The Stone Age

" Stone & Ancient Civilizations "

Edited by Paul F. Kisak

Contents

Chapter 1

Stone Age

For other uses, see Stone Age (disambiguation).

The **Stone Age** is a broad prehistoric period during which

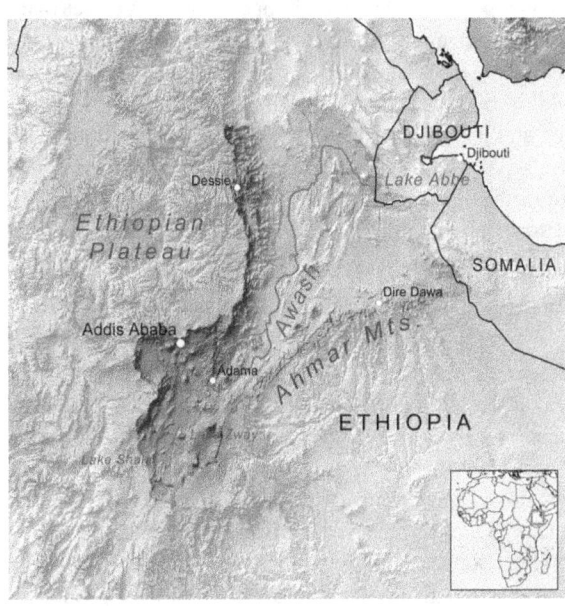

Modern Awash River, Ethiopia, descendant of the Palaeo-Awash, source of the sediments in which the oldest Stone Age tools have been found

stone was widely used to make implements with a sharp edge, a point, or a percussion surface. The period lasted roughly 3.4 million years, and ended between 6000 BCE and 2000 BCE with the advent of metalworking.[1] Stone Age artifacts include tools used by modern humans and by their predecessor species in the genus *Homo*, as well as the earlier partly contemporaneous genera *Australopithecus* and *Paranthropus*. Bone tools were used during this period as well but are rarely preserved in the archaeological record. The Stone Age is further subdivided by the types of stone tools in use.

The Stone Age is the first of the three-age system of archaeology, which divides human technological prehistory into three periods:

- The **Stone Age**
- The Bronze Age
- The Iron Age

1.1 Historical significance

The Stone Age is contemporaneous with the evolution of the genus *Homo*, the only exception possibly being at the very beginning, when species prior to *Homo* may have manufactured tools. According to the age and location of the current evidence, the cradle of the genus is the East African Rift System, especially toward the north in Ethiopia, where it is bordered by grasslands. The closest relative among the other living Primates, the genus *Pan*, represents a branch that continued on in the deep forest, where the primates evolved. The rift served as a conduit for movement into southern Africa and also north down the Nile into North Africa and through the continuation of the rift in the Levant to the vast grasslands of Asia.

Starting from about 3 million years ago (mya) a single biome established itself from South Africa through the rift, North Africa, and across Asia to modern China, which has been called "transcontinental 'savannahstan'" recently.[2] Starting in the grasslands of the rift, *Homo erectus*, the predecessor of modern humans, found an ecological niche as a toolmaker and developed a dependence on it, becoming a "tool equipped savanna dweller."[3]

1.2 The Stone Age in archaeology

1.2.1 Beginning of the Stone Age

During 2010, fossilised animal bones bearing marks from stone tools were found in the Lower Awash Valley in Ethiopia. Discovered by an international team led by Shannon McPherron, at 3.4 million years old they are the oldest

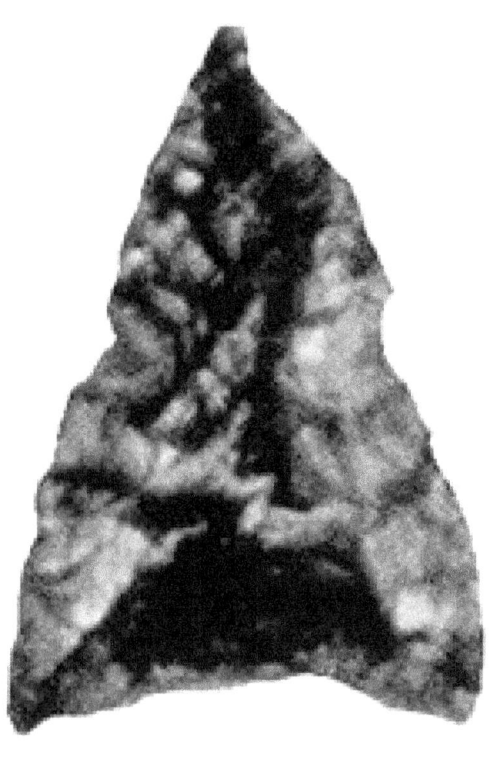

Obsidian projectile point

indirect evidence of stone tool use ever found anywhere in the world.[1] Archaeological discoveries in Kenya in 2015, identifying possibly the oldest known evidence of hominin use of tools to date, have indicated that Kenyanthropus platyops (a 3.2 to 3.5-million-year-old Pliocene hominin fossil discovered in Lake Turkana, Kenya in 1999) may have been the earliest tool-users known.[4]

The oldest known stone tools have been excavated from the site of Lomekwi 3 in West Turkana, northwestern Kenya, and date to 3.3 million years old.[5] Prior to the discovery of these "Lomekwian" tools, the oldest known stone tools had been found several sites at Gona, Ethiopia, on the sediments of the paleo-Awash River, which serve to date them. All the tools come from the Busidama Formation, which lies above a disconformity, or missing layer, which would have been from 2.9 to 2.7 mya. The oldest sites containing tools are dated to 2.6–2.55 mya.[6] One of the most striking circumstances about these sites is that they are from the Late Pliocene, where previous to their discovery tools were thought to have evolved only in the Pleistocene. Rogers and Semaw, excavators at the locality, point out that:[7]

> "...the earliest stone tool makers were skilled flintknappers .... The possible reasons behind this seeming abrupt transition from the absence

of stone tools to the presence thereof include ... gaps in the geological record."

The species who made the Pliocene tools remains unknown. Fragments of *Australopithecus garhi*, *Australopithecus aethiopicus*[8] and *Homo*, possibly *Homo habilis*, have been found in sites near the age of the Gona tools.[9]

1.2.2   End of the Stone Age

Innovation of the technique of smelting ore ended the Stone Age and began the Bronze Age. The first most significant metal manufactured was bronze, an alloy of copper and tin, each of which was smelted separately. The transition from the Stone Age to the Bronze Age was a period during which modern people could smelt copper, but did not yet manufacture bronze, a time known as the Copper Age, or more technically the Chalcolithic, "copper-stone" age. The Chalcolithic by convention is the initial period of the Bronze Age and is unquestionably part of the Age of Metals. The Bronze Age was followed by the Iron Age. During this entire time stone remained in use in parallel with the metals for some objects, including those also used in the Neolithic, such as stone pottery.

The transition out of the Stone Age occurred between 6000 BCE and 2500 BCE for much of humanity living in North Africa and Eurasia. The first evidence of human metallurgy dates to between the 5th and 6th millennium BCE in the archaeological sites of Majdanpek, Yarmovac and Pločnik (a copper axe from 5500 BCE belonging to the Vinca culture), though not conventionally considered part of the Chalcolithic or "Copper Age", this provides the earliest known example of copper metallurgy.[10] Note the Rudna Glava mine in Serbia. Ötzi the Iceman, a mummy from about 3300 BCE carried with him a copper axe and a flint knife.

In regions such as Subsaharan Africa, the Stone Age was followed directly by the Iron Age. The Middle East and southeastern Asian regions progressed past Stone Age technology around 6000 BCE. Europe, and the rest of Asia became post–Stone Age societies by about 4000 BCE. The proto-Inca cultures of South America continued at a Stone Age level until around 2000 BCE, when gold, copper and silver made their entrance, the rest following later. Australia remained in the Stone Age until the 17th century. Stone tool manufacture continued. In Europe and North America, millstones were in use until well into the 20th century, and still are in many parts of the world.

1.2.3   The concept of Stone Age

The terms "Stone Age", "Bronze Age", and "Iron Age" were never meant to suggest that advancement and time periods in prehistory are only measured by the type of tool material, rather than, for example, social organization, food sources exploited, adaptation to climate, adoption of agriculture, cooking, settlement and religion. Like pottery, the typology of the stone tools combined with the relative sequence of the types in various regions provide a chronological framework for the evolution of man and society. They serve as diagnostics of date, rather than characterizing the people or the society.

Lithic analysis is a major and specialised form of archaeological investigation. It involves the measurement of the stone tools to determine their typology, function and the technology involved. It includes scientific study of the lithic reduction of the raw materials, examining how the artifacts were made. Much of this study takes place in the laboratory in the presence of various specialists. In experimental archaeology, researchers attempt to create replica tools, to understand how they were made. Flintknappers are craftsmen who use sharp tools to reduce flintstone to flint tool.

A variety of stone tools

In addition to lithic analysis, the field prehistorian utilizes a wide range of techniques derived from multiple fields. The work of the archaeologist in determining the paleocontext and relative sequence of the layers is supplemented by the efforts of the geologic specialist in identifying layers of rock over geologic time, of the paleontological specialist in identifying bones and animals, of the palynologist in discovering and identifying plant species, of the physicist and chemist in laboratories determining dates by the carbon-14, potassium-argon and other methods. Study of the Stone Age has never been mainly about stone tools and archaeology, which are only one form of evidence. The chief focus has always been on the society and the physical people who belonged to it.

Useful as it has been, the concept of the Stone Age has its limitations. The date range of this period is ambiguous, disputed, and variable according to the region in question. While it is possible to speak of a general 'stone age' period

for the whole of humanity, some groups never developed metal-smelting technology, so remained in a 'stone age' until they encountered technologically developed cultures. The term was innovated to describe the archaeological cultures of Europe. It may not always be the best in relation to regions such as some parts of the Indies and Oceania, where farmers or hunter-gatherers used stone for tools until European colonisation began.

The archaeologists of the late 19th and early 20th centuries CE, who adapted the three-age system to their ideas, hoped to combine cultural anthropology and archaeology in such a way that a specific contemporaneous tribe can be used to illustrate the way of life and beliefs of the people exercising a specific Stone-Age technology. As a description of people living today, the term *stone age* is controversial. The Association of Social Anthropologists discourages this use, asserting:[11]

> "To describe any living group as 'primitive' or 'Stone Age' inevitably implies that they are living representatives of some earlier stage of human development that the majority of humankind has left behind. For some, this could be a positive description, implying, for example, that such groups live in greater harmony with nature .... For others, ... 'primitive' is a negative characterisation. For them, 'primitive' denotes irrational use of resources and absence of the intellectual and moral standards of 'civilised' human societies.... From the standpoint of anthropological knowledge, both these views are equally one-sided and simplistic."

1.2.4 The three-stage system

In the 1920s, South African archaeologists organizing the stone tool collections of that country observed that they did not fit the newly detailed Three-Age System. In the words of J. Desmond Clark,[12]

> "It was early realized that the threefold division of culture into Stone, Bronze and Iron Ages adopted in the nineteenth century for Europe had no validity in Africa outside the Nile valley."

Consequently, they proposed a new system for Africa, the Three-stage System. Clark regarded the Three-age System as valid for North Africa; in sub-Saharan Africa, the Three-stage System was best.[13] In practice, the failure of African archaeologists either to keep this distinction in mind, or to explain which one they mean, contributes to the considerable equivocation already present in the literature. There are in effect two Stone Ages, one part of the Three-age and the other constituting the Three-stage. They refer to one and the same artifacts and the same technologies, but vary by locality and time.

The Three-stage System was proposed in 1929 by Astley John Hilary Goodwin, a professional archaeologist, and Clarence van Riet Lowe, a civil engineer and amateur archaeologist, in an article titled "Stone Age Cultures of South Africa" in the journal *Annals of the South African Museum*. By then, the dates of the Early Stone Age, or Paleolithic, and Late Stone Age, or Neolithic (*neo* = new), were fairly solid and were regarded by Goodwin as absolute. He therefore proposed a relative chronology of periods with floating dates, to be called the Earlier and Later Stone Age. The Middle Stone Age would not change its name, but it would not mean Mesolithic.[14]

The duo thus reinvented the Stone Age. In Sub-Saharan Africa, however, it was ended by the intrusion of the Iron Age from the north. The Neolithic and the Bronze Age never occurred. Moreover, the technologies included in those 'stages', as Goodwin called them, were not exactly the same. Since then, the original relative terms have become identified with the technologies of the Paleolithic and Mesolithic, so that they are no longer relative. Moreover, there has been a tendency to drop the comparative degree in favor of the positive: resulting in two sets of Early, Middle and Late Stone Ages of quite different content and chronologies.

By voluntary agreement, archaeologists respect the decisions of the Pan-African Congress of Prehistory, which meets every four years to resolve archaeological business brought before it. Delegates are actually international; the organization takes its name from the topic. Louis Leakey hosted the first one in Nairobi in 1947. It adopted Goodwin and Lowe's 3-stage system at that time, the stages to be called Early, Middle and Later.

1.2.5 The problem of the transitions

The problem of the transitions in archaeology is a branch of the general philosophic continuity problem, which examines how discrete objects of any sort that are contiguous in any way can be presumed to have a relationship of any sort. In archaeology, the relationship is one of causality. If Period B can be presumed to descend from Period A, there must be a boundary between A and B, the A–B boundary. The problem is in the nature of this boundary. If there is no distinct boundary, then the population of A suddenly stopped using the customs characteristic of A and suddenly started using those of B, an unlikely scenario in the process of evolution. More realistically, a distinct border period, the A/B transition, existed, in which the customs of A were

gradually dropped and those of B acquired. If transitions do not exist, then there is no proof of any continuity between A and B.

The Stone Age of Europe is characteristically in deficit of known transitions. The 19th and early 20th-century innovators of the modern three-age system recognized the problem of the initial transition, the "gap" between the Paleolithic and the Neolithic. Louis Leakey provided something of an answer by proving that man evolved in Africa. The Stone Age must have begun there to be carried repeatedly to Europe by migrant populations. The different phases of the Stone Age thus could appear there without transitions. The burden on African archaeologists became all the greater, because now they must find the missing transitions in Africa. The problem is difficult and ongoing.

After its adoption by the First Pan African Congress in 1947, the Three-Stage Chronology was amended by the Third Congress in 1955 to include a First Intermediate Period between Early and Middle, to encompass the Fauresmith and Sangoan technologies, and the Second Intermediate Period between Middle and Later, to encompass the Magosian technology and others. The chronologic basis for definition was entirely relative. With the arrival of scientific means of finding an absolute chronology, the two intermediates turned out to be will-of-the-wisps. They were in fact Middle and Lower Paleolithic. Fauresmith is now considered to be a facies of Acheulean, while Sangoan is a facies of Lupemban.[15] Magosian is "an artificial mix of two different periods."[16]

Once seriously questioned, the intermediates did not wait for the next Pan African Congress two years hence, but were officially rejected in 1965 (again on an advisory basis) by Burg Wartenstein Conference #29, *Systematic Investigation of the African Later Tertiary and Quaternary*,[17] a conference in anthropology held by the Wenner-Gren Foundation, at Burg Wartenstein Castle, which it then owned in Austria, attended by the same scholars that attended the Pan African Congress, including Louis Leakey and Mary Leakey, who was delivering a pilot presentation of her typological analysis of Early Stone Age tools, to be included in her 1971 contribution to *Olduvai Gorge*, "Excavations in Beds I and II, 1960–1963."[18]

However, although the intermediate periods were gone, the search for the transitions continued.

1.3 Chronology

In 1859 Jens Jacob Worsaae first proposed a division of the Stone Age into older and younger parts based on his work with Danish kitchen middens that began in 1851.[19] In the subsequent decades this simple distinction developed into

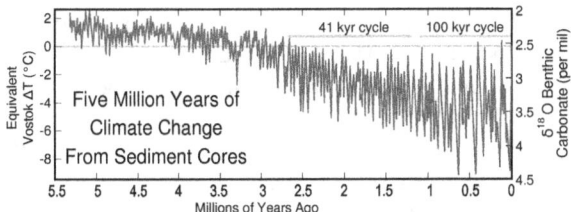

Time series plot of temperature over the previous 5 million years

the archaeological periods of today. The major subdivisions of the Three-age Stone Age cross two epoch boundaries on the geologic time scale:

- The geologic Pliocene–Pleistocene boundary (highly glaciated climate)
    - The Paleolithic period of archaeology
- The geologic Pleistocene–Holocene boundary (modern climate)
    - Mesolithic or Epipaleolithic period of archaeology
    - Neolithic period of archaeology

The succession of these phases varies enormously from one region (and culture) to another.

1.3.1 Three-age chronology

Main articles: Paleolithic, Human evolution and Three-age system

The Paleolithic or Palaeolithic (from Greek: παλαιός, *palaios*, "old"; and λίθος, *lithos*, "stone" lit. "old stone," coined by archaeologist John Lubbock and published in 1865) is the earliest division of the Stone Age. It covers the greatest portion of humanity's time (roughly 99% of "human technological history,"[20] where "human" and "humanity" are interpreted to mean the genus *Homo*), extending from 2.5 or 2.6 million years ago, with the first documented use of stone tools by hominans such as *Homo habilis*, to the end of the Pleistocene around 10,000 BCE.[20] The Paleolithic era ended with the Mesolithic, or in areas with an early neolithisation, the Epipaleolithic.

Lower Paleolithic

Main article: Lower Paleolithic

At sites dating from the Lower Paleolithic Period (about 2,500,000 to 200,000 years ago), simple pebble tools have

been found in association with the remains of what may have been the earliest human ancestors. A somewhat more sophisticated Lower Paleolithic tradition, known as the Chopper chopping-tool industry, is widely distributed in the Eastern Hemisphere. This tradition is thought to have been the work of the hominin species named Homo erectus. Although no such fossil tools have yet been found, it is believed that H. erectus probably made tools of wood and bone as well as stone. About 700,000 years ago, a new Lower Paleolithic tool, the hand ax, appeared. The earliest European hand axes are assigned to the Abbevillian industry, which developed in northern France in the valley of the Somme River; a later, more refined hand-ax tradition is seen in the Acheulian industry, evidence of which has been found in Europe, Africa, the Middle East, and Asia. Some of the earliest known hand axes were found at Olduvai Gorge (Tanzania) in association with remains of H. erectus. Alongside the hand-axe tradition there developed a distinct and very different stone-tool industry, based on flakes of stone: special tools were made from worked (carefully shaped) flakes of flint. In Europe, the Clactonian industry is one example of a flake tradition. The early flake industries probably contributed to the development of the Middle Paleolithic flake tools of the Mousterian industry, which is associated with the remains of Neanderthal man.[21]

This is a Mode 1, or Oldowan, stone tool from the western Sahara.

Oldowan in Africa   Main article: Oldowan

The earliest documented stone tools have been found in eastern Africa, manufacturers unknown, at the 3.3 million year old site of Lomekwi 3 in Kenya.[5] Better known are the later tools belonging to an industry known as Oldowan, after the type site of Olduvai Gorge in Tanzania.

The tools were formed by knocking pieces off a river pebble, or stones like it, with a hammerstone to obtain large and small pieces with one or more sharp edges. The original stone is called a core; the resultant pieces, flakes. Typically, but not necessarily, small pieces are detached from a larger piece, in which case the larger piece may be called the core and the smaller pieces the flakes. The prevalent usage, however, is to call all the results flakes, which can be confusing. A split in half is called bipolar flaking.

Consequently, the method is often called "core-and-flake". More recently, the tradition has been called "small flake" since the flakes were small compared to subsequent Acheulean tools.[22]

> "The essence of the Oldowan is the making and often immediate use of small flakes."

Another naming scheme is "Pebble Core Technology (PBC)":[23]

> "Pebble cores are ... artifacts that have been shaped by varying amounts of hard-hammer percussion."

Various refinements in the shape have been called choppers, discoids, polyhedrons, subspheroid, etc. To date no reasons for the variants have been ascertained:[24]

> "From a functional standpoint, pebble cores seem designed for no specific purpose."

However, they would not have been manufactured for no purpose:[24]

> "Pebble cores can be useful in many cutting, scraping or chopping tasks, but ... they are not particularly more efficient in such tasks than a sharp-edged rock ...."

The whole point of their utility is that each is a "sharp-edged rock" in locations where nature has not provided any. There is additional evidence that Oldowan, or Mode 1, tools were utilized in "percussion technology"; that is, they were designed to be gripped at the blunt end and strike something with the edge, from which use they were given the name of choppers. Modern science has been able to detect mammalian blood cells on Mode 1 tools at Sterkfontein, Member 5 East, in South Africa. As the blood must have come from a fresh kill, the tool users are likely to have done the killing and used the tools for butchering. Plant residues bonded to the silicon of some tools confirm the use to chop plants.[25]

Although the exact species authoring the tools remains unknown, Mode 1 tools in Africa were manufactured and used predominantly by *Homo habilis*. They cannot be said to

have developed these tools or to have contributed the tradition to technology. They continued a tradition of yet unknown origin. As chimpanzees sometimes naturally use percussion to extract or prepare food in the wild, and may use either unmodified stones or stones that they have split, creating an Oldowan tool, the tradition may well be far older than its current record.

Towards the end of Oldowan in Africa a new species appeared over the range of *Homo habilis*: *Homo erectus*. The earliest "unambiguous" evidence is a whole cranium, KNM-ER 3733 (a find identifier) from Koobi Fora in Kenya, dated to 1.78 mya.[26] An early skull fragment, KNM-ER 2598, dated to 1.9 mya, is considered a good candidate also.[27] Transitions in paleoanthropology are always hard to find, if not impossible, but based on the "long-legged" limb morphology shared by *H. habilis* and *H. rudolfensis* in East Africa, an evolution from one of those two has been suggested.[28]

The most immediate cause of the new adjustments appears to have been an increasing aridity in the region and consequent contraction of parkland savanna, interspersed with trees and groves, in favor of open grassland, dated 1.8–1.7 mya. During that transitional period the percentage of grazers among the fossil species increased from 15–25% to 45%, dispersing the food supply and requiring a facility among the hunters to travel longer distances comfortably, which *H. erectus* obviously had.[29] The ultimate proof is the "dispersal" of *H. erectus* "across much of Africa and Asia, substantially before the development of the Mode 2 technology and use of fire ...."[28] *H. erectus* carried Mode 1 tools over Eurasia.

According to the current evidence (which may change at any time) Mode 1 tools are documented from about 2.6 mya to about 1.5 mya in Africa,[30] and to 0.5 mya outside of it.[31] The genus Homo is known from *H. habilis* and *H. rudolfensis* from 2.3 to 2.0 mya, with the latest habilis being an upper jaw from Koobi Fora, Kenya, from 1.4 mya. *H. erectus* is dated 1.8–0.6 mya.[32]

According to this chronology Mode 1 was inherited by *Homo* from unknown Hominans, probably *Australopithecus* and *Paranthropus*, who must have continued on with Mode 1 and then with Mode 2 until their extinction no later than 1.1 mya. Meanwhile, living contemporaneously in the same regions *H. habilis* inherited the tools around 2.3 mya. At about 1.9 mya *H. erectus* came on stage and lived contemporaneously with the others. Mode 1 was now being shared by a number of Hominans over the same ranges, presumably subsisting in different niches, but the archaeology is not precise enough to say which.

Oldowan out of Africa   Tools of the Oldowan tradition first came to archaeological attention in Europe, where, being intrusive and not well defined, compared to the Acheulean, they were puzzling to archaeologists. The mystery would be elucidated by African archaeology at Olduvai, but meanwhile, in the early 20th century, the term "Pre-Acheulean" came into use in climatology. C.E.P, Brooks, a British climatologist working in the United States, used the term to describe a "chalky boulder clay" underlying a layer of gravel at Hoxne, central England, where Acheulean tools had been found.[33] Whether any tools would be found in it and what type was not known. Hugo Obermaier, a contemporary German archaeologist working in Spain, quipped:

> "Unfortunately, the stage of human industry which corresponds to these deposits cannot be positively identified. All we can say is that it is pre-Acheulean...."

This uncertainty was clarified by the subsequent excavations at Olduvai; nevertheless, the term is still in use for pre-Acheulean contexts, mainly across Eurasia, that are yet unspecified or uncertain but with the understanding that they are or will turn out to be pebble-tool.[34]

There are ample associations of Mode 2 with *H. erectus* in Eurasia. *H. erectus* – Mode 1 associations are scantier but they do exist, especially in the Far East. One strong piece of evidence prevents the conclusion that only *H. erectus* reached Eurasia: at Yiron, Israel, Mode 1 tools have been found dating to 2.4 mya,[35] about 0.5 my earlier than the known *H. erectus* finds. If the date is correct, either another Hominan preceded *H. erectus* out of Africa or the earliest *H. erectus* has yet to be found.

After the initial appearance at Gona in Ethiopia at 2.7 mya, pebble tools date from 2.0 mya at Sterkfontein, Member 5, South Africa, and from 1.8 mya at El Kherba, Algeria, North Africa. The manufacturers had already left pebble tools at Yiron, Israel, at 2.4 mya, Riwat, Pakistan, at 2.0 mya, and Renzidong, South China, at over 2 mya.[36] The identification of a fossil skull at Mojokerta, Pernung Peninsula on Java, dated to 1.8 mya, as *H. erectus*, suggests that the African finds are not the earliest to be found in Africa, or that, in fact, erectus did not originate in Africa after all but on the plains of Asia.[28] The outcome of the issue waits for more substantial evidence. Erectus was found also at Dmanisi, Georgia, from 1.75 mya in association with pebble tools.

Pebble tools are found the latest first in southern Europe and then in northern. They begin in the open areas of Italy and Spain, the earliest dated to 1.6 mya at Pirro Nord, Italy. The mountains of Italy are rising at a rapid rate in the framework of geologic time; at 1.6 mya they were lower and covered with grassland (as much of the highlands still are). Europe was otherwise mountainous and covered over with dense forest, a formidable terrain for warm-weather

savanna dwellers. Similarly there is no evidence that the Mediterranean was passable at Gibraltar or anywhere else to *H. erectus* or earlier hominans. They might have reached Italy and Spain along the coasts.

In northern Europe pebble tools are found earliest at Happisburgh, United Kingdom, from 0.8 mya. The last traces are from Kent's Cavern, dated 0.5 mya. By that time *H. erectus* is regarded as having been extinct; however, a more modern version apparently had evolved, *Homo heidelbergensis*, who must have inherited the tools.[37] He also explains the last of the Acheulean in Germany at 0.4 mya.

In the late 19th and early 20th centuries archaeologists worked on the assumptions that a succession of Hominans and cultures prevailed, that one replaced another. Today the presence of multiple hominans living contemporaneously near each other for long periods is accepted as proved true; moreover, by the time the previously assumed "earliest" culture arrived in northern Europe, the rest of Africa and Eurasia had progressed to the Middle and Upper Palaeolithic, so that across the earth all three were for a time contemporaneous. In any given region there was a progression from Oldowan to Acheulean, Lower to Upper, no doubt.

Acheulean in Africa Main article: Acheulean

The end of Oldowan in Africa was brought on by the ap-

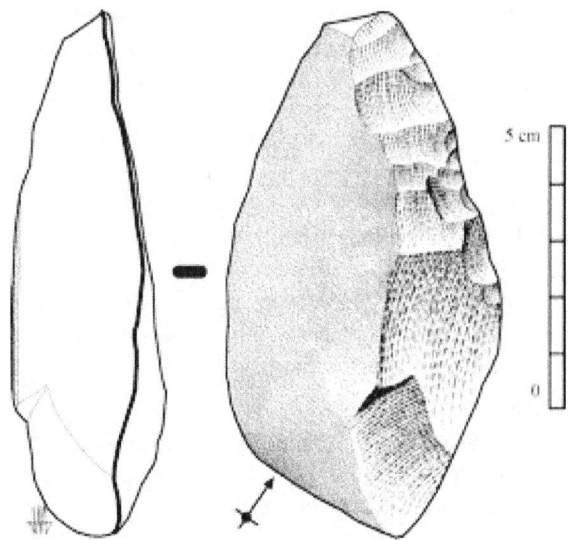

An Acheulean tool, not worked over the entire surface

pearance of Acheulean, or Mode 2, stone tools. The earliest known instances are in the 1.7–1.6 mya layer at Kokiselei, West Turkana, Kenya.[27] At Sterkfontein, South Africa, they are in Member 5 West, 1.7–1.4 mya.[25] The 1.7 is a fairly certain, fairly standard date. Mode 2 is often found in association with *H. erectus*. It makes sense that the most advanced tools should have been innovated by the most advanced Hominan; consequently, they are typically given credit for the innovation.

A Mode 2 tool is a biface consisting of two concave surfaces intersecting to form a cutting edge all the way around, except in the case of tools intended to feature a point. More work and planning go into the manufacture of a Mode 2 tool. The manufacturer hits a slab off a larger rock to use as a blank. Then large flakes are struck off the blank and worked into bifaces by hard-hammer percussion on an anvil stone. Finally the edge is retouched: small flakes are hit off with a bone or wood soft hammer to sharpen or resharpen it. The core can be either the blank or another flake. Blanks are ported for manufacturing supply in places where nature has provided no suitable stone.

Although most Mode 2 tools are easily distinguished from Mode 1, there is a close similarity of some Oldowan and some Acheulean, which can lead to confusion. Some Oldowan tools are more carefully prepared to form a more regular edge. One distinguishing criterion is the size of the flakes. In contrast to the Oldowan "small flake" tradition, Acheulean is "large flake:" "The primary technological distinction remaining between Oldowan and the Acheulean is the preference for large flakes (>10 cm) as blanks for making large cutting tools (handaxes and cleavers) in the Acheulean."[38] "Large Cutting Tool (LCT)" has become part of the standard terminology as well.[24]

In North Africa, the presence of Mode 2 remains a mystery, as the oldest finds are from Thomas Quarry in Morocco at 0.9 mya.[36] Archaeological attention, however, shifts to the Jordan Rift Valley, an extension of the East African Rift Valley (the east bank of the Jordan is slowly sliding northward as East Africa is thrust away from Africa). Evidence of use of the Nile Valley is in deficit, but Hominans could easily have reached the palaeo-Jordan river from Ethiopia along the shores of the Red Sea, one side or the other. A crossing would not have been necessary, but it is more likely there than over a theoretical but unproven land bridge through either Gibraltar or Sicily.

Meanwhile, Acheulean went on in Africa past the 1.0 mya mark and also past the extinction of *H. erectus* there. The last Acheulean in East Africa is at Olorgesailie, Kenya, dated to about 0.9 mya. Its owner was still *H. erectus*,[36] but in South Africa, Acheulean at Elandsfontein, 1.0–0.6 mya, is associated with Saldanha man, classified as *H. heidelbergensis*, a more advanced, but not yet modern, descendant most likely of *H. erectus*. The Thoman Quarry Hominans in Morocco similarly are most likely Homo rhodesiensis,[39] in the same evolutionary status as *H. heidelbergensis*.

Acheulean out of Africa   Mode 2 is first known out of Africa at 'Ubeidiya, Israel, a site now on the Jordan River, then frequented over the long term (hundreds of thousands

of years) by Homo on the shore of a variable-level palaeo-lake, long since vanished. The geology was created by successive "transgression and regression" of the lake[40] resulting in four cycles of layers. The tools are located in the first two, Cycles Li (Limnic Inferior) and Fi (Fluviatile Inferior), but mostly in Fi. The cycles represent different ecologies and therefore different cross-sections of fauna, which makes it possible to date them. They appear to be the same faunal assemblages as the Ferenta Faunal Unit in Italy, known from excavations at Selvella and Pieterfitta, dated to 1.6–1.2 mya.[41]

At 'Ubeidiya the marks on the bones of the animal species found there indicate that the manufacturers of the tools butchered the kills of large predators, an activity that has been termed "scavenging."[42] There are no living floors, nor did they process bones to obtain the marrow. These activities cannot be understood therefore as the only or even the typical economic activity of Hominans. Their interests were selective: they were primarily harvesting the meat of Cervids,[43] which is estimated to have been available without spoiling for up to four days after the kill.

The majority of the animals at the site were of "Palaearctic biogeographic origin."[44] However, these overlapped in range on 30–60% of "African biogeographic origin."[45] The biome was Mediterranean, not savanna. The animals were not passing through; there was simply an overlap of normal ranges. Of the Hominans, *H. erectus* left several cranial fragments. Teeth of undetermined species may have been *H. ergaster*.[46] The tools are classified as "Lower Acheulean" and "Developed Oldowan." The latter is a disputed classification created by Mary Leakey to describe an Acheulean-like tradition in Bed II at Olduvai. It is dated 1.53–1.27 mya. The date of the tools therefore probably does not exceed 1.5 mya; 1.4 is often given as a date. This chronology, which is definitely later than in Kenya, supports the "out of Africa" hypothesis for Acheulean, if not for the Hominans.

From Southwest Asia, as the Levant is now called, the Acheulean extended itself more slowly eastward, arriving at Isampur, India, about 1.2 mya. It does not appear in China and Korea until after 1mya and not at all in Indonesia. There is a discernible boundary marking the furthest extent of the Acheulean eastward before 1 mya, called the Movius Line, after its proposer, Hallam L. Movius. On the east side of the line the small flake tradition continues, but the tools are additionally worked Mode 1, with flaking down the sides. In Athirampakkam at Chennai in Tamil Nadu the Acheulean age started at 1.51 mya and it is also prior than North India and Europe.[47]

The cause of the Movius Line remains speculative, whether it represents a real change in technology or a limitation of archeology, but after 1 mya evidence not available to

Movius indicates the prevalence of Acheulean. For example, the Acheulean site at Bose, China, is dated 0.803±3K mya.[48] The authors of this chronologically later East Asian Acheulean remain unknown, as does whether it evolved in the region or was brought in.

There is no named boundary line between Mode 1 and Mode 2 on the west; nevertheless, Mode 2 is equally late in Europe as it is in the Far East. The earliest comes from a rock shelter at Estrecho de Quípar in Spain, dated to greater than 0.9 mya. Teeth from an undetermined Hominan were found there also.[49] The last Mode 2 in Southern Europe is from a deposit at Fontana Ranuccio near Anagni in Italy dated to 0.45 mya, which is generally linked to *Homo cepranensis*, a "late variant of *H. erectus*," a fragment of whose skull was found at Ceprano nearby, dated 0.46 mya.[50]

Middle Paleolithic

Main article: Middle Paleolithic

This period is best known as the era during which the Neanderthals lived in Europe and the Near East (c. 300,000–28,000 years ago). Their technology is mainly the Mousterian, but Neanderthal physical characteristics have been found also in ambiguous association with the more recent Châtelperronian archeological culture in Western Europe and several local industries like the Szeletian in Eastern Europe/Eurasia. There is no evidence for Neanderthals in Africa, Australia or the Americas.

Neanderthals nursed their elderly and practised ritual burial indicating an organised society. The earliest evidence (Mungo Man) of settlement in Australia dates to around 40,000 years ago when modern humans likely crossed from Asia by island-hopping. Evidence for symbolic behavior such as body ornamentation and burial is ambiguous for the Middle Paleolithic and still subject to debate. The Bhimbetka rock shelters exhibit the earliest traces of human life in India, some of which are approximately 30,000 years old.

Upper Paleolithic

Main article: Upper Paleolithic

From 50,000 to 10,000 years ago in Europe, the Upper Paleolithic ends with the end of the Pleistocene and onset of the Holocene era (the end of the last ice age). Modern humans spread out further across the Earth during the period known as the Upper Paleolithic. The Upper Paleolithic is marked by a relatively rapid succession of often complex stone artifact technologies and a large increase in the cre-

ation of art and personal ornaments. During period between 35 and 10 kya evolved: from 38 to 30 kya Châtelperronian, 40–28 Aurignacian, 28–22 Gravettian, 22–17 Solutrean, and 18–10 Magdalenian. All of these industries except the Châtelperronian are associated with anatomically modern humans. Authorship of the Châtelperronian is still the subject of much debate.

The Americas were colonised via the Bering land bridge which was exposed during this period by lower sea levels. These people are called the Paleo-Indians, and the earliest accepted dates are those of the Clovis culture sites, some 13,500 years ago. Globally, societies were hunter-gatherers but evidence of regional identities begins to appear in the wide variety of stone tool types being developed to suit very different environments.

Epipaleolithic/Mesolithic

Main articles: Epipaleolithic, Mesolithic

The period starting from the end of the last ice age, 10,000 years ago, to around 6,000 years ago was characterized by rising sea levels and a need to adapt to a changing environment and find new food sources. The development of Mode 5 (microlith) tools began in response to these changes. They were derived from the previous Paleolithic tools, hence the term Epipaleolithic, or were intermediate between the Paleolithic and the Neolithic, hence the term Mesolithic (Middle Stone Age). The choice of a word depends on exact circumstances and the inclination of the archaeologists excavating the site. Microliths were used in the manufacture of more efficient composite tools, resulting in an intensification of hunting and fishing and with increasing social activity the development of more complex settlements, such as Lepenski Vir. Domestication of the dog as a hunting companion probably dates to this period.

The earliest known battle occurred during the Mesolithic period at a site in Egypt known as Cemetery 117.

Neolithic

Main article: Neolithic
The Neolithic, New Stone Age, was approximately characterized by the adoption of agriculture, the shift from food gathering to food producing in itself is one of the most revolutionary changes in human history so-called Neolithic Revolution, the development of pottery, polished stone tools and more complex, larger settlements such as Göbekli Tepe and Çatal Hüyük. Some of these features began in certain localities even earlier, in the transitional Mesolithic. The first Neolithic cultures started around 7000 BCE in the fertile crescent and spread concentrically to other areas of

Ġgantija temples, Gozo. Some of the world's oldest free-standing structures.

Skara Brae, Scotland. Europe's most complete Neolithic village

the world; however, the Near East was probably not the only nucleus of agriculture, the cultivation of maize in Meso-America and of rice in the Far East being others.

Due to the increased need to harvest and process plants, ground stone and polished stone artifacts became much more widespread, including tools for grinding, cutting, and chopping. Skara Brae located on Orkney island off Scotland is one of Europe's best examples of a Neolithic village. The community contains stone beds, shelves and even an indoor toilet linked to a stream. The first large-scale constructions were built, including settlement towers and walls, e.g., Jericho and ceremonial sites, e.g.: Stonehenge. The Ġgantija temples of Gozo in the Maltese archipelago are the oldest surviving free standing structures in the world, erected c. 3600–2500 BCE. The earliest evidence for established trade exists in the Neolithic with newly settled people importing exotic goods over distances of many hundreds of miles.

These facts show that there were sufficient resources and cooperation to enable large groups to work on these projects.

To what extent this was a basis for the development of elites and social hierarchies is a matter of ongoing debate.[51] Although some late Neolithic societies formed complex stratified chiefdoms similar to Polynesian societies such as the Ancient Hawaiians, based on the societies of modern tribesmen at an equivalent technological level, most Neolithic societies were relatively simple and egalitarian.[52] A comparison of art in the two ages leads some theorists to conclude that Neolithic cultures were noticeably more hierarchical than the Paleolithic cultures that preceded them.[53]

1.3.2   Three-stage chronology

The Earlier or Early Stone Age (ESA)

Main articles: Paleolithic and Lower Paleolithic
This period is not to be identified with "Old Stone Age",

Acheulean biface from Lake Langano area, Ethiopia.

a translation of Paleolithic, or with Paleolithic, or with the "Earlier Stone Age" that originally meant what became the Paleolithic and Mesolithic. In the initial decades of its definition by the Pan-African Congress of Prehistory, it was parallel in Africa to the Upper and Middle Paleolithic. However, since then Radiocarbon dating has shown that the Middle Stone Age is in fact contemporaneous with the Middle Paleolithic.[54] The Early Stone Age therefore is contemporaneous with the Lower Paleolithic and happens to include the same main technologies, Oldowan and Acheulean, which produced Mode 1 and Mode 2 stone tools respectively. A distinct regional term is warranted, however, by the location and chronology of the sites and the exact typology.

The Middle Stone Age (MSA)

Main article: Middle Stone Age

The Middle Stone Age was a period of African prehistory between Early Stone Age and Late Stone Age. It began around 300,000 years ago and ended around 50,000 years ago.[55] It is considered as an equivalent of European Middle Paleolithic.[56] It is associated with anatomically modern or almost modern *Homo sapiens*. Early physical evidence comes from Omo [57] and Herto,[58] both in Ethiopia and dated respectively at c. 195 ka and at c. 160 ka.

The Later Stone Age (LSA)

Main article: Later Stone Age

The Later Stone Age (LSA, sometimes also called the **Late Stone Age**) refers to a period in African prehistory. Its beginnings are roughly contemporaneous with the European Upper Paleolithic. It lasts until historical times and this includes cultures corresponding to Mesolithic and Neolithic in other regions.

1.4 Material culture

1.4.1 Tools

Stone tools were made from a variety of stones. For example, flint and chert were shaped (or *chipped*) for use as cutting tools and weapons, while basalt and sandstone were used for ground stone tools, such as quern-stones. Wood, bone, shell, antler (deer) and other materials were widely used, as well. During the most recent part of the period, sediments (such as clay) were used to make pottery. Agriculture was developed and certain animals were domesticated as well.

Some species of non-primates are able to use stone tools, such as the sea otter, which breaks abalone shells with them. Primates can both use and manufacture stone tools. This combination of abilities is more marked in apes and men, but only men, or more generally Hominans, depend on tool use for survival.[59] The key anatomical and behavioral features required for tool manufacture, which are possessed only by Hominans, are the larger thumb and the ability to hold by means of an assortment of grips.[60]

1.4.2 Food and drink

Main articles: Paleolithic diet and Paleolithic diet and nutrition

Food sources of the Palaeolithic hunter-gatherers were wild plants and animals harvested from the environment. They liked animal organ meats, including the livers, kidneys and brains. Large seeded legumes were part of the human diet long before the agricultural revolution, as is evident from archaeobotanical finds from the Mousterian layers of Kebara Cave, in Israel.[61] Moreover, recent evidence indicates that humans processed and consumed wild cereal grains as far back as 23,000 years ago in the Upper Paleolithic.[62]

Near the end of the Wisconsin glaciation, 15,000 to 9,000 years ago, mass extinction of Megafauna such as the Wooly mammoth occurred in Asia, Europe, North America and Australia. This was the first Holocene extinction event. It possibly forced modification in the dietary habits of the humans of that age and with the emergence of agricultural practices, plant-based foods also became a regular part of the diet. A number of factors have been suggested for the extinction: certainly over-hunting, but also deforestation and climate change.[63] The net effect was to fragment the vast ranges required by the large animals and extinguish them piecemeal in each fragment.

1.4.3 Shelter and habitat

Around 2 million years ago, *Homo habilis* is believed to have constructed the first man-made structure in East Africa, consisting of simple arrangements of stones to hold branches of trees in position. A similar stone circular arrangement believed to be around 380,000 years old was discovered at Terra Amata, near Nice, France. (Concerns about the dating have been raised, see Terra Amata). Several human habitats dating back to the Stone Age have been discovered around the globe, including:

- A tent-like structure inside a cave near the Grotte du Lazaret, Nice, France.

- A structure with a roof supported with timber, discovered in Dolni Vestonice, the Czech Republic, dates to around 23,000 BCE. The walls were made of packed clay blocks and stones.

- Many huts made of mammoth bones were found in Eastern Europe and Siberia. The people who made these huts were expert mammoth hunters. Examples have been found along the Dniepr river valley of Ukraine, including near Chernihiv, in Moravia, Czech Republic and in southern Poland.

- An animal hide tent dated to around 15000 to 10000 BCE, in the Magdalenian, was discovered at Plateau Parain, France.

1.4.4 Art

Prehistoric art is visible in the artifacts. Prehistoric music is inferred from found instruments, while parietal art can be found on rocks of any kind. The latter are petroglyphs and rock paintings. The art may or may not have had a religious function.

Petroglyphs

Main article: Petroglyph

Petroglyphs appeared in the Neolithic. A Petroglyph is an intaglio abstract or symbolic image engraved on natural stone by various methods, usually by prehistoric peoples. They were a dominant form of pre-writing symbols. Petroglyphs have been discovered in different parts of the world, including Asia (Bhimbetka, India), North America (Death Valley National Park), South America (Cumbe Mayo, Peru), and Europe (Finnmark, Norway).

Rock paintings

Rock painting at Bhimbetka, India, a World heritage site

Main article: Cave painting

In paleolithic times, mostly animals were painted, in theory ones that were used as food or represented strength, such as the rhinoceros or large cats (as in the Chauvet Cave). Signs

such as dots were sometimes drawn. Rare human representations include handprints and half-human/half-animal figures. The Cave of Chauvet in the Ardèche *département*, France, contains the most important cave paintings of the paleolithic era, dating from about 31,000 BCE. The Altamira cave paintings in Spain were done 14,000 to 12,000 BCE and show, among others, bisons. The hall of bulls in Lascaux, Dordogne, France, dates from about 15,000 to 10,000 BCE.

The meaning of many of these paintings remains unknown. They may have been used for seasonal rituals. The animals are accompanied by signs that suggest a possible magic use. Arrow-like symbols in Lascaux are sometimes interpreted as calendar or almanac use, but the evidence remains interpretative.[64]

Some scenes of the Mesolithic, however, can be typed and therefore, judging from their various modifications, are fairly clear. One of these is the battle scene between organized bands of archers. For example, "the marching Warriors," a rock painting at Cingle de la Mola, Castellón in Spain, dated to about 7,000–4,000 BCE, depicts about 50 bowmen in two groups marching or running in step toward each other, each man carrying a bow in one hand and a fistful of arrows in the other. A file of five men leads one band, one of whom is a figure with a "high crowned hat." In other scenes elsewhere, the men wear head-dresses and knee ornaments but otherwise fight nude. Some scenes depict the dead and wounded, bristling with arrows.[65] One is reminded of Ötzi the Iceman, a Copper Age mummy revealed by an Alpine melting glacier, who collapsed from loss of blood due to an arrow wound in the back.

1.4.5 Stone Age rituals and beliefs

Main articles: Paleolithic religion, Prehistoric religion and Mother goddess
Modern studies and the in-depth analysis of finds dating from the Stone Age indicate certain rituals and beliefs of the people in those prehistoric times. It is now believed that activities of the Stone Age humans went beyond the immediate requirements of procuring food, body coverings, and shelters. Specific rites relating to death and burial were practiced, though certainly differing in style and execution between cultures.

- Megalithic tombs, multichambered, and dolmens, single-chambered, were graves with a huge stone slab stacked over other similarly large stone slabs; they have been discovered all across Europe and Asia and were built in the Neolithic and the Bronze Age.

Poulnabrone dolmen in County Clare, Ireland

Monte Bubbonia *dolmen (single-chambered tomb), Sicily*[66]

1.5 Modern popular culture and the Stone Age

Imaginative depiction of the Stone Age, by Viktor Vasnetsov

The image of the caveman is commonly associated with the Stone Age. For example, the 2003 documentary series showing the evolution of humans through the Stone

Age was called *Walking with Cavemen*, although only the last programme showed humans living in caves. While the idea that human beings and dinosaurs coexisted is sometimes portrayed in popular culture in cartoons, films and computer games, such as *The Flintstones*, *One Million Years B.C.* and *Chuck Rock*, the notion of hominids and non-avian dinosaurs co-existing is not supported by any scientific evidence.

Other depictions of the Stone Age include the best-selling *Earth's Children* series of books by Jean M. Auel, which are set in the Paleolithic and are loosely based on archaeological and anthropological findings. The 1981 film *Quest for Fire* by Jean-Jacques Annaud tells the story of a group of neanderthals searching for their lost fire. A twenty first century series, *Chronicles of Ancient Darkness* by Michelle Paver tells of two New Stone Age children fighting to fulfil a prophecy and save their clan.

1.6 See also

- Megalith
- Prehistoric warfare
- Ice Age
- Pleistocene
- *Homo*
- List of Stone Age art
- Timeline of the Stone Age

1.7 Notes

[1] http://www.nhm.ac.uk/about-us/news/2010/august/oldest-tool-use-and-meat-eating-revealed75831.html

[2] Barham & Mitchell 2008, p. 106

[3] Barham & Mitchell 2008, p. 147

[4] BBC News, 21/05/2015: Oldest stone tools pre-date earliest humans

[5] Harmand, Sonia; et al. (21 May 2015). "3.3-million-year-old stone tools from Lomekwi 3, West Turkana, Kenya". *Nature* **521**: 310–315. doi:10.1038/nature14464.

[6] Rogers & Semaw 2009, pp. 162–163

[7] Rogers & Semaw 2009, p. 155

[8] As to whether aethiopicus is the genus *Australopithecus* or the genus *Paranthropus*, broken out to include the more robust forms, anthropological opinion is divided and both usages occur in the professional sources.

[9] Rogers & Semaw 2009, p. 164

[10] "Neolithic Vinca was a metallurgical culture". Archaeo News. Reuters. 17 November 2007. Retrieved 25 January 2011.

[11] "ASA Statement on the use of 'primitive' as a descriptor of contemporary human groups". *ASA News* (Association of Social Anthropologists of the UK and Commonwealth). 27 August 2007.

[12] Clark 1970, p. 22

[13] Clark 1970, pp. 18–19

[14] Deacon & Deacon 1999, pp. 5–6

[15] Isaac, Glynn (1982). "The Earliest Archaeological Traces". In Clark, J. Desmond. *The Cambridge History of Africa*. Volume. I: From the Earliest Times to C. 500 BC. Cambridge: Cambridge University Press. p. 246.

[16] Willoughby, Pamela R. (2007). *The evolution of modern humans in Africa: a comprehensive guide*. Lanham, MD: AltaMira Press. p. 54.

[17] Barham & Mitchell 2008, p. 477

[18] "History: Systematic Investigation of the African Later Tertiary and Quaternary". The Wenner-Gren Foundation. Retrieved 3 March 2011.

[19] "Worsaae, Jens Jacob Asmussen". *Encyclopædia Britannica*.

[20] Toth, Nicholas; Schick, Kathy (2007). "21 Overview of Paleolithic Archaeology". In Henke, H.C. Winfried; Hardt, Thorolf; Tattersall, Ian. *Handbook of Paleoanthropology*. Volume **3**. Berlin; Heidelberg; New York: Springer-Verlag. p. 1944. ISBN 978-3-540-32474-4

[21] http://www.britannica.com/EBchecked/topic/439507/Paleolithic-Period

[22] Barham & Mitchell 2008, p. 130.

[23] Shea 2010, p. 49

[24] Shea 2010, p. 50

[25] Barham & Mitchell 2008, p. 132

[26] Barham & Mitchell 2008, pp. 126–127.

[27] Barham & Mitchell 2008, p. 128

[28] Barham & Mitchell 2008, p. 145

[29] Barham & Mitchell 2008, p. 146.

[30] Barham & Mitchell 2008, p. 112

[31] Shea 2010, p. 57

[32] Barham & Mitchell 2008, p. 73

[33] Brooks, Charles E.P. (1919), "The Correlation of the Quaternary Deposits of the British Isles with Those of the Continent of Europe", *Annual Report of the Board of Regents of the Smithsonian Institution 1917*, Washington: Government Pronting Office, p. 277

[34] Hugo Obermaier; Christine Matthew; Henry Osborne (1924). *Fossil Man in Spain*. New Haven: Yale University Press for the Hispanic Society of America. p. 272.

[35] Barham & Mitchell 2008, pp. 106–107

[36] Shea 2010, pp. 55–57

[37] Barham & Mitchell 2008, p. 24

[38] Barham & Mitchell 2008, p. 130

[39] Jean-Paul Raynal; et al. (2010). "Hominid Cave at Thomas Quarry I (Casablanca, Morocco): Recent findings and their context" (PDF). *Quaternary International* (223–224): 369–382.

[40] Belmaker 2006, p. 9

[41] Belmaker 2006, pp. 119–120

[42] Belmaker 2006, p. 149

[43] Belmaker 2006, p. 147

[44] Belmaker 2006, p. 67

[45] Belmaker 2006, p. 21

[46] Belmaker 2006, p. 20

[47] "Acheulian stone tools discovered near Chennai". The Hindu.

[48] "Bose, China". *What Does It Mean to be Human?*. Smithsonian National Museum of Natural History.

[49] Dalton, Rex (2 September 2009). "Europe's oldest axes discovered". *Nature News* (Nature). doi:10.1038/news.2009.878.

[50] Giovanni Muttoni; et al. (2009). "Pleistocene magnetochronology of early hominid sites at Ceprano and Fontana Ranuccio, Italy" (PDF). *Earth and Planetary Science Letters* **286**: 255–268. doi:10.1016/j.epsl.2009.06.032.

[51] Kuijt, Ian (2000). "Chapter 13: Near Eastern Neolithic Research: Directions and Trends". In Kuijt, Ian. *Life in Neolithic Farming Communities: Social Organization, Identity, and differentiation*. Fundamental Issues in Archaeology. New York: Kluwer Academic/Plenum Publishers. p. 317

[52] Boehm, Christopher (2000). "The Origin of Morality as Social Control". In Katz, Leonard D. *Evolutionary Origins of Morality: Cross-disciplinary Perspectives*. Journal of Consciousness Studies Volume 7. Thorverton: Imprint Academic. p. 158. ISBN 0-7190-5612-8

[53] Guthrie, Russell Dale (2005). *The Nature of Paleolithic Art*. Chicago: University of Chicago Press. pp. 419–420. ISBN 978-0-226-31126-5.

[54] Clark, J. Desmond (1982). "The Culture of the Middle Paleolithic/MIddle Stone Age". In Clark, J. Desmond. *The Cambridge History of Africa*. Volume. I: From the Earliest Times to C. 500 BC. Cambridge: Cambridge University Press. p. 248.

[55] McBrearty and Brooks 2000

[56] Biological origins of modern humans

[57] McDougall et al. 2005

[58] White et al. 2003

[59] Barham & Mitchell 2008, p. 74

[60] Barham & Mitchell 2008, p. 108

[61] Efraim Lev; Mordechai E. Kislev; Ofer Bar-Yosef (March 2005). "Mousterian vegetal food in Kebara Cave, Mt. Carmel". *Journal of Archaeological Science* **32** (3): 475–484. doi:10.1016/j.jas.2004.11.006.

[62] Dolores R. Piperno; Ehud Weiss; Irene Holst; Dani Nadel (5 August 2004). "Processing of wild cereal grains in the Upper Palaeolithic revealed by starch grain analysis" (PDF). *Nature* **430** (7000): 670–3. doi:10.1038/nature02734. PMID 15295598.

[63] Turvey, Samuel T. (2009). "Chapter 2: In the shadow of the megafauna: prehistoric mammal and bird extinctions across the Holocene". In Turvey, Samuel T. *Holocene Extinctions*. Oxford Biology. Oxford: Oxford University Press. pp. 16–17

[64] Aczel, Amir D. (2000). *The Cave and the Cathedral: How a Real-Life Indiana Jones and a Research Scholar Decoded the Ancient Art of Man*. Hoboken: John Wiley & Sons Inc. pp. 157–158.

[65] Martínez, Antonio Beltrán (1982) [1979]. *Rock art of the Spanish Levant*. The Imprint of Man. Cambridge: Cambridge University Press. pp. 48–51.

[66] Salvatore Piccolo, *Ancient Stones...*, op. cit.

1.8 References

- Barham, Lawrence; Mitchell, Peter (2008). *The First Africans: African Archaeology from the Earliest Toolmakers to Most Recent Foragers*. Cambridge World Archaeology. Oxford: Oxford University Press.

- Belmaker, Miriam (March 2006). *Community Structure through Time: 'Ubeidiya, a Lower Pleistocene Site as a Case Study (Thesis)* (PDF). Paleoanthropology Society.

- Clark, J. Desmond (1970). *The Prehistory of Africa*. Ancient People and Places, Volume 72. New York; Washington: Praeger Publishers.

- Deacon, Hilary John; Deacon, Janette (1999). *Human beginnings in South Africa: uncovering the secrets of the Stone Age*. Walnut Creek, Calif. [u.a.]: Altamira Press.

- Piccolo, Salvatore (2013). *Ancient Stones: The Prehistoric Dolmens of Sicily*. Thornham/Norfolk (UK): Brazen Head Publishing.

- Rogers, Michael J.; Semaw, Sileshi (2009). "From Nothing to Something: The Appearance and Context of the Earliest Archaeological Record". In Camps i Calbet, Marta; Chauhan, Parth R. *Sourcebook of paleolithic transitions: methods, theories, and interpretations*. New York: Springer.

- Schick, Kathy D.; Nicholas Toth (1993). *Making Silent Stones Speak: Human Evolution and the Dawn of Technology*. New York: Simon & Schuster. ISBN 0-671-69371-9.

- Shea, John J. (2010). "Stone Age Visiting Cards Revisited: a Strategic Perspective on the Lithic Technology of Early Hominin Dispersal". In Fleagle, John G.; Shea, John J.; Grine, Frederick E.; Boden, Andrea L.; Leakey, Richard E,. *Out of Africa I: the First Hominin Colonization of Eurasia*. Dordrecht; Heidelberg; London; New York: Springer. pp. 47–64.

1.9 Further reading

- Scarre, Christopher (ed.) (1988). *Past Worlds: The Times Atlas of Archaeology*. London: Times Books. ISBN 0-7230-0306-8.

1.10 External links

- Giusepi, Robert A. (2000). "The Stone Age". History World International. Retrieved 22 February 2011.

- Kowalski, D.R. "Stone Age Hand-axes". AerobiologicalEngineering.com. Retrieved 22 February 2011.

- Kowalski, D.R. "Stone Age Habitats". AerobiologicalEngineering.com. Retrieved 22 February 2011.

- "PanAfrican Archaeological Association". Retrieved 28 February 2011.

- "Society of Africanist Archaeologists". Retrieved 3 March 2011.

- "The ASA". Association of Social Anthropologists of
the UK and Commonwealth.

Chapter 2

Paleolithic

The **Paleolithic** (American spelling; British spelling: **Palaeolithic**; pronunciation: /ˌpæliəˈlɪθɪk/ or /ˌpeɪl-/) **Age**, **Era** or **Period** is a prehistoric period of human history distinguished by the development of the most primitive stone tools discovered (Grahame Clark's Modes I and II), and covers roughly 95%[1] of human technological prehistory. It extends from the earliest known use of stone tools, probably by hominins such as australopithecines, 2.6 million years ago, to the end of the Pleistocene around 10,000 BP.[2]

The Paleolithic era is followed by the Mesolithic. The date of the Paleolithic–Mesolithic boundary may vary by locality as much as several thousand years. During the Paleolithic period, humans grouped together in small societies such as bands, and subsisted by gathering plants and fishing, hunting or scavenging wild animals.[3] The Paleolithic is characterized by the use of knapped stone tools, although at the time humans also used wood and bone tools. Other organic commodities were adapted for use as tools, including leather and vegetable fibers; however, due to their nature, these have not been preserved to any great degree. Surviving artifacts of the Paleolithic era are known as paleoliths. Humankind gradually evolved from early members of the genus *Homo* such as *Homo habilis* – who used simple stone tools – into fully behaviorally and anatomically modern humans (*Homo sapiens*)during the Paleolithic era.[4] During the end of the Paleolithic, specifically the Middle and or Upper Paleolithic, humans began to produce the earliest works of art and engage in religious and spiritual behavior such as burial and ritual.[5][6] The climate during the Paleolithic consisted of a set of glacial and interglacial periods in which the climate periodically fluctuated between warm and cool temperatures.

The term "Paleolithic" was coined by archaeologist John Lubbock in 1865.[7] It derives from Greek: παλαιός, *palaios*, "old"; and λίθος, *lithos*, "stone", meaning "old age of the stone" or "Old Stone Age."

2.1 Human evolution

Main article: Human evolution

Human evolution is the part of biological evolution con-

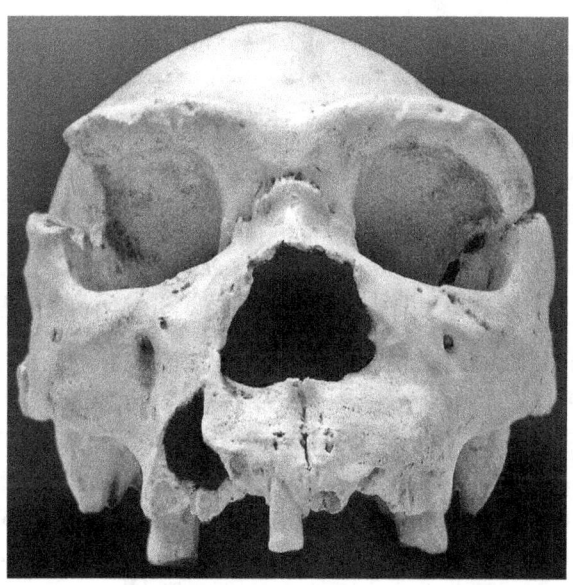

This cranium, of Homo heidelbergensis, *a Lower Paleolithic predecessor to* Homo neanderthalensis *and possibly* Homo sapiens, *dates to sometime between 500,000 and 400,000 BP.*

cerning the emergence of humans as a distinct species.

2.2 Paleogeography and climate

Main articles: Pleistocene § Paleogeography and climate, Pliocene_climate and Pliocene § Paleogeography

The Paleolithic Period coincides almost exactly with the Pleistocene epoch of geologic time, which lasted from 2.6 million years ago to about 12,000 years ago.[8] This epoch experienced important geographic and climatic changes that affected human societies.

During the preceding Pliocene, continents had continued to

The Paleolithic climate consisted of a set of glacial and interglacial periods.

retreats. The minor excursion is a "stadial"; times between stadials are "interstadials". Each glacial advance tied up huge volumes of water in continental ice sheets 1500–3000 m deep, resulting in temporary sea level drops of 100 m or more over the entire surface of the Earth. During interglacial times, such as at present, drowned coastlines were common, mitigated by isostatic or other emergent motion of some regions.

Many great mammals such as woolly mammoths, woolly rhinoceros, and cave lions inhabited places like Siberia during the Pleistocene.

Paleoindians hunting a glyptodon. Glyptodons were hunted to extinction within two millennia after humans' arrival to South America.

drift from possibly as far as 250 km from their present locations to positions only 70 km from their current location. South America became linked to North America through the Isthmus of Panama, bringing a nearly complete end to South America's distinctive marsupial fauna. The formation of the Isthmus had major consequences on global temperatures, because warm equatorial ocean currents were cut off, and the cold Arctic and Antarctic waters lowered temperatures in the now-isolated Atlantic Ocean. Most of Central America formed during the Pliocene to connect the continents of North and South America, allowing fauna from these continents to leave their native habitats and colonize new areas.[9] Africa's collision with Asia created the Mediterranean Sea, cutting off the remnants of the Tethys Ocean. During the Pleistocene, the modern continents were essentially at their present positions; the tectonic plates on which they sit have probably moved at most 100 km from each other since the beginning of the period.[10]

Climates during the Pliocene became cooler and drier, and seasonal, similar to modern climates. Ice sheets grew on Antarctica. The formation of an Arctic ice cap around three million years ago is signaled by an abrupt shift in oxygen isotope ratios and ice-rafted cobbles in the North Atlantic and North Pacific ocean beds.[11] Mid-latitude glaciation probably began before the end of the epoch. The global cooling that occurred during the Pliocene may have spurred on the disappearance of forests and the spread of grasslands and savannas.[9]

The Pleistocene climate was characterized by repeated glacial cycles during which continental glaciers pushed to the 40th parallel in some places. Four major glacial events have been identified, as well as many minor intervening events. A major event is a general glacial excursion, termed a "glacial". Glacials are separated by "interglacials". During a glacial, the glacier experiences minor advances and

The effects of glaciation were global. Antarctica was ice-bound throughout the Pleistocene and the preceding Pliocene. The Andes were covered in the south by the Patagonian ice cap. There were glaciers in New Zealand and Tasmania. The now decaying glaciers of Mount Kenya, Mount Kilimanjaro, and the Ruwenzori Range in east and central Africa were larger. Glaciers existed in the mountains of Ethiopia and to the west in the Atlas mountains. In the northern hemisphere, many glaciers fused into one. The Cordilleran ice sheet covered the North American northwest; the Laurentide covered the east. The Fenno-Scandian ice sheet covered northern Europe, including Great Britain; the Alpine ice sheet covered the Alps. Scattered domes stretched across Siberia and the Arctic shelf. The northern seas were frozen. During the late Upper Paleolithic

(Latest Pleistocene) *c.* 18,000 BP, the Beringia land bridge between Asia and North America was blocked by ice,[10] which may have prevented early Paleo-Indians such as the Clovis culture from directly crossing Beringa to reach the Americas.

According to Mark Lynas (through collected data), the Pleistocene's overall climate could be characterized as a continuous El Niño with trade winds in the south Pacific weakening or heading east, warm air rising near Peru, warm water spreading from the west Pacific and the Indian Ocean to the east Pacific, and other El Niño markers.[12]

The Paleolithic is often held to finish at the end of the ice age (the end of the Pleistocene epoch), and Earth's climate became warmer. This may have caused or contributed to the extinction of the Pleistocene megafauna, although it is also possible that the late Pleistocene extinctions were (at least in part) caused by other factors such as disease and overhunting by humans.[13][14] New research suggests that the extinction of the woolly mammoth may have been caused by the combined effect of climatic change and human hunting.[14] Scientists suggest that climate change during the end of the Pleistocene caused the mammoths' habitat to shrink in size, resulting in a drop in population. The small populations were then hunted out by Paleolithic humans.[14] The global warming that occurred during the end of the Pleistocene and the beginning of the Holocene may have made it easier for humans to reach mammoth habitats that were previously frozen and inaccessible.[14] Small populations of wooly mammoths survived on isolated Arctic islands, Saint Paul Island and Wrangel Island, till circa 3700 and 1700 BCE respectively. The Wrangel Island population went extinct around the same time the island was settled by prehistoric humans.[15] There's no evidence of prehistoric human presence on Saint Paul island (though early human settlements dating as far back as 6500 BCE were found on nearby Aleutian Islands).[16]

2.3 Human way of life

Nearly all of our knowledge of Paleolithic human culture and way of life comes from archaeology and ethnographic comparisons to modern hunter-gatherer cultures such as the !Kung San who live similarly to their Paleolithic predecessors.[18] The economy of a typical Paleolithic society was a hunter-gatherer economy.[19] Humans hunted wild animals for meat and gathered food, firewood, and materials for their tools, clothes, or shelters.[19] Human population density was very low, around only one person per square mile.[3] This was most likely due to low body fat, infanticide, women regularly engaging in intense endurance exercise,[20] late weaning of infants and a nomadic lifestyle.[3] Like contemporary hunter-gatherers, Paleolithic humans enjoyed an

An artist's rendering of a temporary wood house, based on evidence found at Terra Amata (in Nice, France) and dated to the Lower Paleolithic (c. 400,000 BP)

abundance of leisure time unparalleled in both Neolithic farming societies and modern industrial societies.[19][21] At the end of the Paleolithic, specifically the Middle and or Upper Paleolithic, humans began to produce works of art such as cave paintings, rock art and jewellery and began to engage in religious behavior such as burial and ritual.[22]

2.3.1 Distribution

At the beginning of the Paleolithic, hominids were found primarily in eastern Africa, east of the Great Rift Valley. Most known hominid fossils dating earlier than one million years before present are found in this area, particularly in Kenya, Tanzania, and Ethiopia.

By 1.5-2 million years before present, groups of hominids began leaving Africa and settling southern Europe and Asia. Southern Caucasus was occupied by 1.7 million years BP, and northern China was reached by 1.66 million years BP. By the end of the Lower Paleolithic, members of the hominid family were living in what is now China, western Indonesia, and, in Europe, around the Mediterranean and as far north as England, southern Germany, and Bulgaria. Their further northward expansion may have been limited by the lack of control of fire: studies of cave settlements in Europe indicate no regular use of fire prior to 300,000-400,000 BP.[23] East Asian fossils from this period are typically placed in the genus Homo erectus. Very little fossil evidence is available at known Lower Paleolithic sites in Europe, but it is believed that hominids who inhabited these sites were likewise *Homo erectus*. There is no evidence of hominids in America, Australia, or almost anywhere in Oceania during this time period.

Fates of these early colonists, and their relationships to modern humans, are still subject to debate. According to current archeological and genetic models, there were at least two notable expansion events subsequent to peopling of Eurasia 2-1.5 million years BP. Around 500,000 BP, a group of early humans, frequently called Homo heidelbergensis, came to Europe from Africa and eventually evolved into Neanderthals. Both *Homo erectus* and Neanderthals went extinct by the end of the Paleolithic, having been replaced by a new wave of humans, the anatomically modern Homo sapiens, which emerged in eastern Africa circa 200,000 BP, left Africa around 50,000 BP and expanded throughout the planet. It is likely that multiple groups co-existed for some time in certain locations. Neanderthals were still found in parts of Eurasia 30,000 years before present, and engaged in a limited degree of interbreeding with *Homo sapiens*. Hominid fossils not belonging either to *Homo neanderthalensis* or to *Homo sapiens* geni, found in Altai and Indonesia, were radiocarbon dated to 30,000-40,000 BP and 17,000 BP respectively.

Two Lower Paleolithic bifaces

The technological revolution of the Middle and Upper Paleolithic allowed humans to reach places that weren't accessible earlier. In the Middle Paleolithic, Neanderthals were present in Poland. By 40,000-50,000 BP, first humans set foot in Australia. By 45,000 BP, humans lived at 61° north latitude in Europe.[24] By 30,000 BP, Japan was reached, and by 27,000 BP humans were present in Siberia above the Arctic Circle.[24] At the end of the Upper Paleolithic, a group of humans crossed the Bering land bridge and quickly expanded throughout North and South America. Northern Eurasia became depopulated during the last Glacial Maximum (27,000 to 16,000 BP), but was repopulated as the climate got warmer and glaciers retreated.

For the duration of the Paleolithic, human populations remained low, especially outside the equatorial region. The entire population of Europe between 16,000-11,000 BP likely averaged some 30,000 individuals, and, between 40,000-16,000 BP, it was even lower, at 4,000-6,000 individuals.[25]

Stone ball from a set of Paleolithic bolas

2.3.2 Technology

Tools

Paleolithic humans made tools of stone, bone, and wood.[19] The early paleolithic hominids, Australopithecus, were the first users of stone tools. Excavations in Gona, Ethiopia have produced thousands of artifacts, and through radioisotopic dating and magnetostratigraphy, the sites can be firmly dated to 2.6 million years ago. Evidence shows these early hominids intentionally selected raw materials with good flaking qualities and chose appropriate sized stones for their needs to produce sharp-edged tools for cutting.[26] The earliest Paleolithic stone tool industry, the Olduwan,

began around 2.6 million years ago.[27] It contained tools such as choppers, burins and awls. It was completely replaced around 250,000 years ago by the more complex Acheulean industry, which was first conceived by *Homo ergaster* around 1.8 or 1.65 million years ago.[28] The most recent Lower Paleolithic (Acheulean) implements completely vanished from the archeological record around 100,000 years ago and were replaced by more complex Middle Paleolithic/Middle Stone Age tool kits such as the Mousterian and the Aterian industries.[29]

Lower Paleolithic humans used a variety of stone tools, including hand axes and choppers. Although they appear to have used hand axes often, there is disagreement about their use. Interpretations range from cutting and chopping tools, to digging implements, flake cores, the use in traps and a purely ritual significance, maybe in courting behavior. William H. Calvin has suggested that some hand axes could have served as "killer Frisbees" meant to be thrown at a herd of animals at a water hole so as to stun one of them. There are no indications of hafting, and some artifacts are far too large for that. Thus, a thrown hand axe would not usually have penetrated deeply enough to cause very serious injuries. Nevertheless, it could have been an effective weapon for defense against predators. Choppers and scrapers were likely used for skinning and butchering scavenged animals and sharp ended sticks were often obtained for digging up edible roots. Presumably, early humans used wooden spears as early as five million years ago to hunt small animals, much as their relatives, chimpanzees, have been observed to do in Senegal, Africa.[30] Lower Paleolithic humans constructed shelters such as the possible wood hut at Terra Amata.

Fire use

Fire was used by the Lower Paleolithic hominid *Homo erectus*/*Homo ergaster* as early as 300,000 or 1.5 million years ago and possibly even earlier by the early Lower Paleolithic (Oldowan) hominid *Homo habilis* and/or by robust australopithecines such as *Paranthropus*.[3] However, the use of fire only became common in the societies of the following Middle Stone Age/Middle Paleolithic Period.[2] Use of fire reduced mortality rates and provided protection against predators.[31] Early hominids may have begun to cook their food as early as the Lower Paleolithic (*c.* 1.9 million years ago) or at the latest in the early Middle Paleolithic (*c.* 250,000 years ago).[32] Some scientists have hypothesized that Hominids began cooking food to defrost frozen meat, which would help ensure their survival in cold regions.[32]

Rafts

The Lower Paleolithic hominid *Homo erectus* possibly invented rafts (*c.* 800,000 or 840,000 BP) to travel over large bodies of water, which may have allowed a group of *Homo erectus* to reach the island of Flores and evolve into the small hominid *Homo floresiensis*. However, this hypothesis is disputed within the anthropological community.[33][34] The possible use of rafts during the Lower Paleolithic may indicate that Lower Paleolithic Hominids such as *Homo erectus* were more advanced than previously believed, and may have even spoken an early form of modern language.[33] Supplementary evidence from Neanderthal and Modern human sites located around the Mediterranean Sea such as Coa de sa Multa (*c.* 300,000 BP) has also indicated that both Middle and Upper Paleolithic humans used rafts to travel over large bodies of water (i.e. the Mediterranean Sea) for the purpose of colonizing other bodies of land.[33][35]

Advanced tools

Around 200,000 BP, Middle Paleolithic Stone tool manufacturing spawned a tool making technique known as the prepared-core technique, that was more elaborate than previous Acheulean techniques.[4] This technique increased efficiency by allowing the creation of more controlled and consistent flakes.[4] It allowed Middle Paleolithic humans to create stone tipped spears, which were the earliest composite tools, by hafting sharp, pointy stone flakes onto wooden shafts. In addition to improving tool making methods, the Middle Paleolithic also saw an improvement of the tools themselves that allowed access to a wider variety and amount of food sources. For example, microliths or small stone tools or points were invented around 70,000 or 65,000 BP and were essential to the invention of bows and spear throwers in the following Upper Paleolithic period.[31] Harpoons were invented and used for the first time during the late Middle Paleolithic (c.90,000 years ago); the invention of these devices brought fish into the human diets, which provided a hedge against starvation and a more abundant food supply.[35][36] Thanks to their technology and their advanced social structures, Paleolithic groups such as the Neanderthals who had a Middle Paleolithic level of technology, appear to have hunted large game just as well as Upper Paleolithic modern humans[37] and the Neanderthals in particular may have likewise hunted with projectile weapons.[38] Nonetheless, Neanderthal use of projectile weapons in hunting occurred very rarely (or perhaps never) and the Neanderthals hunted large game animals mostly by ambushing them and attacking them with mêlée weapons such as thrusting spears rather than attacking them from a distance with projectile weapons.[22][39]

Other inventions

During the Upper Paleolithic, further inventions were made, such as the net (*c.* 22,000 or 29,000 BP)[31] bolas,[40] the spear thrower (c.30,000 BP), the bow and arrow (*c.* 25,000 or 30,000 BP)[3] and the oldest example of ceramic art, the Venus of Dolní Věstonice (*c.* 29,000–25,000 BCE).[3] Early dogs were domesticated, sometime between 30,000 BP and 14,000 BP, presumably to aid in hunting.[41] However, the earliest instances of successful domestication of dogs may be much more ancient than this. Evidence from canine DNA collected by Robert K. Wayne suggests that dogs may have been first domesticated in the late Middle Paleolithic around 100,000 BP or perhaps even earlier.[42] Archeological evidence from the Dordogne region of France demonstrates that members of the European early Upper Paleolithic culture known as the Aurignacian used calendars (*c.* 30,000 BP). This was a lunar calendar that was used to document the phases of the moon. Genuine solar calendars did not appear until the following Neolithic period.[43] Upper Paleolithic cultures were probably able to time the migration of game animals such as wild horses and deer.[44] This ability allowed humans to become efficient hunters and to exploit a wide variety of game animals.[44] Recent research indicates that the Neanderthals timed their hunts and the migrations of game animals long before the beginning of the Upper Paleolithic.[37]

2.3.3 Social organization

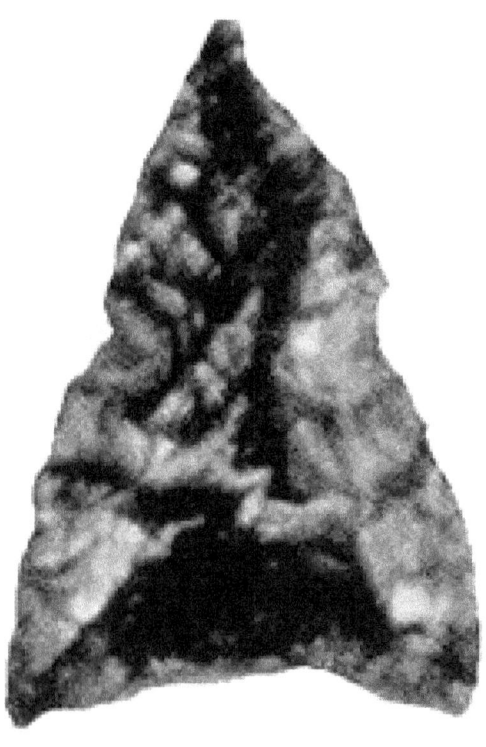

Humans may have taken part in long-distance trade between bands for rare commodities and raw materials (such as stone needed for making tools) as early as 120,000 years ago in Middle Paleolithic.

The social organization of the earliest Paleolithic (Lower Paleolithic) societies remains largely unknown to scientists, though Lower Paleolithic hominids such as *Homo habilis* and *Homo erectus* are likely to have had more complex social structures than chimpanzee societies.[45] Late Oldowan/Early Acheulean humans such as *Homo ergaster/Homo erectus* may have been the first people to invent central campsites or home bases and incorporate them into their foraging and hunting strategies like contemporary hunter-gatherers, possibly as early as 1.7 million years

ago;[4] however, the earliest solid evidence for the existence of home bases or central campsites (hearths and shelters) among humans only dates back to 500,000 years ago.[4]

Similarly, scientists disagree whether Lower Paleolithic humans were largely monogamous or polygynous.[45] In particular, the Provisional model suggests that bipedalism arose in Pre Paleolithic australopithecine societies as an adaptation to monogamous lifestyles; however, other researchers note that sexual dimorphism is more pronounced in Lower Paleolithic humans such as *Homo erectus* than in Modern humans, who are less polygynous than other primates, which suggests that Lower Paleolithic humans had a largely polygynous lifestyle, because species that have the most pronounced sexual dimorphism tend more likely to be polygynous.[46]

Human societies from the Paleolithic to the early Neolithic farming tribes lived without states and organized governments. For most of the Lower Paleolithic, human societies were possibly more hierarchical than their Middle and Upper Paleolithic descendants, and probably were not grouped into bands,[47] though during the end of the Lower Paleolithic, the latest populations of the hominid *Homo erectus* may have begun living in small-scale (possibly egalitarian) bands similar to both Middle and Upper Paleolithic societies and modern hunter-gatherers.[47]

Middle Paleolithic societies, unlike Lower Paleolithic and early Neolithic ones, consisted of bands that ranged from 20 to 30 or 25 to 100 members and were usually nomadic.[3][47] These bands were formed by several families. Bands sometimes joined together into larger "macrobands" for activities such as acquiring mates and celebrations or where resources were abundant.[3] By the end of the Paleolithic era, about 10,000 BP people began to settle down into permanent locations, and began to rely on agriculture for sustenance in many locations. Much evidence exists that humans took part in long-distance trade between bands for rare commodities (such as ochre, which was often used for religious purposes such as ritual[48][49]) and raw materials, as early as 120,000 years ago in Middle Paleolithic.[22] Inter-band trade may have appeared during the Middle Paleolithic because trade between bands would have helped ensure their survival by allowing them to exchange resources and commodities such as raw materials during times of relative scarcity (i.e. famine, drought).[22] Like in modern hunter-gatherer societies, individuals in Paleolithic societies may have been subordinate to the band as a whole.[18][19] Both Neanderthals and modern humans took care of the elderly members of their societies during the Middle and Upper Paleolithic.[22]

Some sources claim that most Middle and Upper Paleolithic societies were possibly fundamentally egalitarian[3][19][35][50] and may have rarely or never engaged in organized violence between groups (i.e. war).[35][51][52][53] Some Upper Paleolithic societies in resource-rich environments (such as societies in Sungir, in what is now Russia) may have had more complex and hierarchical organization (such as tribes with a pronounced hierarchy and a somewhat formal division of labor) and may have engaged in endemic warfare.[35][54] Some argue that there was no formal leadership during the Middle and Upper Paleolithic. Like contemporary egalitarian hunter-gatherers such as the Mbuti pygmies, societies may have made decisions by communal consensus decision making rather than by appointing permanent rulers such as chiefs and monarchs.[6] Nor was there a formal division of labor during the Paleolithic. Each member of the group was skilled at all tasks essential to survival, regardless of individual abilities. Theories to explain the apparent egalitarianism have arisen, notably the Marxist concept of primitive communism.[55][56] Christopher Boehm (1999) has hypothesized that egalitarianism may have evolved in Paleolithic societies because of a need to distribute resources such as food and meat equally to avoid famine and ensure a stable food supply.[57] Raymond C. Kelly speculates that the relative peacefulness of Middle and Upper Paleolithic societies resulted from a low population density, cooperative relationships between groups such as reciprocal exchange of commodities and collaboration on hunting expeditions, and because the invention of projectile weapons such as throwing spears provided less incentive for war, because they increased the damage done to the attacker and decreased the relative amount of territory attackers could gain.[53] However, other sources claim that most Paleolithic groups may have been larger, more complex, sedentary and warlike than most contemporary hunter-gatherer societies, due to occupying more resource-abundant areas than most modern hunter-gatherers who have been pushed into more marginal habitats by agricultural societies.[58]

Anthropologists have typically assumed that in Paleolithic societies, women were responsible for gathering wild plants and firewood, and men were responsible for hunting and scavenging dead animals.[3][35] However, analogies to existent hunter-gatherer societies such as the Hadza people and the Australian aborigines suggest that the sexual division of labor in the Paleolithic was relatively flexible. Men may have participated in gathering plants, firewood and insects, and women may have procured small game animals for consumption and assisted men in driving herds of large game animals (such as woolly mammoths and deer) off cliffs.[35][52] Additionally, recent research by anthropologist and archaeologist Steven Kuhn from the University of Arizona is argued to support that this division of labor did not exist prior to the Upper Paleolithic and was invented relatively recently in human pre-history.[59][60] Sexual divi-

sion of labor may have been developed to allow humans to acquire food and other resources more efficiently.[60] Possibly there was approximate parity between men and women during the Middle and Upper Paleolithic, and that period may have been the most gender-equal time in human history.[51][61][62] Archeological evidence from art and funerary rituals indicates that a number of individual women enjoyed seemingly high status in their communities, and it is likely that both sexes participated in decision making.[62] The earliest known Paleolithic shaman (c. 30,000 BP) was female.[63] Jared Diamond suggests that the status of women declined with the adoption of agriculture because women in farming societies typically have more pregnancies and are expected to do more demanding work than women in hunter-gatherer societies.[64] Like most contemporary hunter-gatherer societies, Paleolithic and the Mesolithic groups probably followed mostly matrilineal and ambilineal descent patterns; patrilineal descent patterns were probably rarer than in the following Neolithic period.[31][49]

2.3.4   Art and music

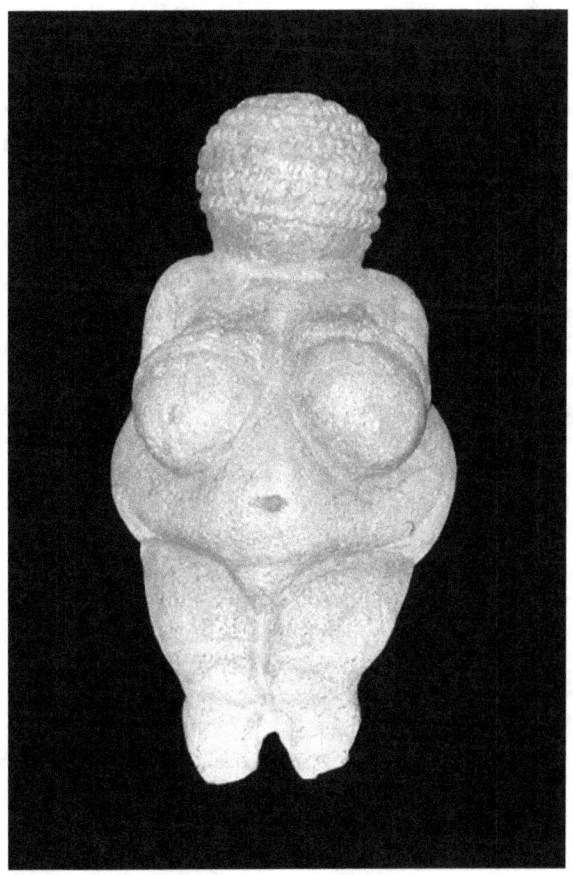

The Venus of Willendorf is one of the most famous Venus figurines.

Early examples of artistic expression, such as the Venus of

Tan-Tan and the patterns found on elephant bones from Bilzingsleben in Thuringia, may have been produced by Acheulean tool users such as *Homo erectus* prior to the start of the Middle Paleolithic period. However, the earliest undisputed evidence of art during the Paleolithic period comes from Middle Paleolithic/Middle Stone Age sites such as Blombos Cave –South Africa– in the form of bracelets,[65] beads,[66] rock art,[48] and ochre used as body paint and perhaps in ritual.[35][48] Undisputed evidence of art only becomes common in the following Upper Paleolithic period.[67]

According to Robert G. Bednarik, Lower Paleolithic Acheulean tool users began to engage in symbolic behavior such as art around 850,000 BP and decorated themselves with beads and collected exotic stones for aesthetic rather than utilitarian qualities.[68] According to Bednarik, traces of the pigment ochre from late Lower Paleolithic Acheulean archeological sites suggests that Acheulean societies, like later Upper Paleolithic societies, collected and used ochre to create rock art.[68] Nevertheless, it is also possible that the ochre traces found at Lower Paleolithic sites is naturally occurring.[69]

Vincent W. Fallio interprets Lower and Middle Paleolithic marking on rocks at sites such as Bilzingsleben (such as zig zagging lines) as accounts or representation of altered states of consciousness[70] though some other scholars interpret them as either simple doodling or as the result of natural processes.

Upper Paleolithic humans produced works of art such as cave paintings, Venus figurines, animal carvings and rock paintings.[71] Upper Paleolithic art can be divided into two broad categories: figurative art such as cave paintings that clearly depicts animals (or more rarely humans); and non-figurative, which consists of shapes and symbols.[71] Cave paintings have been interpreted in a number of ways by modern archeologists. The earliest explanation, by the pre-historian Abbe Breuil, interpreted the paintings as a form of magic designed to ensure a successful hunt.[72] However, this hypothesis fails to explain the existence of animals such as saber-toothed cats and lions, which were not hunted for food, and the existence of half-human, half-animal beings in cave paintings. The anthropologist David Lewis-Williams has suggested that Paleolithic cave paintings were indications of shamanistic practices, because the paintings of half-human, half-animal paintings and the remoteness of the caves are reminiscent of modern hunter-gatherer shamanistic practices.[72] Symbol-like images are more common in Paleolithic cave paintings than are depictions of animals or humans, and unique symbolic patterns might have been trademarks that represent different Upper Paleolithic ethnic groups.[73] Venus figurines have evoked similar controversy. Archeologists and anthropologists have described the figurines as representations of goddesses, pornographic imagery, apotropaic amulets used for sympathetic magic, and even as self-portraits of women themselves.[35][74]

R. Dale Guthrie[75] has studied not only the most artistic and publicized paintings, but also a variety of lower-quality art and figurines, and he identifies a wide range of skill and ages among the artists. He also points out that the main themes in the paintings and other artifacts (powerful beasts, risky hunting scenes and the over-sexual representation of women) are to be expected in the fantasies of adolescent males during the Upper Paleolithic.

The Venus figurines have sometimes been interpreted as representing a mother goddess; the abundance of such female imagery has led some to believe that Upper Paleolithic (and later Neolithic) societies had a female-centered religion and a female-dominated society. For example, this was proposed by the archeologist Marija Gimbutas and the feminist scholar Merlin Stone who was the author of the 1978 book *When God Was a Woman*.[76][77] Various other explanations for the purpose of the figurines have been proposed, such as Catherine McCoid and LeRoy McDermott's hypothesis that the figurines were created as self-portraits of actual women[74] and R.Dale Gutrie's hypothesis that the venus figurines represented a kind of "stone age pornography".

The origins of music during the Paleolithic are unknown, since the earliest forms of music probably did not use musical instruments but instead used the human voice and or natural objects such as rocks, which leave no trace in the archaeological record. However, the anthropological and archeological designation suggests that human music first arose when language, art and other modern behaviors developed in the Middle or the Upper Paleolithic period. Music may have developed from rhythmic sounds produced by daily activities such as cracking nuts by hitting them with stones, because maintaining a rhythm while working may have helped people to become more efficient at daily activities.[78] An alternative theory originally proposed by Charles Darwin explains that music may have begun as a hominid mating strategy as many birds and some other animals produce music like calls to attract mates.[79] This hypothesis is generally less accepted than the previous hypothesis, but it nonetheless provides a possible alternative. Another explanation is that humans began to make music simply because of the pleasure it produced.

Upper Paleolithic (and possibly Middle Paleolithic)[80] humans used flute-like bone pipes as musical instruments,[35][81] and music may have played a large role in the religious lives of Upper Paleolithic hunter-gatherers. As with modern hunter-gatherer societies, music may have been used in ritual or to help induce trances. In particular, it appears that animal skin drums may have been used in religious events by Upper Paleolithic shamans, as shown by the remains of drum-like instruments from some Upper Paleolithic graves of shamans and the ethnographic record of contemporary hunter-gatherer shamanic and ritual practices.[71][63]

2.3.5 Religion and beliefs

Main article: Paleolithic Religion
According to James B. Harrod humankind first devel-

Picture of a half-human, half-animal being in a Paleolithic cave painting in Dordogne. France. Archeologists believe that cave paintings of half-human, half-animal beings may be evidence for early shamanic practices during the Paleolithic.

oped religious and spiritual beliefs during the Middle Paleolithic or Upper Paleolithic.[82] Controversial scholars of prehistoric religion and anthropology, James Harrod and Vincent W. Fallio, have recently proposed that religion and spirituality (and art) may have first arisen in Pre-Paleolithic chimpanzees[83] or Early Lower Paleolithic (Oldowan) societies.[70][84] According to Fallio, the common ancestor of chimpanzees and humans experienced altered states of consciousness and partook in ritual, and ritual was used in their societies to strengthen social bonding and group cohesion.[70]

Middle Paleolithic humans' use of burials at sites such as Krapina, Croatia (*c.* 130,000 BP) and Qafzeh, Israel (*c.* 100,000 BP) have led some anthropologists and archeologists, such as Philip Lieberman, to believe that Middle Paleolithic humans may have possessed a belief in an afterlife and a "concern for the dead that transcends daily life".[5] Cut marks on Neanderthal bones from various sites, such as

Combe-Grenal and Abri Moula in France, suggest that the Neanderthals like some contemporary human cultures may have practiced ritual defleshing for (presumably) religious reasons. According to recent archeological findings from *H. heidelbergensis* sites in Atapuerca, humans may have begun burying their dead much earlier, during the late Lower Paleolithic; but this theory is widely questioned in the scientific community.

Likewise, some scientists have proposed that Middle Paleolithic societies such as Neanderthal societies may also have practiced the earliest form of totemism or animal worship, in addition to their (presumably religious) burial of the dead. In particular, Emil Bächler suggested (based on archeological evidence from Middle Paleolithic caves) that a bear cult was widespread among Middle Paleolithic Neanderthals.[85] A claim that evidence was found for Middle Paleolithic animal worship c 70,000 BCE originates from the Tsodilo Hills in the African Kalahari desert has been denied by the original investigators of the site.[86] Animal cults in the following Upper Paleolithic period, such as the bear cult, may have had their origins in these hypothetical Middle Paleolithic animal cults.[87] Animal worship during the Upper Paleolithic was intertwined with hunting rites.[87] For instance, archeological evidence from art and bear remains reveals that the bear cult apparently involved a type of sacrificial bear ceremonialism, in which a bear was sliced with arrows, finished off by a blast in the lungs, and ritualistically worshipped near a clay bear statue covered by a bear fur with the skull and the body of the bear buried separately.[87] Barbara Ehrenreich controversially theorizes that the sacrificial hunting rites of the Upper Paleolithic (and by extension Paleolithic cooperative big-game hunting) gave rise to war or warlike raiding during the following Epi-Paleolithic/Mesolithic or late Upper Paleolithic period.[52]

The existence of anthropomorphic images and half-human, half-animal images in the Upper Paleolithic period may further indicate that Upper Paleolithic humans were the first people to believe in a pantheon of gods or supernatural beings,[88] though such images may instead indicate shamanistic practices similar to those of contemporary tribal societies.[72] The earliest known undisputed burial of a shaman (and by extension the earliest undisputed evidence of shamans and shamanic practices) dates back to the early Upper Paleolithic era (*c.* 30,000 BP) in what is now the Czech Republic.[63] However, during the early Upper Paleolithic it was probably more common for all members of the band to participate equally and fully in religious ceremonies, in contrast to the religious traditions of later periods when religious authorities and part-time ritual specialists such as shamans, priests and medicine men were relatively common and integral to religious life.[19] Additionally, it is also possible that Upper Paleolithic religions, like

contemporary and historical animistic and polytheistic religions, believed in the existence of a single creator deity
in addition to other supernatural beings such as animistic
spirits.[89]

Vincent W. Fallio writes that ancestor cults first emerged
in complex Upper Paleolithic societies. He argues that the
elites of these societies (like the elites of many more contemporary complex hunter-gatherers such as the Tlingit)
may have used special rituals and ancestor worship to so-
lidify control over their societies, by convincing their subjects that they possess a link to the spirit world that also
gives them control over the earthly realm.[70] Secret societies may have served a similar function in these complex
quasi-theocratic societies, by dividing the religious practices of these cultures into the separate spheres of Popular
Religion and Elite Religion.[70]

Religion was possibly apotropaic; specifically, it may have
involved sympathetic magic.[35] The Venus figurines, which
are abundant in the Upper Paleolithic archeological record,
provide an example of possible Paleolithic sympathetic
magic, as they may have been used for ensuring success in hunting and to bring about fertility of the land
and women.[3] The Upper Paleolithic Venus figurines have
sometimes been explained as depictions of an earth goddess similar to Gaia, or as representations of a goddess who
is the ruler or mother of the animals.[87][90] James Harrod
has described them as representative of female (and male)
shamanistic spiritual transformation processes.[91]

*People may have first fermented grapes in animal skin pouches to
create wine during the Paleolithic.[92]*

2.3.6 Diet and nutrition

Paleolithic hunting and gathering people ate varying proportions of leafy vegetables, fruit, nuts and insects, meat,
fish, and shellfish.[93][94] However, there is little direct evidence of the relative proportions of plant and animal
foods.[95] Although the term "paleolithic diet", without references to a specific timeframe or locale, is sometimes used
with an implication that most humans shared a certain diet
during the entire era, that is not entirely accurate. The Paleolithic was an extended period of time, during which multiple technological advances were made, many of which had
impact on human dietary structure. For example, humans
probably did not possess the control of fire until the Middle Paleolithic,[96] or tools necessary to engage in extensive
fishing. On the other hand, both these technologies are generally agreed to have been widely available to humans by
the end of the Paleolithic (consequently, allowing humans
in some regions of the planet to rely heavily on fishing and
hunting). In addition, the Paleolithic involved a substantial
geographical expansion of human populations. During the
Lower Paleolithic, ancestors of modern humans are thought
to have been constrained to Africa east of the Great Rift

Valley. During the Middle and Upper Paleolithic, humans
greatly expanded their area of settlement, reaching ecosystems as diverse as New Guinea and Alaska, and adapting
their diets to whatever local resources available.

Another view is that until the Upper Paleolithic, humans
were frugivores (fruit eaters) who supplemented their meals
with carrion, eggs, and small prey such as baby birds and
mussels, and only on rare occasions managed to kill and
consume big game such as antelopes.[97] This view is supported by studies of higher apes, particularly chimpanzees.
Chimpanzees are the closest to humans genetically, sharing more than 96% of their DNA code with humans, and
their digestive tract is functionally very similar to that of
humans.[98] Chimpanzees are primarily frugivores, but they
could and would consume and digest animal flesh, given
the opportunity. In general, their actual diet in the wild is
about 95% plant-based, with the remaining 5% filled with
insects, eggs, and baby animals.[99][100] In some ecosystems,
however, chimpanzees are predatory, forming parties to
hunt monkeys.[101] Some comparative studies of human and
higher primate digestive tracts do suggest that humans have

evolved to obtain greater amounts of calories from sources such as animal foods, allowing them to shrink the size of the gastrointestinal tract relative to body mass and to increase the brain mass instead.[102][103]

A difficulty with the frugivore point of view is that humans are established to conditionally require certain long-chain polyunsaturated fatty acids (LC-PUFAs), such as AA and DHA, from the diet.[104] Humans' LC-PUFA requirements are much greater than chimpanzees' because of humans' larger brain mass, and humans' abilities to synthesize them from other nutrients are poor, suggesting readily available external sources.[105] Pregnant and lactating females require 100 mg of DHA per day. However, LC-PUFAs are almost nonexistent in plants and in most tissues of warm-climate animals.

Anthropologists have diverse opinions about the proportions of plant and animal foods consumed. Just as with still existing hunters and gatherers, there were many varied "diets" - in different groups - and also varying through this vast amount of time. Some paleolithic hunter-gatherers consumed a significant amount of meat and possibly obtained most of their food from hunting,[106] while others are shown as a primarily plant-based diet,[59] Most, if not all, are believed to have been opportunistic omnivores.[107] One hypothesis is that carbohydrate tubers (plant underground storage organs) may have been eaten in high amounts by pre-agricultural humans.[108][109][110][111] It is thought that the Paleolithic diet included as much as 1.65–1.9 kilograms per day of fruit and vegetables.[112] The relative proportions of plant and animal foods in the diets of Paleolithic people often varied between regions, with more meat being necessary in colder regions (which weren't populated by anatomically modern humans until 30,000-50,000 BP).[113] It is generally agreed that many modern hunting and fishing tools, such as fish hooks, nets, bows, and poisons, weren't introduced until the Upper Paleolithic and possibly even Neolithic.[31] The only hunting tools widely available to humans during any significant part of the Paleolithic period were hand-held spears and harpoons. There's evidence of Paleolithic people killing and eating seals and elands as far as 100,000 years BP. On the other hand, buffalo bones found in African caves from the same period are typically of very young or very old individuals, and there's no evidence that pigs, elephants or rhinos were hunted by humans at the time.[114]

Paleolithic peoples suffered less famine and malnutrition than the Neolithic farming tribes that followed them.[18][115] This was partly because Paleolithic hunter-gatherers accessed to a wider variety natural foods, which allowed them a more nutritious diet and a decreased risk of famine.[18][20][64] Many of the famines experienced by Neolithic (and some modern) farmers were caused or amplified by their dependence on a small number of crops.[18][20][64] It is thought that wild foods can have a significantly different nutritional profile than cultivated foods.[116] The greater amount of meat obtained by hunting big game animals in Paleolithic diets than Neolithic diets may have also allowed Paleolithic hunter-gatherers to enjoy a more nutritious diet than Neolithic agriculturalists.[115] It has been argued that the shift from hunting and gathering to agriculture resulted in an increasing focus on a limited variety of foods, with meat likely taking a back seat to plants.[117] It is also unlikely that Paleolithic hunter-gatherers were affected by modern diseases of affluence such as Type 2 diabetes, coronary heart disease and cerebrovascular disease, because they ate mostly lean meats and plants and frequently engaged in intense physical activity,[118][119] and because the average lifespan was shorter than the age of common-onset of these conditions.[120][121]

Large-seeded legumes were part of the human diet long before the Neolithic agricultural revolution, as evident from archaeobotanical finds from the Mousterian layers of Kebara Cave, in Israel.[122] There is evidence suggesting that Paleolithic societies were gathering wild cereals for food use at least as early as 30,000 years ago.[123] However, seeds, such as grains and beans, were rarely eaten and never in large quantities on a daily basis.[124] Recent archeological evidence also indicates that winemaking may have originated in the Paleolithic, when early humans drank the juice of naturally fermented wild grapes from animal-skin pouches.[92] Paleolithic humans consumed animal organ meats, including the livers, kidneys and brains. Upper Paleolithic cultures appear to have had significant knowledge about plants and herbs and may have, albeit very rarely, practiced rudimentary forms of horticulture.[125] In particular, bananas and tubers may have been cultivated as early as 25,000 BP in southeast Asia.[58] Late Upper Paleolithic societies also appear to have occasionally practiced pastoralism and animal husbandry, presumably for dietary reasons. For instance, some European late Upper Paleolithic cultures domesticated and raised reindeer, presumably for their meat or milk, as early as 14,000 BP.[41] Humans also probably consumed hallucinogenic plants during the Paleolithic period.[3] The Australian Aborigines have been consuming a variety of native animal and plant foods, called bushfood, for an estimated 60,000 years, since the Middle Paleolithic.

People during the Middle Paleolithic, such as the Neanderthals and Middle Paleolithic Homo sapiens in Africa, began to catch shellfish for food as revealed by shellfish cooking in Neanderthal sites in Italy about 110,000 years ago and Middle Paleolithic *Homo sapiens* sites at Pinnacle Point, in Africa around 164,000 BP.[35][126] Although fishing only became common during the Upper Paleolithic,[35][127] fish have been part of human diets long before the dawn of the

Large game animals such as deer were an important source of protein in Middle and Upper Paleolithic diets.

Upper Paleolithic and have certainly been consumed by humans since at least the Middle Paleolithic.[44] For example, the Middle Paleolithic *Homo sapiens* in the region now occupied by the Democratic Republic of the Congo hunted large 6-foot (1.8 m)-long catfish with specialized barbed fishing points as early as 90,000 years ago.[35][44] The invention of fishing allowed some Upper Paleolithic and later hunter-gatherer societies to become sedentary or semi-nomadic, which altered their social structures.[81] Example societies are the Lepenski Vir as well as some contemporary hunter-gatherers such as the Tlingit. In some instances (at least the Tlingit) they developed social stratification, slavery and complex social structures such as chiefdoms.[31]

Anthropologists such as Tim White suggest that cannibalism was common in human societies prior to the beginning of the Upper Paleolithic, based on the large amount of "butchered human" bones found in Neanderthal and other Lower/Middle Paleolithic sites.[128] Cannibalism in the Lower and Middle Paleolithic may have occurred because of food shortages.[129] However, it may have been for religious reasons, and would coincide with the development of religious practices thought to have occurred during the Upper Paleolithic.[87][130] Nonetheless, it remains possible that Paleolithic societies never practiced cannibalism, and

that the damage to recovered human bones was either the result of ritual post-mortem bone cleaning or predation by carnivores such as saber tooth cats, lions and hyenas.[87]

2.4 Events

- By 11,000 BCE - Paleo-Indians reach the Tierra del Fuego.

2.5 See also

- Abbassia Pluvial
- Caveman
- Japanese Paleolithic
- Lascaux
- Late Glacial Maximum
- List of archaeological sites by continent and age#Palaeolithic
- Luzia Woman
- Models of migration to the New World
- Mousterian Pluvial
- Origins of society
- Palaeoarchaeology
- Paleolithic lifestyle
- Turkana Boy

2.6 References

[1] Christian, David (2014). *Big History: Between Nothing and Everything*. New York, New York: McGraw Hill Education. p. 93.

[2] Toth, Nicholas; Schick, Kathy (2007). "Handbook of Paleoanthropology". In Henke, H.C. Winfried; Hardt, Thorolf; Tatersall, Ian. *Handbook of Paleoanthropology*. Volume 3. Berlin; Heidelberg; New York: Springer-Verlag. p. 1944. (PRINT: ISBN 978-3-540-32474-4 ONLINE: ISBN 978-3-540-33761-4)

[3] McClellan (2006). *Science and Technology in World History: An Introduction*. Baltimore, Maryland: JHU Press. ISBN 0-8018-8360-1. Pages 6–12

[4] "Human Evolution," Microsoft Encarta Online Encyclopedia 2007 Contributed by Richard B. Potts, B.A., Ph.D.

[5] Philip Lieberman (1991). *Uniquely Human*. Cambridge, Mass.: Harvard University Press. ISBN 0-674-92183-6.

[6] Kusimba, Sibel (2003). *African Foragers: Environment, Technology, Interactions* Rowman Altamira. p. 285. ISBN 0-7591-0154-X.

[7] Lubbock, John (1872). *Pre-Historic Times, as Illustrated by Ancient Remains, and the Manners and Customs of Modern Savages*, Williams and Norgate (p 75) ISBN 978-1421270395.

[8] "The Pleistocene Epoch". University of California Museum of Paleontology. Retrieved 22 August 2014.

[9] "University of California Museum of Paleontology website the Pliocene epoch(accessed March 25)". Ucmp.berkeley.edu. Retrieved 2010-01-31.

[10] Christopher Scotese. "Paleomap project". *The Earth has been in an Ice House Climate for the last 30 million years.* Retrieved 2008-03-23.

[11] Van Andel, Tjeerd H. (1994). *New Views on an Old Planet: A History of Global Change*. Cambridge: Cambridge University Press. pp. 454 pp. ISBN 0-521-44243-5.

[12] National Geographic Channel, *Six Degrees Could Change The World*, Mark Lynas interview. Retrieved February 14, 2008.

[13] "University of California Museum of Paleontology website the Pleistocene epoch(accessed March 25)". Ucmp.berkeley.edu. Retrieved 2010-01-31.

[14] Kimberly Johnson. "National geographic news". *Climate Change, Then Humans, Drove Mammoths Extinct from National geographic*. Retrieved 2008-04-04.

[15] Nowak, Ronald M. (1999). *Walker's Mammals of the World*. Baltimore: Johns Hopkins University Press. ISBN 0-8018-5789-9.

[16] "Phylogeographic Analysis of the mid-Holocene Mammoth from Qagnax Cave, St. Paul Island, Alaska" (PDF).

[17] Gamble, Clive (1990), El poblamiento Paleolítico de Europa, Barcelona: Editorial Crítica. ISBN 84-7423-445-X.

[18] Leften Stavros Stavrianos (1997). *Lifelines from Our Past: A New World History*. New Jersey, USA: M.E. Sharpe. ISBN 0-13-357005-3. Pages 9–13Page 70

[19] Leften Stavros Stavrianos (1991). *A Global History from Prehistory to the Present*. New Jersey, USA: Prentice Hall. ISBN 0-13-357005-3. Pages 9–13

[20] "The Consequences of Domestication and Sedentism by Emily Schultz, et al". Primitivism.com. Retrieved 2010-01-31.

[21] Felipe Fernandez Armesto (2003). *Ideas that changed the world*. Newyork: Dorling Kindersley limited. p. 400. ISBN 978-0-7566-3298-4.; Page 10

[22] Hillary Mayell. "When Did "Modern" Behavior Emerge in Humans?". *National Geographic News*. Retrieved 2008-02-05.

[23] "On the earliest evidence for habitual use of fire in Europe", Wil Roebroeks et al, PNAS, 2011

[24] John Weinstock. "Sami Prehistory Revisited: transactions, admixture and assimilation in the phylogeographic picture of Scandinavia".

[25] Jean-Pierre Bocquet-Appel; et al. (2005). "Estimates of Upper Palaeolithic meta-population size in Europe from archaeological data" (PDF). *Journal of Archaeological Science* **32**: 1656–1668. doi:10.1016/j.jas.2005.05.006.

[26] Semaw, Sileshi (2000). "The World's Oldest Stone Artefacts from Gona, Ethiopia: Their Implications for Understanding Stone Technology and Patterns of Human Evolution Between 2.6-1.5 Million Years Ago". *Journal of Archaeological Science* **27** (12): 1197–1214. doi:10.1006/jasc.1999.0592.

[27] Klein, R. (1999). *The Human Career*. University of Chicago Press.

[28] Roche H et al., 2002, *Les sites archaéologiques pioplÉistocènes de la formation de Nachuku 663–673, qtd in Scarre, 2005*

[29] Clark, JD, *Variability in primary and secondary technologies of the Later Acheulian in Africa* in Milliken, S and Cook, J (eds), 2001

[30] Rick Weiss, "Chimps Observed Making Their Own Weapons", *The Washington Post*, February 22, 2007

[31] Marlowe FW (2005). "Hunter-gatherers and human evolution" (PDF). *Evolutionary Anthropology* **14** (2): 15294. doi:10.1002/evan.20046.

[32] Wrangham R, Conklin-Brittain N. (September 2003). "Cooking as a biological trait" (PDF). *Comp Biochem Physiol a Mol Integr Physiol* **136** (1): 35–46. doi:10.1016/S1095-6433(03)00020-5. PMID 14527628.

[33] "First Mariners Project Photo Gallery 1". Mc2.vicnet.net.au. Retrieved 2010-01-31.

[34] "First Mariners - National Geographic project 2004". Mc2.vicnet.net.au. 2004-10-02. Retrieved 2010-01-31.

[35] Miller, Barbra; Bernard Wood; Andrew Balansky; Julio Mercader; Melissa Panger (2006). *Anthropology*. Boston Massachusetts: Allyn and Bacon. p. 768. ISBN 0-205-32024-4.

[36] "Human Evolution," Microsoft Encarta Online Encyclopedia 2007 Contributed by Richard B. Potts, B.A., Ph.D.

[37] Ann Parson. "Neanderthals Hunted as Well as Humans, Study Says". *National Geographic News*. Retrieved 2008-02-01.

[38] Boëda, E.; Geneste, J.M.; Griggo, C.; Mercier, N.; Muhesen, S.; Reyss, J.L.; Taha, A.; Valladas, H. (1999). "A Levallois point embedded in the vertebra of a wild ass (Equus africanus): Hafting, projectiles and Mousterian hunting". *Antiquity* **73**: 394–402.

[39] Cameron Balbirnie (2005-02-10). "The icy truth behind Neanderthals". *BBC News*. Retrieved 2008-04-01.

[40] J. Chavaillon, D. Lavallée, « Bola », in *Dictionnaire de la Préhistoire*, PUF, 1988.

[41] Lloyd, J & Mitchinson, J: "The Book of General Ignorance". Faber & Faber, 2006.

[42] Christine Mellot. "stalking the ancient dog" (PDF). *Science news*. Retrieved 2008-01-03.

[43] Felipe Fernandez Armesto (2003). *Ideas that changed the world*. New York: Dorling Kindersley limited. p. 400. ISBN 978-0-7566-3298-4.;

[44] "Stone Age," Microsoft Encarta Online Encyclopedia 2007 Contributed by Kathy Schick, B.A., M.A., Ph.D. and Nicholas Toth, B.A., M.A., Ph.D.

[45] Nancy White. "Intro to archeology The First People and Culture". *Introduction to archeology*. Retrieved 2008-03-20.

[46] James Urquhart (2007-08-08). "Finds test human origins theory". *BBC News*. Retrieved 2008-03-20.

[47] Christopher Boehm (1999) "Hierarchy in the Forest: The Evolution of Egalitarian Behavior" page 198–208 Harvard University Press

[48] Sean Henahan. "Blombos Cave art". *Science News*. Retrieved 2008-03-12.

[49] Felipe Fernandez Armesto (2003). *Ideas that changed the world*. Newyork: Dorling Kindersley limited. p. 400. ISBN 978-0-7566-3298-4.;

[50] Christopher Boehm (1999) "Hierarchy in the Forest: The Evolution of Egalitarian Behavior" page 198 Harvard University Press

[51] R Dale Gutrie (2005). *The Nature of Paleolithic art*. Chicago: University of Chicago Press. ISBN 0-226-31126-0. Pages 420-422

[52] Barbara Ehrenreich (1997). *Blood Rites: Origins and History of the Passions of War*. London, United Kingdom: Macmillan. ISBN 0-8050-5787-0. Page 123

[53] Kelly, Raymond (October 2005). "The evolution of lethal intergroup violence". *PNAS* **102** (43): 15294–8. doi:10.1073/pnas.0505955102. PMC 1266108. PMID 16129826.

[54] Kelly, Raymond C. Warless societies and the origin of war. Ann Arbor : University of Michigan Press, 2000.

[55] Marx, Karl; Friedrich Engels (1848). *The Communist Manifesto*. London. p. 87. ISBN 978-1-59986-995-7. Page 71

[56] Rigby, Stephen Henry (1999). *Marxism and History: A Critical Introduction*, p111 & p314 Manchester University Press, (ISBN 0-7190-5612-8).

[57] Christopher Boehm (1999) "Hierarchy in the Forest: The Evolution of Egalitarian Behavior" page 192 Harvard university press

[58] Thomas M. Kiefer (Spring 2002). "Anthropology E-20". *Lecture 8 Subsistence, Ecology and Food production*. Harvard University. Retrieved 2008-03-11.

[59] Dahlberg, Frances (1975). *Woman the Gatherer*. London: Yale university press. ISBN 0-300-02989-6.

[60] Stefan Lovgren. "Sex-Based Roles Gave Modern Humans an Edge, Study Says". *National Geographic News*. Retrieved 2008-02-03.

[61] Leften Stavros Stavrianos (1991). *A Global History from Prehistory to the Present*. New Jersey, USA: Prentice Hall. ISBN 0-13-357005-3. the sexes were more equal during Paleolithic millennia than at any time since. Page 9

[62] Museum of Antiquites web site . Retrieved February 13, 2008.

[63] Tedlock, Barbara. 2005. The Woman in the Shaman's Body: Reclaiming the Feminine in Religion and Medicine. New York: Bantam.

[64] Jared Diamond. "The Worst Mistake in the History of the Human Race". *Discover*. Retrieved 2008-01-14.

[65] Jonathan Amos (2004-04-15). "Cave yields 'earliest jewellery'". *BBC News*. Retrieved 2008-03-12.

[66] Hillary Mayell. "Oldest Jewelry? "Beads" Discovered in African Cave". *National Geographic News*. Retrieved 2008-03-03.

[67] "Human Evolution," Microsoft Encarta Online Encyclopedia 2007 Contributed by Richard B. Potts, B.A., Ph.D.

[68] Robert G. Bednarik. "Beads and the origins of symbolism". Retrieved 2008-04-05.

[69] Richard G. Klein, "The Dawn of Human Culture" ISBN 0-471-25252-2

[70] Vincent W. Fallio (2006). *New Developments in Consciousness Research*. New York, United States: Nova Publishers. ISBN 1-60021-247-6. Pages 98 to 109

[71] "Paleolithic Art," Microsoft Encarta Online Encyclopedia 2007 http://encarta.msn.com/encyclopedia_761578676/Paleolithic_Art.html

[72] Jean Clottes. "Shamanism in Prehistory". *Bradshaw foundation*. Retrieved 2008-03-11.

[73] "Paleolithic Art," Microsoft Encarta Online Encyclopedia 2007 http://encarta.msn.com/encyclopedia_761578676/Paleolithic_Art.html Microsoft Encarta

[74] McDermott, LeRoy. "Self-Representation in Upper Paleolithic Female Figurines". Current Anthropology, Vol. 37, No. 2, April., 1996. pp. 227–275.

[75] R. Dale Guthrie, *The Nature of Paleolithic Art*. University Of Chicago Press, 2006. ISBN 978-0-226-31126-5. Preface.

[76] Merlin Stone. (1978). *When God Was a Woman*. Harcourt Brace Jovanovich. p. 265. ISBN 0-15-696158-X.

[77] Marija Gimbutas 1991. The Civilization of the Goddess

[78] Karl Bücher. **Trabajo y ritmo**. Biblioteca Científico-Filosófica, Madrid.

[79] Charles Darwin. **The origin of man**. Edimat books, S. A. ISBN 84-8403-034-2.

[80] Nelson, D.E., *Radiocarbon dating of bone and charcoal from Divje babe I cave*, cited by Morley, p. 47

[81] Bahn, Paul (1996) "The atlas of world archeology" Copyright 2000 The Brown Reference Group PLC

[82] "About OriginsNet by James Harrod". Originsnet.org. Retrieved 2010-01-31.

[83] "Appendices for chimpanzee spirituality by James Harrod" (PDF). Retrieved 2010-01-31.

[84] "Oldowan Art, Religion, Symbols, Mind by James Harrod". Originsnet.org. Retrieved 2010-01-31.

[85] Wunn, Ina (2000). "Beginning of Religion", Numen, **47**(4), pp. 434–435

[86] Robbins, Lawrence H.; AlecC. Campbell; George A. Brook; Michael L. Murphy (June 2007). "World's Oldest Ritual Site? The "Python Cave" at Tsodilo Hills World Heritage Site, Botswana" (PDF). *NYAME AKUMA, the Bulletin of the Society of Africanist Archaeologists* (67). Retrieved 1 December 2010.

[87] Karl J. Narr. "Prehistoric religion". *Britannica online encyclopedia 2008*. Retrieved 2008-03-28.

[88] Steven Mithen (1996). *The Prehistory of the Mind: The Cognitive Origins of Art, Religion and Science*. Thames & Hudson. ISBN 0-500-05081-3.

[89] Lerro, Bruce (2000). *From earth spirits to sky gods Socioecological Origins of Monotheism*. Lanham MD: Lexington Press. p. 327. ISBN 0-7391-0098-X. pages 17–20

[90] Christopher L. C. E. Witcombe, "Women in the Stone Age," in the essay "The Venus of Willendorf" . Retrieved March 13, 2008.

[91] "Upper Paleolithic Art, Religion, Symbols, Mind By James Harrod". Originsnet.org. Retrieved 2010-01-31.

[92] William Cocke. "First Wine? Archaeologist Traces Drink to Stone Age". *National Geographic News*. Retrieved 2008-02-03.

[93] Gowlett JAJ (2003). "What actually was the Stone Age Diet?" (PDF). *J Nutr Environ Med* **13** (3): 143–7. doi:10.1080/13590840310001619338.

[94] Weiss E, Wetterstrom W, Nadel D, Bar-Yosef O (June 29, 2004). "The broad spectrum revisited: Evidence from plant remains". *Proc Natl Acad Sci USA* **101** (26): 9551–5. doi:10.1073/pnas.0402362101. PMC 470712. PMID 15210984.

[95] Richards, MP (December 2002). "A brief review of the archaeological evidence for Palaeolithic and Neolithic subsistence". *Eur J Clin Nutr* **56** (12): 1270–1278. doi:10.1038/sj.ejcn.1601646. PMID 12494313.

[96] Johanson, Donald; Blake, Edgar (2006). *From Lucy to Language: Revised, Updated, and Expanded*. Berlin: Simon & Schuster. pp. 96–97. ISBN 0743280644.

[97] Donna Hart, Robert W. Sussman. *Man the Hunted*. ISBN 0-8133-3936-7.

[98] Lovgren, Stefan (31 August 2005). "Chimps, Humans 96 Percent the Same, Gene Study Finds". Retrieved 23 December 2013.

[99] "Chimp hunting and flesh-eating".

[100] "Chimpanzees 'hunt using spears'". *BBC News*. February 22, 2007.

[101] "The Predatory Behavior and Ecology of Wild Chimpanzees".

[102] Milton, Katharine (1999). "A hypothesis to explain the role of meat-eating in human evolution" (PDF). *Evolutionary Anthropology* **8** (1): 11–21. doi:10.1002/(SICI)1520-6505(1999)8:1<11::AID-EVAN6>3.0.CO;2-M.

[103] Leslie C. Aiello, Peter Wheeler (1995). "The expensive-tissue hypothesis" (PDF). *Current Anthropology*.

[104] Kris-Etherton, PM; Harris, WS; Appel, LJ; Nutrition, Committee (2003). "Fish Consumption, Fish Oil, Omega-3 Fatty Acids, and Cardiovascular Disease". *Arteriosclerosis, thrombosis, and vascular biology* **23** (2): e20–30. doi:10.1161/01.ATV.0000038493.65177.94. PMID 12588785.

[105] Crawford, M. A.; et al. (1999). "Evidence for the Unique Function of Docosahexaenoic Acid (DHA) During the Evolution of the Modern Hominid Brain" (PDF). *Lipids* **34**: S39–S47. doi:10.1007/bf02562227.

[106] Cordain L. Implications of Plio-Pleistocene Hominin Diets for Modern Humans. In: Early Hominin Diets: The Known, the Unknown, and the Unknowable. Ungar, P (Ed.), Oxford University Press, Oxford, 2006, pp 363–83.

[107] Nature's Magic: Synergy in Evolution and the Fate of Humankind By Peter Corning

[108] Laden G, Wrangham R (October 2005). "The rise of the hominids as an adaptive shift in fallback foods: plant underground storage organs (USOs) and australopith origins" (PDF). *J. Hum. Evol.* **49** (4): 482–98. doi:10.1016/j.jhevol.2005.05.007. PMID 16085279.

[109] Wrangham RW, Jones JH, Laden G, Pilbeam D, Conklin-Brittain N (December 1999). "The Raw and the Stolen. Cooking and the Ecology of Human Origins". *Curr Anthropol* **40** (5): 567–94. doi:10.1086/300083. PMID 10539941.

[110] Yeakel JD, Bennett NC, Koch PL, Dominy NJ (July 2007). "The isotopic ecology of African mole rats informs hypotheses on the evolution of human diet" (PDF). *Proc Biol Sci.* **274** (1619): 1723–30. doi:10.1098/rspb.2007.0330. PMC 2493578. PMID 17472915.

[111] Hernandez-Aguilar RA, Moore J, Pickering TR (December 2007). "Savanna chimpanzees use tools to harvest the underground storage organs of plants" (PDF). *Proc Natl Acad Sci U S A.* **105** (49): 19210–13. doi:10.1073/pnas.0707929104. PMC 2148269. PMID 18032604.

[112] S. Boyd Eaton, Stanley B. Eaton III, Andrew J. Sinclair, Loren Cordain, Neil J. Mann (1998). "Dietary intake of long-chain polyunsaturated fatty acids during the Paleolithic" (PDF). *World Rev Nutr Diet.*

[113] J. A. J. Gowlet (September 2003). "What actually was the stone age diet?" (PDF). *Journal of environmental medicine* **13** (3): 143–147. doi:10.1080/13590840310001619338. Retrieved 2008-05-04.)

[114] Diamond, Jared. *The third chimpanzee: the evolution and future of the human animal.*

[115] Sharman Apt Russell (2006). *Hunger an unnatural history.* Basic books. ISBN 0-465-07165-1. Pages 2

[116] Milton, Katharine (2002). "Hunter-gatherer diets: wild foods signal relief from diseases of affluence (PDF)" (PDF). In Ungar, Peter S. & Teaford, Mark F. *Human Diet: Its Origins and Evolution.* Westport, CT: Bergin and Garvey. pp. 111–22. ISBN 0-89789-736-6.

[117] Larsen, Clark Spencer (1 November 2003). "Animal source foods and human health during evolution". *Journal of Nutrition* **133** (11, Suppl 2): 3893S–3897S. PMID 14672287.

[118] Cordain L, Eaton SB, Sebastian A, Mann N, Lindeberg S, Watkins BA, O'Keefe JH, Brand-Miller J (2005). "Origins and evolution of the Western diet: health implications for the 21st century". *Am. J. Clin. Nutr.* **81** (2): 341–54. PMID 15699220.

[119] Thorburn AW, Brand JC, Truswell AS. (1 January 1987). "Slowly digested and absorbed carbohydrate in traditional bushfoods: a protective factor against diabetes?". *Am J Clin Nutr* **45** (1): 98–106. PMID 3541565.

[120] Hillard Kaplan, Kim Hill, Jane Lancaster, and A. Magdalena Hurtado (2000). "A Theory of Human Life History Evolution: Diet, Intelligence and Longevity" (PDF). *Evolutionary Anthropology* **9** (4): 156–185. doi:10.1002/1520-6505(2000)9:4<156::AID-EVAN5>3.0.CO;2-7. Retrieved 12 September 2010

[121] Caspari, Rachel & Lee, Sang-Hee (July 27, 2004). "Older age becomes common late in human evolution". *Proceedings of the National Academy of Sciences* **101** (20): 10895–10900. doi:10.1073/pnas.0402857101. PMC 503716. PMID 15252198. Retrieved 12 September 2010

[122] Efraim Lev, Mordechai E. Kislev, Ofer Bar-Yosef (March 2005). "Mousterian vegetal food in Kebara Cave, Mt. Carmel". *Journal of Archaeological Science* **32** (3): 475–484. doi:10.1016/j.jas.2004.11.006.

[123] Revedin, Anna; Aranguren, B; Becattini, R; Longo, L; Marconi, E; Lippi, MM; Skakun, N; Sinitsyn, A; et al. (2010). "Thirty thousand-year-old evidence of plant food processing". *Proc Natl Acad Sci U S A* **107** (44): 18815–9. doi:10.1073/pnas.1006993107. PMC 2973873. PMID 20956317.

[124] Lindeberg, Staffan (June 2005). "Palaeolithic diet ("stone age" diet)". *Scandinavian Journal of Food & Nutrition* **49** (2): 75–77. doi:10.1080/11026480510032043.

[125] Academic American Encyclopedia By Grolier Incorporated (1994). *Academic American Encyclopedia By Grolier Incorporated.* University of Michigan: Grolier Academic Reference.; p 61

[126] John Noble Wilford (2007-10-18). "Key Human Traits Tied to Shellfish Remains". *New York times.* Retrieved 2008-03-11.

[127] African Bone Tools Dispute Key Idea About Human Evolution National Geographic News article.

[128] Tim D. White (2006-09-15). *Once were Cannibals. Evolution: A Scientific American Reader.* ISBN 978-0-226-74269-4. Retrieved 2008-02-14.

[129] James Owen. "Neandertals Turned to Cannibalism, Bone Cave Suggests". *National Geographic News.* Retrieved 2008-02-03.

[130] Pathou-Mathis M (2000). "Neanderthal subsistence behaviours in Europe". *International Journal of Osteoarchaeology* **10** (5): 379–395. doi:10.1002/1099-1212(200009/10)10:5<379::AID-OA558>3.0.CO;2-4.

Chapter 3

Lower Paleolithic

Four views of an Acheulean handaxe

The **Lower Paleolithic** (or **Lower Palaeolithic**) is the earliest subdivision of the Paleolithic or Old Stone Age. It spans the time from around 3.3 million years ago when the first evidence for stone tool production and use by hominins appears in the current archaeological record,[1] until around 300,000 years ago, spanning the Oldowan ("mode 1") and Acheulean ("mode 2") lithics industries.

In African archaeology, this time period roughly corresponds to the Early Stone Age, beginning approximately 3.3 million years ago with Lomekwian stone tool technology, spanning Mode 1 stone tool technology which begins roughly 2.6 million years ago, and ending between 400,000 and 250,000 years ago with Mode 2 technology.[1][2][3]

The Lower Paleolithic is followed by the Middle Paleolithic, which sees the appearance of the more advanced prepared-core tool-making technologies such as the Mousterian. Whether the earliest control of fire by hominids dates to the Lower or to the Middle Paleolithic remains an open question.

Further information: Gelasian, Homo habilis and Olduvai Gorge

The Lower Paleolithic begins with the appearance of the oldest stone tools in the world, roughly 3.3 million ago in eastern Africa, which were produced by an as-yet undetermined hominin.[1] The Gelasian (Lower Pleistocene),

some 2.5 million years ago, sees the appearance of the *Homo* genus (*Homo habilis*), possibly developing out of australopithecine forebears (such as *Australopithecus garhi*). These early members of the *Homo* genus had primitive tools, summarized under the Oldowan or Mode 1 horizon, which remained dominant for the best part of a million years, from about 2.5 to 1.7 million years ago. *Homo habilis* is assumed to have lived primarily on scavenging, using the tools to cleave meat off carrion or to break bones in order to extract the marrow.

The move from the mostly frugivorous or omnivorous diet of *Australopithecus* to the carnivorous scavenging lifestyle of early *Homo* has been explained by the climate changes in East Africa associated with the Quaternary glaciation. Decreasing oceanic evaporation resulted in a drier climate and an expansion of the savannah at the expense of forests. Reduced availability of fruits forced some Australopithecine to unlock new food sources found in the drier savannah climate. Derek Bickerton has placed to this period the move from simple animal communication systems as they are found in all great apes to the earliest form of symbolic communication systems capable of displacement (referring to items not currently within sensory perception), motivated for the need for "recruitment" of group members for scavenging large carcasses.[4]

Homo erectus appears by about 1.8 million years ago, via the transitional variety *Homo ergaster*.

3.1 Calabrian

Main articles: Calabrian (stage) and Homo

Homo erectus moved from scavenging to hunting, developing the hunting-gathering lifestyle that would remain dominant throughout the Paleolithic into the Mesolithic. The unlocking of the new niche of hunting-gathering subsistence drove a number of further changes, behavioral and physiological, leading to the appearance of *Homo heidelbergensis* by some 600,000 years ago.

Homo erectus migrated out of Africa and dispersed throughout Eurasia. Stone tools in Malaysia have been dated to be 1.83 million years old.[5] The Peking Man fossil, discovered in 1929, is roughly 700,000 years old.

In Europe, the Olduwan tradition (known in Europe as Abbevillian) split into two parallel traditions, the Clactonian, a flake tradition, and the Acheulean, a hand-axe tradition. The Levallois technique for knapping flint developed during this time.

The carrier species from Africa to Europe undoubtedly was *Homo erectus*. This type of human is more clearly linked to the flake tradition, which spread across southern Europe through the Balkans to appear relatively densely in southeast Asia. Many Mousterian finds in the Middle Paleolithic have been knapped using a Levallois technique, suggesting that Neanderthals evolved from *Homo erectus* (but, perhaps, *Homo heidelbergensis* (see below)).

At the site of Monte Poggiolo, near Forlì, in Italy, thousands of stone handaxes have been found that date from 800,000 years ago.

The appearance of *Homo heidelbergensis* about 600,000 years ago heralds a number of other new varieties, such as *Homo rhodesiensis* and *Homo cepranensis* about 400,000 years ago. *Homo heidelbergensis* is a candidate for first developing an early form of symbolic language. Whether control of fire and earliest burials date to this period or only appear during the Middle Paleolithic is an open question.

Also in Europe, there appeared a type of human intermediate between *Homo erectus* and *Homo sapiens*, sometimes summarized under archaic *Homo sapiens*, typified by such fossils as those found at Swanscombe, Steinheim, Tautavel, and Vertesszollos (*Homo palaeohungaricus*). The hand-axe tradition originates in the same period. The intermediate may have been *Homo heidelbergensis*, held responsible for the manufacture of improved *Mode 2* Acheulean tool types, in Africa, after 600,000 years BP. Flakes and axes coexisted in Europe, sometimes at the same site. The axe tradition, however, spread to a different range in the east. It appears in Arabia and India, but more importantly, it does not appear in southeast Asia.

3.2 Transition to the Middle Paleolithic

Further information: Homo rhodesiensis and Anatomically modern humans

From about 300,000 years ago, technology, social struc-

tures and behaviour appear to grow more complex, with prepared-core technique lithics, earliest instances of burial and hunting-gathering subsistence. *Homo sapiens* first appear about 200,000 years ago.

3.3 See also

- Control of fire by early humans

3.4 References

[1] Harmand, Sonia; et al. (21 May 2015). "3.3-million-year-old stone tools from Lomekwi 3, West Turkana, Kenya". *Nature* **521**: 310–315. doi:10.1038/nature14464.

[2] "Early Stone Age Tools". *What does it mean to be human?*. Smithsonian Institution. 2014-09-29. Retrieved 2014-09-30.

[3] Barham, Lawrence; Mitchell, Peter (2008). *The First Africans: African Archaeology from the Earliest Toolmakers to Most Recent Foragers*. New York: Cambridge. p. 16. ISBN 978-0-521-61265-4.

[4] Derek Bickerton, *Adam's Tongue: How Humans Made Language, How Language Made Humans*, New York: Hill and Wang 2009.

[5] Malaysian scientists find stone tools 'oldest in Southeast Asia'

3.5 External links

- The First People and Culture at Indiana University Bloomington

Chapter 4

Oldowan

Oldowan tradition chopper

The **Oldowan**, sometimes spelled **Olduwan**, is the archaeological term used to refer to the earliest stone tool archaeological industry in prehistory. Oldowan tools were used during the Lower Paleolithic period, 2.6 million years ago up until 1.7 million years ago, by ancient hominins across much of Africa, South Asia, the Middle East and Europe. This technological industry was followed by the more sophisticated Acheulean industry.

The term "Oldowan" is taken from the site of Olduvai Gorge in Tanzania, where the first Oldowan lithics were discovered by the archaeologist Louis Leakey in the 1930s. However, some contemporary archaeologists and palaeoanthropologists prefer to use the term "Mode 1" tools to designate pebble tool industries (including Oldowan), with "Mode 2" designating bifacially-worked tools (including Acheulean handaxes), "Mode 3" designating prepared-core tools, and so forth.[1]

Classification of Oldowan tools is still somewhat contended. Mary Leakey was the first to create a system to clas-sify Oldowan assemblages, and built her system based on prescribed use. The system included choppers, scrapers, and pounders.[2][3] However, more recent classifications of Oldowan assemblages have been made that focus primar-liy on manufacture due to the problematic nature of as-suming use from stone artefacts. An example is Isaac et al.'s tri-modal categories of "Flaked Pieces" (cores/ chop-pers), "Detached Pieces" (flakes and fragments), "Pounded Pieces" (cobbles utilized as hammerstones, etc.) and "Un-modified Pieces" (manuports, stones transported to sites)[4] Oldowan tools are sometimes called "pebble tools," so named because the blanks chosen for their production al-ready resemble, in pebble form, the final product.[5]

It is not known for sure which hominin species actually cre-ated and used Oldowan tools. Its emergence is often associ-ated with the species *Australopithecus garhi*[6] and its flour-ishing with early species of *Homo* such as *H. habilis* and *H. ergaster*. Early *Homo erectus* appears to inherit Oldowan technology and refines it into the Acheulean industry begin-ning 1.7 million years ago.[7]

4.1 Dates and ranges

The oldest known Oldowan tools have been found in Gona, Ethiopia, and are dated to about 2.6 mya.[8]

The use of tools by apes including chimpanzees[9] and orangutans[10] can be used to argue in favour of tool-use as an ancestral feature of the hominin family. Tools made from bone, wood, or other organic materials were there-fore in all probability used before the Oldowan.[11] Oldowan stone tools are simply the oldest recognisable tools which have been preserved in the archaeological record.

There is a flourishing of Oldowan tools in eastern Africa, spreading to southern Africa, between 2.4 and 1.7 mya. At 1.7 mya., the first Acheulean tools appear even as Oldowan assemblages continue to be produced. Both technologies are occasionally found in the same areas, dating to the same time periods. This realisation required a rethinking of old cultural sequences in which the more "advanced" Acheulean was supposed to have succeeded the Oldowan. The different traditions may have been used by different species of hominins living in the same area, or multiple techniques may have been used by an individual species in response to different circumstances.

Sometime before 1.8 mya *Homo erectus* had spread outside of Africa, reaching as far east as Java by 1.8 mya.[12] and in Northern China by 1.66 mya.[13] In these newly colonised areas, no Acheulean assemblages have been found. In China, only "Mode 1" Oldowon assemblages were produced, while in Indonesia stone tools from this age are unknown.

By 1.8 mya early *Homo* was present in Europe, as shown by the discovery of fossil remains and Oldowan tools in Dmanisi, Georgia.[14] Remains of their activities have also been excavated in Spain at sites in the Guadix-Baza basin[15] and near Atapuerca.[16] Most early European sites yield "Mode 1" or Oldowan assemblages. The earliest Acheulean sites in Europe only appear around 0.5 mya. In addition, the Acheulean tradition does not seem to spread to Eastern Asia.[17] It is unclear from the archaeological record when the production of Oldowon technologies ended. Other tool-making traditions seem to have supplanted Oldowon technologies by 0.25 mya.

4.2   The tools

4.2.1   Manufacture of the tools

To obtain an Oldowan tool, a roughly spherical hammerstone is struck on the edge, or striking platform, of a suitable core rock to produce a conchoidal fracture with sharp edges useful for various purposes. The process is often called lithic reduction. The chip removed by the blow is the flake. Below the point of impact on the core is a characteristic bulb with fine fissures on the fracture surface. The flake evidences ripple marks.

The materials of the tools were for the most part quartz, quartzite, basalt, or obsidian, and later flint and chert. Any rock that can hold an edge will do. The main source of these rocks is river cobbles, which provide both hammer stones and striking platforms. The earliest tools were simply split cobbles. It is not always clear which is the flake. Later tool-makers clearly identified and reworked flakes. Complaints that artifacts could not be distinguished from naturally fractured stone have helped spark careful studies of Oldowon techniques. These techniques have now been duplicated many times by archaeologists and other knappers, making misidentification of archaeological finds less likely.

Use of bone tools by hominins also producing Oldowon tools is known from Swartkrans, where a bone shaft with a polished point was discovered in Member (layer) I, dated 1.8–1.5 mya. The Osteodontokeratic industry, the "bone-tooth-horn" industry hypothesized by Raymond Dart, is less certain.

4.2.2   Shapes and uses of the tools

Mary Leakey classified the Oldowan tools as Heavy Duty, Light Duty, Utilized Pieces and Debitage, or waste.[18] Heavy-duty tools are mainly cores. A chopper has an edge on one side. It is unifacial if the edge was created by flaking on one face of the core, or bifacial if on two. Discoid tools are roughly circular with a peripheral edge. Polyhedral tools are edged in the shape of a polyhedron. In addition there are spheroidal hammer stones.

Light-duty tools are mainly flakes. There are scrapers, awls (with points for boring) and burins (with points for engraving). Some of these functions belong also to heavy-duty tools. For example, there are heavy-duty scrapers.

Utilized pieces are tools that began with one purpose in mind but were utilized opportunistically.

Oldowan tools were probably used for many purposes, which have been discovered from observation of modern apes and hunter-gatherers. Nuts and bones are cracked by hitting them with hammer stones on a stone used as an anvil. Battered and pitted stones testify to this possible use.

Heavy-duty tools could be used as axes for woodworking. Both choppers and large flakes were probably used for this purpose. Once a branch was separated, it could be scraped clean with a scraper, or hollowed with pointed tools. Such uses are attested by characteristic microscopic alterations of edges used to scrape wood. Oldowon tools could also have been used for preparing hides. Hides must be cut by slicing, piercing and scraping it clean of residues. Flakes are most suitable for this purpose.

Lawrence Keeley, following in the footsteps of Sergei Semenov, conducted microscopic studies (with a high-powered optical microscope) on the edges of tools manufactured de novo and used for the originally speculative purposes described above. He found that the marks were characteristic of the use and matched marks on prehistoric tools. Studies of the cut marks on bones using an electron microscope produce a similar result.

4.2.3   Abbevillian

Abbevillian is a currently obsolescent name for a tool tradition that is increasingly coming to be called Oldowan. The label Abbevillian prevailed until the Leakey family discovered older (yet similar) artifacts at Olduvai Gorge (a.k.a. Oldupai Gorge) and promoted the African origin of man. Oldowan soon replaced Abbevillian in describing African and Asian lithics. The term Abbevillian is still used but is now restricted to Europe. The label, however, continues to lose popularity as a scientific designation.

In the late 20th century, discovery of the discrepancies in date caused a crisis of definition. If Abbevillian did not necessarily precede Acheulean and both traditions had flakes and bifaces, how was the difference to be defined? It was in this spirit that many artifacts formerly considered Abbevillian were labeled Acheulean. In consideration of the difficulty, some preferred to name both phases Acheulean. When the topic of Abbevillian came up, it was simply put down as a phase of Acheulean. Whatever was from Africa was Oldowan, and whatever from Europe, Acheulean.

The solution to the definition problem is stated in the article on Acheulean. The difference is to be defined in terms of complexity. Simply struck tools are Oldowan. Retouched, or reworked tools are Acheulean. Retouching is a second working of the artifact. The manufacturer first creates an Oldowan tool. Then he reworks or retouches the edges by removing very small chips so as to straighten and sharpen the edge. Typically but not necessarily the reworking is accomplished by pressure flaking.

The pictures in the introduction to this article are mainly labeled Acheulean, but this is the now false Acheulean, which also includes Abbevillian. The artifacts shown are clearly in the Oldowan tradition. One or two of the more complex bifaces may have edges made straighter by a large percussion or two, but there is no sign of pressure flaking as depicted. The pictures included with this subsection show the difference.

4.3 The tool users

Current anthropological thinking is that Oldowan tools were made by late *Australopithecus* and early *Homo*. *Homo habilis* was named "skillful" because it was considered the earliest tool-using human ancestor. Indeed, the genus *Homo* was in origin intended to separate tool-using species from their tool-less predecessors, hence the name of *Australopithecus garhi*, *garhi* meaning "surprise", a tool-using Australopithecine discovered in 1996 and described as the "missing link" between the *Australopithecus* and *Homo* genera. There is also evidence that some species of *Paranthropus* utilized the tools associated with this culture.[19]

There is presently no evidence to show that Oldowan tools were the sole property of the *Homo* line or that the ability to produce them was the special characteristic of only our

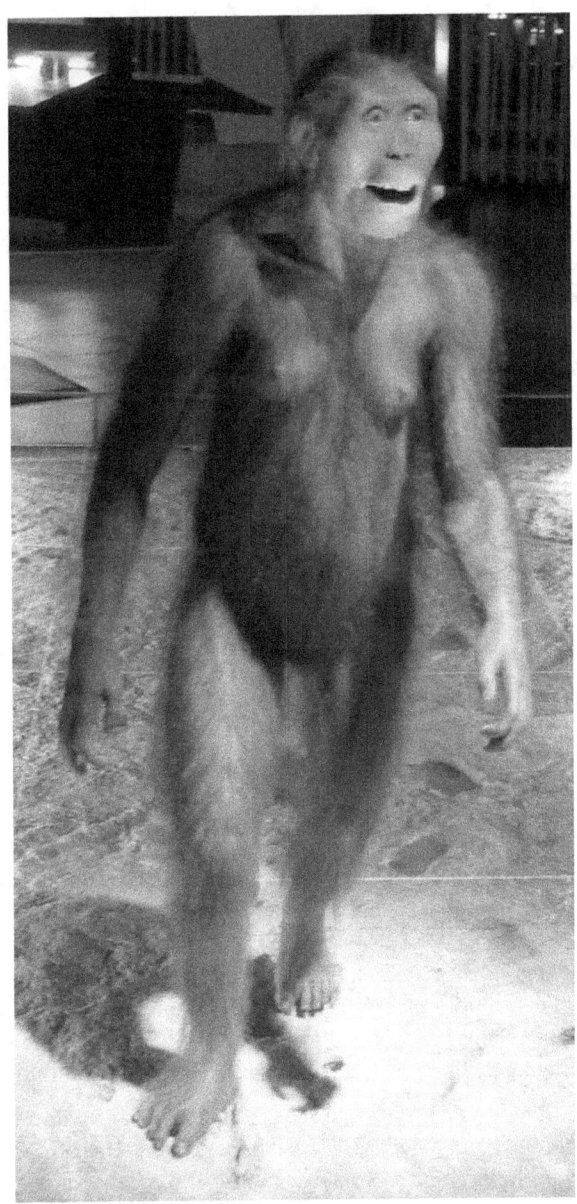

A reconstruction of a female member of the Australopithecus genus. The emergence of Oldowan tools is often associated with the species Australopithecus garhi.[6]

ancestors. A research of tool use by modern wild chimpanzees in West Africa shows there is an operational sequences when chimpanzees use lithic implements to crack nuts. In the course of nut cracking, sometimes they will create unintentional flakes. Although the morphology of chimpanzees' hammer is different from Oldowan hammer, the ability of tool use of chimpanzees indicates that first lithic industries were probably not produced by only one kind of hominin species.[20]

The makers of Oldowan tools were mainly right-handed.[21] "Handedness" (lateralization) had thus already evolved, though it is not clear how related to modern lateralization it was, since other animals show handedness as well.

In the mid-1970s, Glynn Isaac touched off a debate by proposing that human ancestors of this period had a "place of origin" and that they foraged outward from this home base, returning with high quality food to share and to be processed. Over the course of the last 30 years, a variety of competing theories about how foraging occurred have been proposed, each one implying certain kinds of social strategies. The available evidence from the distribution of tools and remains is not enough to decide which theories are the most probable. However, three main groups of theories predominate.

- Glynn Isaac's model became the [Central Forage Point]- as he responded to critics that accused him of attributing too much 'modern' behavior to early hominins with relatively free-form searches outward.

- A second group of models took modern chimpanzee behavior as a starting point, having the hominids use relatively fixed routes of foraging, and leaving tools where it was best to do so on a constant track.

- A third group of theories had relatively loose bands scouring the range, taking care to move carcasses from dangerous death sites and leaving tools more or less at random.

Each group of models implies different grouping and social strategies, from the relative altruism of central base models to the relatively disjointed search models. (See also central foraging theory, scavenging station model, Lewis Binford)

Most models rely on social and communication networks to hold the band together. These social networks range from requiring no more communication than modern primates, to requiring more sophisticated sharing and teaching. At present, no evidence has been found that sharply divides these theories.

Hominins probably lived in social groups that had contact with others. This conclusion is supported by the large number of bones at many sites, too large to be the work of one individual, and all of the scatter patterns implying many different individuals. Since modern primates in Africa have fluid boundaries between groups, as individuals enter, become the focus of bands, and others leave, it is also probable that the tools we find are the result of many overlapping groups working the same territories, and perhaps competing over them. Because of the huge expanse of time and the multiplicity of species associated with possible Oldowan tools, it is difficult to be more precise than this, since it is almost certain that different social groupings were used at different times and in different places.

There is also the question of what mix of hunting, gathering and scavenging the tool users employed. Early models focused on the tool users as hunters. The animals butchered by the tools include waterbuck, hartebeest, Springbok, pig and zebra. However, the disposition of the bones allows some question about hominin methods of obtaining meat. That they were omnivores is unquestioned, as the digging implement and the probable use of hammer stones to smash nuts indicate. Lewis Binford first noticed that the bones at Olduvai contained a disproportionately high incidence of extremities, which are low in food substance. He concluded other predators had taken the best meat, and the hominins had only scavenged. The counter view is that while hunting many large animals would be beyond the reach of an individual human, groups could bring down larger game, as pack hunting animals are capable of doing. Moreover, since many animals both hunt and scavenge, it is possible that hominins hunted smaller animals, but were not above driving carnivores from larger kills, as they probably were driven from kills themselves from time to time.

4.4 Sites and archaeologists

A complete catalog of Oldowan sites would be too extensive for listing here. Some of the better-known sites include the following:

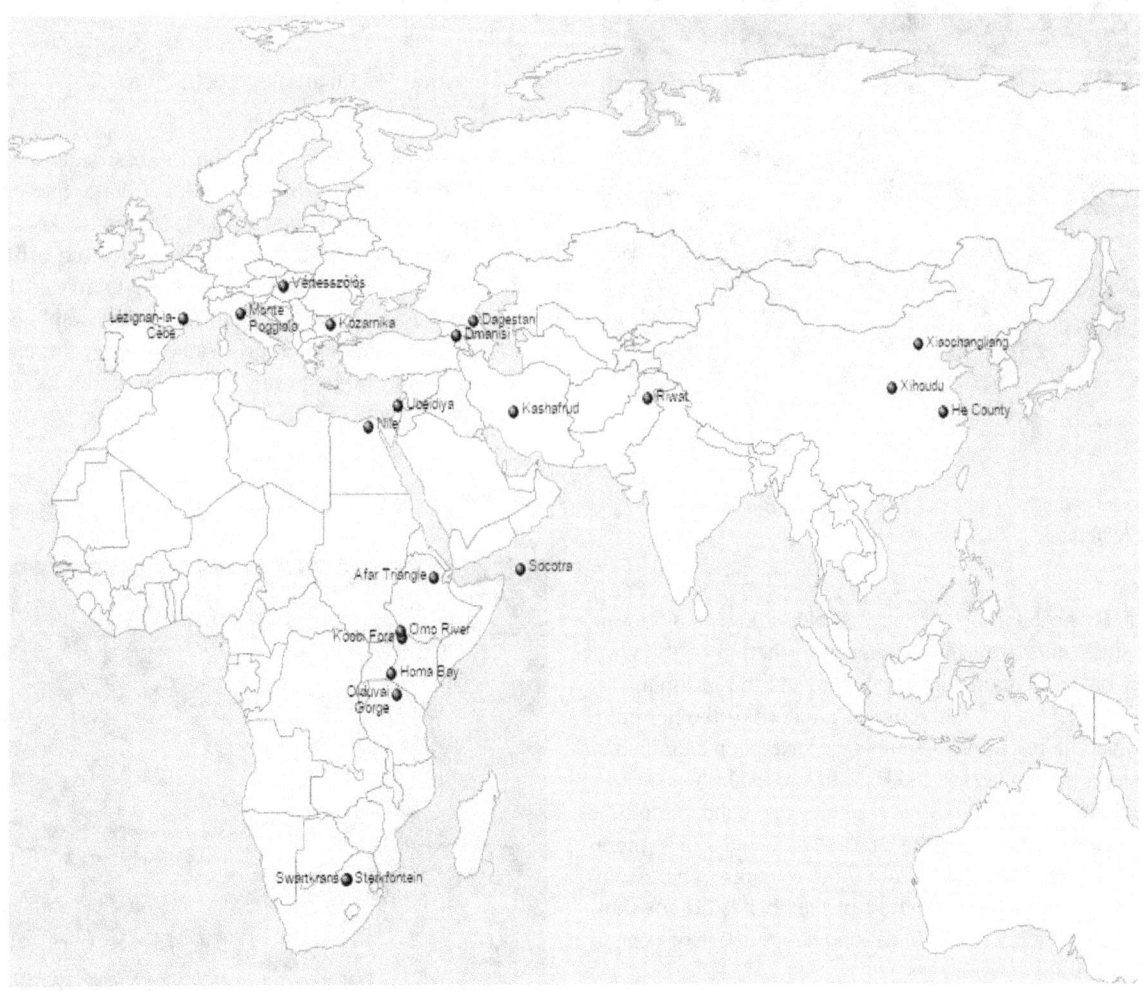

4.4.1 Africa

Ethiopia

Afar Triangle Sites in the Gona river system in the Hadar region of the Afar triangle, excavated by Helene Roche, J. W. Harris and Sileshi Semaw, yielded some of the oldest known Oldowan assemblages, dating to about 2.6 million years ago. Raw material analysis done by Semaw showed that some assemblages in this region are biased towards a certain material (e.g.: 70% of the artifacts at sites EG10 and EG12 were composed of trachyte) indicating a selectivity in the quality of stone used.[22] Recent excavations have yielded tools in association with cut-marked bones, indicating that Oldowan were used in meat-processing or -acquiring activities.

Oldowan choppers dating to 1.7 million years BP, from Melka Kunture, Ethiopia

Omo River basin The second oldest known Oldowan tool site comes from the Shungura formation of the Omo River basin. This formation documents the sediments of the Plio-Pleistocene and provides a record of the hominins that lived there. Lithic assemblages have been classified as Oldowan in members E and F in the lower Omo basin. Although there have been lithic assemblages found in multiple sites in these areas, only the Omo sites 57 and 123 in member F are accepted as hominin lithic remains. The assemblages at Omo sites 71 and 84 in member E do not show evidence of hominin modification and are therefore classified as natural assemblages.[23]

The tools are never found in direct association with the hominins, but archaeologists believe that they would be the strongest candidates for tool manufacture. There are no hominins in those layers, but the same layers elsewhere in the Omo valley contain *Paranthropus* and early *Homo* fossils. *Paranthropus* occurs in the preceding layers. In the last layer at 1.4 million years ago is only *Homo erectus*.

Egypt

Along the Nile River, within the 100 foot terrace, evidence of Chellean or Oldowan cultures has been found.[24]

Kenya

Kanjera South, part of the Kanjera site complex, is located on the Homa Peninsula.[25] The site is estimated around 2 million years old.[26] One of the significant excavations, in the area, is Leakey's expedition in 1932-35.[25] In 1995, Oldowan and Plio-Pleistocene faunal remains surfaced from the site.[25] There has been fieldwork to understand the geochronology of Kanjera.[25]

East Turkana   Main article. Koobi Fora

The numerous Koobi Fora sites on the east side of Lake Turkana are now part of Sibiloi National Park. Sites were initially excavated by Richard Leakey, Meave Leakey, Jack Harris, Glynn Isaac and others. Currently the artifacts found are classified as Oldowan or KBS Oldowan dated from 1.9–1.7 mya, Karari (or "advanced Oldowan") dated to 1.6–1.4 mya, and some early Acheulean at the end of the Karari. Over 200 hominins have been found, including *Australopithecus* and *Homo*.

Tanzania

Chopper from Olduvai Gorge, some 1.8 million years old

Olduvai Gorge The Oldowan industry is named after discoveries made in the Olduvai Gorge of Tanzania in east Africa by the Leakey family, primarily Mary Leakey, but also her husband Louis and their son, Richard. Mary Leakey organized a typology of Early Pleistocene stone tools, which developed Oldowan tools into three chronological variants, A, B and C.[27] Developed Oldowan B is of particular interest due to changes in morphology that appear to have been driven mostly by the short term availability of a chert resource from 1.65 mya to 1.53 mya. The flaking properties of this new resource resulted in considerably more core reduction and a higher prevalence of flake retouch.[28] Similar tools had already been found in various locations in Europe and Asia for some time, where they were called Chellean and Abbevillian.

The oldest tool sites are in the East African Rift system, on the sediments of ancient streams and lakes. This is consistent with what we surmise of the evolution of man. Genetic studies tell us that the human line (the hominins) possibly diverged from the chimpanzee line (the hominids), and the native territory of the latter is the forests of Central Africa nearby. Fossil chimpanzees have been found in Kenya.

The forests of central and western Africa are a stable environment containing food in abundance for animals such as chimpanzees, and any species living in such an environment would have been under little pressure to evolve further. Eastern Africa is a land of often harsh and unstable environments, and resources are correspondingly scarcer and more difficult to get. Species living in the latter environment would be under greater pressure to evolve and change as needed to survive. A facility for tool-using would contribute to the species' chances of survival.

Even though Olduvai Gorge is the type site, Oldowan tools from here are not the oldest known examples. They occur in Beds I–IV. Bed I, dated 1.85 mya to 1.7 mya, contains Oldowan and fossils of *Paranthropus boisei* as well as *Homo habilis*, as does Bed II, 1.7–1.2 mya. *H. habilis* gives way to *Homo erectus* at about 1.6 mya but *P. boisei* goes on. Oldowan continues to Bed IV at 800,000 to 600,000 before present (BP).

South Africa

Abbé Breuil was the first recognized archaeologist to go on record to assert the existence of Oldowan tools. While his description was for "Chello-Abbevillean" tools, and postdated Leakey's finds at Olduvai Gorge by at least ten years, his descriptions nonetheless represented the scholarly acceptance of this technology as legitimate. These findings were cited as being from the location of the Vaal River, at Vereeniging, and Breuil noted the distinct absence of a significant number of cores, suggesting a "portable culture". At the time, this was considered very significant, as portability supported the conclusion that the Oldowan toolmakers were capable of planning for future needs, by creating the tools in a location which was distant from their use.[29]

Swartkrans The Swartkrans site is a cave filled with layered fossil-bearing limestone deposits. Oldowan is found in Members (layers) I–III, 1.8–.5 mya, in association with *Paranthropus robustus* and *Homo habilis*. The Member I assemblage also includes a shaft of pointed bone polished at the pointed end.

Member I contained a high percentage of primate remains compared to other animal remains, which did not fit the hypothesis that *H. habilis* or *P. robustus* lived in the cave. C. K. Brain conducted a more detailed study and discovered the cave had been the abode of leopards, who preyed on the hominins.[30]

Sterkfontein Another site of limestone caves is Sterkfontein, found in South Africa. This site contains a large number of not only Oldowan tools, but also early Acheulean technology. [31]

4.4.2 Europe

Georgia

Reconstruction of the Homo erectus georgicus *fossil Dmanisi skull believed to be about 1.8 million years old, discovered at the Oldowan site of Dmanisi. Reconstruction by Élisabeth Daynès at the Musée de Préhistoire in Quinson, France*

Stone tool (Oldowan style) from Dmanisi paleontological site (right, 1.8 mya, replica), to be compared with the more "modern" Acheulean style (left)

In 1999 and 2002, two *Homo erectus* skulls (*H. georgicus*) were discovered at Dmanisi in southern Georgia. The archaeological layer in which the human remains, hundreds of Oldowan stone tools, and numerous animal bones were unearthed is dated approximately 1.6-1.8 million years ago.

The site yields the earliest unequivocal evidence for presence of early humans outside the African continent.[32]

Bulgaria

At Kozarnika, in the ground layers, dated to 1,4-1,6 millions BP archaeologists have discovered a human molar tooth, lower palaeolithic assemblages that belong to a core-and-flake non-Acheulian industry and incised bones that may be the earliest example of human symbolic behaviour.

Russia

Ainikab-1 and Muhkay-2 (North Caucasus, Daghestan) are the extraordinary sites in relation to date and the culture. Geological and geomorphological data, palynological studies and paleomagnetic testing unequivocally point to Early Pleistocene (Eopleistocene), indicating the age of the sites as being within the range of 1.8 – 1.2 mln years ago.[33][34]

Spain

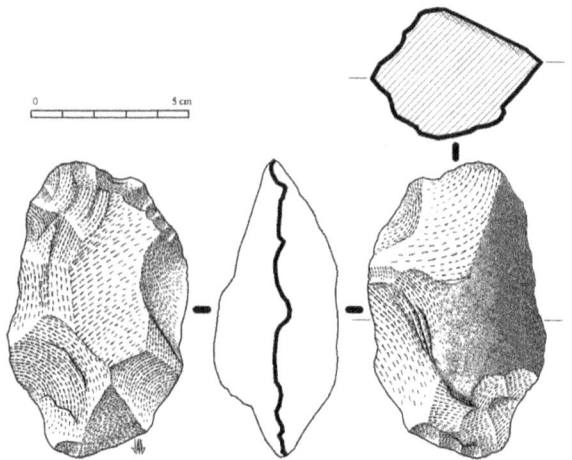

Handaxe extremely archaic from the Quaternary fluvial terraces of Duero river (Valladolid, Spain) and it is dated in Oldowan/Abbevillian period (Lower Paleolithic).

Oldowan tools have been found at the following sites: Fuente Nueva 3, Barranco del Leon, Sima del Elefante, Atapuerca TD 6

France

Oldowan tools have been found at: Lézignan-la-Cèbe, 1.5 mya; Abbeville, 1–.5 mya; Vallonet cave, French Riviera; Soleihac, open-air site in Massif Central. Oldowan tools have also been found at Tautavel in the foothills of the

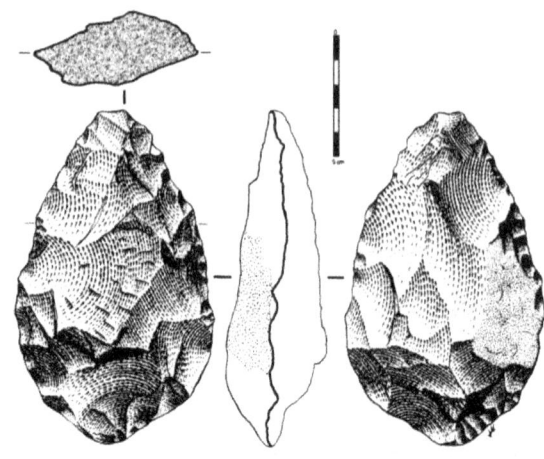

Typical Acheulean handhaxe, it has tear form and proceeds from a superficial site in the Zamora province (Spain), in the Duero river.

Pyrenees. These were discovered by Henry de Lumbley alongside human remains (cranium). The tools are of limestone and quartz.

Elsewhere

Oldowan tools have been found in Italy at the Monte Poggiolo open air site dated to approximately 850,000 BP, making them the oldest evidence of human habitation in Italy. In Germany tools have been found in river gravels at Kärlich dating from .3 mya. In the Czech Republic tools have been found in ancient lake deposits at Przeletice and a cave site at Stranska Skala, dated no later than .5 mya. In Hungary tools have been found at a spring site at Vértesszőlős dating from .5 mya.

4.4.3   Asia

Oldowan tools have been found at sites in South Asia and Southwest Asia. In November 2008 tens of sites of Oldowan tools industry have been found on the island of Socotra (Yemen).[35][36][37]

In Pakistan tools have been found dating from 1.9 mya at Riwat. In Iran tools have been found dating from the Late Pliocene or Early Pleistocene at Kashafrud. In Israel tools have been found dating from 1.4 mya at el 'Ubeidiya.

4.5   Notes

[1] Clark, J. G. D. (1969). *World prehistory: a new outline.* Cambridge: Cambridge University Press.

[2] Clark, J.; de Heinzelin, J.; Schick, K.; Hart, W.; White, T.; WoldeGabriel, G.; Walter, R.; Suwa, G.; Asfaw, B.; Vrba, E.; et al. (1994). "African Homo erectus: Old radiometric ages and young Oldowan assemblages in the middle Awash Valley, Ethiopia". *Science* **264** (5167): 1907–1909. doi:10.1126/science.8009220. PMID 8009220.

[3] Leakey, Mary (1971). *A Summary and Discussion of the Archaeological Evidence from Bed I and Bed II, Olduvai Gorge, Tanzania. Human Origins.* pp. 431–460.

[4] Isaac, G. Ll., Harris, J. W. K. & Marshall, F. 1981. "Small is informative: the application of the study of mini-sites and least effort criteria in the interpretation of the Early Pleistocene archaeological record at Koobi Fora, Kenya." in "Inter-nacional de Ciencias Prehistoricas Y Protohistoricas", Mexico City. Mexico, pp. 101–119.

[5] Napier, John. 1960. "Fossil Hand Bones from Olduvai Gorge." in *Nature", December 17th edition.*

[6] De Heinzelin, J; Clark, JD; White, T; Hart, W; Renne, P; Woldegabriel, G; Beyene, Y; Vrba, E (1999). "Environment and behavior of 2.5-million-year-old Bouri hominids". *Science* **284** (5414): 625–9. doi:10.1126/science.284.5414.625. PMID 10213682.

[7] Richards, M.P. (December 2002). "A brief review of the archaeological evidence for Palaeolithic and Neolithic subsistence". *European Journal of Clinical Nutrition.* 56 Supplement 1, March 2002 (12): 1270–1278. doi:10.1038/sj.ejcn.1601646. PMID 12494313 .

[8] Semaw, S.; Rogers, M. J.; Quade, J.; Renne, P. R.; Butler, R. F.; Domínguez-Rodrigo, M.; Stout, D.; Hart, W. S.; Pickering, T.; et al. (2003). "2.6-Million-year-old stone tools and associated bones from OGS-6 and OGS-7, Gona, Afar, Ethiopia". *Journal of Human Evolution* **45** (2): 169–177. doi:10.1016/S0047-2484(03)00093-9. PMID 14529651.

[9] Whiten, A.; Goodall, J.; McGrew, W. C.; Nishida, T.; Reynolds, V.; Sugiyama, Y.; Tutin, C. E. G.; Wrangham, R. W.; Boesch, C.; et al. (1999). "Cultures in Chimpanzees". *Nature* **399** (6737): 682–685. doi:10.1038/21415. PMID 10385119.

[10] Schaik, CP; Ancrenaz, M.; Borgen, G.; Galdikas, B.; Knott, C. D.; Singleton, I.; Suzuki, A.; Utami, S. S.; Merril, M.; et al. (2003). "Orangutan cultures and the evolution of material culture". *Science* **299** (5603): 102–105. doi:10.1126/science.1078004. PMID 12511649.

[11] Panger, M. A.; Brooks, A. S.; Richmond, B. G.; Wood, B. (2002). "Older than the Oldowan? Rethinking the emergence of hominin tool use". *Evolutionary Anthropology: Issues, News, and Reviews* **11** (6): 235–245. doi:10.1002/evan.10094.

[12] Swisher, C. C.; Curtis, G. H.; Jacob, T.; Getty, A. G.; Suprijo, A.; Widiasmoro (1994). "Age of the earliest known hominids in Java, Indonesia". *Science* **263** (5150): 1118–1121. doi:10.1126/science.8108729. PMID 8108729.

[13] Zhu, R. X.; Potts, R. R.; Xie, F.; Hoffman, K. A.; Shi, C. D.; Pan, Y. X.; Wang, H. Q.; Shi, R. P.; Wang, Y. C.; et al. (2004). "New evidence on the earliest human presence at high northern latitudes in northeast Asia". *Nature* **431** (7008): 559–562. doi:10.1038/nature02829. PMID 15457258.

[14] http://news.nationalgeographic.com/news/2002/07/0703_020704_georgianskull.html

[15] Oms, O.; Pares, J. M.; Martinez-Navarro, B.; Agusti, J.; Toro, I.; Martinez-Fernandez, G.; Turq, A. (2000). "Early human occupation of Western Europe: Paleomagnetic dates for two paleolithic sites in Spain". *Proceedings of the National Academy of Sciences* **97** (19): 10666–10670. doi:10.1073/pnas.180319797.

[16] Pares, J. M.; Perez-Gonzalez, A.; Rosas, A.; Benito, A.; Carbonell, E.; Huguet, R.; Huguet, R (2006). "Matuyama-age lithic tools from the Sima del Elefante site, Atapuerca (northern Spain)". *Journal of Human Evolution* **50** (2): 163–169. doi:10.1016/j.jhevol.2005.08.011. PMID 16249015.

[17] Ambrose, S. H. (2001). "Paleolithic technology and human evolution". *Science* **291** (5509): 1748–1753. doi:10.1126/science.1059487. PMID 11249821.

[18] There is a good online summary of Mary's classification on Effland's site for Anthropology ASB22 at Mesa Community College in Arizona, apparently written by Effland.

[19] Susman, R.L, "Who made the Oldowan Stone Tools?" 1991

[20] Carvalho, S., Cunha, E., Sousa, C., Matsuzawa, T. 2008.Chaînes opératoire and resource-exploitation strategies in chimpanzee (Pan troglodytes) nut cracking. Journal of Archaeological Science.

[21] Klein, Richard G. (22 April 2009). *The Human Career: Human Biological and Cultural Origins* (Third ed.). Chicago: University of Chicago Press. pp. 258–259. ISBN 978-0-226-02752-4. Retrieved 20 May 2014.

[22] Semaw, Sileshi (2000). "The World's Oldest Stone Artefacts from Gona, Ethiopia: Their Implications for Understanding Stone Technology and Patterns of Human Evolution Between 2·6–1·5 Million Years Ago". *Journal of Archaeological Science* **27**: 1197–1214. doi:10.1006/jasc.1999.0592.

[23] de la Torre, Ignacio; deBeaune, S.; Davidson, I.; Gowlett, J.; Hovers, E.; Kimura, Y.; Mercader, J.; de la Torre, I. (2004). "Omo Revisited: Evaluating the Technological Skills of Pliocene Hominids". *Current Anthropology* **45** (4): 439–465. doi:10.1086/422079.

[24] Langer, William L., ed. (1972). *An Encyclopedia of World History* (5th ed.). Boston, MA: Houghton Mifflin Company. p. 9. ISBN 0-395-13592-3.

[25] Bishop, L. C.; Plummer, T. W.; Ferraro, J. V.; Braun, D.; Ditchfield, P. W.; Hertel, F.; Kingston, J. D.; Hicks, J.; Potts, R. (Mar–Jun 2006). "Recent Research into

Oldowan Hominin Activities at Kanjera South, Western Kenya". *The African Archaeological Review* **23** (1/2): 31–40. doi:10.1007/s10437-006-9006-1. JSTOR 25470615.

[26] Thomas Plummer (2005) Stahl, Ann Brower, ed. *African Archaeology*. Malden, MA: Blackwee Publishing Ltd. pp. 55–92.

[27] Barham & Mitchell, Lawernce & Peter (2008). *The First Africans*. Cambridge World Archaeology. pp. 126–127.

[28] Kimura, Yuki (2002). "Examining time trends in the Oldowan technology at Beds I and II, Olduvai Gorge". *Journal of Human Evolution* **43**: 291–321. doi:10.1006/jhev.2002.0576.

[29] Breuil, Abbé Henri; "A Preliminary Survey of Work in South Africa"; *The South African Archaeological Bulletin*; v.1, no.1, Dec. 1945, pp. 5-7

[30] Many scientists had drawn the erroneous conclusion that *Homo habilis* was the predator responsible for these remains, using Oldowan tools. The higher percentage of primate bones was interpreted as a kind of cannibalism, feeding the imagination of Raymond Dart. Brain examined the bones and concluded that the marks resulting from stripping and chewing the bones were made by a leopard.

[31] Petraglia, edited by Michael D.; Korisettar, Ravi (1998). *Early human behaviour in global context the rise and diversity of the Lower Palaeolithic record*. London: Routledge. ISBN 0203203275.

[32] Vekua, A.; Lordkipanidze, D.; Rightmire, G. P.; Agusti, J.; Ferring, R.; Maisuradze, G.; et al. (2002). "A new skull of early Homo from Dmanisi, Georgia". *Science* **297** (5578): 85–9. doi:10.1126/science.1072953. PMID 12098694.

[33] Taymazov A.I. (2011) Main characteristics of the industry at Ainikab I multilayer Early Paleolithic site (based on the data from the 2005–2009 investigations). *Russian Archaeology*, #1, 1-9.

[34] Chepalyga A.L., Amirkhanov Kh.A., Trubikhin V.M., Sadchikova T.A., Pirogov A.N., Taimazov A.I. Geoarchaeology of the earliest paleolithic sites (Oldowan) in the North Caucasus and the East (2012). International Conference GEOMORPHIC PROCESSES AND GEOARCHAEOLOGY: From Landscape Archaeology to Archaeotourism. Moscow-Smolensk, 20–24 August.

[35] Амирханов Х.А., Жуков В.А., Наумкин В.В., Седов А.В. "Эпоха олдована открыта на острове Сокотра"."Природа", № 7/2009

[36] http://www.ihae.ru/konfer/simpozium.htm

[37] Zhukov, Valery A. (2014) The Results of Research of the Stone Age Sites in the Island of Socotra (Yemen) in 2008-2012. - Moscow: Triada Ltd. 2014, pps 114, ill. 134 (in Russian)ISBN 978-5-89282-591-7

4.6 Sources

- Braidwood, Robert J., *Prehistoric Men*, many editions.

- Domínguez-Rodrigo, M.; Pickering, T. R.; Semaw, S.; Rogers, M. J. (2005). "Cutmarked bones from Pliocene archaeological sites at Gona, Afar, Ethiopia: Implications for the function of the world's oldest stone tools". *Journal of Human Evolution* **48** (2): 109–121. doi:10.1016/j.jhevol.2004.09.004. PMID 15701526.

- Edey, Maitland A., *The Missing Link*, Time-Life Books, 1972.

- Schick, Kathy D.; Toth, Nicholas, *Making Silent Stones Speak'*, Simon & Schuster, 1993, ISBN 0-671-69371-9

- Semaw, Sileshi (2000). "The worlds oldest stone artefacts from Gona Ethiopia: Their implications for understanding stone technology and patterns of human evolution between 2.6–1.5 million years ago". *Journal of Archaeological Science* **27** (12): 1197–1214. doi:10.1006/jasc.1999.0592.

- Isaac, Glynn and Harris, JWK *The Scatter between the Patches* 1975

- Isaac, Glynn (1978). "The Food Sharing Behavior of Protohuman Hominids". *Scientific American* **238** (4): 90–108. doi:10.1038/scientificamerican0478-90. PMID 418504.

- Binford, Lewis (1987) *Searching for Camps and Missing the Evidence: Another Look at the Lower Paleolithic*

- Toth, Nicholas (1985) *The Oldowan reassessed: a close look at early stone artifacts Journal of Archaeological Science*

- Susman, Randall L, *Journal of Anthropological Research*, Vol. 47, No. 2, A Quarter Century of Paleoanthropology: Views from the U.S.A. (Summer, 1991), pp. 129–151

4.7 External links

- Oldowan Pebble Tools of Europe
- Oldowan Pebble Tools of Africa

- Oldowan Flake Tool

- Stone Age Hand-axes at the Wayback Machine (archived February 4, 2007)

- Early Palaeolithic

- Stone Age Reference Collection

- Microwear polishes on early stone tools from Koobi Fora, Kenya, article in *Nature* 293, 464–465 (8 October 1981). The summary and the references are displayed at no charge at the *Nature* site.

- Geoarchaeology of the earliest paleolithic sites (Oldowan) in the north Caucasus and the East Europe

- *An Ape's View of the Oldowan* at the Wayback Machine (archived May 21, 2008), T. Wynn and W.C. McGrew, *Man* 24:383–398; 1989.

- *Flaked Stones and Old Bones: Biological and Cultural Evolution at the Dawn of Technology*, Thomas Plummer, *Yearbook of Physical Anthropology* 47:118–164 (2004).

Coordinates: 36°12′03″N 5°39′16″E / 36.2009°N 5.6544°E

Chapter 5

Acheulean

A reconstruction of Homo habilis *at the Westfälisches Museum für Archäologie, Herne. It is thought that Acheulean technologies first developed in Africa out of the more primitive Oldowan technology as long ago as 1.76 million years ago, by* Homo habilis.

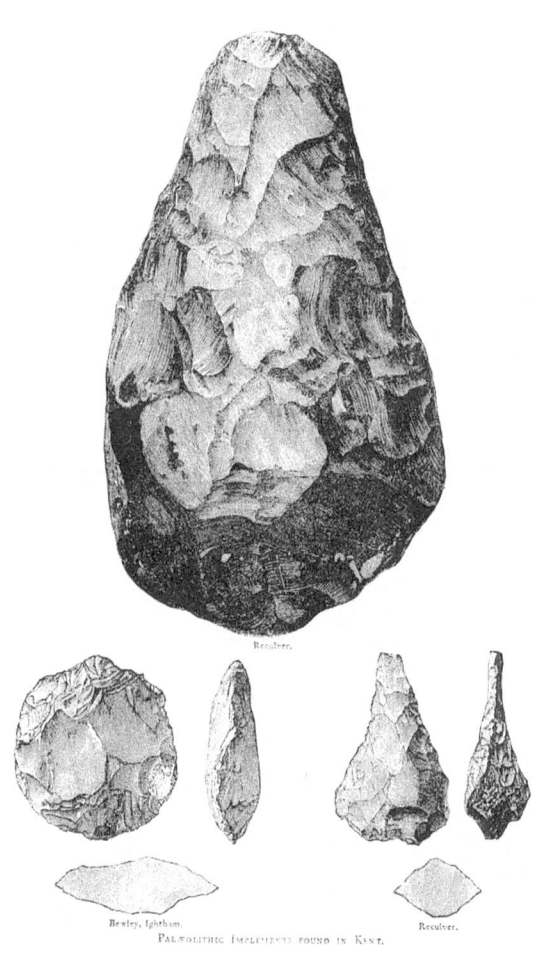

Acheulean hand-axes from Kent. The types shown are (clockwise from top) cordate, ficron, and ovate.

Acheulean (/əˈʃuːliən/; also **Acheulian**), from the French *acheuléen*, is an archaeological industry of stone tool manufacture characterized by distinctive oval and pear-shaped "hand-axes" associated with early humans. Acheulean tools were produced during the Lower Palaeolithic era across Africa and much of West Asia, South Asia, and Europe, and are typically found with *Homo erectus* remains. It is thought that Acheulean technologies first developed in Africa out of the more primitive Oldowan technology as long ago as 1.76 million years ago, by *Homo habilis*. Acheulean tools were the dominant technology for the vast majority of human history.[3][4][5]

Depiction of a Terra Amata hut in Nice, France as postulated by Henry de Lumley dated to 400 thousand years ago. Shelter construction has been discovered in Japan dating back to 500 thousand years ago.[1]

Pigments in Zambia may have been used for body painting as far back as 400 thousand years ago.[2]

5.1 History of research

The type site for the Acheulean is Saint-Acheul, a suburb of Amiens, the capital of the Somme department in Picardy, where artifacts were found in 1859.[6]

John Frere is generally credited as being the first to suggest a very ancient date for Acheulean hand-axes. In 1797, he sent two examples to the Royal Academy in London from Hoxne in Suffolk. He had found them in prehistoric lake deposits along with the bones of extinct animals and concluded that they were made by people "who had not the use of metals" and that they belonged to a "very ancient period indeed, even beyond the present world".[7] His ideas were, however, ignored by his contemporaries, who subscribed to a pre-Darwinian view of human evolution.

Later, Jacques Boucher de Crèvecœur de Perthes, work-

ing between 1836 and 1846, collected further examples of hand-axes and fossilised animal bone from the gravel river terraces of the Somme near Abbeville in northern France. Again, his theories attributing great antiquity to the finds were spurned by his colleagues, until one of de Perthe's main opponents, Dr Marcel Jérôme Rigollot, began finding more tools near Saint Acheul. Following visits to both Abbeville and Saint Acheul by the geologist Joseph Prestwich, the age of the tools was finally accepted.

In 1872, Louis Laurent Gabriel de Mortillet described the characteristic hand-axe tools as belonging to *L'Epoque de St Acheul*. The industry was renamed as the Acheulean in 1925.

5.2 Dating the Acheulean

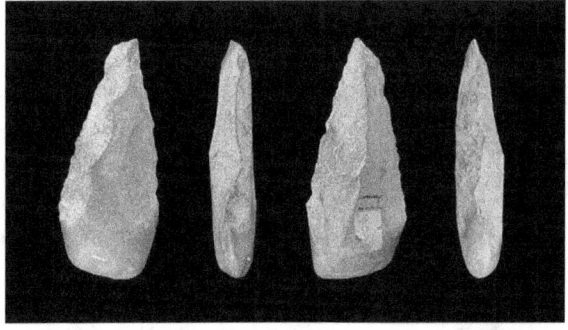

An Acheulean handaxe, Haute-Garonne France – MHNT

Providing calendrical dates and ordered chronological sequences in the study of early stone tool manufacture is often accomplished through one or more geological techniques, such as radiometric dating, often potassium-argon dating, and magnetostratigraphy. From the Konso Formation of Ethiopia, Acheulean hand-axes are dated to about 1.5 million years ago using radiometric dating of deposits containing volcanic ashes.[8] Acheulean tools in South Asia have also been found to be dated as far as 1.5 million years ago.[9] However, the earliest accepted examples of the Acheulean currently known come from the West Turkana region of Kenya and were first described by a French-led archaeology team.[10] These particular Acheulean tools were recently dated through the method of magnetostratigraphy to about 1.76 million years ago, making them the oldest not only in Africa but the world.[11] The earliest user of Acheulean tools was *Homo ergaster*, who first appeared about 1.8 million years ago. Not all researchers use this formal name, and instead prefer to call these users *early Homo erectus*.[5]

From geological dating of sedimentary deposits, it appears that the Acheulean originated in Africa and spread to Asian, Middle Eastern, and European areas sometime between 1.5

million years ago and about 800 thousand years ago.[12][13] In individual regions, this dating can be considerably refined; in Europe for example, it was thought that Acheulean methods did not reach the continent until around 500,000 years ago. However more recent research demonstrated that hand-axes from Spain were made more than 900,000 years ago.[13]

Relative dating techniques (based on a presumption that technology progresses over time) suggest that Acheulean tools followed on from earlier, cruder tool-making methods, but there is considerable chronological overlap in early prehistoric stone-working industries, with evidence in some regions that Acheulean tool-using groups were contemporary with other, less sophisticated industries such as the Clactonian[14] and then later with the more sophisticated Mousterian, as well. It is therefore important not to see the Acheulean as a neatly defined period or one that happened as part of a clear sequence but as one tool-making technique that flourished especially well in early prehistory. The enormous geographic spread of Acheulean techniques also makes the name unwieldy as it represents numerous regional variations on a similar theme. The term Acheulean does not represent a common culture in the modern sense, rather it is a basic method for making stone tools that was shared across much of the Old World.

The very earliest Acheulean assemblages often contain numerous Oldowan-style flakes and core forms and it is almost certain that the Acheulean developed from this older industry. These industries are known as the Developed Oldowan and are almost certainly transitional between the Oldowan and Acheulean.

5.3 Acheulean stone tools

5.3.1 Stages

In the four divisions of prehistoric stone-working,[15] Acheulean artefacts are classified as Mode 2, meaning they are more advanced than the (usually earlier) Mode 1 tools of the Clactonian or Oldowan/Abbevillian industries but lacking the sophistication of the (usually later) Mode 3 Middle Palaeolithic technology, exemplified by the Mousterian industry.

The Mode 1 industries created rough flake tools by hitting a suitable stone with a hammerstone. The resulting flake that broke off would have a natural sharp edge for cutting and could afterwards be sharpened further by striking another smaller flake from the edge if necessary (known as "retouch"). These early toolmakers may also have worked the stone they took the flake from (known as a core) to create chopper cores although there is some debate over whether

these items were tools or just discarded cores.[16]

The Mode 2 Acheulean toolmakers also used the Mode 1 flake tool method but supplemented it by using bone, antler, or wood to shape stone tools. This type of hammer, compared to stone, yields more control over the shape of the finished tool. Unlike the earlier Mode 1 industries, it was the core that was prized over the flakes that came from it. Another advance was that the Mode 2 tools were worked symmetrically and on both sides indicating greater care in the production of the final tool.

Mode 3 technology emerged towards the end of Acheulean dominance and involved the Levallois technique, most famously exploited by the Mousterian industry. Transitional tool forms between the two are called Mousterian of Acheulian Tradition, or MTA types. The long blades of the Upper Palaeolithic Mode 4 industries appeared long after the Acheulean was abandoned.

As the period of Acheulean tool use is so vast, efforts have been made to classify various stages of it such as John Wymer's division into Early Acheulean, Middle Acheulean, Late Middle Acheulean and Late Acheulean[17] for material from Britain. These schemes are normally regional and their dating and interpretations vary.[18]

In Africa, there is a distinct difference in the tools made before and after 600,000 years ago with the older group being thicker and less symmetric and the younger being more extensively trimmed. This may be connected with the appearance of *Homo heidelbergensis* in the archaeological record at this time who may have contributed this more sophisticated approach.

5.3.2 Manufacture

The primary innovation associated with Acheulean hand-axes is that the stone was worked symmetrically and on both sides. For the latter reason, handaxes are, along with cleavers, bifacially worked tools that could be manufactured from the large flakes themselves or from prepared cores.[19]

Tool types found in Acheulean assemblages include pointed, cordate, ovate, ficron, and bout-coupé hand-axes (referring to the shapes of the final tool), cleavers, retouched flakes, scrapers, and segmental chopping tools. Materials used were determined by available local stone types; flint is most often associated with the tools but its use is concentrated in Western Europe; in Africa sedimentary and igneous rock such as mudstone and basalt were most widely used, for example. Other source materials include chalcedony, quartzite, andesite, sandstone, chert, and shale. Even relatively soft rock such as limestone could be exploited.[20] In all cases the toolmakers worked their handaxes close to the source of their raw materials, suggesting that the Acheulean was a set of skills passed between individual groups.[21]

Some smaller tools were made from large flakes that had been struck from stone cores. These flake tools and the distinctive waste flakes produced in Acheulean tool manufacture suggest a more considered technique, one that required the toolmaker to think one or two steps ahead during work that necessitated a clear sequence of steps to create perhaps several tools in one sitting.

A hard hammerstone would first be used to rough out the shape of the tool from the stone by removing large flakes. These large flakes might be re-used to create tools. The tool maker would work around the circumference of the remaining stone core, removing smaller flakes alternately from each face. The scar created by the removal of the preceding flake would provide a striking platform for the removal of the next. Misjudged blows or flaws in the material used could cause problems, but a skilled toolmaker could overcome them.

Once the roughout shape was created, a further phase of flaking was undertaken to make the tool thinner. The thinning flakes were removed using a softer hammer, such as bone or antler. The softer hammer required more careful preparation of the striking platform and this would be abraded using a coarse stone to ensure the hammer did not slide off when struck.

Final shaping was then applied to the usable cutting edge of the tool, again using fine removal of flakes. Some Acheulean tools were sharpened instead by the removal of a tranchet flake. This was struck from the lateral edge of the hand-axe close to the intended cutting area, resulting in the removal of a flake running along (parallel to) the blade of the axe to create a neat and very sharp working edge. This distinctive tranchet flake can been identified amongst flint-knapping debris at Acheulean sites.

5.3.3 Use

Loren Eiseley calculated[22] that Acheulean tools have an average useful cutting edge of 20 centimetres (8 inches), making them much more efficient than the 5-centimetre (2 in) average of Oldowan tools.

Use-wear analysis on Acheulean tools suggests there was generally no specialization in the different types created and that they were multi-use implements. Functions included hacking wood from a tree, cutting animal carcasses as well as scraping and cutting hides when necessary. Some tools, however, could have been better suited to digging roots or butchering animals than others.

Alternative theories include a use for ovate hand-axes as a kind of hunting discus to be hurled at prey.[23] Puzzlingly, there are also examples of sites where hundreds of hand-axes, many impractically large and also apparently unused, have been found in close association together. Sites such as Melka Kunturé in Ethiopia, Olorgesailie in Kenya, Isimila in Tanzania, and Kalambo Falls in Zambia have produced evidence that suggests Acheulean hand-axes might not always have had a functional purpose.

Recently, it has been suggested[24] that the Acheulean tool users adopted the handaxe as a social artifact, meaning that it embodied something beyond its function of a butchery or wood cutting tool. Knowing how to create and use these tools would have been a valuable skill and the more elaborate ones suggest that they played a role in their owners' identity and their interactions with others. This would help explain the apparent over-sophistication of some examples which may represent a "historically accrued social significance".[25]

One theory goes further and suggests that some special hand-axes were made and displayed by males in search of mate, using a large, well-made hand-axe to demonstrate that they possessed sufficient strength and skill to pass on to their offspring. Once they had attracted a female at a group gathering, it is suggested that they would discard their axes, perhaps explaining why so many are found together.[26]

Hand-axe as a left over core

Stone knapping with limited digital dexterity makes the center of gravity the required direction of flake removal. Physics then dictates a circular or oval end pattern, similar to the handaxe, for a leftover core after flake production. This would explain the abundance, wide distribution, proximity to source, consistent shape, and lack of actual use, of these artifacts.[27]

Money

Mimi Lam, a researcher from the University of British Columbia, has suggested that Acheulean hand-axes became "the first commodity: A marketable good or service that has value and is used as an item for exchange."[28]

5.3.4 Distribution

Acheulean Biface from Saint Acheul

Map of Afro-Eurasia showing important sites of the Acheulean industry.

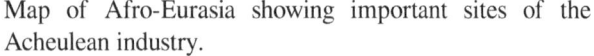

Map of Afro-Eurasia showing important sites of the Acheulean industry.

The geographic distribution of Acheulean tools – and thus the peoples who made them – is often interpreted as being the result of palaeo-climatic and ecological factors, such as glaciation and the desertification of the Sahara Desert.[29]

Acheulean stone tools have been found across the continent of Africa, save for the dense rainforest around the River Congo which is not thought to have been colonized by humans until later. It is thought that from Africa their use spread north and east to Asia: from Anatolia, through the Arabian Peninsula, across modern day Iran[30] and Pakistan, and into India, and beyond. In Europe their users reached the Pannonian Basin and the western Mediterranean regions, modern day France, the Low Countries, western Germany, and southern and central Britain. Areas further north did not see human occupation until much later, due to glaciation. In Athirampakkam at Chennai in Tamil Nadu the Acheulean age started at 1.51 mya and it is also prior than North India and Europe.[31]

Until the 1980s, it was thought that the humans who arrived in East Asia abandoned the hand-axe technology of their ancestors and adopted chopper tools instead. An apparent division between Acheulean and non-Acheulean tool industries was identified by Hallam L. Movius, who drew the Movius Line across northern India to show where the traditions seemed to diverge. Later finds of Acheulean tools at Chongokni in South Korea and also in Mongolia and China, however, cast doubt on the reliability of Movius's distinction.[32] Since then, a different division known as the Roe Line has been suggested. This runs across North Africa to Israel and thence to India, separating two different techniques used by Acheulean toolmakers. North and east of the Roe Line, Acheulean hand-axes were made directly from large stone nodules and cores; while, to the south and west, they were made from flakes struck from these nodules.[33]

5.4 Acheulean tool users

For further details of the known environment and people during the time when Acheulean tools were being made, see Palaeolithic and Lower Palaeolithic.

Acheulean tools were not made by fully modern humans – that is, *Homo sapiens* – although the early or non-modern (transitional) *Homo sapiens idaltu* did use Late Acheulean tools, as did the proto-Neanderthal species.[34] Most notably, however, it is *Homo ergaster* (sometimes called early *Homo erectus*), whose assemblages are almost exclusively Acheulean, who used the technique. Later, the related species *Homo heidelbergensis* (the common ancestor of both Neanderthals and *Homo sapiens*) used it extensively.

The symmetry of the hand-axes has been used to suggest that Acheulean tool users possessed the ability to use language;[35] the parts of the brain connected with fine control and movement are located in the same region that controls speech. The wider variety of tool types compared to earlier industries and their aesthetically as well as functionally pleasing form could indicate a higher intellectual level in Acheulean tool users than in earlier hominines.[36] Others argue that there is no correlation between spatial abilities in tool making and linguistic behaviour, and that language is not learned or conceived in the same manner as artefact manufacture.[37]

Lower Palaeolithic finds made in association with Acheulean hand-axes, such as the Venus of Berekhat Ram,[38] have been used to argue for artistic expression amongst the tool users. The incised elephant tibia from Bilzingsleben[39] in Germany, and ochre finds from Kapthurin in Kenya[40] and Duinefontein in South Africa,[41] are sometimes cited as being some of the earliest examples of an aesthetic sensibility in human history. There are numerous other explanations put forward for the creation of these artefacts, however; and there is no unequivocal evidence of human art until around 50,000 years ago, after the emergence of modern *Homo sapiens*.[42]

The kill site at Boxgrove in England is another famous Acheulean site. Up until the 1970s these kill sites, often at waterholes where animals would gather to drink, were interpreted as being where Acheulean tool users killed game, butchered their carcasses, and then discarded the tools they had used. Since the advent of zooarchaeology, which has placed greater emphasis on studying animal bones from archaeological sites, this view has changed. Many of the animals at these kill sites have been found to have been killed by other predator animals, so it is likely that humans of the period supplemented hunting with scavenging from already dead animals.[43]

Excavations at the Bnot Ya'akov Bridge site, located along the Dead Sea rift in the southern Hula Valley of northern Israel, have revealed evidence of human habitation in the area from as early as 750,000 years ago.[44] Archaeologists from the Hebrew University of Jerusalem claim that the site provides evidence of "advanced human behavior" half a million years earlier than has previously been estimated. Their report describes an Acheulean layer at the site in which numerous stone tools, animal bones, and plant remains have been found.[45]

Only limited artefactual evidence survives of the users of Acheulean tools other than the stone tools themselves. Cave sites were exploited for habitation, but the hunter-gatherers of the Palaeolithic also possibly built shelters such as those identified in connection with Acheulean tools at Grotte du Lazaret[46] and Terra Amata near Nice in France. The presence of the shelters is inferred from large rocks at the sites, which may have been used to weigh down the bottoms of tent-like structures or serve as foundations for huts or windbreaks. These stones may have been naturally deposited. In any case, a flimsy wood or animal skin structure would leave few archaeological traces after so long. Fire was seemingly being exploited by *Homo ergaster*, and would have been a necessity in colonising colder Eurasia from Africa. Conclusive evidence of mastery over it this early is, however, difficult to find.

5.5 See also

- Oldowan

- Lithic reduction

- Lower Palaeolithic

- Palaeolithic

- Stone Age

- Stone tools

- Synoptic table of the principal old world prehistoric cultures

5.6 References

[1] Hadfield, Peter, Gimme Shelter

[2] Earliest evidence of art found

[3] Bar-Yosef, O and Belfer-Cohen, A, 2001, *From Africa to Eurasia — Early Dispersals*, Quaternary International 75, 19–28, Abstract

[4] Ian Tattersall, *Masters of the Planet, the search for our human origins*, 2012, Palgrave Macmillan, chapter 6, pp. 124-125, ISBN 978-0-230-10875-2

[5] Wood, B, 2005, p87.

[6] *Oxford English Dictionary* 2nd Ed. (1989)

[7] Frere, John. "Account of Flint Weapons Discovered at Hoxne in Suffolk.". Archaeologia 13 (1800): 204-205 [reprinted in Grayson (1983), 55-56, and Heizer (1962), 70-71].

[8] (Asfaw, B. et al. The earliest Acheulean from Konso-Gardula. Nature 360, 732–735, 1992, http://www.nature.com/nature/journal/v360/n6406/abs/360732a0.html)

[9] http://www.sciencemag.org/content/331/6024/1596.abstract

[10] (Roche, H. et al. Les sites arche´ologiques plio-ple´istoce`nes de la formation de Nachukui, Ouest-Turkana, Kenya: bilan synthe´tique 1997–2001. C. R. Palevol 2, 663–673, 2003, http://journals2.scholarsportal.info/details.xqy?uri=/16310683/v02i0008/663_lsapdlnokbs1.xml)

[11] (Lepre, C. J., Roche, H., Kent, D. V., Harmand, S., Quinn, R. L., Brugal, J.-P., Texier, P.-J., Lenoble, A., & Feibel, C. S. (2011). An earlier origin for the Acheulian. Nature, 477, 82–85, http://www.nature.com/nature/journal/v477/n7362/abs/nature10372.html)

[12] Goren-Inbar, N.; Feibel, C. S.; Verosub, K. L.; Melamed, Y.; Kislev, M. E.; Tchernov, E.; Saragusti, I. (2000). "Pleistocene Milestones on the Out-of-Africa Corridor at Gesher Benot Ya'aqov, Israel". *Science* **289** (5481): 944–947. doi:10.1126/science.289.5481.944. PMID 10937996.

[13] Scott, G. R.; Gibert, L. (2009). "The oldest handaxes in Europe". *Nature* **461** (7260): 82–85. doi:10.1038/nature08214. PMID 19727198.

[14] Ashton, N, McNabb, J, Irving, B, Lewis, S and Parfitt, S *Contemporaneity of Clactonian and Acheulian flint industries at Barnham, Suffolk* Antiquity 68, 260, p585–589 Abstract

[15] Barton, RNE, *Stone Age Britain* English Heritage/BT Batsford:London 1997 qtd in Butler, 2005. See also Wymer, JJ, *The Lower Palaeolithic Occupation of Britain*, Wessex Archaeology and English Heritage, 1999.

[16] Ashton, NM, McNabb, J, and Parfitt, S, *Choppers and the Clactonian, a reinvestigation*, Proceedings of the Prehistoric Society 58, pp21–28, qtd in Butler, 2005

[17] Wymer, JJ, 1968, *Lower Palaeolithic Archaeology in Britain: as represented by the Thames Valley*, qtd in Adkins, L and R, 1998

[18] Collins, D, 1978, *Early Man in West Middlesex*, qtd in Adkins, L and R, 1998

[19] Barham, Lawrence; Mitchell, Peter (2008). *The First Africans* (1st ed.). Cambridge University Press. p. 16. ISBN 978-0-521-61265-4.

[20] Paddayya, K, Jhaldiyal, R and Petraglia, MD, *Excavation of an Acheulian workshop at Isampur, Karnataka (India)* Antiquity 74, 286, pp 751–752 Abstract

[21] Gamble, C and Steele, J, 1999, *Hominid ranging patterns and dietary strategies* in Ullrich, H (ed.), Hominid evolution: lifestyles and survival strategies, pp 396–409, Gelsenkirchen: Edition Archaea.

[22] Unattributed citation in Renfrew and Bahn, 1991, p277

[23] O'Brien, E, 1981, *The projectile capabilities of an Acheulian handaxe from Olorgesailie*, Current Anthropology 22: 76–9. See also Calvin, W, 1993, *The unitary hypothesis: a common neural circuitry for novel manipulations, language, plan-ahead and throwing*, in K.R. Gibson & T. Ingold (ed.), Tools, language and cognition in human evolution: 230–50. Cambridge: Cambridge University Press.

[24] Gamble, C, 1997, *Handaxes and palaeolithic individuals*, in N. Ashton, F. Healey & P.Pettitt (ed.), Stone Age archaeology: 105–9. Oxford: Oxbow Books. Monograph 102.

[25] White, MJ, 1998, On the significance of Acheulian biface variability in southern Britain, *Proceedings of the Prehistoric Society 64: 15–44.*

[26] Kohn, M and Mithen, S, 1999, *Handaxes: products of sexual selection?*, Antiquity 73, 518–26 Abstract

[27] [citation http://www.ele.net/acheulean/handaxe.htm]

[28] http://www.livescience.com/18751-hand-axe-tools-money.html

[29] Todd, L, Glantz, M and Kappelman, J, *Chilga Kernet: an Acheulean landscape on Ethiopia's western plateau* Antiquity 76, 293 pp 611–612 Abstract

[30] Biglari, F. and Shidrang, S. 2006 The Lower Paleolithic Occupation of Iran, Near Eastern Archaeology 69(3–4): 160-168

[31] http://www.hindu.com/2011/03/25/stories/2011032564021300.htm

[32] Hyeong Woo Lee, *The Palaeolithic industries of Korea: chronology and related new findspots* in Milliken, S and Cook, J (eds), 2001

[33] Gamble, C and Marshall, G, *The shape of handaxes, the structure of the Acheulian world*, in Milliken, S and Cook, J (eds), 2001

[34] Clarke, JD et al., 2003, *Stratigraphic, chronological and behavioural contexts of Pleistocene Homo sapiens from Middle Awash, Ethiopia*, Nature 423, 747–52, Abstract

[35] Isaac, GL, 1976, *Stages of cultural elaboration in the Pleistocene: possible archaeological indicators of the development of language capabilities*, in Origins and Evolution of Languages and Speech (SR Harbard et al. eds.), 276–88, Annals of the New York Academy of Sciences 280, qtd in Renfrew and Bahn, 1991

[36] Wynn, T, 1995, *Handaxe enigmas*, World Archaeology 27, 10–24, qtd in Scarre, 2005

[37] Dibble, HL, 1989, *The implications of stone tool types for the presence of language during the Lower and Middle Palaeolithic*, in The Human Revolution (P Mellars and C Stringer eds) Edinburgh University Press, qtd in Renfrew and Bahn, 1991.

[38] Goren-Inbar, N and Peltz, S, 1995, *Additional remarks on the Berekhat Ram figure*, Rock Art Research 12, 131–132, qtd in Scarre, 2005

[39] Mania, D and Mania, U, 1988, *Deliberate engravings on bone artefacts of Homo Erectus*, Rock Art Research 5, 919–7, qtd in Scarre, 2005

[40] Tryon, CA and McBrearty, S, 2002, *Tephrostatigraphy and the Acheulean to Middle Stone Age transition in the Kapthurin Formation, Kenya*, Journal of Human Evolution 42, 211–35, qtd in Scarre, 2005 Abstract

[41] Cruz-Uribe, K et al., 2003, *Excavation of buried late Acheulean (mid-Quaternary) land surfaces at Duinefontein 2, West Cape Province, South Africa*, Journal of Archaeological Science 30, 559–75, qtd in Scarre, 2005

[42] Scarre, 2005, chapter 3, p118 "However, objects whose artistic meaning is unequivocal become commonplace only after 50,000 years ago, when they are associated with the origins and spread of fully modern humans from Africa.

[43] *...the most conservative conclusion today is that Acheulean people and their contemporaries definitely hunted big animals, though their success rate is not clear* ibid, p 120.

[44] Gesher Benot Ya'aqov, Hebrew University, Retrieved 2010-01-05

[45] Evidence of advanced human life half a million years earlier than previously thought (Dec 22, 2009) in *The Jerusalem Post* Retrieved 2010-01-05

[46] De Lumley, 1975, *Cultural evolution in France in its palaeoecological setting during the middle Pleistocene*, in After the Australopithecines, Butzer, KW and Issac, G Ll. (eds) 745–808. The Hague:Mouton, qtd in Scarre, 2005

5.7 Bibliography

- Adkins, L; and R (1998). *The Handbook of British Archaeology*. London: Constable. ISBN 0-09-478330-6.

- Butler, C (2005). *Prehistoric Flintwork*. Tempus, Stroud. ISBN 0-7524-3340-7.

- Milliken, S; and J Cook (eds) (2001). *A Very Remote Period Indeed. Papers on the Palaeolithic presented to Derek Roe*. Oxford: Oxbow. ISBN 1-84217-056-2.

- Renfrew, C; and P Bahn (1991). *Archaeology, Theories Methods and Practice*. London: Thames and Hudson. ISBN 0-500-27605-6.

- Scarre, C (ed.) (2005). *The Human Past*. London: Thames and Hudson. ISBN 0-500-28531-4.

- Wood, B (2005). *Human Evolution A Very Short Introduction*. Oxford: Oxford University Press. ISBN 0-19-280360-3.

5.8 External links

- Media related to Acheulean at Wikimedia Commons

Chapter 6

Homo

"Pithecanthropus" redirects here. For Pithecanthropus alalus, see Ernst Haeckel. For Pithecanthropus erectus, see Homo erectus. For Pithecanthropus erectus erectus, see Java Man.

"Genus Homo" redirects here. For the novel by L. Sprague de Camp and P. Schuyler Miller, see Genus Homo (novel). For other uses, see Homo (disambiguation).

Homo is the genus that comprises the species *Homo sapiens*, which includes modern humans, as well as several extinct species classified as ancestral to or closely related to modern humans—as for examples *Homo habilis* and *Homo neanderthalensis*. The genus is about 2.8 million years old;[1][2][3][4][5] it first appeared as its earliest species *Homo habilis*, which emerged from the genus *Australopithecus*, which itself had previously split from the lineage of the genus *Pan*, the chimpanzees.[6] *Homo* is the only genus assigned to the subtribe **Hominina** which, with the subtribes Australopithecina and Panina, comprise the tribe Hominini (see evolutionary tree below). All species of the genus *Homo* plus those species of the australopithecines that arose *after* the split from *Pan* are called hominins.

The line to the earliest members of *Homo* made final separation from the lineage of *Pan* by late Miocene or early Pliocene times—with date estimates by several specialists ranging from 13 million years ago [7] to as recent as four million years ago—which (latter) date was soon rejected by some;[8] [9] see current estimates regarding complex speciation. *Homo erectus* appeared about two million years ago in East Africa (where it is dubbed *Homo ergaster*) and, in several early migrations, it spread throughout Africa and Eurasia. It was likely the first hominin to live in a hunter-gatherer society and to control fire. An adaptive and successful species, *Homo erectus* persisted for almost 2 million years before suddenly becoming extinct about 70,000 years ago (0.07 Ma)—perhaps a casualty of the Toba supereruption catastrophe.

Homo sapiens sapiens, or *anatomically* modern humans, emerged about 200,000 years ago (0.2 Ma) in East Africa (see Omo remains). There is division among scholars as to when *H. s. sapiens* became *behaviorally* modern; the debate is: modern behavior developed 1) simultaneously with anatomical development, or 2) separately, and was complete by 50,000 years ago (see Modern human behavior). *Homo sapiens sapiens* is the only surviving species of the genus Homo; all others have become extinct.

Modern humans migrated from Africa as recently as 60,000 years ago, and during Upper Paleolithic times they spread throughout Africa and Eurasia, Oceania, and the Americas; and they encountered archaic humans en route of their migrations. Some archaic humans outside Africa survived alongside modern humans until about 40,000 years ago (*see H. neanderthalensis*),[10] and possibly until as late as the times of the Epipaleolithic culture (about 12,000 years ago). DNA analysis provides evidence of interbreeding between archaic and modern humans.

6.1 Name

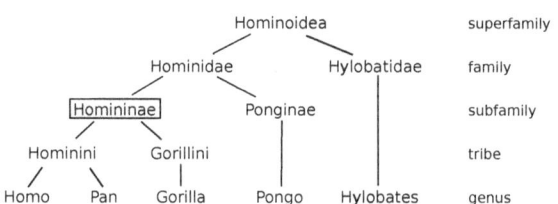

Evolutionary tree chart emphasizing the subfamily Homininae and the tribe Hominini. After diverging from the line to Ponginae the early Homininae split into the tribes Hominini and Gorillini. The early Hominini split further, separating the line to Homo *from the lineage of* Pan. *Currently, tribe* Hominini *designates the subtribes* Hominina, *containing genus* Homo; Panina, *genus* Pan; *and* Australopithecina, *with several extinct genera—the subtribes are not labelled on this chart.*

Further information: List of alternative names for the human species

See Hominidae for an overview of taxonomy.

The Latin noun *homō* (genitive *hominis*) means "human being" or "man" in the generic sense of "human being, mankind".[11] The binomial name *Homo sapiens* was coined by Carl Linnaeus (1758).[12] Names for other species of the genus were introduced beginning in the second half of the 19th century (*H. neanderthalensis* 1864, *H. erectus* 1892).

6.1.1 Taxonomy

Even today, the *Homo* genus has not been properly defined.[13][14][15] Since the early human fossil record began to slowly emerge from the earth, the boundaries and definitions of the *Homo* genus have been poorly defined and constantly in flux. Because there was no reason to think it would ever have any additional members, Carl Linnaeus did not even bother to define the *Homo* genus when he first created it for humans in the 1700s. The discovery of Neanderthal brought the first addition.

The genus *Homo* was given its taxonomic name to suggest that its member species can be classified as human. And, over the decades of the 20th century, fossil finds of pre-human and early human species from late Miocene and early Pliocene times produced a rich mix for debating classifications. There is continuing debate on delineating *Homo* from *Australopithecus*—or, indeed, delineating *Homo* from *Pan*, as one body of scientists argue that the two species of chimpanzee should be classed with genus *Homo* rather than *Pan*. Even so, classifying the fossils of *Homo* coincides with evidences of: 1) competent human bipedalism in *Homo habilis* inherited from the earlier Australopithecus of more than four million years ago, (see Laetoli); and 2) human tool culture having begun by 2.5 million years ago.

From the late-19th to mid-20th century, a number of new taxonomic names including new generic names were proposed for early human fossils; most have since been merged with *Homo* in recognition that *Homo erectus* was a single and singular species with a large geographic spread of early migrations. Many such names are now dubbed as "synonyms" with *Homo*, including *Pithecanthropus*,[16] *Protanthropus*,[17] *Sinanthropus*,[18] *Cyphanthropus*,[19] *Africanthropus*,[20] *Telanthropus*,[21] *Atlanthropus*,[22] and *Tchadanthropus*.[23]

Classifying the genus *Homo* into species and subspecies is subject to incomplete information and remains poorly done. This has led to using common names ("Neanderthal" and "Denisovan") in even scientific papers to avoid trinomial names or the ambiguity of classifying groups as *incertae sedis* (uncertain placement)—for example, *H. neanderthalensis* vs. *H. sapiens neanderthalensis*, or *H. georgicus* vs. *H. erectus georgicus*.[24] Some recently extinct species in the genus *Homo* are only recently discovered and do not as yet have consensus binomial names (see Denisova

hominin and Red Deer Cave people).

John Edward Gray (1825) was an early advocate of classifying taxa by designating tribes and families.[25] Wood and Richmond (2000) proposed that Hominini ("hominins") be designated as a tribe that comprised all species of early humans and pre-humans ancestral to humans back to *after* the chimpanzee-human last common ancestor; and that *Hominina* be designated a subtribe of Hominini to include *only* the genus *Homo*—that is, *not* including the earlier upright walking hominins of the Pliocene such as *Australopithecus*, *Orrorin tugenensis*, *Ardipithecus*, or *Sahelanthropus*.[26] Designations alternative to Hominina existed, or were offered: *Australopithecinae* (Gregory & Hellman 1939) and *Preanthropinae* (Cela-Conde & Altaba 2002);[27] and later, Cela-Conde and Ayala (2003) proposed that the four genera *Australopithecus*, *Ardipithecus*, *Praeanthropus*, and *Sahelanthropus* be grouped with *Homo* within Hominina.[28]

6.2 Evolution

Further information: Timeline of human evolution, Hominina, Archaic humans and Australopithecus

Several   species,   including   *Australopithecus garhi*,

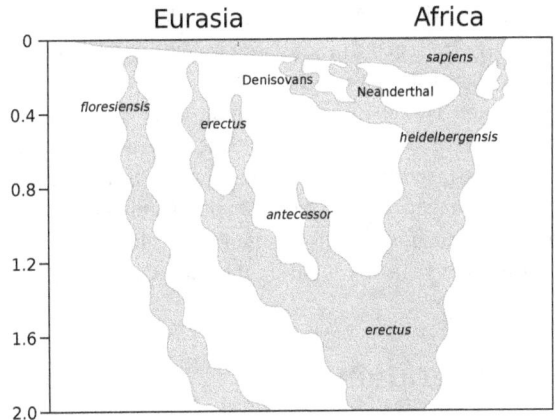

A model of the evolution of the genus Homo *over the last 2 million years (vertical axis). The rapid "Out of Africa" expansion of* H. sapiens *is indicated at the top of the diagram, with admixture indicated with Neanderthals, Denisovans, and unspecified archaic African hominins.*[29]

Australopithecus sediba, *Australopithecus africanus*, and *Australopithecus afarensis*, have been proposed as the direct ancestor of the *Homo* lineage.[30][31] These species have morphological features that align them with *Homo*, but there is no consensus as to which gave rise to *Homo*. The advent of *Homo* was traditionally taken to coincide with the first use of stone tools (the Oldowan industry), and thus by definition with the beginning of the Lower Palaeolithic.[32] The emergence of *Homo* also coincides roughly with the onset of Quaternary glaciation, the beginning of the current ice age.

A fossil jawbone dated to 2.8 million years ago which may represent an intermediate stage between *Australopithecus* and *Homo* was discovered in 2015 in Afar, Ethiopia.[33] Some authors would push the development of *Homo* past 3 Mya, by including *Kenyanthropus* (a fossil dated 3.2 to 3.5 Mya, usually classified as an australopithecine species) into the *Homo* genus.[34]

The most salient physiological development between the earlier australopithecine species and *Homo* is the increase in cranial capacity, from about 450 cm^3 (27 cu in) in *A. garhi* to 600 cm^3 (37 cu in) in *H. habilis*. Within the *Homo* genus, cranial capacity again doubled from *H. habilis* through *Homo ergaster* or *H. erectus* to *Homo heidelbergensis* by 0.6 million years ago. The cranial capacity of *H. heidelbergensis* overlaps with the range found in modern humans.

Homo erectus has often been assumed to have developed anagenetically from *Homo habilis* from about 2 million years ago. This scenario was strengthened with the discovery of *Homo erectus georgicus*, early specimens of *H. erectus* found in the Caucasus, which seemed to exhibit transitional traits with *H. habilis*. As the earliest evidence for *H. erectus* was found outside of Africa, it was considered plausible that *H. erectus* developed in Eurasia and then migrated back to Africa. Based on fossils from the Koobi Fora Formation, east of Lake Turkana in Kenya, Spoor et al. (2007) argued that *H. habilis* may have survived beyond the emergence of *H. erectus*, so that the evolution of *H. erectus* would not have been anagenetically, and *H. erectus* would have existed alongside with *H. habilis* for about half a million years (1.9 to 1.4 million years ago), during the early Calabrian.[35]

6.3   Migration

See also: Human evolution and Archaic human admixture with modern humans

Some of *H. ergaster* migrated to Asia, where they are named *Homo erectus*, and to Europe with *Homo georgicus*. *H. ergaster* in Africa and *H. erectus* in Eurasia evolved separately for almost two million years and presumably separated into two different species. *Homo rhodesiensis*, who were descended from *H. ergaster*, migrated from Africa to Europe and became *Homo heidelbergensis* and later (about 250,000 years ago) *Homo neanderthalensis* and the Denisova hominin in Asia. The first *Homo sapiens*, descendants of *H. rhodesiensis*, appeared in Africa about 250,000 years

ago. About 100,000 years ago, some *H. sapiens sapiens* migrated from Africa to the Levant and met with resident Neanderthals, with some admixture.[36] Later, about 70,000 years ago, perhaps after the Toba catastrophe, a small group left the Levant to populate Eurasia, Australia and later the Americas. A subgroup among them met the Denisovans[37] and, after further admixture, migrated to populate Melanesia. In this scenario, non-African people living today are mostly of African origin ("Out of Africa model"). However, there was also some admixture with Neanderthals and Denisovans, who had evolved locally (the "multiregional hypothesis"). Recent genomic results from the group of Svante Pääbo also show that 30,000 years ago at least three major subspecies coexisted: Denisovans, Neanderthals and anatomically modern humans.[38] Today, only *H. sapiens* remains, with no other extant species.

6.4   List of species

See also: List of human evolution fossils

The species status of *Homo rudolfensis*, *Homo ergaster*, *H. georgicus*, *H. antecessor*, *H. cepranensis*, *H. rhodesiensis*, *Homo neanderthalensis*, Denisova hominin, Red Deer Cave people and *Homo floresiensis* remains under debate. *H. heidelbergensis* and *H. neanderthalensis* are closely related to each other and have been considered to be subspecies of *H. sapiens*. Recently, nuclear DNA from a Neanderthal specimen from Vindija Cave has been sequenced using two different methods that yield similar results regarding Neanderthal and *H. sapiens* lineages, with both analyses suggesting a date for the split between 460,000 and 700,000 years ago, though a population split of around 370,000 years is inferred. The nuclear DNA results indicate about 30% of derived alleles in *H. sapiens* are also in the Neanderthal lineage. This high frequency may suggest some gene flow between ancestral human and Neanderthal populations due to mating between the two.[39]

Homo naledi was discovered near Johannesburg, South Africa in 2013 and announced on 10 September 2015. Fossils indicate the hominid was 1.45-1.5 meters tall and had a small brain.[40] The fossils have yet to be dated but are estimated to be roughly 2.5 million years old.[41]

6.5   See also

- *Dawn of Humanity* (2015 PBS film)

- List of human evolution fossils (*with images*)

6.6 References

[1] Stringer, C.B. (1994). "Evolution of early humans". In Steve Jones, Robert Martin & David Pilbeam (eds.). *The Cambridge Encyclopedia of Human Evolution*. Cambridge: Cambridge University Press. p. 242. ISBN 0-521-32370-3. Also ISBN 0-521-46786-1 (paperback)

[2] McHenry, H.M (2009). "Human Evolution". In Michael Ruse & Joseph Travis. *Evolution: The First Four Billion Years*. Cambridge, Massachusetts: The Belknap Press of Harvard University Press. p. 265. ISBN 978-0-674-03175-3.

[3] Wilford, John Noble (2015-03-04). "Jawbone Fossil Fills a Gap in Early Human Evolution". *The New York Times*. ISSN 0362-4331. Retrieved 2015-05-30.

[4] Spoor, Fred; Gunz, Philipp; Neubauer, Simon; Stelzer, Stefanie; Scott, Nadia; Kwekason, Amandus; Dean, M. Christopher (March 5, 2015). "Reconstructed Homo habilis type OH 7 suggests deep-rooted species diversity in early Homo". *Nature* **519** (7541): 83–86. doi:10.1038/nature14224. ISSN 0028-0836.

[5] Villmoare, Brian; Kimbel, William H.; Seyoum, Chalachew; Campisano, Christopher J.; DiMaggio, Erin N.; Rowan, John; Braun, David R.; Arrowsmith, J. Ramón; Reed, Kaye E. (2015-03-20). "Early Homo at 2.8 Ma from Ledi-Geraru, Afar, Ethiopia". *Science* **347** (6228): 1352–1355. doi:10.1126/science.aaa1343. ISSN 0036-8075. PMID 25739410.

[6] Schuster, Angela M. H. (1997). "Earliest Remains of Genus Homo". *Archaeology* **50** (1). Retrieved 5 March 2015.

[7] Arnason U, Gullberg A, Janke A (December 1998). "Molecular timing of primate divergences as estimated by two nonprimate calibration points". J. Mol. Evol. 47 (6): 718–27. doi:10.1007/PL00006431. PMID 9847414.

[8] Patterson N, Richter DJ, Gnerre S, Lander ES, Reich D (June 2006). "Genetic evidence for complex speciation of humans and chimpanzees". Nature 441 (7097): 1103–8. doi:10.1038/nature04789. PMID 16710306

[9] Wakeley J (March 2008). "Complex speciation of humans and chimpanzees". Nature 452 (7184): E3–4; discussion E4. doi:10.1038/nature06805. PMID 18337768. "Patterson et al. suggest that the apparently short divergence time between humans and chimpanzees on the X chromosome is explained by a massive interspecific hybridization event in the ancestry of these two species. However, Patterson et al. do not statistically test their own null model of simple speciation before concluding that speciation was complex, and—even if the null model could be rejected—they do not consider other explanations of a short divergence time on the X chromosome. These include natural selection on the X chromosome in the common ancestor of humans and chimpanzees, changes in the ratio of male-to-female mutation rates over time, and less extreme versions of divergence with gene flow. I therefore believe that their claim of hybridization is unwarranted."

[10] , BBC

[11] The word "human" itself is from Latin *humanus*, an adjective formed on the root of *homo*, thought to derive from a Proto-Indo-European word for "earth" reconstructed as **dhghem-*. dhghem The American Heritage Dictionary of the English Language: Fourth Edition. 2000.

[12] Linné, Carl von (1758). *Systema naturæ. Regnum animale*. (10 ed.). pp. 18, 20. Retrieved 19 November 2012.. Note: In 1959, Linnaeus was designated as the lectotype for *Homo sapiens* (Stearn, W. T. 1959. "The background of Linnaeus's contributions to the nomenclature and methods of systematic biology", *Systematic Zoology* 8 (1): 4-22, p. 4) which means that following the nomenclatural rules, *Homo sapiens* was validly defined as the animal species to which Linnaeus belonged.

[13] Schwartz, Jeffrey H.; Tattersall, Ian (28 August 2015). "Defining the genus Homo". *Science* **349** (6251): 931–932. doi:10.1126/science.aac6182. Retrieved 2015-11-02.

[14] Lents, Nathan (4 October 2014). "Homo naledi and the Problems with the Homo Genus". *The Wildernist*. Retrieved 2015-11-02.

[15] Wood, B.; Collard, M. (2 April 1999). "The human genus". *Science* **284** (5411): 65–71. PMID 10102822. Retrieved 2015-11-03.

[16] "ape-man", from *Pithecanthropus erectus* (Java Man), Eugène Dubois, *Pithecanthropus erectus : eine menschenähnliche Übergangsform aus Java* (1894), identified with the *Pithecanthropus alalus* (i.e. "non-speaking ape-man") hypothesized earlier by Ernst Haeckel

[17] "early man", *Protanthropus primigenius* Ernst Haeckel, *Systematische Phylogenie* vol. 3 (1895), p. 625

[18] "Sinic man", from *Sinanthropus pekinensis* (Peking Man), Davidson Black (1927).

[19] "crooked man", from *Cyphanthropus rhodesiensis* (Rhodesian Man) William Plane Pycraft (1928).

[20] "African man", used by T. F. Dreyer (1935) for the Florisbad Skull he found in 1932 (also *Homo florisbadensis* or *Homo helmei*). Also the genus suggested for a number of archaic human skulls found at Lake Eyasi by Weinert (1938). Leaky, *Journal of the East Africa Natural History Society' (1942), p. 43*.

[21] "remote man"; from *Telanthropus capensis* (Broom and Robinson 1949), see (1961), p. 487.

[22] from *Atlanthropus mauritanicus*, name given to the species of fossils (three lower jaw bones and a parietal bone of a skull) discovered in 1954 to 1955 by Camille Arambourg

in Tighennif, Algeria. C. Arambourg, "A recent discovery in human paleontology: Atlanthropus of ternifine (Algeria)", *American Journal of Physical Anthropology* 13.2 (June 1955), 191–201, doi: 10.1002/ajpa.1330130203.

[23] Y. Coppens, "L'Hominien du Tchad", *Actes V Congr. PPEC* I (1965), 329f.; "Le Tchadanthropus", *Anthropologia* 70 (1966), 5–16.

[24] Alexandra Vivelo (2013), Characterization of Unique Features of the Denisovan Exome

[25] J. E. Gray, "An outline of an attempt at the disposition of Mammalia into Tribes and Families, with a list of genera apparently appertaining to each Tribe", *Annals of Philosophy', new series (1825), pp. 337–344.*

[26] Wood and Richmond; Richmond, BG (2000). "Human evolution: taxonomy and paleobiology". *Journal of Anatomy* **197** (Pt 1): 19–60. doi:10.1046/j.1469-7580.2000.19710019.x. PMC 1468107. PMID 10999270.

[27] Brunet, M. *et al.* 2002: A new hominid from the upper Miocene of Chad, central Africa. *Nature* (London), **418**: 145-151. Cela-Conde, C.J. and Ayala, F.J., 2003: Genera of the human lineage. *PNAS*, **100**(13): 7684-7689. Wood, B.; Lonergan, N., 2008: The hominin fossil record: taxa, grades and clades. *J. Anat.*, **212**: 354–376. PDF

[28] C. J. Cela-Conde and F. J. Ayala. 2003. "Genera of the human lineage". *Proceedings of the National Academy of Sciences* 100(13):7684-7689.

[29] Stringer, C. (2012). "What makes a modern human". *Nature* **485** (7396): 33–35. doi:10.1038/485033a. PMID 22552077.

[30] Pickering, R.; Dirks, P. H.; Jinnah, Z.; De Ruiter, D. J.; Churchill, S. E.; Herries, A. I.; Berger, L. R. (2011). "Australopithecus sediba at 1.977 Ma and implications for the origins of the genus Homo". *Science* **333** (6048): 1421–1423. doi:10.1126/science.1203697.

[31] Asfaw, B.; White, T.; Lovejoy, O.; Latimer, B.; Simpson, S.; Suwa, G. (1999). "Australopithecus garhi: a new species of early hominid from Ethiopia". *Science* **284** (5414): 629–635. doi:10.1126/science.284.5414.629.

[32] In 2010, evidence was presented that seems to attribute the use of stone tools to *Australopithecus afarensis*, close to a million years before the first appearance of *Homo*. McPherron, S. P.; Alemseged, Z.; Marean, C. W.; Wynn, J. G.; Reed, D.; Geraads, D.; Bobe, R.; Bearat, H. A. (2010). "Evidence for stone-tool-assisted consumption of animal tissues before 3.39 million years ago at Dikika, Ethiopia". *Nature* **466**: 857–860. doi:10.1038/nature09248. "The oldest direct evidence of stone tool manufacture comes from Gona (Ethiopia) and dates to between 2.6 and 2.5 million years (Myr) ago. [...] Here we report stone-tool-inflicted marks on bones found during recent survey work in Dikika, Ethiopia [... showing] unambiguous stone-tool cut marks for flesh removal [..., dated] to between 3.42 and 3.24 Myr ago [...] Our discovery extends by approximately 800,000 years the antiquity of stone tools and of stone-tool-assisted consumption of ungulates by hominins; furthermore, this behaviour can now be attributed to Australopithecus atarensis."

[33] Erin N. DiMaggio EN, Campisano CJ, Rowan J, Dupont-Nivet G, Deino AL; et al. "Late Pliocene fossiliferous sedimentary record and the environmental context of early *Homo* from Afar, Ethiopia". *Science*. doi:10.1126/science.aaa1415. See also: "Oldest known member of human family found in Ethiopia". *New Scientist*. 4 March 2015. Retrieved 7 March 2015., Ghosh, Pallab (4 March 2015). "'First human' discovered in Ethiopia". *BBC News*. Retrieved 5 March 2015.

[34] Cela-Conde and Ayala (2003) recognize five genera within *Hominina*: *Ardipithecus*, *Australopithecus* (including *Paranthropus*), *Homo* (including *Kenyanthropus*), *Praeanthropus* (including *Orrorin*), and *Sahelanthropus*. C. J. Cela-Conde and F. J. Ayala. 2003. "Genera of the human lineage". *Proceedings of the National Academy of Sciences* 100(13):7684-7689.

[35] "A partial maxilla assigned to H. habilis reliably demonstrates that this species survived until later than previously recognized, making an anagenetic relationship with H. erectus unlikely. The discovery of a particularly small calvaria of H. erectus indicates that this taxon overlapped in size with H. habilis, and may have shown marked sexual dimorphism. The new fossils confirm the distinctiveness of H. habilis and H. erectus, independently of overall cranial size, and suggest that these two early taxa were living broadly sympatrically in the same lake basin for almost half a million years." Spoor, F; Leakey, M.G; Gathogo, P.N; Brown, F.H; Antón, S.C; McDougall, I; Kiarie, C; Manthi, F.K.; Leakey, L.N. (2007). "Implications of new early Homo fossils from Ileret, east of Lake Turkana, Kenya". *Nature* **448** (7154): 688–691. doi:10.1038/nature05986.

[36] Green, RE; Krause, J; et al. (2010). "A draft sequence of the Neanderthal genome". *Science* **328** (5979): 710–22. doi:10.1126/science.1188021. PMID 20448178.

[37] Reich, D; Green, RE; Kircher, M; et al. (December 2010). "(December 2010). "Genetic history of an archaic hominin group from Denisova Cave in Siberia"". *Nature* **468** (7327): 1053–60. doi:10.1038/nature09710. PMID 21179161.

[38] Reich; et al. (October 2011). "Denisova admixture and the first modern human dispersals into southeast Asia and Oceania". *Am J Hum Genet* **89** (4): 516–28. doi:10.1016/j.ajhg.2011.09.005. PMC 3188841. PMID 21944045.

[39] Biological Anthropology: 2nd Edition. 2009. Craig Stanford et al.

[40] Shaun Smillie,"Homo naledi--New human ancestor buried its dead," *Times Live,* 10 Sept 2015.

[41] Berger, Lee R.; et al. (10 September 2015). "*Homo naledi*, a new species of the genus *Homo* from the Dinaledi Chamber,

South Africa". *eLife* **4**. doi:10.7554/eLife.09560. Retrieved 10 September 2015. Lay summary.

> Full list of authors: Lee R Berger, John Hawks, Darryl J de Ruiter, Steven E Churchill, Peter Schmid, Lucas K Delezene, Tracy L Kivell, Heather M Garvin, Scott A Williams, Jeremy M DeSilva, Matthew M Skinner, Charles M Musiba, Noel Cameron, Trenton W Holliday, William Harcourt-Smith, Rebecca R Ackermann, Markus Bastir, Barry Bogin, Debra Bolter, Juliet Brophy, Zachary D Cofran, Kimberly A Congdon, Andrew S Deane, Mana Dembo, Michelle Drapeau, Marina C Elliott, Elen M Feuerriegel, Daniel Garcia-Martinez, David J Green, Alia Gurtov, Joel D Irish, Ashley Kruger, Myra F Laird, Damiano Marchi, Marc R Meyer, Shahed Nalla, Enquye W Negash, Caley M Orr, Davorka Radovcic, Lauren Schroeder, Jill E Scott, Zachary Throckmorton, Matthew W Tocheri, Caroline VanSickle, Christopher S Walker, Pianpian Wei, Bernhard Zipfel.

[42] Schrenk, Friedemann; Kullmer, Ottmar; Bromage, Timothy (2007). "The Earliest Putative *Homo* Fossils". In Henke, Winfried; Tattersall, Ian. *Handbook of Paleoanthropology* **1**. In collaboration with Thorolf Hardt. Berlin, Heidelberg: Springer. pp. 1611–1631. doi:10.1007/978-3-540-33761-4_52. ISBN 978-3-540-32474-4. Confirmed *H. habilis* fossils are dated to between 2.1 and 1.5 million years ago. This date range overlaps with the emergence of *Homo erectus*. Wilford, John Noble (August 9, 2007). "Fossils in Kenya Challenge Linear Evolution". *The New York Times*. Retrieved 2015-05-04.

> • DiMaggio, Erin N.; Campisano, Christopher J.; Rowan, John; et al. (March 20, 2015). "Late Pliocene fossiliferous sedimentary record and the environmental context of early *Homo* from Afar, Ethiopia". *Science* (Washington, D.C.: American Association for the Advancement of Science) **347** (6228): 1355–1359. doi:10.1126/science.aaa1415. ISSN 0036-8075. PMID 25739409. Hominins with "proto-Homo" traits may have lived as early as 2.8 million years ago, as suggested by a fossil jawbone classified as transitional between *Australopithecus* and *Homo* discovered in 2015.

[43] Haviland, William A.; Walrath, Dana; Prins, Harald E. L.; McBride, Bunny (2007). *Evolution and Prehistory: The Human Challenge* (8th ed.). Belmont, CA: Thomson Wadsworth. p. 162. ISBN 978-0-495-38190-7. *H. erectus* may have appeared some 2 million years ago. Fossils dated to as much as 1.8 million years ago have been found both in Africa and in Southeast Asia, and the oldest fossils by a narrow margin (1.85 to 1.77 million years ago) were found in the Caucasus, so that it is unclear whether *H. erectus* emerged in Africa and migrated to Eurasia, or if, conversely, it evolved in Eurasia and migrated back to Africa.

> • Ferring, R.; Oms, O.; Agusti, J.; Berna, F.; Nioradze, M.; Shelia, T.; Tappen, M.; Vekua, A.; Zhvania, D.; Lordkipanidze, D. (2011). "Earliest human occupations at Dmanisi (Georgian Caucasus) dated to 1.85-1.78 Ma". *Proceedings of the National Academy of Sciences* **108** (26): 10432. doi:10.1073/pnas.1106638108.

> • "New discovery suggests Homo erectus originated from Asia". *Daily News and Analysis* (Mumbai, India: Diligent Media Corporation Ltd.). Asian News International. June 8, 2011. Retrieved 2015-05-04.

> • Frazier, Kendrick (November–December 2006). "Leakey Fights Church Campaign to Downgrade Kenya Museum's Human Fossils". *Skeptical Inquirer* (Amherst, NY: Committee for Skeptical Inquiry) **30** (6). ISSN 0194-6730. Retrieved 2015-05-04.

[44] Now also included in *H. erectus* are Peking Man (formerly *Sinanthropus pekinensis*) and Java Man (formerly *Pithecanthropus erectus*). *H. erectus* is now grouped into various subspecies, including *Homo erectus erectus*, *Homo erectus yuanmouensis*, *Homo erectus lantianensis*, *Homo erectus nankinensis*, *Homo erectus pekinensis*, *Homo erectus palaeojavanicus*, *Homo erectus soloensis*, *Homo erectus tautavelensis*, *Homo erectus georgicus*. The distinction from descendant species such as *Homo ergaster*, *Homo floresiensis*, *Homo antecessor*, *Homo heidelbergensis* and indeed *Homo sapiens* is not entirely clear.

[45] Curnoe, Darren (June 2010). "A review of early *Homo* in southern Africa focusing on cranial, mandibular and dental remains, with the description of a new species (*Homo gautengensis* sp. nov.)". *HOMO - Journal of Comparative Human Biology* (Amsterdam, the Netherlands: Elsevier) **61** (3): 151–177. doi:10.1016/j.jchb.2010.04.002. ISSN 0018-442X. PMID 20466364. A species proposed in 2010 based on the fossil remains of three individuals dated between 1.9 and 0.6 million years ago. The same fossils were also classified as *H. habilis*, *H. ergaster* or *Australopithecus* by other anthropologists.

[46] Hazarika, Manjil (2007). "*Homo erectus/ergaster* and Out of Africa: Recent Developments in Paleoanthropology and Prehistoric Archaeology" (PDF). *EAA Summer School eBook* **1**. European Anthropological Association. pp. 35–41. Retrieved 2015-05-04. "Intensive Course in Biological Anthrpology, 1st Summer School of the European Anthropological Association, 16–30 June, 2007, Prague, Czech Republic"

[47] The type fossil is Mauer 1, dated to ca. 0.6 million years ago. The transition from *H. heidelbergensis* to *H. neanderthalensis* at about 0.35 to 0.25 million years ago is largely conventional. Relevant examples are fossils found at Bilzingsleben (also classified as *Homo erectus bilzingslebensis*).

[48] Bischoff, James L.; Shamp, Donald D.; Aramburu, Arantza; et al. (March 2003). "The Sima de los Huesos Hominids Date to Beyond U/Th Equilibrium (>350 kyr) and Perhaps to 400–500 kyr: New Radiometric Dates". *Journal*

of Archaeological Science (Amsterdam, the Netherlands: Elsevier) **30** (3): 275–280. doi:10.1006/jasc.2002.0834. ISSN 0305-4403. The first humans with "proto-Neanderthal traits" lived in Eurasia as early as 0.6 to 0.35 million years ago (classified as *H. heidelbergensis*, also called a chronospecies because it represents a chronological grouping rather than being based on clear morphological distinctions from either *H. erectus* or *H. neanderthalensis*), with the first "true Neanderthals" appearing between 0.25 and 0.2 million years ago.

  • Papagianni, Dmitra; Morse, Michael A. (2013). *The Neanderthals Rediscovered: How Modern Science is Rewriting Their Story*. New York: Thames & Hudson. ISBN 978-0-500-05177-1.

[49] Chang, Chun-Hsiang; Kaifu, Yousuke; Takai, Masanaru; Kono, Reiko T.; Grün, Rainer; Matsu'ura, Shuji; Kinsley, Les; Lin, Liang-Kong (2015). "The first archaic *Homo* from Taiwan". *Nature Communications* **6**: 6037. doi:10.1038/ncomms7037.

[50] "Fossil Reanalysis Pushes Back Origin of *Homo sapiens*". *Scientific American* (Stuttgart: Georg von Holtzbrinck Publishing Group). February 17, 2005. ISSN 0036-8733. Retrieved 2015-05-04. The oldest fossil remains of anatomically modern humans are the Omo remains, which date to 195,000 (±5,000) years ago and include two partial skulls as well as arm, leg, foot and pelvis bones.

  • McDougall, Ian; Brown, Francis H.; Fleagle, John G. (February 17, 2005). "Stratigraphic placement and age of modern humans from Kibish, Ethiopia". *Nature* (London: Nature Publishing Group) **433** (7027): 733–736. Bibcode:2005Natur.433..733M. doi:10.1038/nature03258. ISSN 0028-0836. PMID 15716951. *H. sapiens idaltu* is a confirmed subspecies, based on 3 craniums dated 0.16 – 0.15 Mya found in Ethiopia (1997/2003).

6.7 Further reading

• Serre; Langaney, André; Chech, Mario; Teschler-Nicola, Maria; Paunovic, Maja; Mennecier, Philippe; Hofreiter, Michael; Possnert, Göran; Pääbo, Svante; et al. (2004). "No evidence of Neandertal mtDNA contribution to early modern humans". *PLoS Biology* **2** (3): 313–7. doi:10.1371/journal.pbio.0020057. PMC 368159. PMID 15024415.

6.8 External links

• Exploring the Hominid Fossil Record (Center for the Advanced Study of Hominid Paleobiology at George Washington University)

• Hominid species

• Prominent Hominid Fossils

• Mikko's Phylogeny archive

• *Homo* at the Encyclopedia of Life

Chapter 7

Control of fire by early humans

"Control of fire" redirects here. For the process of suppressing or more extinguishing a fire, see Fire control. For components that assist weapon systems, see Fire-control system.
The control of fire by early humans was a turning point in

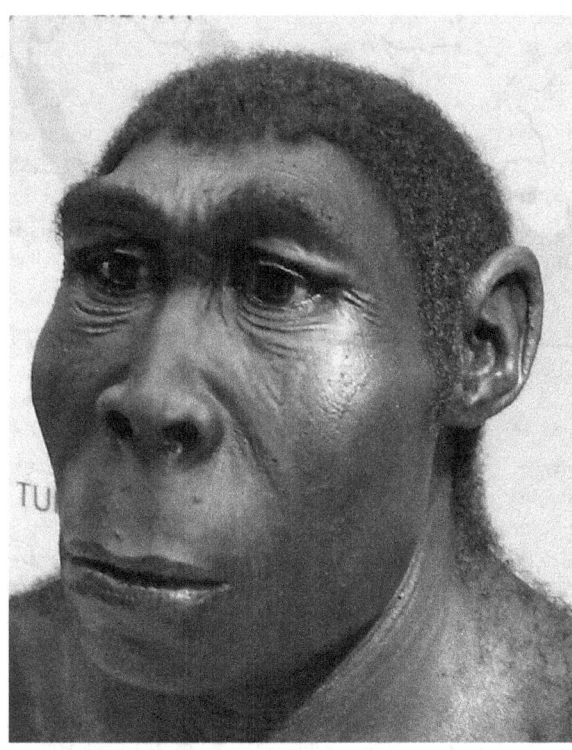

A reconstruction of Homo erectus, *the earliest human species that is known to have controlled fire (reconstruction shown in Westfälisches Landesmuseum, Herne, Germany, in a 2006 exhibition)*

the cultural aspect of human evolution that allowed humans to cook food and obtain warmth and protection. Making fire also allowed the expansion of human activity into the dark and colder hours of the night, and provided protection from predators and insects.

Evidence of widespread control of fire dates to approximately 125,000 years ago and earlier.[1] Evidence for the controlled use of fire by *Homo erectus* beginning

some 400,000 years ago has wide scholarly support, with claims regarding earlier evidence finding increasing scientific support.[2][3]

Claims for the earliest definitive evidence of control of fire by a member of *Homo* range from 0.2 to 1.7 million years ago (Mya).[4]

7.1 Lower Paleolithic evidence

All evidence of control of fire during the Lower Paleolithic is uncertain and has at best limited scholarly support. In fact, definitive evidence of controlled use of fire is one of the factors characteristic of the transition from the Lower to the Middle Paleolithic in the period of 400,000 to 200,000 Before Present (BP).

East African sites, such as Chesowanja near Lake Baringo, Koobi Fora, and Olorgesailie in Kenya, show some possible evidence that fire was utilized by early humans. At Chesowanja, archaeologists found red clay shards dated to be 1.42 Mya.[4] Reheating these shards show that the clay must have been heated to 400 °C (752 °F) to harden. At Koobi Fora, sites FxJjzoE and FxJj50 show evidence of control of fire by *Homo erectus* at 1.5 Mya, with the reddening of sediment that can only come from heating at 200–400 °C (392–752 °F).[4] A "hearth-like depression" exists at a site in Olorgesailie, Kenya. Some microscopic charcoal was found, but it could have resulted from a natural brush fire.[4] In Gadeb, Ethiopia, fragments of welded tuff that appeared to have been burned were found in Locality 8E, but re-firing of the rocks might have occurred due to local volcanic activity.[4] These have been found amongst *H. erectus*–produced Acheulean artifacts. In the Middle Awash River Valley, cone-shaped depressions of reddish clay were found that could have been formed by temperatures of 200 °C (392 °F). These features are thought to be burned tree stumps such that they would have fire away from their habitation site.[4] Burnt stones are also found in the Awash Valley, but volcanic welded tuff is also found in the area.

61

A site at Bnot Ya'akov Bridge, Israel, has been claimed to show that *H. erectus* or *H. ergaster* made fires between 790,000 and 690,000 BP.[5] To date this has been the most widely accepted claim, although recent reanalysis of burnt bone fragments and plant ashes from the Wonderwerk Cave have sparked claims of evidence supporting human control of fire by 1 Ma.[6]

In Xihoudu in Shanxi Province, China, there is evidence of burning by the black, gray, and grayish-green discoloration of mammalian bones found at the site. Another site in China is Yuanmou in Yunnan Province, where blackened mammal bones were found in 1985 and dated to 1.7 Ma BP.[4]

At Trinil, Java, burned wood has been found in layers that carried *H. erectus* fossils dated from 500,000 to 830,000 BP, but the charring may have resulted from natural fires because Central Java is a volcanic region.[4]

Based on the feeding time comparison between human and nonhuman primates (4.7% versus predicted 48% of daily activity), researchers have inferred that this is due to an evolutionary consequence of food processing dating back to 1.9 million years ago. This may imply control of fire as early as 1.9 million years ago by the *Homo* genus.[7]

7.2 Middle Paleolithic evidence

7.2.1 Africa

The earliest definitive evidence of human control of fire was found at Swartkrans, South Africa.[8] Several burnt bones were found among Acheulean tools, bone tools, and bones with hominid-inflicted cut marks.[4] This site also shows some of the earliest evidence of carnivory in *H. erectus*. The Cave of Hearths in South Africa has burned deposits dated from 200,000 to 700,000 BP, as do various other sites such as Montagu Cave (58,000 to 200,000 BP) and at the Klasies River Mouth (120,000 to 130,000 BP).[4]

The strongest evidence comes from Kalambo Falls in Zambia where several artifacts related to the use of fire by humans had been recovered including charred logs, charcoal, reddened areas, carbonized grass stems and plants, and wooden implements which may have been hardened by fire. The site was dated through radiocarbon dating to be at 61,000 BP and 110,000 BP through amino acid racemization.[4]

Fire was used to heat treat silcrete stones to increase their workability before they were knapped into tools by Stillbay culture.[9][10][11] This research identifies this not only with Stillbay sites that date back to 72,000 BP but sites that could be as old as 164,000 BP.[9]

7.2.2 Asia

Zhoukoudian Caves, a World Heritage Site and an early site of human use of fire in China

At Qesem Cave 12 km east of Tel-Aviv evidence exists of the regular use of fire from before 382,000 BP to around 200,000 BP at the end of Lower Pleistocene. The large quantities of burnt bone and moderately heated soil lumps suggest butchering and prey-defleshing took place near fireplaces.[12]

At Zhoukoudian in China, evidence of fire is as early as 230,000 to 460,000 BP.[1] Fire in Zhoukoudian is suggested by the presence of burned bones, burned chipped-stone artifacts, charcoal, ash, and hearths alongside *H. erectus* fossils in Layer 10 at Locality 1.[4][13] This evidence comes from Locality 1 at Zhoukoudian where several bones were found to be uniformly black to grey. The extracts from the bones were determined to be characteristic of burned bone rather than manganese staining. These residues also showed IR spectra for oxides, and a bone that was turquoise was reproduced in the laboratory by heating some of the other bones found in Layer 10. At the site, the same effect might have been due to natural heating, as the effect was produced on white, yellow, and black bones.[13] Layer 10 itself is described as ash with biologically produced silicon, aluminum, iron, and potassium, but wood ash remnants such as siliceous aggregates are missing.

Among these are possible hearths "represented by finely laminated silt and clay interbedded with reddish-brown and yellow brown fragments of organic matter, locally mixed with limestone fragments and dark brown finely laminated silt, clay and organic matter."[13] The site itself does not show that fires were made in Zhoukoudian, but the association of blackened bones with stone artifacts at least shows that humans did control fire at the time of the habitation of the Zhoukoudian cave.

7.2.3 Europe

Multiple sites in Europe have also shown evidence of use of fire by later versions of *H. erectus*. The oldest has been found in England at the site of Beeches Pit, Suffolk; Uranium series dating and thermoluminescence dating place the use of fire at 415,000 BP.[14] At Vértesszőlős, Hungary, where evidence of burned bones, but no charcoal, had been found, dating from c. 350,000 years ago. At Torralba and Ambrona, Spain, show charcoal and wood, Acheulean stone tools dated 300,000 to 500,000 BP.[4]

At Saint-Estève-Janson in France, there is evidence of five hearths and reddened earth in the Escale Cave. These hearths have been dated to 200,000 BP.[4]

7.3 Changes to behavior

An important change in the behavior of humans was brought about by the control of fire and its accompanying light.[15] Activity was no longer restricted to the daylight hours. In addition, some mammals and biting insects avoid fire and smoke.[4] Fire also led to improved nutrition from cooked proteins.[13][16] Fire could also be used to fire-harden spear points and collect melted birch pitch to use as a "glue" to affix flint points to spears.

Cooking plant foods might trigger brain expansion by allowing complex carbohydrates in starchy foods to become more digestible and allow humans to absorb more food energy per unit of food consumed.[17][18][19] The human digestive system has evolved to deal with cooked foods, and so it is thought that cooking may explain the increase in hominid brain sizes, shorter digestive tracts, smaller teeth and jaws and the decrease in sexual dimorphism that occurred roughly 1.8 million years ago.[20][21]

Other anthropologists argue that the evidence suggests that cooking fires began in earnest only 250,000 BP, when ancient hearths, earth ovens, burnt animal bones, and flint appear across Europe and the Middle East.[22] Two million years ago, the only sign of fire is burnt earth with human remains, which most other anthropologists consider to be mere coincidence rather than evidence of intentional fire.[23] The mainstream view among anthropologists is that the increases in human brain-size occurred well before the advent of cooking, due to a shift away from the consumption of nuts and berries to the consumption of meat.[24][25]

7.3.1 Changes to diet

Because of the indigestible components of plants such as raw cellulose and starch, some parts of the plant such as stems, mature leaves, enlarged roots, and tubers would not have been part of the hominid diet prior to the advent of fire.[26] Instead, the consumption of plants would be limited to parts that were made of simpler sugars and carbohydrates such as seeds, flowers, and fleshy fruits. The incorporation of toxins into the seeds and similar carbohydrate sources also affected the diet, as cyanogenic glycosides such as those found in linseed and cassava are made non-toxic through cooking.[26]

The teeth of *H. erectus* over time showed gradual shrinking, suggesting that later members of the species converted from eating crunchier foods such as crisp root vegetables to softer foods such as meat and various cooked foods.[27][28] The evidence of cooking of meat comes from burned and blackened animal bones found at various archaeological sites.[29]

Cooking meat may have acted as a form of "pre-digestion", allowing less food energy intake to be spent on digesting the tougher proteins such as collagen. The digestive tract shrank, allowing more energy to be given to the growing brain of *H. erectus*.[30] If they ate only raw, unprocessed food, humans would need to eat for 9.3 hours per day in order to fuel their brains; which use about twice as much resting energy by percentage as other primates.[31]

Dental evidence suggests that the advent of cooking and preparing food (making it easier to chew) led to the issues of malocclusion in modern humans and a steady decrease in the size of teeth. Because the food was made smaller, but not small tough, the human jaw shrank too much to adequately fit all of the teeth.[32]

7.4 See also

- Hunting hypothesis

- Savanna theory

7.5 References

[1] "First Control of Fire by Human Beings—How Early?". Retrieved 2007-11-12.

[2] Luke, Kim. "Evidence That Human Ancestors Used Fire One Million Years Ago". Retrieved 2013-10-27. An international team led by the University of Toronto and Hebrew University has identified the earliest known evidence of the use of fire by human ancestors. Microscopic traces of wood ash, alongside animal bones and stone tools, were found in a layer dated to one million years ago

[3] http://discovermagazine.com/2013/may/09-archaeologists-find-earliest-evidence-of-humans-cooking-with-fire

[4] James, Steven R. (February 1989). "Hominid Use of Fire in the Lower and Middle Pleistocene: A Review of the Evidence" (PDF). *Current Anthropology* (University of Chicago Press) **30** (1): 1–26. doi:10.1086/203705. Retrieved 2012-04-04.

[5] Rincon, Paul (29 April 2004). "Early human fire skills revealed". BBC News. Retrieved 2007-11-12.

[6] Pringle, Heather (2 April 2012), "Quest for Fire Began Earlier Than Thought", *ScienceNOW* (American Association for the Advancement of Science), retrieved 2012-04-04

[7] "Phylogenetic rate shifts in feeding time during the evolution of Homo". Retrieved 2012-11-03.

[8] Renfrew and Bahn (2004). *Archaeology: Theories, Methods and Practice* (Fourth Edition). Thames and Hudson, p341.

[9] Brown, KS; Marean, CW; Herries, AI; Jacobs, Z; Tribolo, C; Braun, D; Roberts, DL; Meyer, MC; Bernatchez, J (2009). "Fire As an Engineering Tool of Early Modern Humans". *Science* **325**: 859–862. doi:10.1126/science.1175028.

[10] Webb, J. Domanski M. (2009). "Fire and Stone". *Science* **325**: 820–821. doi:10.1126/science.1178014.

[11] Callaway. E. (13 August 2009) Earliest fired knives improved stone age tool kit New Scientist, online

[12] Karkanas P, Shahack-Gross R, Ayalon A; et al. (August 2007). "Evidence for habitual use of fire at the end of the Lower Paleolithic: site-formation processes at Qesem Cave, Israel" (PDF). *J. Hum. Evol.* **53** (2): 197–212. doi:10.1016/j.jhevol.2007.04.002. PMID 17572475.

[13] Weiner, S.; Q. Xu; P. Goldberg; J. Liu; O. Bar-Yosef (1998). "Evidence for the Use of Fire at Zhoukoudian, China". *Science* **281** (5374): 251–253. doi:10.1126/science.281.5374.251. PMID 9657718.

[14] Preece, R. C. (2006). "Humans in the Hoxnian: habitat, context and fire use at Beeches Pit, West Stow, Suffolk, UK". *Journal of Quaternary Science* (Wiley): 485–496.

[15] Stone, Linda; Paul F. Lurquin; Luigi Luca Cavalli-Sforza (2007). *Genes, Culture, And Human Evolution: A Synthesis.* Blackwell Publishing. p. 33.

[16] Eisley, Loren C. (1955). "Fossil Man and Human Evolution". *Yearbook of Anthropology* (University of Chicago Press): 61–78.

[17] William R. Leonard. "Food for Thought: Into the Fire". *Scientific american.* Retrieved 2008-02-22.

[18] Wrangham R, Conklin-Brittain N. (Sep 2003). "Cooking as a biological trait" (PDF). *Comp Biochem Physiol a Mol Integr Physiol* **136** (1): 35–46. doi:10.1016/S1095-6433(03)00020-5. PMID 14527628.

[19] Lambert, Craig (May–June 2004). "The Way We Eat Now". Harvard Magazine.

[20] Clement, Brian (2006). "The Cooking Enigma". In Ungar, Peter S. *Evolution of the Human Diet: The Known, the Unknown, and the Unknowable.* Oxford, USA: Oxford University Press. pp. 308–23. ISBN 0-19-518346-0.

[21] Wrangham, R; Conklin-Brittain, N (2003). "'Cooking as a biological trait'." (PDF). *Comparative biochemistry and physiology. Part A, Molecular & integrative physiology* **136** (1): 35–46. doi:10.1016/S1095-6433(03)00020-5. PMID 14527628.

[22] Pennisi, Elizabeth (26 March 1999). "Human evolution: Did Cooked Tubers Spur the Evolution of Big Brains?". *Science* **283** (5410): 2004–2005. doi:10.1126/science.283.5410.2004. PMID 10206901.

[23] Gorman, RM (2008). "Cooking up bigger brains.". *Scientific American* **298** (1): 102, 104–5. doi:10.1038/scientificamerican0108-102.

[24] "Meat-eating was essential for human evolution, says UC Berkeley anthropologist specializing in diet". 14 June 1999. Retrieved 2010-12-06.

[25] Mann, Neil (15 August 2007). "Meat in the human diet: An anthropological perspective". *Nutrition & Dietetics* **64** (Supplement s4): 102–107. doi:10.1111/j.1747-0080.2007.00194.x.

[26] Stahl, Ann Brower (April 1984). "Hominid Dietary Selection Before Fire". *Current Anthropology* (University of Chicago Press) **25** (2): 151–168. doi:10.1086/203106.

[27] Viegas, Jennifer (22 November 2005). "Homo erectus ate crunchy food". *News in Science* (abc.net.au). Retrieved 2007-11-12.

[28] "Early Human Evolution: Homo ergaster and erectus". Retrieved 2007-11-12.

[29] "What evidence is there that Homo erectus used fire? Why did they use it?". Retrieved 2007-11-12.

[30] Gibbons, Ann (15 June 2007). "Food for Thought" (pdf). *Science* **316** (5831): 1558–60. doi:10.1126/science.316.5831.1558. PMID 17569838. Retrieved 2007-11-12.

[31] Ann Gibbons (2012-10-22). "Raw Food Not Enough to Feed Big Brains". *ScienceNow.* American Association for the Advancement of Science. Retrieved 23 October 2012.

[32] Pickrell, John (2005-02-19). "Human 'dental chaos' linked to evolution of cooking". Retrieved 2011-10-07.

7.6 Further reading

- Goudsblom, J (1992): *Fire and Civilization,* Allen Lane.

- Bejan, Adrian (8 June 2015). "Why humans build fires shaped the same way". (Nature) Scientific Reports 5: 11270. doi:10.1038/srep11270

7.7 External links

- How our pact with fire made us what we are Article by Stephen J Pyne

Chapter 8

Stone tool

Not to be confused with Tool stone.

A **stone tool** is, in the most general sense, any tool made either partially or entirely out of stone. Although stone tool-dependent societies and cultures still exist today, most stone tools are associated with prehistoric, particularly Stone Age cultures that have become extinct. Archaeologists often study such prehistoric societies, and refer to the study of stone tools as lithic analysis. Ethnoarchaeology has been a valuable research field in order to further the understanding and cultural implications of stone tool use and manufacture.[1] Stone has been used to make a wide variety of different tools throughout history, including arrow heads, spearpoints and querns. Stone tools may be made of either ground stone or chipped stone, and a person who creates tools out of the latter is known as a flintknapper.

Chipped stone tools are made from cryptocrystalline materials such as chert or flint, radiolarite, chalcedony, obsidian, basalt, and quartzite via a process known as lithic reduction. One simple form of reduction is to strike stone flakes from a nucleus (core) of material using a hammerstone or similar hard hammer fabricator. If the goal of the reduction strategy is to produce flakes, the remnant lithic core may be discarded once it has become too small to use. In some strategies, however, a flintknapper reduces the core to a rough unifacial or bifacial preform, which is further reduced using soft hammer flaking techniques or by pressure flaking the edges. More complex forms of reduction include the production of highly standardized blades, which can then be fashioned into a variety of tools such as scrapers, knives, sickles and microliths. In general terms, chipped stone tools are nearly ubiquitous in all pre-metal-using societies because they are easily manufactured, the tool stone is usually plentiful, and they are easy to transport and sharpen.

A selection of prehistoric stone tools.

8.1 Evolutionary development of technocomplexes

From the 19th century archaeologists had been turning up prehistoric worked stone tools that appeared to be typologically classifiable into taxa. They referred to these homotaxial groups of stone tools as industries and named them after the type site; for example, Acheulean after Saint Acheul, and later Oldowan from Olduvai Gorge. In the earlier 20th century they became complexes and technologies; in the later, technocomplexes. So, in archaeology a technocomplex is a distinct culture that employed a specific technology.

In 1969 in the 2nd edition of *World Prehistory*, Grahame Clark envisioned an evolutionary progression of flint-knapping in which the "dominant lithic technologies" occurred in a fixed sequence from Mode 1 through Mode 5.[2] He assigned to them relative dates: Modes 1 and 2 to the Lower Palaeolithic, 3 to the Middle Palaeolithic, 4 to the Advanced and 5 to the Mesolithic. They were not to be conceived, however, as either universal—that is, they did not account for all lithic technology; or as synchronous—they were not in effect in different regions simultaneously. Mode 1, for example, was in use in Europe long after it had been replaced by Mode 2 in Africa.

Clarke's scheme was adopted enthusiastically by the archaeological community. One of its advantages was the simplicity of terminology; for example, the Mode 1 / Mode 2 Transition. The transitions are currently of greatest interest. Consequently in the literature the stone tools used in the period of the Palaeolithic are divided into four "modes", each of which designate a different form of complexity, and which in most cases followed a rough chronological order.

8.1.1 Pre-Mode I

The earliest evidence of the use of stone tools dates to 3.4 million years ago. also Archaeological discoveries in Kenya in 2015, identifying possibly the oldest known evidence of hominin use of tools to date, have indicated that Kenyanthropus platyops (a 3.2 to 3.5-million-year-old Pliocene hominin fossil discovered in Lake Turkana, Kenya in 1999) may have been the earliest tool-users known. Grooved, cut and fractured animal bone fossils, made by using stone tools, were found in Dikika in Ethiopia near (200 yards) the remains of *Selam*, a young *Australopithecus afarensis* girl who lived about 3.3 million years ago, the time of the *Australopithecus afarensis* Lucy.[3][4] The oldest known stone tools, found in 2011 at Lake Turkana in Kenya, are dated to be 3.3 million years old, and predate the genus Homo by half a million years. The dating of the stone tools was completed in 2013. The oldest known *Homo* fossil is 2.8 million years old compared to the 3.3 million year old stone tools.[5] The stone tools may have been made by *Australopithecus afarensis*, the species whose best fossil example is Lucy, which inhabited East Africa at the same time as the date of the oldest stone tools.[6][7][8][9][10] NatGeo Video

8.1.2 Mode I: The Oldowan Industry

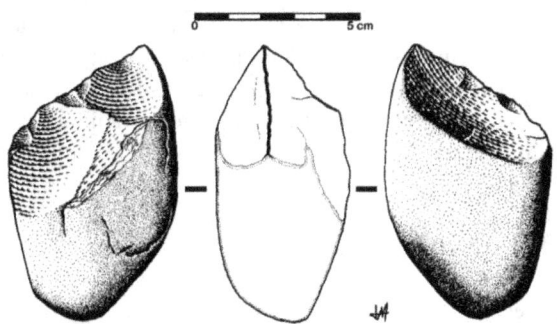

A typical Oldowan simple chopping-tool. This example is from the Duero Valley, Valladolid.

The earliest stone tools in the life span of the genus *Homo* are Mode 1 tools,[11] and come from what has been termed

the Oldowan Industry, named after the type of site (many sites, actually) found in Olduvai Gorge, Tanzania, where they were discovered in large quantities. Oldowan tools were characterised by their simple construction, predominantly using core forms. These cores were river pebbles, or rocks similar to them, that had been struck by a spherical hammerstone to cause conchoidal fractures removing flakes from one surface, creating an edge and often a sharp tip. The blunt end is the proximal surface; the sharp, the distal. Oldowan is a percussion technology. Grasping the proximal surface, the hominid brought the distal surface down hard on an object he wished to detach or shatter, such as a bone or tuber.

The earliest known Oldowan tools yet found date from 2.6 million years ago, during the Lower Palaeolithic period, and have been uncovered at Gona in Ethiopia.[12] After this date, the Oldowan Industry subsequently spread throughout much of Africa, although archaeologists are currently unsure which Hominan species first developed them, with some speculating that it was *Australopithecus garhi*, and others believing that it was in fact *Homo habilis*.[13] *Homo habilis* was the hominin who used the tools for most of the Oldowan in Africa, but at about 1.9-1.8 million years ago Homo erectus inherited them. The Industry flourished in southern and eastern Africa between 2.6 and 1.7 million years ago, but was also spread out of Africa and into Eurasia by travelling bands of *H. erectus*, who took it as far east as Java by 1.8 million years ago and Northern China by 1.6 million years ago.

8.1.3 Mode II: The Acheulean Industry

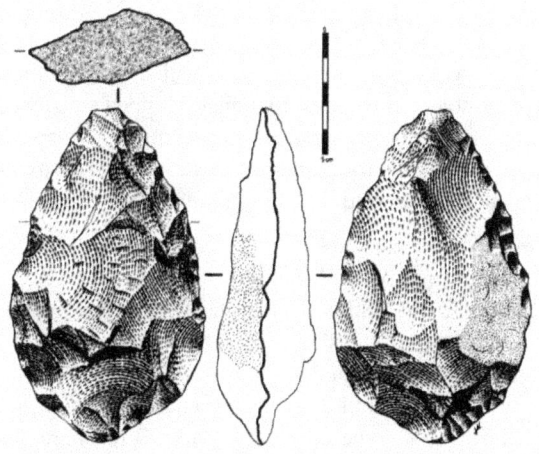

A typical Acheulean handaxe; this example is from the Douro valley, Zamora, Spain. The small chips on the edge are from reworking.

Eventually, more complex, Mode 2 tools began to be developed through the Acheulean Industry, named after the site of Saint-Acheul in France. The Acheulean was characterised not by the core, but by the biface, the most notable form of which was the hand axe.[14] The Acheulean first appears in the archaeological record as early as 1.7 million years ago in the West Turkana area of Kenya and contemporaneously in southern Africa.

The Leakeys, excavators at Olduvai, defined a "Developed Oldowan" Period in which they believed they saw evidence of an overlap in Oldowan and Acheulean. In their species-specific view of the two industries, Oldowan equated to *H. habilis* and Acheulean to *H. erectus*. Developed Oldowan was assigned to *habilis* and Acheulean to *erectus*. Subsequent dates on *H. erectus* pushed the fossils back to well before Acheulean tools; that is, *H. erectus* must have initially used Mode 1. There was no reason to think, therefore, that Developed Oldowan had to be *habilis*; it could have been *erectus*. Opponents of the view divide Developed Oldowan between Oldowan and Acheulean. There is no question, however, that *habilis* and *erectus* coexisted, as *habilis* fossils are found as late as 1.4 million years ago. Meanwhile, African *H. erectus* developed Mode 2. In any case a wave of Mode 2 then spread across Eurasia, resulting in use of both there. *H. erectus* may not have been the only hominin to leave Africa; European fossils are sometimes associated with *Homo ergaster*, a contemporary of *H. erectus* in Africa.

In contrast to an Oldowan tool, which is the result of a fortuitous and probably ex tempore operation to obtain one sharp edge on a stone, an Acheulean tool is a planned result of a manufacturing process. The manufacturer begins with a blank, either a larger stone or a slab knocked off a larger rock. From this blank he or she removes large flakes, to be used as cores. Standing a core on edge on an anvil stone, he or she hits the exposed edge with centripetal blows of a hard hammer to roughly shape the implement. Then he or she works it over again, or retouches it, with a soft hammer of wood or bone to produce a tool finely chipped all over consisting of two convex surfaces intersecting in a sharp edge. Such a tool is used for slicing; concussion would destroy the edge and cut the hand.

Some Mode 2 tools are disk-shaped, others ovoid, others leaf-shaped and pointed, and others elongated and pointed at the distal end, with a blunt surface at the proximal end, obviously used for drilling. Mode 2 tools are used for butchering; not being composite (having no haft) they are not very appropriate killing instruments. The killing must have been done some other way. Mode 2 tools are larger than Oldowan. The blank was ported to serve as an ongoing source of flakes until it was finally retouched as a finished tool itself. Edges were often sharpened by further retouching.

8.1.4 Mode III: The Mousterian Industry

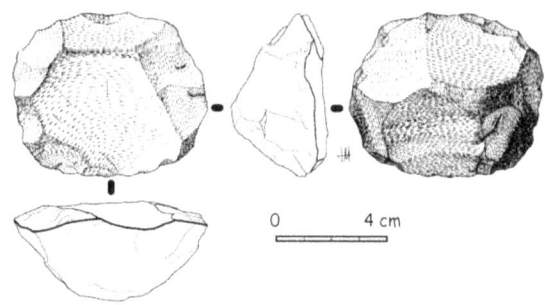

A tool made by the Levallois technique. This example is from La Parrilla (Valladolid, Spain).

Eventually, the Acheulean in Europe was replaced by a lithic technology known as the Mousterian Industry, which was named after the site of Le Moustier in France, where examples were first uncovered in the 1860s. Evolving from the Acheulean, it adopted the Levallois technique to produce smaller and sharper knife-like tools as well as scrapers.[15] The Mousterian Industry was developed and used primarily by the Neanderthals, a native European and Middle Eastern hominin species.[16]

8.1.5 Mode IV: The Aurignacian Industry

The long blades (rather than flakes) of the Upper Palaeolithic Mode 4 industries appeared during the Upper Palaeolithic.[17] The Aurignacian culture is a good example of mode 4 tool production.[18]

8.1.6 Mode V: The Microlithic Industries

Mode 5 stone tools involve the production of microliths, which were used in composite tools, mainly fastened to a haft.[19] Examples include the Magdalenian culture. Such a technology makes much more efficient use of available materials like flint, although required greater skill in manufacturing the small flakes.

8.1.7 Neolithic industries

In prehistoric Japan, ground stone tools appear during the Japanese Paleolithic period.[20] Elsewhere, ground stone tools became important during the Neolithic period. These ground or polished implements are manufactured from larger-grained materials such as basalt, jade and jadeite, greenstone and some forms of rhyolite which are not suitable for flaking. The greenstone industry was important in

An array of Neolithic artifacts, including bracelets, axe heads, chisels, and polishing tools.

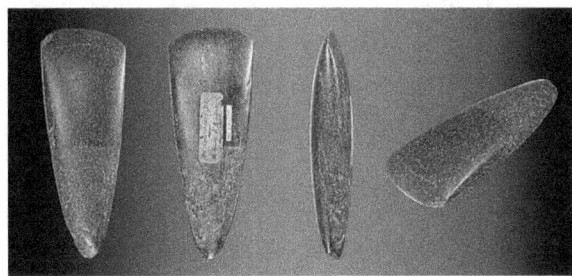

Polished Neolithic jadeitite axe from the Museum of Toulouse

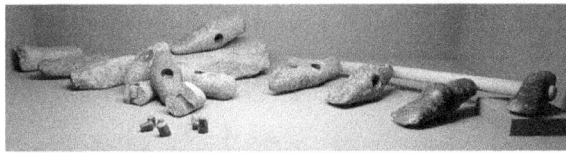

Axe heads found at a 2700 BC Neolithic manufacture site in Switzerland, arranged in the various stages of production from left to right.
Click to see individual images.

the English Lake District, and is known as the Langdale axe industry. Ground stone implements included adzes, celts, and axes, which were manufactured using a labour-intensive, time-consuming method of repeated grinding against an abrasive stone, often using water as a lubricant. Because of their coarse surfaces, some ground stone tools were used for grinding plant foods and were polished not just by intentional shaping, but also by use. Manos are hand stones used in conjunction with metates for grinding corn or grain. Polishing increased the intrinsic mechanical strength of the axe. Polished stone axes were important for the widespread clearance of woods and forest during the Neolithic period, when crop and livestock farming developed on a large scale. They are distributed very widely and were traded over great distances since the best rock types were often very local. They also became venerated objects,

and were frequently buried in long barrows or round barrows with their former owners.

During the Neolithic period, large axes were made from flint nodules by chipping a rough shape, a so-called "rough-out". Such products were traded across a wide area. The rough-outs were then polished to give the surface a fine finish to create the axe head. Polishing not only increased the final strength of the product but also meant that the head could penetrate wood more easily.

Such axe heads were needed in large numbers for forest clearance and the establishment of settlements and farmsteads, a characteristic of the Neolithic period. There were many sources of supply, including Grimes Graves in Suffolk, Cissbury in Sussex and Spiennes near Mons in Belgium to mention but a few. In Britain, there were numerous small quarries in downland areas where flint was removed for local use, for example.

Many other rocks were used to make axes from stones, including the Langdale axe industry as well as numerous other sites such as Penmaenmawr and Tievebulliagh in Co Antrim, Ulster. In Langdale, there many outcrops of the greenstone were exploited, and knapped where the stone was extracted. The sites exhibit piles of waste flakes, as well as rejected rough-outs. Polishing improved the mechanical strength of the tools, so increasing their life and effectiveness. Many other tools were developed using the same techniques. Such products were traded across the country and abroad.

8.2 Modern uses

The invention of the flintlock gun mechanism in the sixteenth century produced a demand for specially shaped gunflints. The gunflint industry survived until the middle of the twentieth century in some places, including in the English town of Brandon.[21]

For specialist purposes glass knives are still made and used today, particularly for cutting thin sections for electron microscopy in a technique known as microtomy. Freshly cut blades are always used since the sharpness of the edge is very great. These knives are made from high-quality manufactured glass, however, not from natural raw materials such as chert or obsidian. Surgical knives made from obsidian are still used in some delicate surgeries.

8.3 Tool Stone

Main article: Tool stone

In archaeology, a tool stone is a type of stone that is used to manufacture stone tools. Alternatively, the term can be used to refer to stones used as the raw material for tools.

8.4 See also

- Chaîne opératoire

- Eccentric flint (archaeology)

- Flint

- Knapping

- Langdale axe industry

- Lithic technology

- Manuport

- Mount William stone axe quarry

- Prismatic blade

8.5 References

[1] Sillitoe, P. and K. Hardy 2003 Living lithics: ethnoarchaeology in highland Papua New Guinea. Antiquity 77:555-566

[2] Clarke, Grahame (1969). *World Prehistory: a New Outline* (2 ed.). Cambridge: Cambridge University Press. p. 31.

[3] Shannon P. McPherron; Zeresenay Alemseged; Curtis W. Marean; Jonathan G. Wynn; Denné Reed; Denis Geraads; René Bobe; Hamdallah A. Béarat (2010). "Evidence for Stone-tool-assisted Consumption of Animal Tissues before 3.39 Million Years Ago at Dikika, Ethiopia". *Nature* **466** (7308): 857–860. doi:10.1038/nature09248. PMID 20703305.

[4] "Scientists Discover Oldest Evidence of Stone Tool Use and Meat-Eating Among Human Ancestors". Retrieved 27 November 2013.

[5] http://www.bbc.co.uk/news/science-environment-32804177

[6] Drake, Nadia; 20, for National Geographic PUBLISHED May. "Wrong Turn Leads to Discovery of Oldest Stone Tools". *National Geographic News*. Retrieved 2015-05-21.

[7] Harmand, Sonia; Lewis, Jason E.; Feibel, Craig S.; Lepre, Christopher J.; Prat, Sandrine; Lenoble, Arnaud; Boës, Xavier; Quinn, Rhonda L.; Brenet, Michel (May 21, 2015). "3.3-million-year-old stone tools from Lomekwi 3, West Turkana, Kenya". *Nature* **521** (7552): 310–315. doi:10.1038/nature14464. ISSN 0028-0836.

[8] Thompson, Helen. "The Oldest Stone Tools Yet Discovered Are Unearthed in Kenya". Retrieved 2015-05-21.

[9] Wilford, John Noble (2015-05-20). "Stone Tools From Kenya Are Oldest Yet Discovered". *The New York Times*. ISSN 0362-4331. Retrieved 2015-05-30.

[10] "Oldest Known Stone Tools Discovered: 3.3 Million Years Old". *video.nationalgeographic.com*. Retrieved 2015-06-15.

[11] Clarke's "chopper tools and flakes."

[12] Semaw, S.; M. J. Rogers; J. Quade; P. R. Renne; R. F. Butler; M. Domínguez-Rodrigo; D. Stout; W. S. Hart; T. Pickering; S. W. Simpson (2003). "2.6-Million-year-old stone tools and associated bones from OGS-6 and OGS-7, Gona, Afar, Ethiopia". *Journal of Human Evolution* **45**: 169–177. doi:10.1016/S0047-2484(03)00093-9. PMID 14529651.

[13] Toth, Nicholas; Schick, Kathy (2009), "African Origins", in Scarre, Chris, *The Human Past: World Prehistory and the Development of Human Societies* (2nd ed.), London: Thames and Hudson, pp. 67–68

[14] Clarke's "bifacially flaked hand axes."

[15] Clarke's "flake tools from prepared cores."

[16] Pettitt, Paul (2009), "The Rise of Modern Humans", in Scarre, Chris, *The Human Past: World Prehistory and the Development of Human Societies* (2nd ed.), London: Thames and Hudson, pp. 149–151

[17] Lewin, R.; Foley, R. A. (2004). *Principles of Human Evolution* (2 ed.). UK: Blackwell Science. p. 311. ISBN 0-632-04704-6.

[18] Clarke's "punch-struck blades with steep retouch."

[19] Clarke's "microlithic components of composite artifacts."

[20] "Prehistoric Japan, New perspectives on insular East Asia", Keiji Imamura, University of Hawaii Press, Honolulu, ISBN 0-8248-1853-9

[21] Clarke, R (1935), The Flint-knapping Industry at Brandon, *Antiquity*, vol. IX

8.6 External links

- Michaels, George H.; Fagan, Brian M. (1990–1998). "Principles of Lithic Technology". University of California. Retrieved 22 January 2011.

- Gunness, Jo Lynn (1998). "Lithic Technologies Notes". University of Hawaii Anthropology Department. Retrieved 22 January 2011.

- Prindle, Tara (1994–2011). "Flaked Stone Tool Technology". Nativetech.org. Retrieved 22 January 2011.

- "Typology". Stone Age Reference Collection (SARC), University of Oslo. Retrieved 22 January 2011.

- "Stone Tools of Texas Indians". Texas Beyond History, University of Texas at Austin. 2001. Retrieved 18 January 2011.

- Prindle, Tara (1994–2011). "Common Stone Types and Northeastern Lithic Technologies". Nativetech.org. Retrieved 18 January 2011.

- Grace, Roger. "Interpreting the Function of Stone Tools". Stone Age Reference Collection (SARC), University of Oslo. Retrieved 18 January 2011.

- "How to recognize prehistoric stone tools". newarchaeology.com. Retrieved 18 January 2011.

- "The World Museum of Man and Prehistory". World Museum of Man. 2004–2011. Retrieved 18 January 2011.

- Beach, Chandler B., ed. (1914). "Flint-Implements". *The New Student's Reference Work*. Chicago: F. E. Compton and Co.

Chapter 9

Middle Paleolithic

Bifacial points, engraved ochre and bone tools from the c. 75,000- to 80,000-year-old M1 & M2 phases at Blombos cave.

The **Middle Paleolithic** (or **Middle Palaeolithic**) is the second subdivision of the Paleolithic or Old Stone Age as it is understood in Europe, Africa and Asia. The term Middle Stone Age is used as an equivalent or a synonym for the Middle Paleolithic in African archeology.[1] The Middle Paleolithic and the Middle Stone Age broadly spanned from 300,000 to 30,000 years ago. There are considerable dating differences between regions. The Middle Paleolithic/Middle Stone Age was succeeded by the Upper Paleolithic subdivision which first began between 50,000 and 40,000 years ago.[1]

According to the Out of Africa Hypothesis, modern humans began migrating out of Africa during the Middle Stone Age/Middle Paleolithic around 100,000 or 70,000 years ago and began to replace earlier pre-existent *Homo* species such as the Neanderthals and *Homo erectus*.[2]

9.1   Origin of behavioral modernity

The earliest evidence of behavioral modernity first appears during the Middle Paleolithic/Middle Stone Age; undisputed evidence of behavioral modernity, however, only becomes common during the following Upper Paleolithic period.[1]

Middle Paleolithic burials at sites such as Krapina, Croatia (c. 130,000 BP) and Qafzeh, Israel (c. 100,000 BP) have led some anthropologists and archeologists, such as Philip Lieberman, to believe that Middle Paleolithic cultures may have possessed a developing religious ideology which included belief in concepts such as an afterlife; other scholars suggest the bodies were buried for secular reasons.[3][4]

According to recent archeological findings from *H. heidelbergensis* sites in Atapuerca the practice of intentional burial may have begun much earlier during the late Lower Paleolithic but this theory is widely questioned in the scientific community. Cut marks on Neanderthal bones from various sites such as Combe-Grenal and Abri Moula in France may imply that the Neanderthals, like some contemporary human cultures, may have practiced ritual defleshing for (presumably) religious reasons.

Also the earliest undisputed evidence of artistic expression during the Paleolithic period comes from Middle Paleolithic/Middle Stone Age sites such as Blombos Cave in the form of bracelets,[5] beads,[6] art rock,[7] ochre used as body paint and perhaps in ritual,[1][7] though earlier examples of artistic expression such as the Venus of Tan-Tan and the patterns found on elephant bones from Bilzingsleben in Thuringia may have been produced by Acheulean tool users such as *Homo erectus* prior to the start of the Middle Paleolithic period.[8] Activities such as catching large fish and hunting large game animals with specialized tools connote increased group-wide cooperation and more elaborate social organization.[1]

In addition to developing other advanced cultural traits such as religion and art, humans also first began to take part in long distance trade between groups for rare commodities (such as ochre, which was often used for religious purposes such as ritual[7][9]) and raw materials during the Middle Paleolithic as early as 120,000 years ago.[1][10] Intergroup trade may have appeared during the Middle Paleolithic because trade between bands would have helped ensure their survival by allowing them to exchange resources and commodities such as raw materials during times of relative scarcity (i.e., famine or drought).[10]

9.2 Social stratification

A model of a Neanderthal male by modern scientists at the Zagros Paleolithic Museum

Evidence from archeology and comparative ethnography indicates that Middle Paleolithic/Middle Stone Age people lived in small egalitarian band societies similar to those of Upper Paleolithic societies and (some) existent Hunter gatherers such as the !Kung San and the Mbuti.[1][11] Both Neanderthal and modern human societies took care of the elderly members of their societies during the Middle Paleolithic.[10] Christopher Boehm (1999) has hypothesized that egalitarianism may have arisen in Middle Paleolithic societies because of a need to distribute resources such as food and meat equally to avoid famine and ensure a stable food supply.[12]

Typically, it has been assumed that women gathered plants and firewood and men hunted and scavenged dead animals through the Paleolithic.[13] However, recent archaeological research done by the anthropologist and archaeologist Steven Kuhn from the University of Arizona suggests that this gender-based division of labor (presumably) did not exist prior to the Upper Paleolithic in Middle Paleolithic societies (Modern humans before 40,000 or 50,000 BCE and Neanderthals) and evolved relatively recently in human prehistory. The gender-based division of labor may have evolved to allow humans to acquire food and other resources more efficiently and thus may have allowed Upper Paleolithic *Homo sapiens* to out-compete the Neanderthals in Europe.[13]

9.3 Nutrition

Although gathering and hunting comprised most of the food supply during the Middle Paleolithic, people began to supplement their diet with seafood and began smoking and drying meat to preserve and store it. For instance the Middle Stone Age inhabitants of the region now occupied by the Democratic Republic of the Congo hunted large 6-foot (1.8 m) long catfish with specialized barbed fishing points as early as 90,000 years ago,[1][14] and Neanderthals and Middle Paleolithic *Homo sapiens* in Africa began to catch shellfish for food as revealed by shellfish cooking in Neanderthal sites in Italy about 110,000 years ago and Middle Paleolithic *Homo sapiens* sites at Pinnacle Point, in Africa.[1][15]

Anthropologists such as Tim D. White suggest that cannibalism was common in human societies prior to the beginning of the Upper Paleolithic, based on the large amount of "butchered human" bones found in Neanderthal and other Middle Paleolithic sites.[16] Cannibalism in the Middle Paleolithic may have occurred because of food shortages.[17]

However it is also possible that Middle Paleolithic cannibalism occurred for religious reasons which would coincide with the development of religious practices thought to have occurred during the Upper Paleolithic.[18][19] Nonetheless it remains possible that Middle Paleolithic societies never practiced cannibalism and that the damage to recovered human bones was either the result of ritual post-mortem bone cleaning or predation by carnivores such as Saber tooth cats, lions and hyenas.[19]

9.4 Technology

Around 200,000 BP Middle Paleolithic Stone tool manufacturing spawned a tool-making technique known as the prepared-core technique, that was more elaborate than previous Acheulean techniques.[20] Wallace and Shea split the core artifacts into two different types: formal cores and expedient cores. Formal cores are designed to extract the maximum amount from the raw material while expedient cores are more based on function need.[21] This method increased efficiency by permitting the creation of more controlled and consistent flakes.[20] This method allowed Middle Paleolithic humans correspondingly to create stone-tipped spears, which were the earliest composite tools, by hafting sharp, pointy stone flakes onto wooden shafts. Paleolithic groups such as the Neanderthals who possessed a Middle Paleolithic level of technology appear to have hunted large game just as well as Upper Paleolithic modern humans[22] and the Neanderthals in particular may have likewise hunted with projectile weapons.[23]

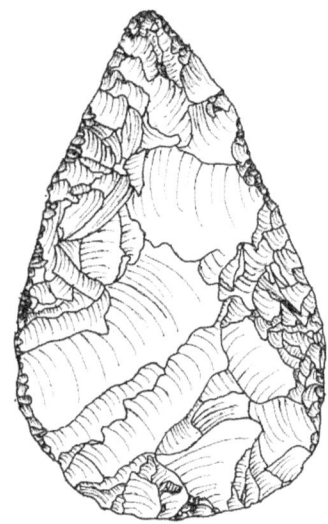

This is a drawing of a replica of an Acheulean hand-axe found during the Lower Paleolithic period. The raw material this tool is made of in this drawing is black Obsidian and is even worked on both sides.

Nonetheless Neanderthal usage of projectile weapons in hunting occurred very rarely (or perhaps never) and the Neanderthals hunted large game animals mostly by ambushing them and attacking them with mêlée weapons such as thrusting spears rather than attacking them from a distance with projectile weapons.[10][24] An ongoing controversy about the nature of Middle Paleolithic tools is whether there were a series of functionally specific and preconceived tool forms or whether there was a simple continuum of tool morphology that reflect the extent of edge maintenance, as Harold Dibble has suggested.[25]

The use of fire became widespread for the first time in human prehistory during the Middle Paleolithic and humans began to cook their food c. 250,000 years ago.[26][27] Some scientists have hypothesized that hominids began cooking food to defrost frozen meat which would help ensure their survival in cold regions.[27] Robert K. Wayne, a molecular biologist, has controversially claimed, based on a comparison of canine DNA, that dogs may have been first domesticated during the Middle Paleolithic around or even before 100,000 BCE.[28] Christopher Boehm (1999) has hypothesized that egalitarianism may have arisen in Middle Paleolithic societies because of a need to distribute resources such as food and meat equally to avoid famine and ensure a stable food supply.[12]

9.5 Sites

9.5.1 Cave sites

- Axlor, Spain

- Petralona, Greece

- Le Moustier, France—*see also* Mousterian

- Neanderthal, Germany

- Grotte de Spy, Spy, Belgium

9.5.2 Open-air sites

- Biache-Saint-Vaast, France

- Maastricht-Belvédère, The Netherlands

- Veldwezelt-Hezerwater, Belgium

9.6 See also

- Early human migrations

- Recent African origin of modern humans

- Timeline of human prehistory

9.7 References

[1] Miller, Barbra; Bernard Wood; Andrew Balansky; Julio Mercader; Melissa Panger (2006). *Anthropology* (PDF). Boston Massachusetts: Allyn and Bacon. p. 768. ISBN 0-205-32024-4.

[2] *Origins of Modern Humans: Multiregional or Out of Africa?* By Donald Johanson

[3] Evolving in their graves: early burials hold clues to human origins - research of burial rituals of Neanderthals

[4] phillip lieberman (1991). *Uniquely Human*. Cambridge, Mass.: Harvard University Press. ISBN 0-674-92183-6.

[5] Jonathan Amos (2004-04-15). "Cave yields 'earliest jewellery'". *BBC News*. Retrieved 2008-03-12.

[6] Hillary Mayell. "Oldest Jewelry? "Beads" Discovered in African Cave". *National Geographic News*. Retrieved 2008-03-03.

[7] Sean Henahan. "Blombos Cave art". *Science news.* Retrieved 2008-03-12.

[8] "Human Evolution," Microsoft Encarta Online Encyclopedia 2007 © 1997–2007 Microsoft Corporation. All Rights Reserved. Contributed by Richard B. Potts, B.A., Ph.D. Archived 2009-11-01.

[9] Felipe Fernandez Armesto (2003). *Ideas that changed the world.* Newyork: Dorling Kindersley limited. p. 400. ISBN 978-0-7566-3298-4.;

[10] Hillary Mayell. "When Did "Modern" Behavior Emerge in Humans?". *National Geographic News.* Retrieved 2008-02-05.

[11] Boehm, Christopher (1999). *Hierarchy in the forest: the evolution of egalitarian behavior.* Cambridge: Harvard University Press. ISBN 0-674-39031-8.; p. 198

[12] Christopher Boehm (1999) "Hierarchy in the Forest: The Evolution of Egalitarian Behavior" page 192 Harvard university press

[13] Stefan Lovgren. "Sex-Based Roles Gave Modern Humans an Edge, Study Says". *National Geographic News.* Retrieved 2008-02-03.

[14] "Human Evolution," Microsoft Encarta Online Encyclopedia 2007 © 1997–2007 Microsoft Corporation. All Rights Reserved. Contributed by Richard B. Potts, B.A., Ph.D. Archived 2009-11-01.

[15] John Noble Wilford (2007-10-18). "Key Human Traits Tied to Shellfish Remains". *New York times.* Retrieved 2008-03-11.

[16] Tim D. White (2006-09-15). *Once were Cannibals. Evolution: A Scientific American Reader* (University of Chicago Press). ISBN 978-0-226-74269-4. Retrieved 2008-02-14.

[17] James Owen. "Neandertals Turned to Cannibalism, Bone Cave Suggests". *National Geographic News.* Retrieved 2008-02-03.

[18] Pathou-Mathis M (2000). "Neanderthal subsistence behaviours in Europe". *International Journal of Osteoarchaeology* **10** (5): 379–395. doi:10.1002/1099-1212(200009/10)10:5<379::AID-OA558>3.0.CO;2-4.

[19] Karl J. Narr. "Prehistoric religion". *Britannica online encyclopedia 2008.* Retrieved 2008-03-28.

[20] "Human Evolution," Microsoft Encarta Online Encyclopedia 2007 © 1997–2007 Microsoft Corporation. All Rights Reserved. Contributed by Richard B. Potts, B.A., Ph.D. Archived 2009-11-01.

[21] Wallace, Ian; Shea, John (2006). "Mobility patterns and core technologies in the Middle Paleolithic of the Levant". *Journal of Archaeological Science* **33**: 1293–1309. doi:10.1016/j.jas.2006.01.005.

[22] Ann Parson. "Neanderthals Hunted as Well as Humans, Study Says". *National Geographic News.* Retrieved 2008-02-01.

[23] Boëda, E.; Geneste, J.M.; Griggo, C.; Mercier, N.; Muhesen, S.; Reyss, J.L.; Taha, A.; Valladas, H. (1999). "A Levallois point embedded in the vertebra of a wild ass (Equus africanus): Hafting, projectiles and Mousterian hunting". *Antiquity* **73**: 394–402.

[24] Cameron Balbirnie (2005-02-10). "The icy truth behind Neanderthals". *BBC News.* Retrieved 2008-04-01.

[25] Dibble, H.L. 1995. Middle paleolithic scraper reduction: Background, clarification, and review of the evidence to date. Journal of Archaeological Method and Theory 2:299-368

[26] Nicholas Toth and Kathy Schick (2007). *Handbook of Paleoanthropology.* Springer Berlin Heidelberg. p. 1963. ISBN 978-3-540-32474-4.

[27] Wrangham, Richard; Conklin-Brittain, NancyLou (September 2003). "'Cooking as a biological trait'" (PDF). *Comparative Biochemistry and Physiology, Part A: Molecular & Integrative Physiology* **136** (1): 35–46. doi:10.1016/S1095-6433(03)00020-5. PMID 14527628. Archived from the original (pdf) on 19 May 2005. Retrieved 5 June 2014.

[28] Christine mellot. "stalking the ancient dog" (PDF). *Science news.* Retrieved 3-1-08. Check date values in: |access-date= (help)

9.8 External links

- Veldwezelt-Hezerwater

- Picture Gallery of the Paleolithic (reconstructional palaeoethnology), Libor Balák at the Czech Academy of Sciences, the Institute of Archaeology in Brno, The Center for Paleolithic and Paleoethnological Research

Chapter 10

Middle Stone Age

This article is about the term as applied to African prehistory. See Mesolithic for the "middle" period of the Stone Age in general. See Middle Paleolithic for the "middle" part of the "Old Stone Age".

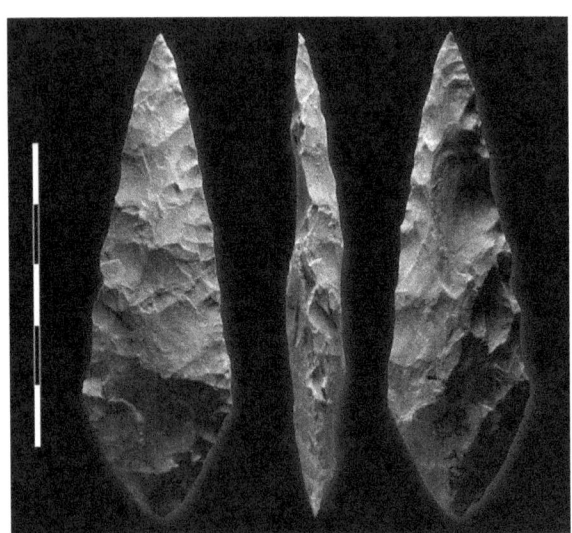

Middle Stone Age tool from Blombos Cave

The **Middle Stone Age** (or **MSA**) was a period of African prehistory between the Early Stone Age and the Later Stone Age. It is generally considered to have begun around 280,000 years ago and ended around 50–25,000 years ago.[1] The beginnings of particular MSA stone tools have their origins as far back as 550–500,000 years ago and as such some researchers consider this to be the beginnings of the MSA.[2] The MSA is often mistakenly understood to be synonymous with the Middle Paleolithic of Europe, especially due to their roughly contemporaneous time span, however, the Middle Paleolithic of Europe represents an entirely different hominin population, *Homo neanderthalensis*, than the MSA of Africa, which did not have Neanderthal populations. Additionally, current archaeological research in Africa has yielded much evidence to suggest that modern human behavior and cogni-

tion was beginning to develop much earlier in Africa during the MSA than it was in Europe during the Middle Paleolithic.<ref name=d'Errico2013>D'Errico, Francesco; Banks, William E. (2013). "Identifying Mechanisms behind Middle Paleolithic and Middle Stone Age Cultural Trajectories". *Current Anthropology* **54** (8): 371–387. doi:10.1086/673388.</ref> The MSA is associated with both anatomically modern humans (*Homo sapiens*) as well as archaic *Homo sapiens*, sometimes referred to as *Homo helmei*. Early physical evidence comes from the Gademotta Formation in Ethiopia, The Kapthurin Formation in Kenya and Kathu Pan in South Africa.[2]

10.1 Regional development

Middle Stone Age Silcrete bifacial points, engraved ochre and bone tools from the c. 75,000- to 80,000-year-old M1 & M2 phases at Blombos cave

It is difficult to discuss the MSA of Africa without first considering the immense size of the continent. There are archaeological sites and evidence from across the continent, and, for the sake of ease, it has often been divided into five regions: northern Africa, which is often, though not without controversy, taken into consideration more so with Southwest Asia and Europe than with the rest of Africa; eastern Africa, stretching roughly from the highlands of Ethiopia to the southern part of Kenya; central Africa, which is arguably the least explored region, stretching from the borders of Tanzania and Kenya to include Angola; southern Africa, which includes the numerous cave sites of South Africa; and western Africa.[3][4]

In northern and western Africa, the desiccation and humectation of the modern Sahara desert has led to very fruitful archaeological sites, followed by completely barren soil, only to once again show evidence of population when the aridity of the region was ameliorated. Preservation in these two regions are alternately superb and lamentable, yet the sites that have been uncovered document the adaptive nature of early hominins to climatically unstable environments. Researchers such as Marean and Assefa consider the historic distinction between northern Africa and the rest of Africa, as though they represent divergent cultural developments, an arbitrary and antiquated distinction.[4]

Eastern Africa represents some of the most reliable dates, due to the use of radiocarbon dating on volcanic ash deposits, as well as some of the earliest MSA sites. Faunal preservation, however, is not spectacular, and standardization in site excavation and lithic classification was, until recently, lacking. Unlike northern Africa, shifts between lithic technologies were not nearly as pronounced, likely due to more favorable climatic conditions that would have allowed for more continuous occupation of sites.[3][4][5] Central Africa reflects similar patterning to eastern Africa, yet more archaeological research of the region is certainly required.

Southern Africa consists of many cave sites, most of which show very punctuated starts and stops in stone tool technology. Research in southern Africa has been continuous and quite standardized, allowing for reliable comparisons between sites in the region. Much of the archaeological evidence for the origins of modern human behavior is traced back to sites in this region, including Blombos Cave, Howiesons Poort, Still Bay, and Pinnacle Point.[3][4]

10.2 Early development

The origins of the MSA are characterized in most regions by the Acheulian to MSA transition. This transition is considered to be a gradual process, rather than a singular event wherein hominin technologies advanced rapidly. Although the dates for this transition vary widely, the oldest reliably dated MSA site is Gademotta in Ethiopia at greater than 276 thousand years ago.[6] The Middle Awash valley of Ethiopia and the Central Rift Valley of Kenya constituted a major center for behavioural innovation.[7] It is likely that the large terrestrial mammal biomass of these regions supported substantial human populations with subsistence and manufac-

The Awash Valley

turing patterns similar to those of ethnographically known foragers.

Archaeological evidence from eastern Africa extending from the Rift Valley from Ethiopia to northern Tanzania represents the largest archaeological evidence of the shift from the Late Acheulian to the Middle Stone Age tool technologies. This transition is characterized by stratigraphic layering of Acheulian stone tools, a bifacial handaxe technology, underneath and even contemporaneous with MSA technologies, such as Levallois tools, flakes, flaked tools, pointed flakes, smaller bifaces that are projectile in form, and, on rare occasions, hafted tools.[4][6] Evidence of the gradual displacement of Acheulian by MSA technologies is further supported by this layering and contemporaneous placement, as well as by the earliest appearance of MSA technologies at Gademotta and the latest Acheulian technologies at the Bouri Formation of Ethiopia, dated to 154 to 160 kya. This suggests a possible overlap of 100-150 thousand years.[6]

South African cave sites have also contributed to the data regarding this shift with accurate dating due to deposits of volcanic ash, which have allowed these sites to be dated to between 999 and 49 thousand years ago. The Cave of Hearths and Montague Cave in South Africa contain evidence of Acheulian technologies, as well as later MSA technologies, however there is no evidence of crossover in this region.[4]

10.3 Lithic technology

Early blades have been documented as far back as 550-500,000 years in the Kapthurin Formation in Kenya and Kathu Pan in South Africa.[2] Backed pieces from the Twin Rivers and Kalambo Falls sites in Zambia, dated at some-

2009 excavations at the Diepkloof Rock Shelter

time between 300 and 140,000 years, likewise indicate a suite of new behaviors.[2][8] A high level of technical competence is also indicated for the c. 280 ka blades recovered from the Kapthurin Formation, Kenya.[9]

The stone tool technology in use during the Middle Stone Age shows a mosaic of techniques. Beginning approximately 300 kya, the large cutting tools of the Achuelian are gradually displaced by Levallois prepared core technologies, also widely used by Neanderthals during the European Middle Palaeolithic.[10] As the MSA progresses, highly varied technocomplexes become common throughout Africa and include pointed artifacts, blades, retouched flakes, end and side scrapers, grinding stones, and even bone tools.[1][4] However, the use of blades (associated mainly with the Upper Palaeolithic in Europe) is seen at many sites as well.[1] In Africa, blades may have been used during the transition from the Early Stone Age to the Middle Stone Age onwards.[11] Finally, during the later part of the Middle Stone Age, microlithic technologies aimed at producing replaceable components of composite hafted tools are seen from at least 70 ka at sites such as Pinnacle Point and Diepkloof Rock Shelter in South Africa.[12][13]

Artifact technology during the Middle Stone Age shows a pattern of innovation followed by disappearance. This occurs with technology such as the manufacture of shell beads,[14] arrows and hide working tools including needles,[15] and gluing technology.[16] These pieces of evidence provide a counterpoint to the classic "Out of Africa" scenario in which increasing complexity accumulated during the Middle Stone Age. Instead, it has been argued that such technological innovations "appear, disappear and reappear in a way that best fits a scenario in which historical contingencies and environmental rather than cognitive changes are seen as main drivers".[15]

10.4 Hominin evolution and migration

Homo erectus *skull, Museum of Natural History, Ann Arbor*

See also: Human Evolution

There have been two migration events out of Africa, the first was the expansion of *H. erectus* into Eurasia approximately 1.9 to 1.7 million years ago, and the second, by *H. sapiens* began during the MSA by 80 – 50 ka MSA out of Africa to Asia, Australia and Europe.[17][18] Perhaps only in small numbers initially, but by 30 ka they had replaced Neanderthals and *H. erectus*.[19] Each of these migrations represent the increased flexibility of the genus *Homo* to survive in widely varied climates. Based on the measurement of a large number of human skulls a recent study supports a central/southern African origin for *Homo sapiens* as this region shows the highest intra-population diversity in phenotypic measurements. Genetic data supports this conclusion.[19] However, there is genetic evidence to suggest that dispersal out of Africa began in eastern Africa. Sites such as the Omo Kibish Formation, the Herto Member of the Bouri Formation, and Mumba Cave contain fossil evidence to support this conclusion as well.[6]

10.5 Evidence for modern human behavior

See also: modern behaviour

There have been a number of theories proposed regarding the development of modern human behavior, but in recent years the mosaic approach has been the most favored perspective in regards to the MSA, especially when taken in consideration with the archaeological evidence.[20] Some scholars including Klein[21] have argued for discontinuity, while others including McBrearty and Brooks have argued that cognitive advances can be detected in the MSA and that the origin of our species is linked with the appearance of Middle Stone Age technology at 250–300 ka.[1]

The earliest remains of Homo sapiens date back to approximately 195 thousand years ago in eastern Africa.[6] In the archaeological record of both eastern Africa and southern Africa, there is immense variability associated with Homo sapiens sites, and it is during this time that we see evidence of the origins of modern human behavior. According to McBrearty and Brooks, there are four features that are characteristic of modern human behavior: abstract thinking, the ability to plan and strategize, "behavioral, economic and technological innovativeness," and symbolic behavior.[1] Many of these aspects of modern human behavior can be broken down into more specific categories, including art, personal adornment, technological advancement, yet these four overarching categories allow for a thorough, albeit significantly overlapping, discussion of behavioral modernity.

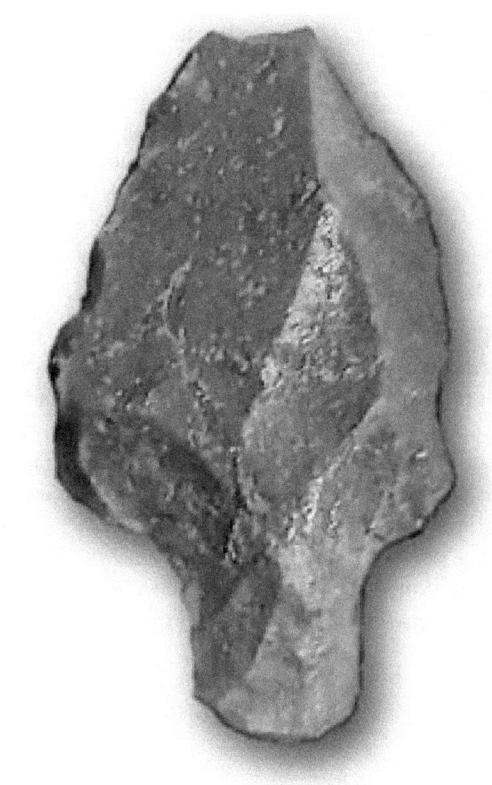

Aterian stone tool

10.5.1 Possible cultural complexes

As early Homo sapiens began to diversify the ecological zones that they inhabited during the MSA, the archaeological record associated with these zones begins to show evidence for regional continuities. These continuities are significant for a number of reasons. The expansion of Homo sapiens into various ecological zones demonstrates an ability to adapt to a variety of environmental contexts including marine environments, savanna grasslands, relatively arid deserts, and forests. This adaptability is reflected in MSA artifacts found in these zones. These artifacts display stylistic variability depending on zone. During the Acheulian, which spanned from 1.5 million years ago to 300 thousand years ago, lithic technology displayed incredible homogeneity throughout all ecological niches. MSA technologies, with their evidence for regional variability and continuity, represent a remarkable advance.[1][5][6] These data have been used to support theories of social and stylistic development throughout the MSA.[22]

In southern Africa, we see the technocomplexes of Howiesons Poort and Stillbay, named after the sites at which they were first discovered. Several others have not been dated or have been dated unreliably; these include the Lupemban technocomplex of central Africa, the Bambatan

in southeast Africa, 70-80ka, and the Aterian technocomplex of northern Africa, 160-90ka.[1][20]

10.5.2 Abstract thinking

Evidence of abstract thinking can be seen in the archaeological record as early as the Acheulean-Middle Stone Age transition, approximately 300,000-250,000 years ago. This transition involves a shift in stone tool technology from Mode 2, Acheulean tools, to Mode 3 and 4, which include blades and microliths. The manufacture of these tools requires planning and the understanding of how striking a stone will produce different flaking patterns.[23] This requires abstract thought, one of the hallmarks of modern human behavior.[1] The shift from large cutting tools in the Acheulian to smaller and more diversified toolkits in the MSA represents a better cognitive and conceptual understanding of flintknapping, as well as the potential functional effects of distinct tool types.

10.5.3 Planning depth

The ability to plan and strategize, much like abstract thinking, can be seen in the more diversified toolkit of the Middle Stone Age, as well as in the subsistence patterns of the period. As MSA hominins began to migrate into a range of different ecological zones, it became necessary to base hunting strategies around seasonally available resources. Awareness of seasonality is evident in the faunal remains found at temporary sites. In less forgiving ecological zones, this awareness would have been essential for survival and the ability to plan subsistence strategies based on this awareness demonstrates an ability to think beyond the present tense and act upon this knowledge.[1]

This planning depth is also seen in the presence of exotic raw materials at a variety of sites throughout the MSA. Procurement of local raw materials would have been a simple task to accomplish, yet MSA sites regularly contain raw materials that were obtained from sources over 100 km away, and sometimes farther than 300 km.[4] Obtaining raw materials from this distance would require an awareness of the resources, a perceived value in the resources, whether it be functional or symbolic, and, possibly, the ability to organize an exchange network in order to obtain the materials.[1][4]

10.5.4 Innovation

The ability to expand into new environments throughout Africa and, ultimately, the world, displays a level of adaptability and, consequently, innovativeness that is often seen as characteristic of behavioral modernity.[1] This, however, is not the only evidence of innovativeness that can be seen in early Homo sapiens. The development of new, regionally relevant tools, such as those used for the collection of marine resources seen at Abdur, Ethiopia, Pinnacle Point Cave, South Africa, and Blombos Cave, South Africa.[1][3] The use of fire demonstrates another innovative aspect of human behavior when it is used in order to create stronger tools, such as the heated silcrete at Blombos, Howiesons Poort and Still Bay,[3][13] and the heat treated bone tools from Still Bay.[20]

Hafted tools are further representative of human innovation. The large cutting tools of the Acheulian technocomplex become smaller, as more complex tools are better suited towards the needs of highly diversified environments. Composite tools represent a new level of innovation in their increased efficacy and more complex manufacturing process. The ability to conceptualize beyond the mere reduction of stone cores demonstrates cognitive flexibility, and the use of glue, which was often processed with ochre, to attach flakes to hafts demonstrates an understanding of chemical changes that can be utilized beyond the simple use of color.[3] Adhesives were used to construct hafted tools by 95ka at Sibudu Cave in South Africa.[1][3]

Zoomorphic pictogram on stone slab from the MSA of Apollo 11 Cave, Namibia

10.5.5 Symbolic behavior

Symbolic behavior is, perhaps, one of the most difficult aspects of modern human behavior to distinguish archaeologically. When searching for evidence of symbolic behavior in the MSA, there are three lines of evidence that can be considered: direct evidence reflecting concrete examples of symbols; indirect evidence reflecting behaviors that would have been used to convey symbolic thought; and technological evidence reflecting the tools and skills that would have been used to produce art. Direct evidence is difficult to find beyond 40ka, and indirect evidence is essentially intangible, thus technological evidence is the most fruitful of the three.[4]

Today there is widespread agreement among archaeologists that the world's first art and symbolic culture dates to the southern African Middle Stone Age. Some of the most striking artifacts, including engraved pieces of red ochre, were manufactured at Blombos Cave in South Africa 70 ka. Pierced and ochred *Nassarius* shell beads were also recovered from Blombos, with even earlier examples (Middle Stone Age, Aterian) from the Taforalt Caves. Arrows and hide working tools have been found at Sibudu Cave[15] as evidence of making weapons with compound heat treated gluing technology.[16]

10.5.6 Complex cognition

A series of innovations have been documented by 170–160,000 years ago at the site of Pinnacle Point 13B on the southern Cape coast of South Africa.[24] This includes the oldest confirmed evidence for the utilization of ochre and marine resources in the form of shellfish exploitation for food. Based on his analysis of the MSA bovid assemblage at Klasies, Milo[25] reports MSA people were formidable hunters and that their social behavior patterns approached those of modern humans. Deacon[26] maintains that the management of plant food resources through deliberate burning of the veld to encourage the growth of plants with corms or tubers in the southern Cape during the Howiesons Poort (c. 70–55 ka) is indicative of modern human behavior. A family basis to foraging groups, color symbolism and the reciprocal exchange of artifacts and the formal organization of living space are, he suggests, further evidence for modernity in the MSA.

Lyn Wadley et al.[16] have argued that the complexity of the skill needed to process the heat-treated compound glue (gum and red ochre) used to haft spears would seem to argue for continuity between modern human cognition and that of humans 70,000 BP at Sibudu Cave.[27]

10.5.7 Evidence for language

Ochre is reported from some early MSA sites, for example at Kapthurin and Twin Rivers, and is common after c. 100 ka.[28] Barham[29] argues that even if some of this ochre was used in a symbolic, color-related role then this abstraction could not have worked without language. Ochre, he suggests, could be one proxy for trying to find the emergence of language.

Formal bone tools are frequently associated with modern behaviour by archaeologists.[30] Sophisticated bone harpoons manufactured at Katanda, West Africa at c. 90 ka[31][32] and bone tools from Blombos Cave dated at c. 77 ka[30] may then also serve as examples of material culture associated with modern language.

Language has been suggested to be necessary to maintain exchange networks. Evidence of some form of exchange networks during the Middle Stone Age is presented in Marwick (2003) in which the distance between the source of raw material and location in which a stone artifact was found was compared throughout sites containing early stone artifacts.[33] Five Middle Stone Age sites contained distances between 140–340 km and have been interpreted, when compared with ethnographic data, that these distances were made possible through exchange networks.[33] Barham[34] also views syntactic language as one aspect of behavior that in fact allowed MSA people to settle in the tropical forest environments of what is now the Democratic Republic of Congo.

Many authors have speculated that at the core of this symbolic explosion, and in tandem, was the development of syntactic language that evolved through a highly specialized social learning system[35] providing the means for semantically unbounded discourse.[36] Syntax would have played a key role in this process and its full adoption could have been a crucial element of the symbolic behavioral package in the MSA.[37]

10.6 Brain change

Although the advent of anatomical physical modernity cannot confidently be linked with palaeoneurological change,[38] it does seem probable that hominid brains evolved through the same selection processes as other body parts.[39] Genes that promoted a capacity for symbolism may have been selected for, suggesting that the foundations for symbolic culture may well be grounded in biology. However, behavior that was mediated by symbolism may have only come later, even though this physical capacity was already in place much earlier. Skoyles and Sagan, for example, argue that human brain expansion by increasing the prefrontal cortex would have created a brain capable of symbolizing its previously non-symbolic cognition, and that this process, slow to begin with, increasingly accelerated during the last 100,000 years.[40] Symbolically mediated behavior may then feed back upon this process by creating a greater ability to manufacture symbolic artifacts and social networks.

10.7 Sites

Numerous sites in southern Africa reflect the four characteristics of behavioral modernity. Blombos Cave, South Africa contains personal ornaments and what are presumed to be the tools used for the production of artistic imagery, as well as bone tools.[20] Still Bay and Howieson's Poort contain variable tool technologies.[41] These different types of assemblages allow researchers to extrapolate behaviors that would likely be associated with such technologies, such as shifts in foraging behaviors, which are further supported by faunal data at these sites.

- Blombos Cave, South Africa

- Klasies River Caves, South Africa

- Sibudu Cave, South Africa

- Diepkloof Rock Shelter, South Africa

- Pinnacle Point, South Africa

- Mumba Cave, Tanzania

- Mumbwa Caves, Zambia

Excavations at Pinnacle Point, South Africa

10.8   See also

- Out of Africa hypothesis

- Symbolic culture

- The Human Revolution (human origins)

- Later Stone Age

10.9   Notes

[1] McBrearty, Sally; Brooks, Alison A. (2000). "The revolution that wasn't: A new interpretation of the origin of modern human behaviour". *Journal of Human Evolution* **39**: 453–563. doi:10.1006/jhev.2000.0435.

[2] Herries, A.I.R. 2011. "A chronological perspective on the Acheulian and its transition to the Middle Stone Age in southern Africa: the question of the Fauresmith" *International Journal of Evolutionary Biology* 961401, doi:10.4061/2011/961401

[3] Lombard, Marlize (2012). "Thinking through the Middle Stone Age of sub-Saharan Africa". *Quaternary International* **270**: 140–155. doi:10.1016/j.quaint.2012.02.033.

[4] Marean, Curtis W. and Zelalem Assefa. 2004. "The Middle and Upper Pleistocene African Record for the Biological and Behavioral Origins of Modern Humans" In *African Archaeology: A Critical Introduction*, edited by Ann B. Stahl, pp.93-129. Wiley-Blackwell, New Jersey

[5] Ambrose, Stanley H (2001). "Paleolithic Technology and Human Evolution". *Science* **291**: 1748–1753. doi:10.1126/science.1059487.

[6] Tryon, Christopher A.; Faith, Tyler (2013). "Variability in the Middle Stone Age of Eastern Africa". *Current Anthropology* **54** (8): 234–254. doi:10.1086/673752.

[7] Brooks, A. S. 2006. "Recent perspectives on the Middle Stone Age of Africa" Paper presented at the African Genesis Symposium on Hominid Evolution in Africa: Johannesburg.

[8] Barham, Lawrence (2002). "Backed tools in Middle Pleistocene central Africa and their evolutionary significance". *Journal of Human Evolution* **43**: 585–603. doi:10.1006/jhev.2002.0597.

[9] Deino, Alan L.; McBrearty, Sally (2002). "40Ar/(39)Ar dating of the Kapthurin Formation, Baringo, Kenya". *Journal of Human Evolution* **42** (1-2): 185–210. doi:10.1006/jhev.2001.0517.

[10] Shea, John (2011). "*Homo sapiens* is as *Homo sapiens* was". *Current Anthropology* **52**: 1–35. doi:10.1086/658067.

[11] Porat, Naomi; Chazan, Michael; Grün, Rainer; Aubert, Maxime; Eisenman, Vera; Kolska Horwitz, Liora (2010). "New radiometric ages for the Fauresmith industry from Kathu Pan, southern Africa: Implications for the Earlier to Middle Stone Age transition". *Journal of Archaeological Science* **37**: 269–283. doi:10.1016/j.jas.2009.09.038.

[12] Rigaud, Jean-Phillipe; Texier, Pierre-Jean; Parkington, John; Poggenpoel, Cedric (2006). "Le mobilier Stillbay et Howiesons Poort de l'abri Diepkloof: La chronologie du *Middle Stone Age* sud-africain et ses implications". *Comptes Rendus Palevol* **5** (6): 839–849. doi:10.1016/j.crpv.2006.02.003.

[13] Brown, Kyle S.; Marean, Curtis W.; Jacobs, Zenobia; Schoville, Benjamin J.; Oestmo, Simen; Fisher, Erich C.; Bernatchez, Jocelyn; Karkanas, Panagiotis; Matthews, Thalassa (2012). "An early and enduring advanced technology originating 71,000 years ago in South Africa". *Nature* **491**: 590–593. doi:10.1038/nature11660.

[14] D'Errico, Francesco; Vanhaeren, Marian; Wadley, Lyn (2008). "Possible shell beads from the Middle Stone Age layers of Sibudu Cave, South Africa". *Journal of Archaeological Science* **35** (10): 2675–2685. doi:10.1016/j.jas.2008.04.023.

[15] Backwell, L; d'Errico, F; Wadley, L (2008). "Middle Stone Age bone tools from the Howiesons Poort layers, Sibudu Cave, South Africa". *Journal of Archaeological Science* **35**: 1566–1580. doi:10.1016/j.jas.2007.11.006.

[16] Wadley, L; Hodgskiss, T; Grant, M (2009). "Implications for complex cognition from the hafting of tools with compound adhesives in the Middle Stone Age, South Africa". *Proceedings of the National Academy of Sciences* **106**: 9590–9594. doi:10.1073/pnas.0900957106. PMC 2700998. PMID 19433786.

[17] Anton, Susan C.; Potts, Richard; Aeillo, Leslie C. (2014). "Evolution of early Homo: An integrated biological perspective". *Science* **345**: 45–59. doi:10.1126/science.1236828.

[18] Mellers, Paul (2006). "A new radiocarbon revolution and the dispersal of modern humans in Eurasia". *Nature* **439**: 931–935. doi:10.1038/nature04521.

[19] Manica, Andrea; Amos, William; Balloux, Francois; Hanihara, Tsunehiko (2007). "The effect of ancient population bottlenecks on human phenotypic variation". *Nature* **448**: 346–348. doi:10.1038/nature05951.

[20] Henshilwood, Christopher S.; d'Errico, Francesco; Marean, Curtis W.; Milo, Richard G.; Yates, Royden (2001). "An early bone tool industry from the Middle Stone Age at Blombos Cave, South Africa: implications for the origins of modern human behaviour, symbolism and language.". *Journal of Human Evolution* **41**: 631–678. doi:10.1006/jhev.2001.0515.

[21] Klein, R. G. (2000). "Archaeology and the evolution of human behavior". *Evolutionary Anthropology* **9**: 17–36. doi:10.1002/(sici)1520-6505(2000)9:1<17::aid-evan3>3.0.co;2-a.

[22] Hogsberg, Anders; Larsson, Lars (2011). "Lithic technology and behavioural modernity: New results from the Still Bay site, Hollow Rock Shelter, Western Cape Province, South Africa". *Journal of Human Evolution* **61**: 133–155. doi:10.1016/j.jhevol.2011.02.006.

[23] Lombard, Marlize (2012). "Thinking through the Middle Stone Age of sub-Saharan Africa". *Quaternary International* **270**: 140–155. doi:10.1016/j.quaint.2012.02.033.

[24] Marean, Curtis W.; Bar-Matthews, Miryam; Bernatchez, Jocelyn; Fisher, Erich; Goldberg, Paul; Herries, Andy I. R.; Jacobs, Zenobia; Jerardino, Antonieta; Karkanas, Panagiotis; Minichillo, Tom; Nilssen, Peter J.; Thompson, Erin; Watts, Ian; Williams, Hope M. (2007). "Early human use of marine resources and pigment in South Africa during the Middle Pleistocene". *Nature* **449**: 905–908. doi:10.1038/nature06204.

[25] Milo, R. G. (1998). "Evidence for hominid predation at Klasies River Mouth, South Africa, and its implications for the behavior of early modern humans". *Journal of Archaeological Science* **25**: 99–133. doi:10.1006/jasc.1997.0233.

[26] Deacon, H. J. 2001. "Modern human emergence: an African archaeological perspective" In *Humanity from African Naissance to Coming Millennia: Colloquia in Human Biology and Palaeoanthropology*, edited by P. V. Tobias, M. A. Raath, J. Maggi-Cecchi, and G. A. Doyle, pp.217-226. Florence University Press, Florence.

[27] Wynn, Thomas (2009). "Hafted spears and the archaeology of mind". *Proceedings of the National Academy of Sciences* **106** (24): 9544–9545. doi:10.1073/pnas.0904369106.

[28] Watts, I (2002). "Ochre in the Middle Stone Age of southern Africa: ritualized display or hide preservative?". *South African Archaeological Bulletin* **57**: 64–74.

[29] Barham, L. S. (2002). "Systematic pigment use in the Middle Pleistocene of south central Africa". *Current Anthropology* **43** (1): 181–190. doi:10.1086/338292.

[30] Henshilwood, C. S.; d'Errico, F.; Marean, C.; Milo, R.; Yates, R. (2001). "An early bone tool industry from the Middle Stone Age at Blombos Cave, South Africa: implications for the origins of modern human behaviour, symbolism and language". *Journal of Human Evolution* **41**: 631–678. doi:10.1006/jhev.2001.0515.

[31] Yellen, J.E.; Brooks, A.S.; Cornelissen, E.; Mehlman, M.J.; Stewart, K. (1995). "A Middle Stone Age worked bone industry from Katanda, Upper Semliki Valley, Zaire". *Science* **268**: 553–556. doi:10.1126/science.7725100.

[32] Brooks, A.S.; Helgren, D.M.; Cramer, J.S.; Franklin, A.; Hornyak, W.; Keating, J.M.; Klein, R.G.; Rink, W.J.; Schwarcz, H.; Smith, J.N.L.; Stewart, K.; Todd, N.E.; Verniers, J.; Yellen, J.E. (1995). "Dating and Context of Three Middle Stone Age Sites with Bone Points in the Upper Semliki Valley, Zaire". *Science* **268**: 548–553. doi:10.1126/science.7725099.

[33] Marwick, Ben (2003). "Pleistocene Exchange Networks as Evidence for the Evolution of Language". *Cambridge Archaeological Journal* **13** (1): 67–81. doi:10.1017/s0959774303000040.

[34] Barham, Lawrence. 2001. "Central Africa and the emergence of regional identity in the Middle Pleistocene" In *Human Roots: Africa and Asia in the Middle Pleistocene*, edited by Lawrence Barham and Kate Robson-Brown, pp.65-80. Western Academic and Specialist Press, Bristol.

[35] Richerson, P. and Boyd, R. 1998. "The Pleistocene and the origins of human culture: built for speed" Paper presented at the 5th Biannual Symposium on the Science of Behaviour: Behaviour, Evolution and Culture. University of Guadalajara, Mexico.

[36] Rappaport, R. A. 1999. *Ritual and Religion in the Making of Humanity* Cambridge University Press, Cambridge.

[37] Bickerton, D. 2003. "Symbol and structure: A comprehensive framework for language evolution" In *Language Evolution*, edited by M. H. Christiansen and S. Kirby, pp. 77-93. Oxford University Press, Oxford.

[38] Holloway, R.L. 1996. "Evolution of the human brain" In *Handbook of Human Symbolic Evolution*, edited by A. Locke and C. Peters, pp. 74-116. Oxford University Press, New York.

[39] Gabora, L. 2001. *Cognitive mechanisms underlying the origin and evolution of culture* Ph.D. dissertation. Center Leo Apostel For interdisciplinary Studies, Vrije Universiteit Brussels, Brussels, Belgium.

[40] Skoyles JR. Sagan D. 2002. *Up from Dragons: The evolution of intelligence* McGraw-Hill.

[41] Henshilwood, Christopher S.; Dubreuil, Benoit (2011). "The Still Bay and Howiesons Poort, 77-59 ka: Symbolic Material Culture and the Evolution of the Mind during the African Middle Stone Age". *Current Anthropology.* doi:10.1086/660022.

Chapter 11

Neanderthal

For other uses, see Neanderthal (disambiguation).
"Neanderthal man" redirects here. For other uses, see Neanderthal man (disambiguation).

The **Neanderthal** or **Neandertal** UK /niˈændərˌtɑːlz/, us also /neɪ/-, -/ˈɑːndər/-, -/ˌtɔːlz/, -/ˌθɔːlz/)[3][4] (named after the Neandertal area) was a species of human in the genus *Homo* that became extinct between 41,000 and 39,000 years ago. They were closely related to modern humans,[5][6] differing in DNA by just 0.12%.[7] Remains left by Neanderthals include bone and stone tools, which are found in Eurasia, from Western Europe to Central and Northern Asia and the Middle East. The Neanderthal is generally classified by biologists as the species *Homo neanderthalensis*, but a minority considers them to be a subspecies of *Homo sapiens* (*Homo sapiens neanderthalensis*).[8][9][10]

Several cultural assemblages have been linked to the Neanderthals in Europe. The earliest, the Mousterian stone tool culture, dates to about 300,000 years ago.[11] Late Mousterian artifacts were found in Gorham's Cave on the south-facing coast of Gibraltar.[12][13]

Neanderthals were large compared to Homo sapiens because they inhabited higher latitudes, in conformance with Bergmann's rule, and their larger stature explains their larger brain size because brain size generally increases with body size.[14] With an average cranial capacity of 1600 cm^3,[15] the cranial capacity of Neanderthals is notably larger than the 1400 cm^3 average for modern humans, indicating that their brain size was larger. Males stood 164–168 cm (65–66 in) and females 152–156 cm (60–61 in) tall.[16]

A 2008 study by the Max Planck Institute for Evolutionary Anthropology in Leipzig suggested Neanderthals probably did not interbreed with anatomically modern humans,[17][18] while the Neanderthal genome project published in 2010 and 2014 suggests that Neanderthals did contribute to the DNA of modern humans, including most non-Africans as well as a few African populations, through interbreeding, likely between 50,000 to 60,000 years ago.[19][20][21]

In December 2013, researchers reported evidence that Neanderthals practiced burial behavior and intentionally buried their dead.[22] In addition, scientists reported having sequenced the entire genome of a Neanderthal for the first time. The genome was extracted from the toe bone of a 50,000-year-old Neanderthal found in a Siberian cave.[23][24][25]

11.1 Name

The species is named after one of the first sites where its fossils were discovered in the 19th-century, about 12 km (7.5 mi) east of Düsseldorf, Germany, in the Feldhofer Cave, located at the Düssel River's Neander valley.[26] *Thal* is the older spelling of the German word *Tal* (with the same pronunciation), which means "valley" (cognate with English *dale*).[lower-alpha 1][27][28]

Neanderthal 1 was known as the "Neanderthal cranium" or "Neanderthal skull" in anthropological literature, and the individual reconstructed on the basis of the skull was occasionally called "the Neanderthal man".[29] The binomial name *Homo neanderthalensis* – extending the name "Neanderthal man" from the individual type specimen to the entire species – was first proposed by the Anglo-Irish geologist William King in 1864, although that same year King changed his mind and thought that the Neanderthal fossil was distinct enough from humans to warrant a separate genus.[30] Nevertheless, King's name had priority over the proposal put forward in 1866 by Ernst Haeckel, *Homo stupidus*.[27] The practice of referring to "the Neanderthals" and "a Neanderthal" emerged in the popular literature of the 1920s.[31]

The German pronunciation of *Neanderthaler* and *Neandertaler* is [neˈandɐˌtʰaːlɐ] in the International Phonetic Alphabet. In British English, "Neanderthal" is pronounced with the /t/ as in German but different vowels (IPA: /niˈændɐtɑːl/).[32][33][34] In layman's American English, "Neanderthal" is pronounced with a /θ/ (the voiceless *th* as in *thin*) and /ɔ/ instead of the longer British /ɑː/ (IPA:

/niːˈændərθɔːl/),[35] although scientists typically use the /t/ as in German.[36][37]

11.2 Classification

For some time, scientists have debated whether Neanderthals should be classified as *Homo neanderthalensis* or *Homo sapiens neanderthalensis*, the latter placing Neanderthals as a subspecies of *H. sapiens*.[38][39] Some morphological studies support the view that *H. neanderthalensis* is a separate species and not a subspecies.[40][41] Others, for example University of Cambridge Professor Paul Mellars, say "no evidence has been found of cultural interaction"[42] and evidence from mitochondrial DNA studies has been interpreted as evidence Neanderthals were not a subspecies of *H. sapiens*.[43]

11.3 Origin

The first humans with proto-Neanderthal[44] traits are believed to have existed in Eurasia as early as 350,000–600,000 years ago[45] with the first "true Neanderthals" appearing between 200,000 and 250,000 years ago.[46] The exact date of their extinction had been disputed but in 2014, a team led by Thomas Higham of the University of Oxford used an improved radiocarbon dating technique on material from 40 archaeological sites to show that Neanderthals died out in Europe between 41,000 and 39,000 years ago. This coincides with the start of a very cold period in Europe and is 5,000 years after *Homo sapiens* reached the continent.[47]

Comparison of the DNA of Neanderthals and *Homo sapiens* suggests that they diverged from a common ancestor between 350,000 and 400,000 years ago. This ancestor was probably *Homo heidelbergensis*. Heidelbergensis originated between 800,000 and 1,300,000 years ago, and continued until about 200,000 years ago. It ranged over Eastern and South Africa, Europe and Western Asia. Between 350,000 and 400,000 years ago the African branch is thought to have started evolving towards modern humans and the Eurasian branch towards Neanderthals. Scientists do not agree when Neanderthals can first be recognised in the fossil record, with dates ranging between 200,000 and 300,000 years BP.[48][49][50][51]

11.4 Discovery

Neander Valley site

The site of Kleine Feldhofer Grotte where the type specimen was unearthed by miners in the 19th century

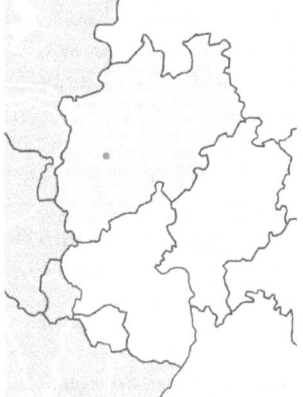

Location of Neander Valley, Germany with the modern federal state of North Rhine-Westphalia highlighted

Neanderthal skulls were first discovered in the Engis Caves (fr), in what is now Belgium (1829) by Philippe-Charles Schmerling and in Forbes' Quarry, Gibraltar, dubbed Gibraltar 1 (1848), both prior to the type specimen discovery in a limestone quarry of the Neander Valley in Erkrath near Düsseldorf in August 1856, three years before Charles Darwin's *On the Origin of Species* was published.[52]

The type specimen, dubbed Neanderthal 1, consisted of a skull cap, two femora, three bones from the right arm, two from the left arm, part of the left ilium, fragments of a scapula, and ribs. The workers who recovered this material originally thought it to be the remains of a bear. They gave the material to amateur naturalist Johann Carl Fuhlrott, who turned the fossils over to anatomist Hermann Schaaffhausen.

To date, the bones of over 400 Neanderthals have been found.[53]

11.4.1 Timeline

Neanderthal fossils

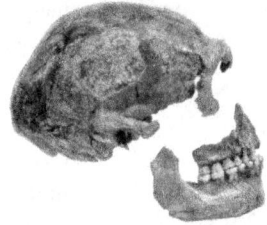

Skull, found in 1886 in Spy, Belgium

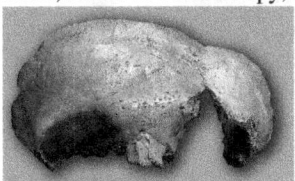

Frontal bone of a neanderthal child from the cave of La Garigüela

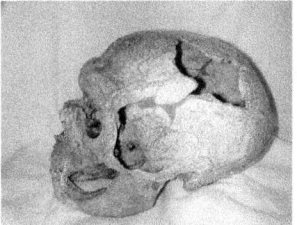

Skull from La Chapelle aux Saints

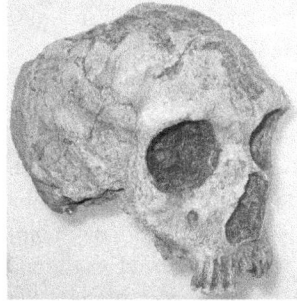

Semi-frontal view of a neanderthal skull from Gibraltar

- 1829: The Engis 2 Neanderthal skulls were discovered in Engis, in present-day Belgium.

- 1848: Neanderthal skull Gibraltar 1 found in Forbes' Quarry, Gibraltar. Called "an ancient human" at the time.

- 1856: Johann Karl Fuhlrott first recognized the fossil called "Neanderthal man", discovered in Neanderthal, a valley near Mettmann in what is now North Rhine-Westphalia, Germany.

- 1864: William King proposed the name *Homo neanderthalensis* at a meeting of the British Association for the Advancement of Science, but then changed

his mind and argued that Neanderthals were different enough from humans to warrant a separate genus, under the assumption that they were likely "incapable of moral and theositic conceptions."[30]

- 1880: The mandible of a Neanderthal child was found in a secure context and associated with cultural debris, including hearths, Mousterian tools, and bones of extinct animals.

- 1886: Two nearly perfect skeletons of a man and woman were found at Spy, Belgium at the depth of 16 ft with numerous Mousterian-type implements.

- 1899: Hundreds of Neanderthal bones were described in stratigraphic position in association with cultural remains and extinct animal bones.

- 1899: Sand excavation workers found bone fragments on a hill in Krapina, Croatia called *Hušnjakovo brdo*. Local Franciscan friar Dominik Antolković requested Dragutin Gorjanović-Kramberger to study the remains of bones and teeth that were found there.

- 1905: During the excavation in Krapina more than 5 000 items were found, of which 874 residue of human origin, including bones of prehistoric man and animals, artifacts.

- 1908: A nearly complete Neanderthal skeleton was discovered in association with Mousterian tools and bones of extinct animals.

- 1925: Francis Turville-Petre finds the 'Galilee Man' or 'Galilee Skull' in the Zuttiyeh Cave in Wadi Amud in The British Mandate of Palestine (now Israel).

- 1926: Skull fragments of Gibraltar 2, a four-year-old Neanderthal girl, discovered by Dorothy Garrod.

- 1953–1957: Ralph Solecki uncovered nine Neanderthal skeletons in Shanidar Cave in the Kurdistan region of northern Iraq.

- 1975: Erik Trinkaus' study of Neanderthal feet confirmed they walked like modern humans.

- 1981: The Bontnewydd Palaeolithic site, Wales yields the most north-western site in Eurasia.

- 1987: Thermoluminescence results from Israeli fossils date Neanderthals at Kebara to 60,000 BP and humans at Qafzeh to 90,000 BP. These dates were confirmed by electron spin resonance (ESR) dates for Qafzeh (90,000 BP) and Es Skhul (80,000 BP).

- 1991: ESR dates showed the Tabun Neanderthal was contemporaneous with modern humans from Skhul and Qafzeh.

- 1993: 127,000-year-old DNA is found on the child of Sclayn, found in Scladina (fr), Belgium.

- 1994: Neanderthal remains inadvertently uncovered inside the Sidrón Cave in Piloña municipality, Asturias, northwestern Spain. Since then the remains of at least 12 individuals: three men, three adolescent boys, three women, and three infants have been found. In 2009 Neanderthal ancient mtDNA was partially sequenced in HVR region for three distinct Neanderthals there.[54][55]

- 1997: Matthias Krings *et al.* are the first to amplify Neanderthal mitochondrial DNA (mtDNA) using a specimen from Feldhofer grotto in the Neander valley.[56]

- 1997–2000: During new excavations in the Neandertal, additional bone fragments are found, some of which fit the fragments found in 1856, thus pinpointing the exact location of the original find. The exact location had previously been unknown, as the site of the find (the „Kleine Feldhofer Grotte") was destroyed by limestone mining.[57]

- 1998: A team led by pre-history archeologist João Zilhão discovered an early Upper Paleolithic human burial in Portugal, at Abrigo do Lagar Velho, which provided evidence of early modern humans from the west of the Iberian Peninsula. The remains, a largely complete skeleton of an approximately 4-year-old child, buried with pierced shell and red ochre, is dated to *ca.* 24,500 years BP.[58] The cranium, mandible, dentition, and postcrania present a mosaic of European early modern human and Neanderthal features.[58]

- 2000: Igor Ovchinnikov *et al.* retrieved mitochondrial DNA of a Neanderthal infant from Mezmaiskaya Cave in the Caucasus.[59]

- 2005: The Max Planck Institute for Evolutionary Anthropology launched a project to reconstruct the Neanderthal genome, working with Connecticut-based 454 Life Sciences. In 2009, the Max Planck Institute announced the "first draft" of a complete Neanderthal genome is completed.[60]

- 2010: Comparison of Neanderthal genome with modern humans from Africa and Eurasia shows that 1–4% of modern non-African human genome might come from the Neanderthals.[19][61]

- 2010: Discovery of Neanderthal tools far away from the influence of *H. sapiens* indicate that the species might have been able to create and evolve tools on its own, and therefore be more intelligent than previously

thought. Furthermore, it was proposed that the Neanderthals might be more closely related to *Homo sapiens* than previously thought and that may in fact be a subspecies of it.[62] Evidence has more recently emerged that these artifacts are probably of *H. sapiens sapiens* origin.[63]

- 2012: Charcoal found next to six paintings of seals in Nerja caves, Malaga, Spain, has been dated to between 42,300 and 43,500 years old. The paintings themselves will be dated in 2013, and if their pigment matches the date of the charcoal, they would be the oldest known cave paintings. José Luis Sanchidrián at the University of Cordoba, Spain believes the paintings are more likely to have been painted by Neanderthals than early modern humans.[64]

- 2013: A jawbone found in Italy had features intermediate between Neanderthals and *Homo sapiens* suggesting it could be a hybrid. The mitochondrial DNA is Neanderthal.[65]

- 2013: An international team of researchers reported evidence that Neanderthals practiced burial behavior and intentionally buried their dead.[22]

- 2014: Researchers at the University of Colorado Museum in Boulder report that Neanderthals were not less intelligent than modern humans and "that single-factor explanations for the disappearance of the Neandertals are not warranted any more."[66]

- 2014: Prof Thomas Higham of the University of Oxford performed the most comprehensive dating of Neanderthal bones and tools ever carried out, which demonstrated that Neanderthals died out in Europe between 41,000 and 39,000 years ago—this coincides with the start of a very cold period in Europe and is 5,000 years after *Homo sapiens* reached the continent.[67][68]

11.5 Habitat and range

Further information: List of Neanderthal sites

Early Neanderthals lived in the last glacial period for a span of about 100,000 years. Because of the damaging effects the glacial period had on the Neanderthal sites, not much is known about the early species. Countries where their remains are known include most of Europe south of the line of glaciation, roughly along the 50th parallel north. This includes most of Western Europe, Central Europe, the Carpathians, and the Balkans,[69] some sites in Ukraine and in western Russia, Central and Northern Asia up to the Altai Mountains, and Western Asia from the Levant

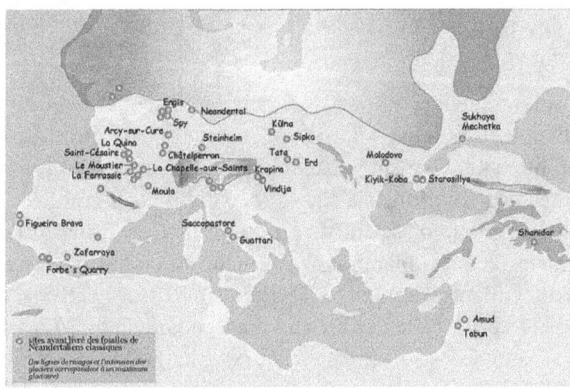

Sites where typical Neanderthal fossils have been found

up to the Indus River. The Bontnewydd Palaeolithic site at Denbighshire, North Wales is the most north-western site of Neanderthal remains and one the oldest remains in Britain (230,000 years ago). It is estimated that the total Neanderthal population across this habitat range numbered at around 70,000 at its peak.[70]

Neanderthal fossils have not been found to date in Africa, but there have been finds close to North Africa, both on Gibraltar and in the Levant. At some Levantine sites, Neanderthal remains date from after the same sites were vacated by modern humans. Mammal fossils of the same time period show cold-adapted animals were present alongside these Neanderthals in this region of the Eastern Mediterranean. This implies Neanderthals were better adapted biologically to cold weather than modern humans and at times displaced them in parts of the Middle East when the climate got cold enough. However this has been disputed recently through studies of their cranial morphology and sinuses. Some scholars have also posited that Neanderthals could be an Asian species which expanded into Europe, making Neanderthals a tropical species rather than cold adaptive.[71][72]

Homo sapiens sapiens appears to have been the only human type in the Nile River Valley during these periods, and Neanderthals are not known to have ever lived south-west of present-day Israel. When climate change caused warmer temperatures, the Neanderthal range likewise retreated to the north, along with the cold-adapted species of mammals. Apparently these weather-induced population shifts took place before modern people secured competitive advantages over the Neanderthal, as these shifts in range took place well over ten thousand years before modern people totally replaced the Neanderthal, despite the recent evidence of some successful interbreeding.[71]

Separate developments in the human line, in other regions such as Southern Africa, somewhat resembled the Eurasian Neanderthals, but these people were not Neanderthals. One

such example is Rhodesian Man (*Homo rhodesiensis*), who existed long before any classic Eurasian Neanderthals, but had a more modern set of teeth. Some *H. rhodesiensis* populations appeared to be on the road to modern *H. sapiens sapiens*. At any rate, the populations in Eurasia underwent more and more "Neanderthalization" as time went on. There is some argument that *H. rhodesiensis* in general was ancestral to both modern humans and Neanderthals, and that at some point the two populations went their separate ways, but this supposes that *H. rhodesiensis* goes back to around 600,000 years ago.

To date, no intimate connection has been found between these similar archaic people and the Eurasian Neanderthals, at least during the same time as *H. rhodesiensis* seems to have lived about 600,000 years ago, long before the time of classic Neanderthals. This said, some researchers think that *H. rhodesiensis* may have lived much later than this period, depending on the method used to date the fossils, leaving this issue open to debate. Some *H. rhodesiensis* features, like the large brow ridge, may have been caused by convergent evolution.

It appears incorrect, based on present research and known fossil finds, to refer to any fossils outside Eurasia as true Neanderthals. They had a known range that possibly extended as far east as the Altai Mountains, but not farther to the east or south, and apparently not into Africa. At any rate, in North-East Africa the land immediately south of the Neanderthal range was possessed by modern humans *Homo sapiens idaltu* or *Homo sapiens*, since at least 160,000 years before the present. 160,000-year-old hominid fossils at Jebel Irhoud in Morocco were previously thought to be Neanderthal, but it is now clear that they are early modern humans.[73]

Classic Neanderthal fossils have been found over a large area, from northern Germany to Israel and Mediterranean countries like Spain[74] and Italy[75] in the south and from England and Portugal in the west to Uzbekistan in the east. This area probably was not occupied all at the same time. The northern border of their range, in particular, would have contracted frequently with the onset of cold periods. On the other hand, the northern border of their range as represented by fossils may not be the real northern border of the area they occupied, since Middle Palaeolithic-looking artifacts have been found even farther north, up to 60° N, on the Russian plain.[76] Recent evidence has extended the Neanderthal range by about 1,250 miles (2,010 km) east into southern Siberia's Altai Mountains.[77][78]

11.6 Anatomy

Main article: Neanderthal anatomy
Neanderthal anatomy differed from modern humans in that

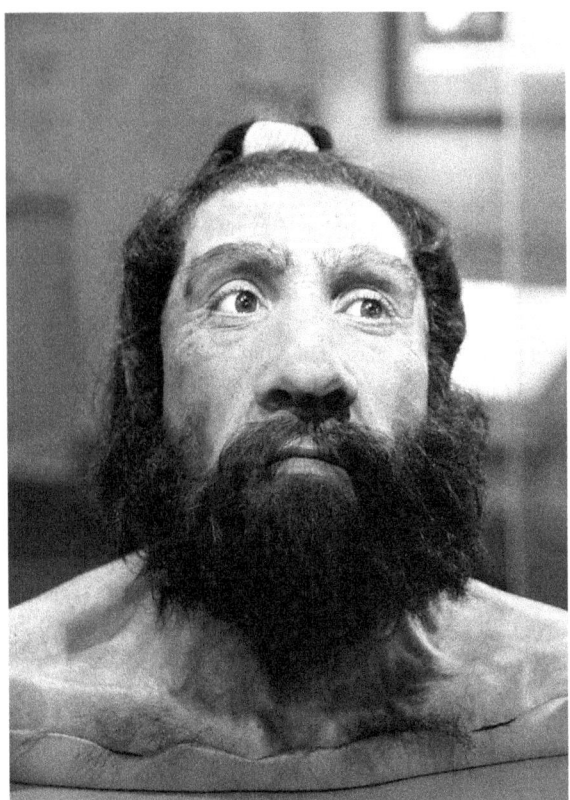

Reconstruction of the head of the Shanidar 1 fossil, a Neanderthal male who lived c. 70,000 years ago (John Gurche 2010)

they had a more robust build and distinctive morphological features, especially on the cranium, which gradually accumulated more derived aspects as it was described by Marcellin Boule,[79] particularly in certain isolated geographic regions. These include shorter limb proportions, a wider, barrel-shaped rib cage, a reduced chin and, perhaps most notably, a large nose, which was much larger in both length and width, and started somewhat higher on the face, than in modern humans.[46] Evidence suggests they were much stronger than modern humans, with particularly strong arms and hands,[80][81] while they were comparable in height; based on 45 long bones from at most 14 males and 7 females, Neanderthal males averaged 164–168 cm (65–66 in) and females 152–156 cm (60–61 in) tall.[16] Samples of 26 specimens in 2010 found an average weight of 77.6 kg (171 lb) for males and 66.4 kg (146 lb) for females.[82] A 2007 genetic study suggested some Neanderthals may have had red hair and blond hair, along with a light skin tone.[83]

A 2013 study of Neanderthal skulls suggests that their eyesight may have been better than that of modern humans,

owing to larger eye sockets and larger areas of the brain devoted to vision.[84]

Neanderthals are known for their large cranial capacity, which at 1600 cm³ is larger on average than that of modern humans. One study has found that Neanderthal brains were more asymmetric than other hominid brains.[85] In 2008, a group of scientists produced a study using three-dimensional computer-assisted reconstructions of Neanderthal infants based on fossils found in Russia and Syria. It indicated that Neanderthal and modern human brains were the same size at birth, but that by adulthood, the Neanderthal brain was larger than the modern human brain.[86] They had almost the same degree of encephalization (i.e. brain to body size ratio) as modern humans.[87][88]

11.7 Behavior

Main article: Neanderthal behavior
Neanderthals made advanced tools,[89] probably had a lan-

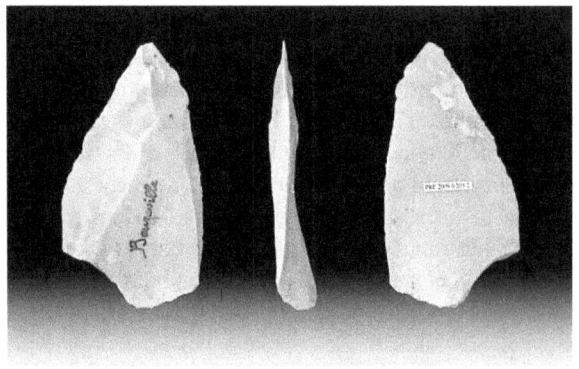

Levallois point—Beuzeville, France

guage (the nature of which is debated and likely unknowable) and lived in complex social groups. The Molodova archaeological site in eastern Ukraine suggests some Neanderthals built dwellings using animal bones. A building was made of mammoth skulls, jaws, tusks and leg bones, and had 25 hearths inside.[90]

Circumstantial evidence suggests Neanderthals may have been building some form of watercraft since the Middle Paleolithic.[91][92] Scientists have speculated that these watercraft may have been similar to dugout canoes, which are among the oldest known boats in the archaeological record.[92] Mousterian stone tools discovered on the southern Ionian Islands suggests that Neanderthals were sailing the Mediterranean Sea as early as 110,000 years BP.[93][94] Quartz hand-axes, three-sided picks, and stone cleavers from Crete have also been recovered that date back about 170,000 years BP.[95]

It was once thought that Neanderthals lacked the sophistication for hunting, perhaps scavenging meat from carcasses,[46] but increasing evidence suggests they were apex predators,[96][97] capable of bringing down a wide range of prey from red deer, reindeer, ibex and wild boar, to larger animals such as aurochs and even, on occasion, mammoth, straight-tusked elephant and rhinoceros.[46][98] However, while largely carnivorous,[99][100] new studies indicate Neanderthals also had cooked vegetables in their diet.[97][101] In 2010, an isotope analysis of Neanderthal teeth found traces of cooked vegetable matter, and more recently a 2014 study of Neanderthal coprolites (fossilized feces) found substantial amounts of plant matter, contradicting the earlier belief they were exclusively (or almost exclusively) carnivorous.[99][102]

The size and distribution of Neanderthal sites, along with genetic evidence, suggests Neanderthals lived in much smaller and more sparsely distributed groups than their *Homo sapiens* contemporaries.[103][104] Some experts suggest that this disparity alone was a major contributing factor to their ultimate replacement by *Homo sapiens*, which may have outnumbered them by as much as 9 to 1 according to some estimates.[103] Their lower population density may have also increased Neanderthal susceptibility to mutations caused by inbreeding.[104]

11.7.1 Violence

The St. Césaire 1 skeleton discovered in 1979 at La Roche à Pierrot, France, showed a healed fracture on top of the skull apparently caused by a deep blade wound. Researchers have taken this as evidence of the presence of interpersonal violence among the Neanderthals.[105]

11.8 Genome

Further information: Neanderthal genome project

11.8.1 Background

Early investigations concentrated on mitochondrial DNA (mtDNA), which, owing to strictly matrilineal inheritance and subsequent vulnerability to genetic drift, is of limited value in evaluating the possibility of interbreeding of Neanderthals with Cro-Magnon people.

In 1997, geneticists were able to extract a short sequence of DNA from Neanderthal bones.[106] The extraction of mtDNA from a second specimen was reported in

2000, and showed no sign of modern human descent from Neanderthals.[59]

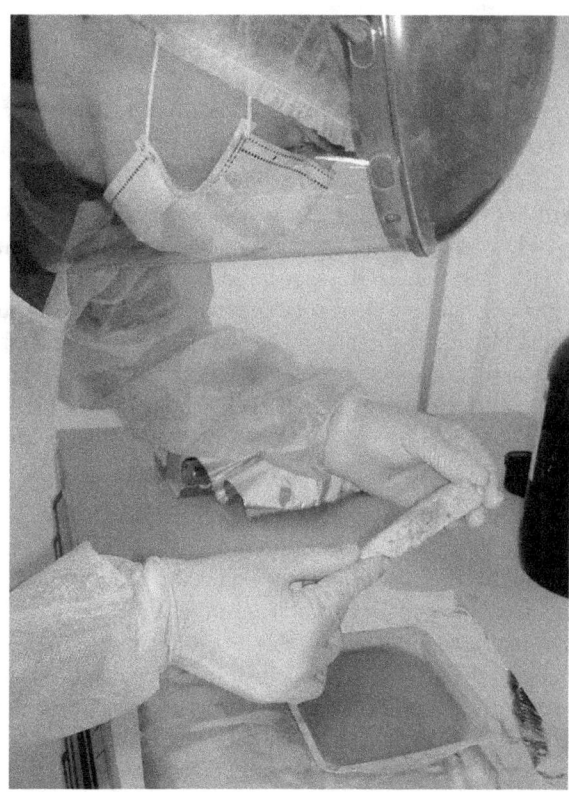

Scientist at the Max Planck Institute for Evolutionary Anthropology extracting the DNA

In July 2006, the Max Planck Institute for Evolutionary Anthropology and 454 Life Sciences announced that they would sequence the Neanderthal genome over the next two years. This genome was expected to be roughly the size of the human genome, three-billion base pairs, and share most of its genes. It was hoped the comparison would expand understanding of Neanderthals, as well as the evolution of humans and human brains.[107]

Svante Pääbo has tested more than 70 Neanderthal specimens. The Neanderthal genome is almost the same size as the human genome and is identical to ours to a level of 99.7% by comparing the accurate order of the nitrogenous bases in the double nucleotide chain.[108] From mtDNA analysis estimates, the two species shared a common ancestor about 500,000 years ago. An article[109] appearing in the journal *Nature* has calculated the species diverged about 516,000 years ago, whereas fossil records show a time of about 400,000 years ago.[110] A 2007 study pushes the point of divergence back to around 800,000 years ago.[111]

Edward Rubin of the Lawrence Berkeley National Laboratory states recent genome testing of Neanderthals suggests human and Neanderthal DNA are some 99.5% to nearly

99.9% identical.[112][113]

11.8.2   Interbreeding with modern humans

On 16 November 2006, Lawrence Berkeley National Laboratory issued a press release suggesting Neanderthals and ancient humans probably did not interbreed.[114] Edward M. Rubin, director of the U.S. Department of Energy's Lawrence Berkeley National Laboratory and the Joint Genome Institute (JGI), sequenced a fraction (0.00002) of genomic nuclear DNA (nDNA) from a 38,000-year-old Vindia Neanderthal femur. They calculated the common ancestor to be about 353,000 years ago, and a complete separation of the ancestors of the species about 188,000 years ago.[115]

Their results show the genomes of modern humans and Neanderthals are at least 99.5% identical, but despite this genetic similarity, and despite the two species having coexisted in the same geographic region for thousands of years, Rubin and his team did not find any evidence of any significant interbreeding between the two. Rubin said, "While unable to definitively conclude that interbreeding between the two species of humans did not occur, analysis of the nuclear DNA from the Neanderthal suggests the low likelihood of it having occurred at any appreciable level."[115]

In 2008 Richard E. Green et al. from Max Planck Institute for Evolutionary Anthropology in Leipzig, Germany, published the full sequence of Neanderthal mitochondrial DNA (mtDNA) and suggested "Neanderthals had a long-term effective population size smaller than that of modern humans."[116] Writing in *Nature* about Green et al.'s findings, James Morgan asserted the mtDNA sequence contained clues that Neanderthals lived in "small and isolated populations, and probably did not interbreed with their human neighbours."[17][18]

In the same publication, it was disclosed by Svante Pääbo that in the previous work at the Max Planck Institute, "Contamination was indeed an issue," and they eventually realized that 11% of their sample was modern human DNA.[117][118] Since then, more of the preparation work has been done in clean areas and 4-base pair 'tags' have been added to the DNA as soon as it is extracted so the Neanderthal DNA can be identified.

With 3 billion nucleotides sequenced, analysis of about ⅓ showed no sign of admixture between modern humans and Neanderthals, according to Pääbo. This concurred with the work of Noonan from two years earlier. The variant of microcephalin common outside Africa, which was suggested to be of Neanderthal origin and responsible for rapid brain growth in humans, was not found in Neanderthals. Nor was the MAPT variant, a very old variant found primarily in Europeans.[117]

However, an analysis of a first draft of the Neanderthal genome by the same team released in May 2010 indicates interbreeding may have occurred.[19][61] "Those of us who live outside Africa carry a little Neanderthal DNA in us," said Pääbo, who led the study. "The proportion of Neanderthal-inherited genetic material is about 1 to 4 percent. It is a small but very real proportion of ancestry in non-Africans today," says Dr. David Reich of Harvard Medical School, who worked on the study. This research compared the genome of the Neanderthals to five modern humans from China, France, sub-Saharan Africa, and Papua New Guinea. The finding is that about 1 to 4 percent of the genes of the non-Africans came from Neanderthals, compared to the baseline defined by the two Africans.[61]

This indicates a gene flow from Neanderthals to modern humans, i.e., interbreeding between the two populations. Since the three non-African genomes show a similar proportion of Neanderthal sequences, the interbreeding must have occurred early in the migration of modern humans out of Africa, perhaps in the Middle East. No evidence for gene flow in the direction from modern humans to Neanderthals was found. Gene flow from modern humans to Neanderthals would not be expected if contact occurred between a small colonizing population of modern humans and a much larger resident population of Neanderthals. A very limited amount of interbreeding could explain the findings, if it occurred early enough in the colonization process.[61]

While interbreeding is viewed as the most parsimonious interpretation of the genetic discoveries, the authors point out they cannot conclusively rule out an alternative scenario, in which the source population of non-African modern humans was already more closely related to Neanderthals than other Africans were, because of ancient genetic divisions within Africa.[61] Other studies carried out since the sequencing of the Neanderthal genome have cast doubt on the level of admixture between Neanderthals and modern humans, or even as to whether the species interbred at all. One study has asserted that the presence of Neanderthal or other archaic human genetic markers can be attributed to shared ancestral traits between the species originating from a 500,000-year-old common ancestor.[119][120][121]

Among the genes shown to differ between present-day humans and Neanderthals were *RPTN*, *SPAG17*, *CAN15*, *TTF1* and *PCD16*.[61]

11.8.3   Epigenetics

In April 2014, a first glimpse into the epigenetics of the Neanderthal was obtained with the publication of the full DNA methylation of the Neanderthal and the Denisovan.[122] The reconstructed DNA methylation map allowed researchers to assess gene activity levels throughout the Neanderthal genome and compare them to modern humans. One of the major findings focused on the limb morphology of Neanderthals. Gokhman et al. found that changes in the activity levels of the HOX cluster of genes were behind many of the morphological differences between Neanderthals and modern humans, including shorter limbs, curved bones and more.[122]

11.9 Extinction hypotheses

Main article: Neanderthal extinction

As the 2014 study by Thomas Higham of Neanderthal bones and tools indicates that Neanderthals died out in Europe between 41,000 and 39,000 years ago, and that *Homo sapiens* arrived in Europe between 45,000 and 43,000 years ago, it is now apparent that the two different human populations shared Europe for as long as 5,000 years.[67] The exact nature of biological and cultural interaction between Neanderthals and other human groups has been contested.[123]

Possible scenarios for the extinction of the Neanderthals are:

1. Neanderthals were a separate species from modern humans, and became extinct (because of climate change or interaction with modern humans) and were replaced by modern humans moving into their habitat between 45,000 and 40,000 years ago.[124] Jared Diamond has suggested a scenario of violent conflict and displacement.[125]

2. Neanderthals were a contemporary subspecies that bred with modern humans and disappeared through absorption (interbreeding theory).

As Paul Jordan notes: "A natural sympathy for the underdog and the disadvantaged lends a sad poignancy to the fate of the Neanderthal folk, however it came about." Jordan, though, does say that there was perhaps interbreeding to some extent, but that populations that remained totally Neanderthal were probably out-competed and marginalized to extinction by the Aurignacians.[71]

11.9.1 Climate change

About 55,000 years ago, the weather began to fluctuate wildly from extreme cold conditions to mild cold and back in a matter of a few decades. Neanderthal bodies were well suited for survival in a cold climate—their barrel chests and

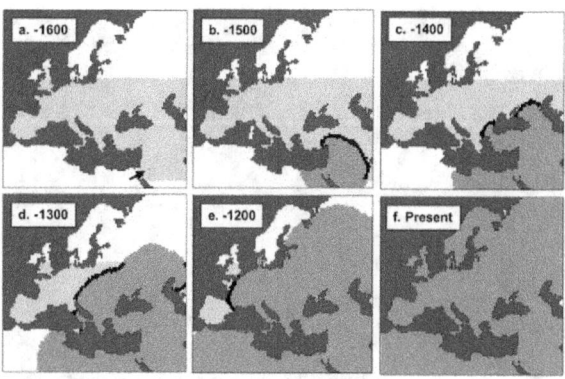

mtDNA-based simulation of modern human expansion in Europe starting 1600 generations ago. Neanderthal range in light grey[126]

stocky limbs stored body heat better than the Cro-Magnons. However, the rapid fluctuations of weather caused ecological changes to which the Neanderthals could not adapt; familiar plants and animals would be replaced by completely different ones within a lifetime. Neanderthals' ambush techniques would have failed as grasslands replaced trees. Neanderthals died out in Europe between 41,000 and 39,000 years ago which coincides with the start of a very cold period.[67][127] Raw material sourcing and the examination of faunal remains by Adler et al. (2006) in the southern Caucasus region suggest that modern humans may have had a survival advantage during this period, being able to use social networks to acquire resources from a greater area. They found that in both the Late Middle Palaeolithic and Early Upper Palaeolithic more than 95% of stone artifacts were drawn from local material, suggesting Neanderthals were restricted to more local resources. Furthermore, excavations at Ortvale Klde Rockshelter discovered that there was a clear break between the Late Middle Paleolithic and the Early Upper Paleolithic lithic assemblages, which were attributed to Neanderthals and modern humans respectively. This would suggest that modern humans came in and replaced Neanderthals, rather than a slow shift or integration occurring in this region. [128]

Studies on Neanderthal body structures have shown that they needed more energy to survive than any other species of hominid. Their energy needs were up to 100 to 350 kcal (420 to 1,460 kJ) more per day comparing to projected anatomically modern human males weighing 68.5 kg (151 lb) and females 59.2 kg (131 lb).[129] When food became scarce, this difference may have played a major role in the Neanderthals' extinction.[127]

11.9.2 Coexistence with *Homo sapiens*

In November 2011 tests conducted at the Oxford Radiocarbon Accelerator Unit in England on what were previ-

Skeleton and reconstruction of the La Ferrassie 1 Neanderthal man from the National Museum of Nature and Science

Christopher Columbus to the New World, which brought and introduced foreign diseases when he and his crew arrived to a native population who had no immunity.

Anthropologist Pat Shipman, of Pennsylvania State University, suggested that domestication of the dog could have played a role in Neanderthals' extinction.[132]

11.10   Interbreeding hypotheses

Main article: Archaic human admixture with modern humans

An alternative to extinction is that Neanderthals were ab-

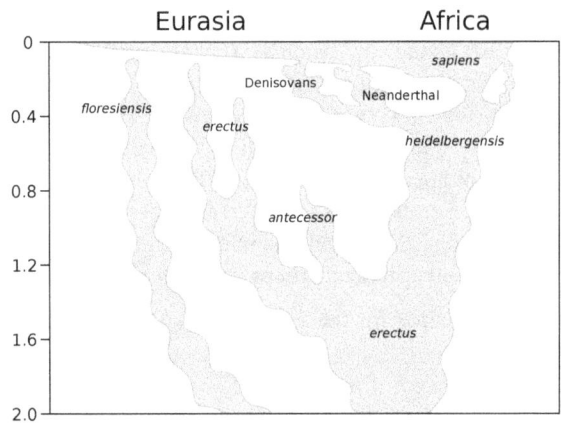

Chris Stringer's hypothesis of the family tree of genus Homo, *published 2012 in* Nature: Homo floresiensis *originated in an unknown location from unknown ancestors and reached remote parts of Indonesia.* Homo erectus *spread from Africa to western Asia, then east Asia and Indonesia; its presence in Europe is uncertain, but it gave rise to* Homo antecessor, *found in Spain.* Homo heidelbergensis *originated from* Homo erectus *in an unknown location and dispersed across Africa, southern Asia and southern Europe (other scientists interpret fossils, here named* heidelbergensis, *as late* erectus). Homo sapiens *spread from Africa to western Asia and then to Europe and southern Asia, eventually reaching Australia and the Americas. In addition to Neanderthals and Denisovans, a third gene flow of archaic Africa origin is indicated at the right.[133]*

ously thought to be Neanderthal baby teeth, which had been unearthed in 1964 from the Grotta del Cavallo in Italy, were identified as the oldest modern human remains discovered anywhere in Europe, dating from between 43,000 and 45,000 years ago.[130] Given that the 2014 study by Thomas Higham of Neanderthal bones and tools indicates that Neanderthals died out in Europe between 41,000 and 39,000 years ago, the two different human populations shared Europe for as long as 5,000 years.[67] The exact nature of biological and cultural interaction between Neanderthals and other human groups has been contested.[123]

Modern humans co-existed with them in Europe starting around 45,000 years ago and perhaps even earlier. Neanderthals inhabited that continent for a long period of time before the arrival of modern humans. H. sapiens may have introduced a disease that contributed to the extinction of Neanderthals, and that may be added to other recent explanations for their extinction. When Neanderthal ancestors left Africa roughly 100,000 years earlier they adapted to the pathogens in their European environment, unlike modern humans who adapted to African pathogens. This transcontinental movement is known as the Out of Africa model. If contact between humans and Neanderthals occurred in Europe and Asia the first contact may have been devastating to the Neanderthal population, because they would have little if any immunity to the African pathogens. More recent historical events in Eurasia and the Americas show a similar pattern, where the unintentional introduction of viral, or bacterial pathogens to unprepared populations has led to mass mortality and local population extinction.[131] The most well known example of this is the arrival of

sorbed into the Cro-Magnon population by interbreeding. This would be counter to strict versions of the Recent African Origin, since it would imply that at least part of the genome of Europeans would descend from Neanderthals.

Hans Peder Steensby, while strongly emphasising that all modern humans are of mixed origins, proposed the interbreeding hypothesis in 1907, in the article *Race studies in Denmark*.[134] He held that this would best fit current observations, and attacked the widespread idea that Neanderthals were ape-like or inferior.

The most vocal proponent of the hybridization hypothesis

is Erik Trinkaus of Washington University.[135] Trinkaus claims various fossils as products of hybridized populations, including the child of Lagar Velho, a skeleton found at Lagar Velho in Portugal.[136][137] In a 2006 publication co-authored by Trinkaus, the fossils found in 1952 in the cave of Peştera Muierii, Romania, are likewise claimed as descendants of previously hybridized populations.[138]

Genetic research has asserted that some admixture took place.[139] The genomes of all non-Africans include portions that are of Neanderthal origin,[140][141] due to interbreeding between Neanderthals and the ancestors of Eurasians in Northern Africa or the Middle East prior to their spread. Rather than absorption of the Neanderthal population, this gene flow appears to have been of limited duration and limited extent. An estimated 1 to 4 percent of the DNA in Europeans and Asians (French, Chinese and Papua probands) is non-modern, and shared with ancient Neanderthal DNA rather than with Sub-Saharan Africans (Yoruba people and San probands).[61] Ötzi the iceman, Europe's oldest preserved mummy, was found to possess an even higher percentage of Neanderthal ancestry.[142] Recent findings suggest there may be even more Neanderthal genes in non-African humans than previously expected: approximately 20% of the Neanderthal gene pool was present in a broad sampling of non-African individuals, though each individual's genome was on average only 2% Neanderthal.[143]

More recent genetic studies seem to suggest that modern humans may have mated with "at least two groups" of ancient humans: Neanderthals and Denisovans.[144] Some researchers suggest admixture of 3.4–7.9% in modern humans of non-African ancestry, rejecting the hypothesis of ancestral population structure.[145] Detractors have argued and continue to argue that the signal of Neanderthal interbreeding may be due to ancient African substructure, meaning that the similarity is only a remnant of a common ancestor of both Neanderthals and modern humans and not the result of interbreeding.[146][147] John D. Hawks has argued that the genetic similarity to Neanderthals may indeed be the result of both structure and interbreeding, as opposed to just one or the other.[148]

While modern humans share some nuclear DNA with the extinct Neanderthals, the two species do not share any mitochondrial DNA,[56] which in primates is always maternally transmitted. This observation has prompted the hypothesis that whereas female humans interbreeding with male Neanderthals were able to generate fertile offspring, the progeny of female Neanderthals who mated with male humans were either rare, absent or sterile.[149] However, some researchers have argued that there is evidence of possible interbreeding between female Neanderthals and male modern humans.[150][151]

11.11 Specimens

Main article: List of human evolution fossils

11.11.1 Notable specimens

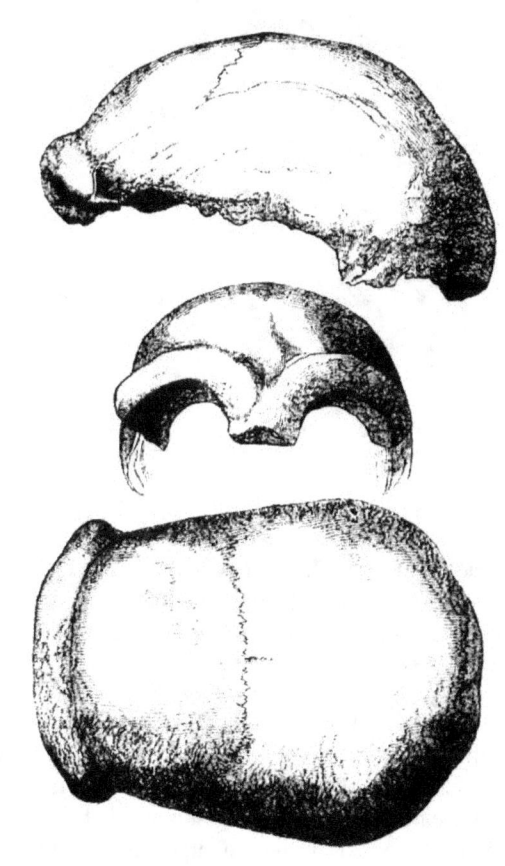

Type Specimen, Neanderthal 1

- Neanderthal 1: The first Neanderthal specimen found during mining in August 1856. It was discovered in a limestone quarry at the Feldhofer grotto in Neanderthal, Germany. The find consisted of a skull cap, two femora, three right arm bones, two left arm bones, ilium, and fragments of a scapula and ribs.

- La Chapelle-aux-Saints 1: Called the Old Man, a fossilized skull discovered in La Chapelle-aux-Saints, France, by A. and J. Bouyssonie, and L. Bardon in 1908. Characteristics include a low vaulted cranium and large browridge typical of Neanderthals. Estimated to be about 60,000 years old, the specimen was severely arthritic and had lost all his teeth, with evidence of healing. For him to have lived on would have required that someone process his food for him, one of the earliest examples of Neanderthal altruism (similar to Shanidar I.)

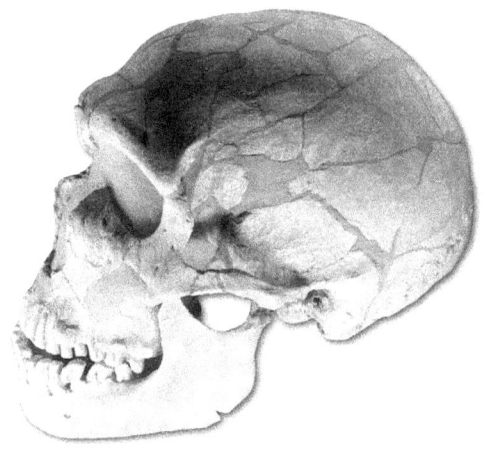

The Ferrassie skull

- La Ferrassie 1: A fossilized skull discovered in La Ferrassie, France, by R. Capitan in 1909. It is estimated to be 70,000 years old. Its characteristics include a large occipital bun, low-vaulted cranium and heavily worn teeth.

- Le Moustier: A fossilized skull, discovered in 1909, at the archaeological site in Peyzac-le-Moustier, Dordogne, France. The Mousterian tool culture is named after Le Moustier. The skull, estimated to be less than 45,000 years old, includes a large nasal cavity and a somewhat less developed brow ridge and occipital bun as might be expected in a juvenile.

- Shanidar Cave: Found in the Zagros Mountains in (Iraqi Kurdistan); a total of nine skeletons found believed to have lived in the Middle Paleolithic. One of the nine remains was missing part of its right arm, which is theorized to have been broken off or amputated. The find is also significant because it shows that stone tools were present among this tribe's culture. One of the skeletons was originally thought to have been buried with flowers, signifying that some type of burial ceremony may have occurred. This is no longer considered to be the case, and Paul B. Pettitt has stated that the "deliberate placement of flowers has now been convincingly eliminated", noting that "A recent examination of the microfauna from the strata into which the grave was cut suggests that the pollen

was deposited by the burrowing rodent Meriones tersicus, which is common in the Shanidar microfauna and whose burrowing activity can be observed today".[152]

- Amud 1: Fossilized remains of an adult Neanderthal, dated to roughly 45,000 years ago, and one of several found in a cave at Nahal Amud, Israel, at least some of which may have been deliberately buried. A particularly notable feature of this find is its cranial capacity, which, at 1,740 cm^3, is among the largest known for any hominid, living or extinct.[46][153]

11.11.2   Chronology

This section describes bones with Neanderthal traits in chronological order.

Mixed with *H. heidelbergensis* traits

- > 350 ka: Sima de los Huesos c. 500:350 ka ago[154][155]

- 350–200 ka: Pontnewydd 225 ka ago.

- 200–135 ka: Atapuerca,[156] Vértesszőlős, Ehringsdorf, Casal de'Pazzi, Biache, La Chaise, Montmaurin, Prince, Lazaret, Fontéchevade

Typical *H. neanderthalensis* traits

- 135–45 ka: Krapina, Saccopastore skulls, Malarnaud, Altamura, Gánovce, Denisova, Okladnikov, Altai Mountains, Pech de l'Azé, Tabun 120–100±5 ka,[157] Qafzeh9 100, Shanidar 1 to 9 80–60 ka, La Ferrassie 1 70 ka, Kebara 60 ka, Régourdou, Mt. Circeo, Combe Grenal, Erd 50 ka, La Chapelle-aux Saints 1 60 ka, Amud I 53±8 ka,[158][159] Teshik-Tash.

- 45–35 ka: Le Moustier 45 ka, Feldhofer 42 ka, La Quina, l'Horus, Hortus, Kulna, Šipka, Saint Césaire, Bacho Kiro, El Castillo, Bañolas, Arcy-sur-Cure.[160]

- < 35 ka: Châtelperron, Figueira Brava, Zafarraya 30 ka,[160] Vogelherd 3?,[161] Vindija 32,400±800 14C B.P.[162] (Vi-208 31,390±220, Vi-207 32,400±1,800 14C B.P.),[162] Velika Pećina,

Homo sapiens with some neanderthal-like archaic traits

- < 35 Pestera cu Oase 35 ka, Mladeč 31 ka, Pestera Muierii 30 ka (n/s),[163] Lapedo Child 24.5 ka.

11.12 Popular culture

Main article: Neanderthals in popular culture

Neanderthals have been portrayed in popular culture including appearances in literature, visual media and comedy, often in an unflattering and inaccurate light.

Early artistic reconstructions mostly presented Neanderthals as beastly creatures, emphasizing hairiness and rough, dark complexion.[164] More recent reconstructions acknowledge that because of the lineage evolution in European latitude there is reason to believe that Neanderthals were fair-skinned and probably with no more facial hair than modern man. Archaeological evidence exists indicating that they probably communicated by speech and used tools. Artist renderings and reconstructions of Neanderthals have become much more intelligent-looking and closely resembling modern humans.[165][166]

11.13 See also

* Abrigo do Lagar Velho—More about "the Lapedo child"

* Almas: wild man of Mongolia

* Altamura Man

* Basajaun

* Biological anthropology

* Caveman

* *Dawn of Humanity* (2015 PBS film)

* Denisova hominin another *Homo* species that may have interbred with Neanderthals

* Early human migrations

* Engis 2

* Max Planck Institute for Evolutionary Anthropology

* Neanderthal 1

* Neanderthal Museum

* Neanderthals of Gibraltar

* Pleistocene megafauna

* Roca dels Bous (archaeological site)

* Species problem

Lists:

* List of fossil sites *(with link directory)*

* List of human evolution fossils *(with images)*

* List of Neanderthal sites

11.14 Notes

[1] *Tal* after the German spelling reform of 1901, whence the German name *Neandertal* for both the valley and species.

11.15 References

[1] Dictionary of Anthropology—Charles Winick - Google Books. Books.google.ca (1956-12-18). Retrieved on 2014-05-24.

[2] Bibliography of Fossil Vertebrates 1954-1958 - C.L. Camp, H.J. Allison, and R.H. Nichols - Google Books. Books.google.ca. Retrieved on 2014-05-24.

[3] "Neanderthal in ODE". *Oxford Dictionaries*.

[4] ""Neanderthal" in Random House Dictionary (US) & Collins Dictionary (UK)". *Dictionary.com*.

[5] Colin P.T. Baillie; *University of California, Berkeley*. "Neandertals: Unique from Humans, or Uniquely Human?" (PDF). berkeley.edu.

[6] *Smithsonian Institution National Museum of American History*. "Ancient DNA and Neanderthals". si.edu.

[7] Gokhman, David; et al. (2014). "Reconstructing the DNA methylation maps of the Neandertal and the Denisovan". *Science* **344** (6183): 523–527. doi:10.1126/science.1250368.

[8] Hublin, J. J. (2009). "The origin of Neandertals". *Proceedings of the National Academy of Sciences* **106** (38): 16022–7. Bibcode:2009PNAS..10616022H. doi:10.1073/pnas.0904119106. JSTOR 40485013. PMC 2752594. PMID 19805257.

[9] Harvati, K.; Frost, S.R.; McNulty, K.P. (2004). "Neanderthal taxonomy reconsidered: implications of 3D primate models of intra- and interspecific differences". *Proc. Natl. Acad. Sci. U.S.A.* **101**: 1147–52. doi:10.1073/pnas.0308085100. PMC 337021. PMID 14745010. Retrieved 2015-02-06.

[10] "Scientists Identify Neanderthal Genes in Modern Human DNA". Sci-News.com. January 30, 2014. Retrieved October 29, 2015.

[11] Skinner, A., B. Blackwell, R. Long, M.R. Seronie-Vivien, A.-M. Tillier and J. Blickstein; New ESR dates for a new bone-bearing layer at Pradayrol, Lot, France; Paleoanthropology Society March 28, 2007

[12] Finlayson, C; Pacheco, FG; Rodríguez-Vidal, J; Fa, DA; Gutierrez, López, JM; Santiago, Pérez, A; Finlayson, G; Allue, E; Baena, Preysler, J; Cáceres, I; Carrión, JS; Fernández, Jalvo, Y; Gleed-Owen, CP; Jimenez, Espejo, FJ; López, P; López, Sáez, JA; Riquelme, Cantal, JA; Sánchez, Marco, A; Guzman, FG; Brown, K; Fuentes, N; Valarino, CA; Villalpando, A; Stringer, CB; Martinez, Ruiz, F; Sakamoto, T (October 2006). "Late survival of Neanderthals at the southernmost extreme of Europe". *Nature* **443** (7113): 850–3. Bibcode:2006Natur.443..850F. doi:10.1038/nature05195. ISSN 0028-0836. PMID 16971951.

[13] Outside Europe, Mousterian tools were made by both Neanderthals and early modern *Homo sapiens*. (Donald Johanson & Blake Edgar (2006) *From Lucy to Language*, Simon & Schuster, p. 272 |url=https://books.google.com/books?id=-VKEjAbpggcC&pg=PT272&lpg=PT2720&focus=viewport&vq=Mousterian+tools&dq=From+Lucy+to+Language+%22Mousterian+tools%22#v=onepage&q=Mousterian%20tools&f=false)

[14] Mellars, Paul (2007). *Rethinking the Human Revolution: New Behavioural and Biological Perspectives on the Origin and Dispersal of Modern Humans*. McDonald Institute for Archaeological Research. p. 143. Brain size of course scales with body size, and more massive human body sizes are found at higher latitudes, in conformance with Bergmann's rule (Ruff et al. 1997; Ruff 2002). As a result Neanderthal brain size and body weight are large compared to Homo sapiens.

[15] "Neanderthal man". infoplease.

[16] Helmuth H (1998). "Body height, body mass and surface area of the Neanderthals". *Zeitschrift Für Morphologie Und Anthropologie* **82** (1): 1–12. PMID 9850627.

[17] Evans PD, Mekel-Bobrov N, Vallender EJ, Hudson RR, Lahn BT (November 2006). "Evidence that the adaptive allele of the brain size gene *microcephalin* introgressed into *Homo sapiens* from an archaic *Homo* lineage". *Proceedings of the National Academy of Sciences* **103** (48): 18178–83. Bibcode:2006PNAS..10318178E. doi:10.1073/pnas.0606966103. PMC 1635020. PMID 17090677.

[18] Evans PD, Gilbert SL, Mekel-Bobrov N, Vallender EJ, Anderson JR, Vaez-Azizi LM, Tishkoff SA, Hudson RR, Lahn BT (September 2005). "Microcephalin, a gene regulating brain size, continues to evolve adaptively in humans". *Science* **309** (5741): 1717–20. Bibcode:2005Sci...309.1717E. doi:10.1126/science.1113722. PMID 16151009.

[19] Rincon, Paul (2010-05-06). "Neanderthal genes 'survive in us'". BBC News. Retrieved 2010-05-07.

[20] "Genome sequence of a 45,000-year-old modern human from western Siberia". *Nature* **514** (7523): 445–449. 23 October 2014. doi:10.1038/nature13810. PMID 25341783.

[21] Brahic, Catherine. "Humanity's forgotten return to Africa revealed in DNA", *The New Scientist* (February 3, 2014).

[22] Wilford, John Noble (December 16, 2013). "Neanderthals and the Dead". *New York Times*. Retrieved December 17, 2013.

[23] http://www.washingtonpost.com/national/health-science/neanderthal-genes-found-in-modern-humans/2014/01/29/f7f81852-8774-11e3-a5bd-844629433ba3_story.html

[24] Zimmer, Carl (December 18, 2013). "Toe Fossil Provides Complete Neanderthal Genome". *New York Times*. Retrieved December 18, 2013.

[25] Prüfer, Kay; et al. (2014). "The complete genome sequence of a Neanderthal from the Altai Mountains". *Nature* **505** (7481): 43–49. doi:10.1038/nature12886.

[26] The valley is named after Joachim Neander, whose Greek last name coincidentally means "new man" in English. *See* Kunzig, Robert. "The Year in Science: Human Origins 1997", *Discover (magazine)* (January 1, 1998) reprinted in *Contemporary Readings in Physical Anthropology*, p. 145 (Alan Almquist ed., Prentice Hall, 2000).

[27] Howell, F. Clark (1957). "The evolutionary significance of variation and varieties of 'Neanderthal' man". *The Quarterly Review of Biology* **32** (4): 330–47. doi:10.1086/401978. JSTOR 2816956. PMID 13506025.

[28] Foley, Tim. TalkOrigins Archive. "Neanderthal or Neandertal?". 2005.

[29] Vogt, Karl C (1864). *Lectures on Man: His Place in Creation, and in the History of the Earth*. London: Longman, Green, Longman and Roberts. pp. 302, 473.

[30] King, William (Jan 1864). "The Reputed Fossil Man of the Neanderthal" (PDF). *The Quarterly Journal of Science* **1**: 96.

[31] *Inter alia, Boys' Life*, p. 18. January 1924.

[32] *The Oxford Illustrated Dictionary*. Great Britain: Oxford University Press. 1976 [1975]. p. 564. (tahl)

[33] "Neanderthal adjective—definition in British English Dictionary & Thesaurus—Cambridge Dictionary Online". Dictionary.cambridge.org. 2013-01-08. Retrieved 2013-01-22.

[34] "Oxford Learner's Dictionaries—Find pronunciation, clear meanings and definitions of words at OxfordLearnersDictionaries.com".

[35] "Neanderthal | Define Neanderthal at Dictionary.com". Dictionary.reference.com. Retrieved 2013-01-22.

[36] Kurtén, Björn (10 October 1995). *Dance of the Tiger: A Novel of the Ice Age*. University of California Press. pp. xxi. ISBN 0-520-20277-5. Retrieved 9 May 2012.

[37] Pollet, Carl J. (September 21, 1991). "...And Etymology". *Science News* **140** (12): 191. doi:10.2307/3975867. JSTOR 3975867.

[38] Tattersall, Ian; Schwartz, Jeffrey H. (1999). "Hominids and hybrids: The place of Neanderthals in human evolution". *Proceedings of the National Academy of Sciences* **96** (13): 7117–9. Bibcode:1999PNAS...96.7117T. doi:10.1073/pnas.96.13.7117. JSTOR 48019. PMC 33580. PMID 10377375.

[39] Duarte, Cidália; Mauricio, João; Pettitt, Paul B.; Souto, Pedro; Trinkaus, Erik; Van Der Plicht, Hans; Zilhao, João (1999). "The early Upper Paleolithic human skeleton from the Abrigo do Lagar Velho (Portugal) and modern human emergence in Iberia". *Proceedings of the National Academy of Sciences* **96** (13): 7604–9. Bibcode:1999PNAS...96.7604D. doi:10.1073/pnas.96.13.7604. JSTOR 48106. PMC 22133. PMID 10377462.

[40] Harvati, K.; Frost, S.R.; McNulty, K.P. (February 2004). "Neanderthal taxonomy reconsidered: Implications of 3D primate models of intra- and interspecific differences". *Proceedings of the National Academy of Sciences* **101** (5): 1147–52. Bibcode:2004PNAS..101.1147H. doi:10.1073/pnas.0308085100. PMC 337021. PMID 14745010.

[41] "Research supports Neanderthals as a separate species". *Archaeology News from Past Horizons*.

[42] "Modern humans, Neanderthals shared earth for 1,000 years". ABC News (Australia). 1 September 2005. Retrieved 19 September 2006.

[43] Hedges SB (December 2000). "Human evolution. A start for population genomics". *Nature* **408** (6813): 652–3. doi:10.1038/35047193. PMID 11130051.

[44] "Palaeontology: How Neanderthals evolved". THE FINANCIAL TIMES LTD 2015. June 27, 2014. Retrieved October 28, 2015.

[45] Bischoff, James L.; Shamp, Donald D.; Aramburu, Arantza; Arsuaga, Juan Luis; Carbonell, Eudald; Bermudez de Castro, J.M. (2003). "The Sima de los Huesos Hominids Date to Beyond U/Th Equilibrium (>350kyr) and Perhaps to 400–500kyr: New Radiometric Dates". *Journal of Archaeological Science* **30** (3): 275–80. doi:10.1006/jasc.2002.0834.

[46] Papagianni, Dmitra; Morse, Michael (2013). *The Neanderthals Rediscovered*. Thames & Hudson. ISBN 978-0-500-05177-1.

[47] Higham, T.; et al. (2014). "The timing and spatiotemporal patterning of Neanderthal disappearance". *Nature* **512** (7514): 306–309. doi:10.1038/nature13621.

[48] "*Homo heidelbergensis*: Evolutionary Tree". Smithsonian National Museum of Natural History. Retrieved 17 March 2013.

[49] Stringer, Chris. "The Ancient Human Occupation of Britain" (PDF). Natural History Museum, London. Retrieved 17 March 2013.

[50] Stringer, Chris (2011). *The Origin of our Species*. Penguin. pp. 26–29, 202. ISBN 978-0-141-03720-2.

[51] Johansson, Donald; Edgar, Blake (2006). *From Lucy to Language*. Simon & Schuster. p. 38. ISBN 978-0-7432-8064-8.

[52] "Homo neanderthalensis". Smithsonian Institution. Retrieved 18 May 2009.

[53] "New Evidence On The Role Of Climate In Neanderthal Extinction". Science Daily.

[54] "Universitat Pompeu Fabra, Barcelona, A paleogenetical study determines the blood group of Neanderthal man"

[55] "Elsevier GmbH, Palaeogenetic research at the El Sidrón Neanderthal site"

[56] Krings, M; Stone, A; Schmitz, RW; Krainitzki, H; Stoneking, M; Pääbo, S (1997). "Neandertal DNA sequences and the origin of modern humans". *Cell* **90** (1): 19–30. doi:10.1016/S0092-8674(00)80310-4. ISSN 0092-8674. PMID 9230299.

[57] Schmitz, Ralf W; et al. (2002). "The Neandertal type site revisited: Interdisciplinary investigations of skeletal remains from the Neander Valley, Germany". *Proceedings of the National Academy of Sciences* **99** (20): 13342–13347. doi:10.1073/pnas.192464099.

[58] Duarte, C; Maurício, J; Pettitt, PB; Souto, P; Trinkaus, E; Van Der Plicht, H; Zilhão, J; et al. (1999). "The early Upper Paleolithic human skeleton from the Abrigo do Lagar Velho (Portugal) and modern human emergence in the Iberian Peninsula". *Proceedings of the National Academy of Sciences of the United States of America* (PNAS) **96** (13): 7604–7609. Bibcode:1999PNAS...96.7604D. doi:10.1073/pnas.96.13.7604. PMC 22133. PMID 10377462. Retrieved 2009-06-21.

[59] Ovchinnikov, IV; Götherström, A; Romanova, GP; Kharitonov, VM; Lidén, K; Goodwin, W (2000). "Molecular analysis of Neanderthal DNA from the northern Caucasus". *Nature* **404** (6777): 490–3. doi:10.1038/35006625. PMID 10761915.

[60] Morgan, James (12 February 2009). "Neanderthals 'distinct from us'". BBC News. Retrieved 22 May 2009.

[61] Green, Richard E.; Krause, Johannes; Briggs, Adrian W.; Maricic, Tomislav; Stenzel, Udo; Kircher, Martin; Patterson, Nick; Li, Heng; Zhai, Weiwei; Fritz, Markus Hsi-Yang; Hansen, Nancy F.; Durand, Eric Y.; Malaspinas, Anna-Sapfo; Jensen, Jeffrey D.; Marques-Bonet, Tomas;

Alkan, Can; Prüfer, Kay; Meyer, Matthias; Burbano, Hernán A.; Good, Jeffrey M.; Schultz, Rigo; Aximu-Petri, Ayinuer; Butthof, Anne; Höber, Barbara; Höffner, Barbara; Siegemund, Madlen; Weihmann, Antje; Nusbaum, Chad; Lander, Eric S.; Russ, Carsten (2010). "A Draft Sequence of the Neandertal Genome". *Science* **328** (5979): 710–22. Bibcode:2010Sci...328..710G. doi:10.1126/science.1188021. PMID 20448178.

[62]

[63] Benazzi, S.; Douka, K.; Fornai, C.; Bauer, C. C.; Kullmer, O.; Svoboda, J. Í.; Pap, I.; Mallegni, F.; Bayle, P.; Coquerelle, M.; Condemi, S.; Ronchitelli, A.; Harvati, K.; Weber, G. W. (2011). "Early dispersal of modern humans in Europe and implications for Neanderthal behaviour". *Nature* **479** (7374): 525–528. doi:10.1038/Nature10617. PMID 22048311.

[64] Fergal MacErlean (10 February 2012). "First Neanderthal cave paintings discovered in Spain". *New Scientist*. Retrieved 10 February 2012.

[65] Viegas, Jennifer (27 March 2013). "First Love Child of Human, Neanderthal Found". *Discovery News*. Retrieved 11 April 2013.

[66] "Neanderthals were not less intelligent than modern humans, scientists find". April 30, 2014. Retrieved April 30, 2014.

[67] "BBC News—New dates rewrite Neanderthal story". *BBC News*.

[68] "The Human Lineage by Matt Cartmill, Fred H. Smith". Google Books. Retrieved September 27, 2015.

[69] "Ancient tooth provides evidence of Neanderthal movement" (Press release). Durham University. 11 February 2008. Retrieved 18 May 2009.

[70] O'Neill, Dennis. "Evolution of Modern Humans: Neanderthals", Palomar College, June 10, 2011, accessed August 21, 2011.

[71] Jordan, P. (2001) *Neanderthal: Neanderthal Man and the Story of Human Origins*. The History Press ISBN 978-0-7509-2676-8.

[72] Bilsborough, A, Rae, TC (2015) [REVISION OF] Hominoid cranial diversity and adaptation. In: Henke, W., Rothe, H. & Tattersall, I. (eds.) Handbook of Palaeoanthropology, New York: Springer

[73] "Fieldwork—Jebel Irhoud". Max Planck Institute, Department of Human Evolution.

[74] Arsuaga, J.L; Gracia, A; Martínez, I; Bermúdez de Castro, J.M; Rosas, A; Villaverde, V; Fumanal, M.P (1989). "The human remains from Cova Negra (Valencia, Spain) and their place in European Pleistocene human evolution". *Journal of Human Evolution* **19**: 55–92. doi:10.1016/0047-2484(89)90023-7.

[75] Mallegni, F., Piperno, M., and Segre, A (1987). "Human remains of *Homo sapiens neanderthalensis* from the Pleistocene deposit of Sants Croce Cave, Bisceglie (Apulia), Italy". *American Journal of Physical Anthropology* **72** (4): 421–429. doi:10.1002/ajpa.1330720402. PMID 3111268.

[76] Pavlov P, Roebroeks W, Svendsen JI (2004). "The Pleistocene colonization of northeastern Europe: a report on recent research". *Journal of Human Evolution* **47** (1–2): 3–17. doi:10.1016/j.jhevol.2004.05.002. PMID 15288521.

[77] Wade, Nicholas (2 October 2007). "Fossil DNA Expands Neanderthal Range". *The New York Times*. Retrieved 18 May 2009.

[78] Ravilious, Kate (1 October 2007). "Neandertals Ranged Much Farther East Than Thought". National Geographic Society. Retrieved 18 May 2009.

[79] L'Homme de Neanderthal par Paul Dardé : L'Homme Primitif https://www.academia.edu/11187487/L_Homme_de_Neanderthal_par_Paul_Dard%C3%A9_L_Homme_Primitif

[80] "Science & Nature—Wildfacts—Neanderthal". BBC. Retrieved 2009-06-21.

[81] "Neanderthal". BBC. Retrieved 18 May 2009.

[82] Froehle, Andrew W; Chruchill, Steven E (2009). "Energetic Competition Between Neandertals and Anatomically Modern Humans" (PDF). *PaleoAnthropology*: 96–116. Retrieved 2011-10-31.

[83] Laleuza-Fox, Carles; Römpler, Holger; et al. (2007-10-25). "A Melanocortin 1 Receptor Allele Suggests Varying Pigmentation Among Neanderthals". *Science* **318** (5855): 1453–5. Bibcode:2007Sci...318.1453L. doi:10.1126/science.1147417. PMID 17962522.; see also Rincon, Paul (25 October 2007). "Neanderthals 'were flame-haired'". BBC News. Retrieved 25 October 2007.

[84] "Neanderthal brains focused on vision and movement leaving less room for social networking". Science Daily. March 19, 2013.

[85] SINC Servicio de Información y Noticias Científicas. "El cerebro neandertal era más asimétrico que el del 'Homo sapiens'".

[86] "Neanderthal Brain Size at Birth Sheds Light on Human Evolution". National Geographic. 2008-09-09. Retrieved 2009-09-19.

[87] Silberman, Neil. *The Oxford Companion to Archaeology*, p. 455 (Oxford University Press 2012): "[I]t is with the Neanderthals that we see the full achievement, for the first time, of the degree of encephalization (brain to body size ratio) that characterizes modern humans."

[88] Abramiuk, Marc. *The Foundations of Cognitive Archaeology*, p. 199 (MIT Press 2012): "the encephalization quotient was slightly smaller".

[89] Moskvitch, Katia (2010-09-24). "Neanderthals were able to 'develop their own tools'". *BBC News* (BBC). Retrieved 2010-10-01.

[90] Gray, Richard (December 18, 2011). "Neanderthals built homes with mammoth bones". *Telegraph.co.uk*.

[91] "Evidence suggests Neanderthals took to boats before modern humans".

[92] "Neanderthals were ancient mariners".

[93] "Neanderthals beat modern humans to the seas by 50,000 years, say scientists".

[94] "Ancient Mariners: Did Neanderthals Sail to Mediterranean?".

[95] "Neanderthals May Have Sailed to Crete".

[96] Bocherens, Hervé; Drucker, Dorothée G.; Billiou, Daniel; Patou-Mathis, Marylène; Vandermeersch, Bernard (2005). "Isotopic evidence for diet and subsistence pattern of the Saint-Césaire I Neanderthal: Review and use of a multi-source mixing model". *Journal of Human Evolution* **49** (1): 71–87. doi:10.1016/j.jhevol.2005.03.003. PMID 15869783.

[97] Ghosh, Pallab. "Neanderthals cooked and ate vegetables." BBC News. December 27, 2010.

[98] Lichfield, John (September 30, 2006). "French dig up Neanderthal 'butcher's shop'". The New Zealand Herald.

[99] Richards, Michael P.; Pettitt, Paul B.; Trinkaus, Erik; Smith, Fred H.; Paunović, Maja; Karavanić, Ivor (2000). "Neanderthal diet at Vindija and Neanderthal predation: The evidence from stable isotopes". *Proceedings of the National Academy of Sciences* **97** (13): 7663–6. Bibcode:2000pnas...97.7663r. doi:10.1073/pnas.120178997. JSTOR 122870. PMC 16602. PMID 10852955.

[100] Fiorenza, Luca; Benazzi, Stefano; Tausch, Jeremy; Kullmer, Ottmar; Bromage, Timothy G.; Schrenk, Friedemann (2011). Rosenberg, Karen, ed. "Molar Macrowear Reveals Neanderthal Eco-Geographic Dietary Variation". *PLoS ONE* **6** (3): e14769. doi:10.1371/journal.pone.0014769. PMC 3060801. PMID 21445243.

[101] Henry, A. G.; Brooks, A. S.; Piperno, D. R. (2010). "Microfossils in calculus demonstrate consumption of plants and cooked foods in Neanderthal diets (Shanidar III, Iraq; Spy I and II, Belgium)". *Proceedings of the National Academy of Sciences* **108** (2): 486–491. Bibcode:2011PNAS..108..486H. doi:10.1073/pnas.1016868108.

[102] Webb, Jonathan (25 June 2014). "Oldest human faeces show Neanderthals ate vegetables". BBC News.

[103] Shaw, Kate (July 29, 2011). "Sheer Numbers Gave Early Humans Edge Over Neanderthals". Wired.com.

[104] Vergano, Dan (April 22, 2014). "Neanderthals Lived in Small, Isolated Populations, Gene Analysis Shows". National Geographic.

[105] Zollikofer, Christoph; Marcia, Ponce; Leon, De; Vandermeersch, Bernard; Leveque, Francois (2002). "Evidence for Interpersonal Violence in the St. Césaire Neanderthal". *PNAS* **99** (9): 6444–448. doi:10.1073/pnas.082111899.

[106] Brown, Cynthia Stokes. Big History. New York, NY: The New Press, 2008. Print.

[107] Moulson, Geir; Associated Press (20 July 2006). "Neanderthal genome project launches". MSNBC. Retrieved 22 August 2006.

[108] Lunine 2013, p. 251: "The Neanderthal genome is about the same size as the human genome, and is identical to ours to a level of 99.7% (this is comparing the ordering of the lettering in the nucleotide bases)."

[109] Green RE, Krause J, Ptak SE, et al. (November 2006). "Analysis of one million base pairs of Neanderthal DNA". *Nature* **444** (7117): 330–6. Bibcode:2006Natur.444..330G. doi:10.1038/nature05336. PMID 17108958.

[110] Wade, Nicholas (15 November 2006). "New Machine Sheds Light on DNA of Neanderthals". *The New York Times*. Retrieved 18 May 2009.

[111] Pennisi, E. (May 2007). "Ancient DNA. No sex please, we're Neandertals". *Science* **316** (5827): 967. doi:10.1126/science.316.5827.967a. PMID 17510332.

[112] "Neanderthal bone gives DNA clues". CNN. Associated Press. 16 November 2006. Archived from the original on 18 November 2006. Retrieved 18 May 2009.

[113] Than, Ker; LiveScience (15 November 2006). "Scientists decode Neanderthal genes". MSNBC. Retrieved 18 May 2009.

[114] "Neanderthal Genome Sequencing Yields Surprising Results And Opens A New Door To Future Studies" (Press release). Lawrence Berkeley National Laboratory. 16 November 2006. Retrieved 31 May 2009.

[115] Hayes, Jacqui (15 November 2006). "DNA find deepens Neanderthal mystery". *Cosmos*. Retrieved 18 May 2009.

[116] Green, RE; Malaspinas, AS; Krause, J; Briggs, Aw; Johnson, PL; Uhler, C; Meyer, M; Good, JM; Maricic, T; Stenzel, U; Prüfer, K; Siebauer, M; Burbano, HA; Ronan, M; Rothberg, JM; Egholm, M; Rudan, P; Brajković, D; Kućan, Z; Gusić, I; Wikström, M; Laakkonen, L; Kelso, J; Slatkin, M; Pääbo, S (2008). "A complete Neandertal mitochondrial genome sequence determined by high-throughput sequencing". *Cell* **134** (3): 416–26. doi:10.1016/j.cell.2008.06.021. ISSN 0092-8674. PMC 2602844. PMID 18692465.

[117] Elizabeth Pennisi (2009). "NEANDERTAL GENOMICS: Tales of a Prehistoric Human Genome". *Science* **323** (5916): 866–871. doi:10.1126/science.323.5916.866. PMID 19213888.

[118] Green RE, Briggs AW, Krause J, Prüfer K, Burbano HA, Siebauer M, Lachmann M, Pääbo S. (2009). "The Neandertal genome and ancient DNA authenticity". *EMBO J.* **28** (17): 2494–502. doi:10.1038/emboj.2009.222. PMC 2725275. PMID 19661919.

[119] Neanderthals did not interbreed with humans, scientists find. Telegraph. Retrieved on 2014-05-24.

[120] Neanderthals 'unlikely to have interbred with human ancestors' | Science. The Guardian. Retrieved on 2014-05-24.

[121] "Neanderthal and Denisova genetic affinities with contemporary humans: Introgression versus common ancestral polymorphisms". *Gene* **530**: 83–94. Nov 2013. doi:10.1016/j.gene.2013.06.005. PMID 23872234. Retrieved 2014-05-24.

[122] Gokhman D, Lavi E, Prüfer K, Fraga MF, Riancho JA, Kelso J, Pääbo S, Meshorer E, Carmel L. (2014). "Reconstructing the DNA methylation maps of the Neandertal and the Denisovan.". *Science* **344** (6183): 523–7. doi:10.1126/science.1250368. PMID 24786081.

[123] Finlayson, C., Carrión, J.S. (April 2007). "Rapid ecological turnover and its impact on Neanderthal and other human populations". *Trends in Ecology & Evolution (Personal Edition)* **22** (4): 213–22. doi:10.1016/j.tree.2007.02.001. PMID 17300854.

[124] "First genocide of human beings occurred 30,000 years ago". *Pravda.* 24 October 2007. Retrieved 18 May 2009.

[125] Diamond, Jared M. (1992). *The third chimpanzee: the evolution and future of the human animal.* New York City: HarperCollins. p. 52. ISBN 0-06-098403-1. OCLC 60088352.

[126] Currat, Mathias; Excoffier, Laurent (2004). "Modern Humans Did Not Admix with Neanderthals during Their Range Expansion into Europe". *PLoS Biology* **2** (12): e421. doi:10.1371/journal.pbio.0020421. PMC 532389. PMID 15562317.

[127] "The Mysterious Downfall of the Neandertals", *Scientific American*, August 2009

[128] Adler, Daniel S.; Bar-Oz, Guy; Belfer-Cohen, Anna; Bar-Yosef, Ofer (2006). "Ahead of the Game: Middle and Upper Palaeolithic Hunting Behaviors in the Southern Caucasus". *Current Anthropology* **47** (1): 89–118. doi:10.1086/432455.

[129] Froehle, Andrew W.; Churchill, Steven E. (2009). "Energetic Competition Between Neandertals and Anatomically Modern Humans" (PDF). *PaleoAnthropology*: 96–116.

[130] Wilford, John Noble (November 2, 2011). "Fossil Teeth Put Humans in Europe Earlier Than Thought". *New York Times.* Retrieved August 27, 2014.

[131] Wolff, H. "Result Filters." National Center for Biotechnology Information. U.S. National Library of Medicine, July 2010. Web. 22 Oct. 2014.

[132] How hunting with wolves helped humans outsmart the Neanderthals, The Guardian, 28 February 2015

[133] Stringer, Chris (2012). "Evolution: What makes a modern human". *Nature* **485** (7396): 33–5. Bibcode:2012Natur.485...33S. doi:10.1038/485033a. PMID 22552077.

[134] http://img.kb.dk/tidsskriftdk/pdf/gto/gto_0019-PDF/gto_0019_67206.pdf[]

[135] Dan Jones: *The Neanderthal within.*, New Scientist 193.2007, H. 2593 (3 March), 28–32. Modern Humans, Neanderthals May Have Interbred; Humans and Neanderthals interbred

[136] ; [http://www.guardian.co.uk/science/story/0,,1871842,00.html[]

[137] Not a lasting last for the Neandertals—John Hawks weblog, September 13, 2006

[138] Soficaru, Andrei; Dobos, Adrian; Trinkaus, Erik (2006). "Early modern humans from the Pestera Muierii, Baia de Fier, Romania". *Proceedings of the National Academy of Sciences* **103** (46): 17196–201. Bibcode:2006PNAS..10317196S. doi:10.1073/pnas.0608443103. JSTOR 30052409. PMC 1859909. PMID 17085588.

[139] "Cousins of Neanderthals Left DNA in Africa, Scientists Report". *The New York Times.* July 26, 2012.

[140] Yotova, V.; Lefebvre, J.-F.; Moreau, C.; Gbeha, E.; Hovhannesyan, K.; Bourgeois, S.; Bédarida, S.; Azevedo, L.; Amorim, A.; Sarkisian, T.; Avogbe, P. H.; Chabi, N.; Dicko, M. H.; Kou' Santa Amouzou, E. S.; Sanni, A.; Roberts-Thomson, J.; Boettcher, B.; Scott, R. J.; Labuda, D. (2011). "An X-Linked Haplotype of Neandertal Origin is Present Among All Non-African Populations". *Molecular Biology and Evolution* **28** (7): 1957–62. doi:10.1093/molbev/msr024. PMID 21266489.

[141] "All Non-Africans Part Neanderthal, Genetics Confirm". *DNews.*

[142] Neandertal ancestry "Iced" - John Hawks weblog, August 15, 2012

[143] "Resurrecting Surviving Neandertal Lineages from Modern Human Genomes". Science. 2014-01-29. Retrieved 2014-02-02.

[144] Mitchell, Alanna (January 30, 2012). "DNA Turning Human Story Into a Tell-All". NYTimes. Retrieved January 31, 2012.

[145] Lohse, Konrad; Frantz, Laurent A. F. (2013). "Maximum likelihood evidence for Neandertal admixture in Eurasian populations from three genomes". *Populations and Evolution* **1307**: 8263. arXiv:1307.8263. Bibcode:2013arXiv1307.8263L.

[146] Jha, Alok (14 August 2012). "Study casts doubt on human-Neanderthal interbreeding theory". *The Guardian*. Retrieved 19 February 2015.

[147] Lowery, Robert K.; Uribe, Gabriel; Jimenez, Eric B.; Weiss, Mark A.; Herrera, Kristian J.; Regueiro, Maria; Herrera, Rene J. (2013). "Neanderthal and Denisova genetic affinities with contemporary humans: Introgression versus common ancestral polymorphisms". *Gene* **530** (1): 83–94. doi:10.1016/j.gene.2013.06.005. PMID 23872234.

[148] Hawks, John (2013). "Significance of Neandertal and Denisovan Genomes in Human Evolution". *Annual Review of Anthropology* **42**: 433–49. doi:10.1146/annurev-anthro-092412-155548.

[149] Mason, Paul H.; Short, Roger V. (2011). "Neanderthal-human Hybrids". *Hypothesis* **9**: e1. doi:10.5779/hypothesis.v9i1.215.

[150] Viegas, Jennifer (27 March 2013). "First Love Child of Human, Neanderthal Found". Discovery.

[151] Condemi, Silvana; Mounier, Aurélien; Giunti, Paolo; Lari, Martina; Caramelli, David; Longo, Laura (2013). Frayer, David, ed. "Possible Interbreeding in Late Italian Neanderthals? New Data from the Mezzena Jaw (Monti Lessini, Verona, Italy)". *PLoS ONE* **8** (3): e59781. doi:10.1371/journal.pone.0059781. PMC 3609795. PMID 23544098.

[152] The Neanderthal Dead, exploring mortuary variability in middle paleolithic eurasia. Paul B. Pettitt (2002)

[153] "*Homo neanderthalensis*–The Neanderthals". Australian Museum. Retrieved 26 June 2014.

[154] Bischoff, J; Shamp, Donald D.; Aramburu, Arantza; Arsuaga, Juan Luis; Carbonell, Eudald; Bermudez De Castro, J.M. (2003). "The Sima de los Huesos Hominids Date to Beyond U/Th Equilibrium (>350kyr) and Perhaps to 400–500kyr: New Radiometric Dates". *Journal of Archaeological Science* **30** (3): 275–280. doi:10.1006/jasc.2002.0834.

[155] Arsuaga JL, Martínez I, Gracia A, Lorenzo C (1997). "The Sima de los Huesos crania (Sierra de Atapuerca, Spain). A comparative study". *Journal of Human Evolution* **33** (2–3): 219–81. doi:10.1006/jhev.1997.0133. PMID 9300343.

[156] Kreger, C. David. "Homo neanderthalensis". ArchaeologyInfo.com. Retrieved 16 May 2009.

[157] Mcdermott, F; Grün, R; Stringer, Cb; Hawkesworth, Cj (May 1993). "Mass-spectrometric U-series dates for Israeli Neanderthal/early modern hominid sites". *Nature* **363** (6426): 252–5. Bibcode:1993Natur.363..252M. doi:10.1038/363252a0. ISSN 0028-0836. PMID 8387643.

[158] Rink, W. Jack, H.P. Schwarcz, H.K. Lee, J. Rees-Jones, R. Rabinovich & E. Hovers (August 2002). "Electron spin resonance (ESR) and thermal ionization mass spectrometric (TIMS) 230Th/234U dating of teeth in Middle Paleolithic layers at Amud Cave, Israel". *Geoarchaeology* **16** (6): 701–717. doi:10.1002/gea.1017.

[159] Valladas, Hélène, N. Merciera, L. Frogeta, E. Hoversb, J.L. Joronc, W.H. Kimbeld & Y. Rak (March 1999). "TL Dates for the Neanderthal Site of the Amud Cave, Israel". *Journal of Archaeological Science* **26** (3): 259–268. doi:10.1006/jasc.1998.0334.

[160] Rincon, Paul (13 September 2006). "Neanderthals' 'last rock refuge'". BBC News. Retrieved 18 May 2009.

[161] Conard, Nj; Grootes, Pm; Smith, Fh (July 2004). "Unexpectedly recent dates for human remains from Vogelherd". *Nature* **430** (6996): 198–201. Bibcode:2004Natur.430..198C. doi:10.1038/nature02690. ISSN 0028-0836. PMID 15241412.

[162] Higham T, Ramsey CB, Karavanić I, Smith FH, Trinkaus E (January 2006). "Revised direct radiocarbon dating of the Vindija G1 Upper Paleolithic Neandertals". *Proceedings of the National Academy of Sciences* **103** (3): 553–7. Bibcode:2006PNAS..103..553H. doi:10.1073/pnas.0510005103. PMC 1334669. PMID 16407102.

[163] Hayes, Jacqui (2 November 2006). "Humans and Neanderthals interbred". *Cosmos*. Retrieved 17 May 2009.

[164] Neanderthal image by Kupka, based on Boule, 1909, in Humanity's Journeys Dr. Kathryn Denning, 2005, retrieved 2012-03-17

[165] Atelier Daynes, Neanderthal reconstructions, retrieved 2012-03-17

[166] Replica of a Neanderthal child, The Ancient Edge Back European Neanderthals Overwhelmed by Sheer Numbers of Invading Homo Sapiens, Genevieve Maul, August 10th 2011, retrieved 2012-03-17

Journals

- Boë, Louis-Jean; Heim, Jean-Louis; Honda, Kiyoshi; Maeda, Shinji (2002). "The potential Neandertal vowel space was as large as that of modern humans". *Journal of Phonetics* **30** (3): 465–84. doi:10.1006/jpho.2002.0170.

- Lieberman, Philip (October 2007). "Current views on Neanderthal speech capabilities: A reply to Boe et al. (2002)". *Journal of Phonetics* **35** (4): 552–63. doi:10.1016/j.wocn.2005.07.002.

- Serre, David; Langaney, André; Chech, Mario; Teschler-Nicola, Maria; Paunovic, Maja; Mennecier, Philippe; Hofreiter, Michael; Possnert, Göran; Pääbo, Svante (2004). "No Evidence of Neandertal mtDNA Contribution to Early Modern Humans". *PLoS Biology* **2** (3): e57. doi:10.1371/journal.pbio.0020057. PMC 368159. PMID 15024415.

- Wild, Eva M.; Teschler-Nicola, Maria; Kutschera, Walter; Steier, Peter; Trinkaus, Erik; Wanek, Wolfgang (2005). "Direct dating of Early Upper Palaeolithic human remains from Mladeč". *Nature* **435** (7040): 332–5. Bibcode:2005Natur.435..332W. doi:10.1038/nature03585. PMID 15902255.

- Zilhão, João; Davis, Simon J. M.; Duarte, Cidália; Soares, António M. M.; Steier, Peter; Wild, Eva (2010). Hawks, John, ed. "Pego do Diabo (Loures, Portugal): Dating the Emergence of Anatomical Modernity in Westernmost Eurasia". *PLoS ONE* **5** (1): e8880. doi:10.1371/journal.pone.0008880. PMC 2811729. PMID 20111705. Lay summary – *ScienceDaily* (January 27, 2010).

Bibliography

- Derev'anko, Anatoliy P.; Powers, William Roger; Shimkin, Demitri Boris (1998). *The Paleolithic of Siberia: new discoveries and interpretations.* Novosibirsk: Institute of Anthropology and Ethnography. ISBN 978-0-252-02052-0. OCLC 36461622.

- Lunine, Jonathan I. (2013). *Earth: Evolution of a Habitable World.* Cambridge University Press. 327. ISBN 978-0-521-85001-8.

11.16 Further reading

- Sankararaman, Sriram; Mallick, Swapan; Dannemann, Michael; Prüfer, Kay; Kelso, Janet; Patterson, Nick; Reich, David (2014). "The genomic landscape of Neanderthal ancestry in present-day humans". *Nature* **507** (7492): 354–357. doi:10.1038/nature12961.

- Vattathil, S.; Akey, J.M. (2015). "Small amounts of archaic admixture provide big insights into human history". *Cell* **163** (2): 281–284. doi:10.1016/j.cell.2015.09.042.

11.17 External links

- Kreger, C. David (30 June 2000). "Homo neanderthalensis". ArchaeologyInfo.com. Retrieved 23 May 2009.

- O'Neil, Dennis (12 May 2009). "Evolution of Modern Humans: Neandertals". Retrieved 23 May 2009.

- "*Homo neanderthalensis*". The Smithsonian Institution.

- "Neanderthal DNA". International Society of Genetic Genealogy.: Includes Neanderthal mtDNA sequences

- Panoramio—'IMG_6922 The Neandertal foot prints' (photo of ~25K years old fossilized footprints discovered in 1970 on volcanic layers near Demirkopru Dam Reservoir, Manisa, Turkey)

- Did better mothering defeat the Neanderthals?

- My Great-great-great Grandfather's a Neanderthal

- Ancient tryst fortified human immune system

- Neanderthal-human hybridisation hypothesis

- Neanderthal hybridization and Haldane's rule

Chapter 12

Archaic Homo sapiens

"Dawn men" redirects here. For the stone circle, see Dawn's Men.

A number of varieties of *Homo* are grouped into the

Homo rhodesiensis *"Broken Hill Cranium": dated to either 130,000 years ago (using amino acid racemization determination) or 800,000 to 600,000 years ago (within the same time as* Homo erectus*), depending on which dating method is used.*

broad category of **archaic humans** in the period beginning 500,000 years ago (or 500ka). It typically includes *Homo neanderthalensis* (40ka-300ka), *Homo rhodesiensis* (125ka-300ka), *Homo heidelbergensis* (200ka-600ka), and may also include *Homo antecessor* (800ka-1200ka).[1] This category is contrasted with anatomically modern humans, which include *Homo sapiens sapiens* and *Homo sapiens idaltu*.

Modern humans are theorized to have evolved from archaic humans, who in turn evolved from *Homo erectus*. Varieties of archaic humans are sometimes included under the binomial name "*Homo sapiens*" because their brain size is very similar to that of modern humans. Archaic humans had a brain size averaging 1200 to 1400 cubic centimeters, which overlaps with the range of modern humans. Archaics are distinguished from anatomically modern humans by having a thick skull, prominent brow ridges and the lack of a prominent chin.[1][2]

Anatomically modern humans appear from about 200,000 years ago and after 70,000 years ago (see Toba catastrophe theory) gradually marginalize the "archaic" varieties. Non-modern varieties of *Homo* are certain to have survived until after 30,000 years ago, and perhaps until as recent as 10,000 years ago. Which of these, if any, are included under the term "archaic human" is a matter of definition and varies among authors. Nonetheless, according to recent genetic studies, modern humans may have bred with "at least two groups" of ancient humans: Neanderthals and Denisovans.[3] Other studies have cast doubt on admixture being the source of the shared genetic markers between archaic and modern humans, pointing to an ancestral origin of the traits originating 500,000 to 800,000 years ago.[4][5][6]

New evidence suggests another group may also have been extant as recently as 11,500 years ago, the *Red Deer Cave people* of China.[7] Chris Stringer of the Natural History Museum in London has suggested that these people could be a result of mating between Denisovans and modern humans.[8] Other scientists remain skeptical, suggesting that the unique features are within the variations expected for human populations.[9]

12.1 Terminology and definition

The category archaic human lacks a single, agreed upon definition.[1] According to one definition, *Homo sapiens* is a single species comprising several subspecies that include the archaics and modern humans. Under this definition, modern humans are referred to as *Homo sapiens sapiens* and archaics are also designated with the prefix "*Homo sapiens*". For example, the Neanderthals are ***Homo sapiens ne-***

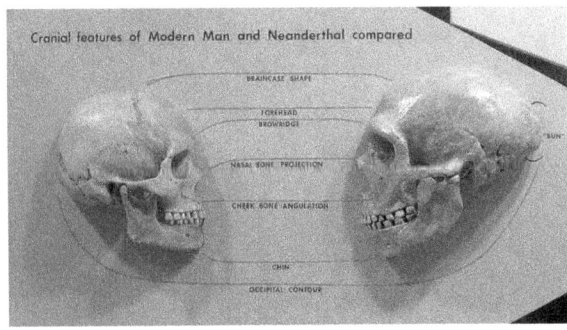

Anatomical comparison of the skulls of anatomically modern humans "wise men" (left) and Homo neanderthalensis *(right)*

anderthalensis, and *Homo heidelbergensis* is ***Homo sapiens heidelbergensis***. Other taxonomists prefer not to consider archaics and modern humans as a single species but as several different species. In this case the standard taxonomy is used, i.e. *Homo rhodesiensis*, or *Homo neanderthalensis*.[1]

The dividing lines that separate modern humans from archaic *Homo sapiens* and archaics from *Homo erectus* are unclear. The earliest known fossils of anatomically modern humans such as the Omo remains from 195,000 years ago, *Homo sapiens idaltu* from 160,000 years ago, and Qafzeh remains from 90,000 years ago are recognizably modern humans. However, these early modern humans do exhibit a mix of some archaic traits, such as moderate, but not prominent, brow ridges.

12.2   Brain size expansion

The emergence of archaic humans is sometimes used as an example of punctuated equilibrium.[10] This occurs when a species undergoes significant biological evolution within a relatively short period. Subsequently, the species undergoes very little change for long periods until the next punctuation. The brain size of archaic humans expanded significantly from 900 cubic centimeters in erectus to 1300 cubic centimeters. Since the peak of human brain size during the archaics, it has begun to decline.[11]

12.3   Origin of language

Main article: Origin of language

Robin Dunbar has argued that archaic humans were the first to use language. Based on his analysis of the relationship between brain size and hominin group size, he concluded that because archaic humans had large brains, they must have lived in groups of over 120 individuals. Dunbar ar-

gues that it was not possible for hominins to live in such large groups without using language, otherwise there could be no group cohesion and the group would disintegrate. By comparison, chimpanzees live in smaller groups of up to 50 individuals.[12][13]

12.4   Fossils

Further information: List of human evolution fossils

- Atapuerca Mountains, *Sima de los Huesos*
- Saldanha Man
- Altamura Man
- Kabwe Skull
- Steinheim Skull

12.5   See also

- Archaic human admixture with modern humans
- *Dawn of Humanity* (2015 PBS film)
- Early human migrations
- Evolution of human intelligence
- *Homo erectus*
- *Homo sapiens idaltu*
- Human evolution
- Middle Paleolithic
- Neanderthal extinction hypotheses
- Recent African origin of modern humans

12.6   References

[1] Dawkins (2005). "Archaic homo sapiens". *The Ancestor's Tale*. Boston: Mariner. ISBN 0-618-61916-X.

[2] Companion encyclopedia of archaeology

[3] Mitchell, Alanna (January 30, 2012). "DNA Turning Human Story Into a Tell-All". NYTimes. Retrieved January 31, 2012.

[4]

[5] http://www.theguardian.com/science/2013/feb/04/ neanderthals-modern-humans-research

[6]

[7] Amos, Jonathan (March 14, 2012). "Human fossils hint at new species". BBC. Retrieved March 14, 2012.

[8] Barras, Colin (2012-03-14). "Chinese human fossils unlike any known species". New Scientist. Retrieved 2012-03-15.

[9] James Owen (2012-03-14). "Cave Fossil Find: New Human Species or "Nothing Extraordinary"?". *National Geographic News*.

[10] Alone in the Universe

[11] http://phys.org/news187877156.html

[12] CO-EVOLUTION OF NEOCORTEX SIZE, GROUP SIZE AND LANGUAGE IN HUMANS

[13] Dunbar (1993). *Grooming, Gossip, and the Evolution of Language*. Harvard University Press. ISBN 978-0-674-36336-6.

12.7 External links

- EARLY AND LATE "ARCHAIC"HOMO SAPIENS AND "ANATOMICALLY MODERN" HOMO SAPIENS

- Origins of Modern Humans: Multiregional or Out of Africa?

- Homo sapiens, Museum of Natural History

Chapter 13

Recent African origin of modern humans

This article is about modern humans. For migrations of early humans, see Out of Africa I.

In paleoanthropology, the **recent African origin of modern humans**, or the **"out of Africa" theory** (**OOA**), is the most widely accepted model of the geographic origin and early migration of anatomically modern humans. The theory is called the "out-of-Africa" theory in the popular press, and the "**recent single-origin hypothesis**" (**RSOH**), "**replacement hypothesis**", or "**recent African origin model**" (**RAO**) by experts in the field. The concept was speculative before it was corroborated in the 1980s by a study of present-day mitochondrial DNA, combined with evidence based on physical anthropology of archaic specimens.

Genetic studies and fossil evidence indicate that archaic *Homo sapiens* evolved to anatomically modern humans solely in Africa between 200,000 and 60,000 years ago,[1] that members of one branch of *Homo sapiens* left Africa at some point between 125,000 and 60,000 years ago, and that over time these humans replaced other populations of the genus Homo such as Neanderthals and *Homo erectus*.[2] The date of the earliest successful "out of Africa" migration (earliest migrants with living descendants) has generally been placed at 60,000 years ago based on mitochondrial genetics, but this model has recently been contested by simulations of mitochrondrial DNA data,[3] 125,000-year-old Arabian archaeological finds of tools in the region[4] and the discovery of *Homo sapiens* teeth in China, dating to at least 80,000 years ago.[5]

The recent single origin of modern humans in East Africa is the predominant position held within the scientific community.[6][7][8][9][10] There are differing theories on whether there was a single exodus or several. An increasing number of researchers think that "long-neglected North Africa"[11][12] may have been the original home of the first modern humans to migrate out of Africa.[13][14]

The major competing hypothesis is the multiregional origin of modern humans, which envisions a wave of *Homo sapiens* migrating earlier from Africa and interbreeding with local *Homo erectus* populations in multiple regions of the globe. Most multiregionalists still view Africa as a major wellspring of human genetic diversity, but allow a much greater role for hybridization.[15][16]

Genetic testing in the last decade has revealed that several now extinct archaic human species may have interbred with modern humans. These species have been claimed to have left their genetic imprint in different regions across the world: Neanderthals in all humans except Sub-Saharan Africans, Denisova hominin in Australasia (for example, Melanesians, Aboriginal Australians and some Negritos) and there could also have been interbreeding between Sub-Saharan Africans and an as-yet-unknown hominin (possibly remnants of the ancient species Homo heidelbergensis). However, the rate of interbreeding was found to be relatively low (1–10%) and other studies have suggested that the presence of Neanderthal or other archaic human genetic markers in modern humans can be attributed to shared ancestral traits originating from a common ancestor 500,000 to 800,000 years ago.[17][18][19][20][21]

13.1 History of the theory

Further information: Timeline of human evolution

With the development of anthropology in the early 19th century, scholars disagreed vigorously about different theories of human development. Those such as Johann Friedrich Blumenbach and James Cowles Prichard held that since the creation, the various human races had developed as different varieties sharing descent from one people (monogenism). Their opponents, such as Louis Agassiz and Josiah C. Nott, argued for polygenism, or the separate development of human races as separate species or had developed as separate species through transmutation of species from apes, with no common ancestor.

Charles Darwin was one of the first to propose common

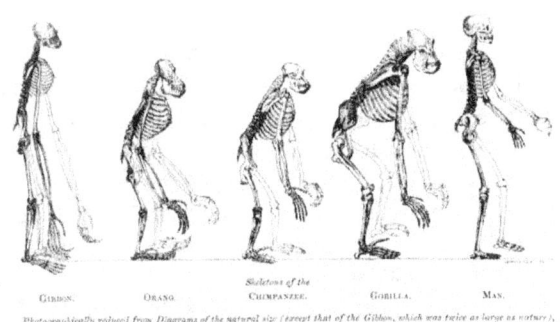

GIBBON. ORANG. CHIMPANZEE. GORILLA. MAN.

Skeletons of the

Photographically reduced from Diagrams of the natural size (except that of the Gibbon, which was twice as large as nature),
drawn by Mr. Waterhouse Hawkins from specimens in the Museum of the Royal College of Surgeons.

The frontispiece to Huxley's Evidence as to Man's Place in Nature *(1863): the image compares the skeletons of apes to humans.*

descent of living organisms, and among the first to suggest that all humans had in common ancestors who lived in Africa.[22] Darwin first suggested the "Out of Africa" hypothesis after studying the behaviour of African apes, one of which was displayed at the London Zoo. The anatomist Thomas Huxley had also supported the hypothesis and suggested that African apes have a close evolutionary relationship with humans.[23] These views were however opposed by Ernst Haeckel the German biologist who was a proponent of the Out of Asia theory. Haeckel argued that humans were more closely related to the primates of Southeast Asia and rejected Darwin's hypothesis of Africa.[24][25]

In the *Descent of Man*, Darwin speculated that humans had descended from apes which still had small brains but walked upright, freeing their hands for uses which favoured intelligence. Further, he thought such apes were African:[26]

> In each great region of the world the living mammals are closely related to the extinct species of the same region. It is, therefore, probable that Africa was formerly inhabited by extinct apes closely allied to the gorilla and chimpanzee; and as these two species are now man's nearest allies, it is somewhat more probable that our early progenitors lived on the African continent than elsewhere. But it is useless to speculate on this subject, for an ape nearly as large as a man, namely the Dryopithecus of Lartet, which was closely allied to the anthropomorphous Hylobates, existed in Europe during the Upper Miocene period; and since so remote a period the earth has certainly undergone many great revolutions, and there has been ample time for migration on the largest scale.
>
> — Charles Darwin, Descent of Man[27]

The prediction was insightful, because in 1871 there were hardly any human fossils of ancient hominids available. Almost fifty years later, Darwin's speculation was supported when anthropologists began finding numerous fossils of ancient small-brained hominids in several areas of Africa (list of hominina fossils).

The debate in anthropology had swung in favour of monogenism by the mid-20th century. Isolated proponents of polygenism held forth in the mid-20th century, such as Carleton Coon, who hypothesized as late as 1962 that *Homo sapiens* arose five times from *Homo erectus* in five places.[28] The "Recent African origin" of modern humans means "single origin" (monogenism) and has been used in various contexts as an antonym to polygenism.

In the 1980s Allan Wilson together with Rebecca L. Cann and Mark Stoneking worked on the so-called "Mitochondrial Eve" hypothesis. In his efforts to identify informative genetic markers for tracking human evolutionary history, he started to focus on mitochondrial DNA (mtDNA) – genes that sit in the cell, but not in the nucleus, and are passed from mother to child. This DNA material is important because it mutates quickly, thus making it easy to plot changes over relatively short time spans. By comparing differences in the mtDNA Wilson thought it was possible to estimate the time, and the place, modern humans first evolved. With his discovery that human mtDNA is genetically much less diverse than chimpanzee mtDNA, he concluded that modern human populations had diverged recently from a single population while older human species such as Neandertals and *Homo erectus* had become extinct. He and his team compared mtDNA in people of different ancestral backgrounds and concluded that all modern humans evolved from one 'lucky mother' in Africa about 150,000 years ago.[29] With the advent of archaeogenetics in the 1990s, scientists were able to date the "out of Africa" migration with some confidence.

In 2000, the mitochondrial DNA (mtDNA) sequence of "Mungo Man 3" (LM3) of ancient Australia was published indicating that Mungo Man was an extinct subspecies that diverged before the most recent common ancestor of contemporary humans. The results, if correct, supports the multiregional origin of modern humans hypothesis.[30][31] This work was later questioned[32][33] and explained by W. James Peacock, leader of the team who sequenced Mungo man's aDNA.[34] In addition, a large-scale genotyping analysis of aboriginal Australians, New Guineans, Southeast Asians and Indians in 2013 showed close genetic relationship between Australian, New Guinean, and the Mamanwa people, with divergence times for these groups estimated at 36,000 y ago. Further, substantial gene flow was detected between the Indian populations and aboriginal Australians, indicating an early "southern route" migration out of Africa, and arrival of other populations in the region by subsequent dispersal. This basically opposes the view that there was an isolated human evolution in Australia.[35]

The question of whether there was inheritance of other typological (not *de facto*) *Homo* subspecies into the *Homo sapiens* genetic pool is debated.

13.2   Early *Homo sapiens*

Main articles: Anatomically modern humans and Archaic Homo sapiens

Anatomically modern humans originated in Africa about

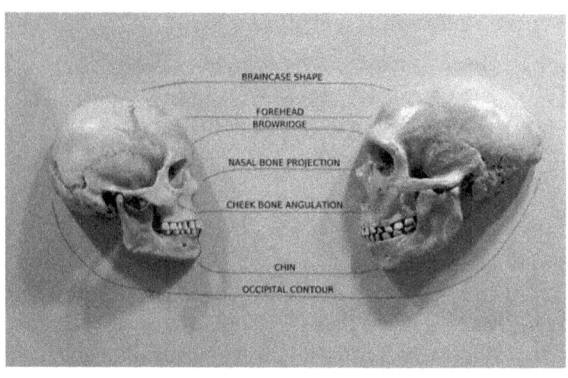

Anatomical comparison of the skulls of a modern human (left) and Homo neanderthalensis *(right).*

250,000 years ago. The trend in cranial expansion and the acheulean elaboration of stone tool technologies which occurred between 400,000 years ago and the second interglacial period in the Middle Pleistocene (around 250,000 years ago) provide evidence for a transition from *H. erectus* to *H. sapiens*.[36] In the Recent African Origin (RAO) scenario, migration within and out of Africa eventually replaced the earlier dispersed *H. erectus*.

Homo sapiens idaltu, found at site Middle Awash in Ethiopia, lived about 160,000 years ago.[37] It is the oldest known anatomically modern human and classified as an extinct subspecies.[38] Fossils of early *Homo sapiens* were found in Qafzeh cave in Israel and have been dated to 80,000 to 100,000 years ago. However these humans seem to have either become extinct or retreated back to Africa 70,000 to 80,000 years ago, possibly replaced by southbound Neanderthals escaping the colder regions of ice age Europe.[39] Hua Liu et al. analyzed autosomal microsatellite markers dates to c. 56,000±5,700 years ago mtDNA evidence. He interprets the paleontological fossil of early modern human from Qafzeh cave as an isolated early offshoot that retracted back to Africa.[40] However, these interpretations have been strongly contested,[3] particularly in the view of discoveries in China.[5]

All other fossils of fully modern humans outside Africa have

been dated to more recent times. The oldest well dated fossil of modern humans found outside Africa is from Manot Cave in Israel, named Manot 1, which have been dated to 54,700 years ago.[41][42] Fossils from Lake Mungo, Australia have been dated to about 42,000 years ago.[43][44] The Tianyuan cave remains in Liujiang region China have a probable date range between 38,000 and 42,000 years ago. More recently, *H. sapiens* teeth in Chine have been dated to at least 80,000 years old.[5] The Tianyuan specimens are most similar in morphology to Minatogawa Man, modern humans dated between 17,000 and 19,000 years ago and found on Okinawa Island, Japan.[45][46] However, others have dated Liujang Man to 111,000 to 139,000 years before the present.[47]

Beginning about 100,000 years ago evidence of more sophisticated technology and artwork begins to emerge and by 50,000 years ago fully modern behaviour becomes more prominent. Stone tools show regular patterns that are reproduced or duplicated with more precision while tools made of bone and antler appear for the first time.[48][49]

13.3   Genetic reconstruction

Further information: Most recent common ancestor and Archaeogenetics

Two pieces of the human genome are quite useful in deciphering human history: mitochondrial DNA and the Y chromosome. These are the only two parts of the genome that are not shuffled about by the evolutionary mechanisms that generate diversity with each generation: instead, these elements are passed down intact. According to the hypothesis, all people alive today have inherited the same mitochondria[50] from a woman who lived in Africa about 160,000 years ago.[51][52] She has been named Mitochondrial Eve. All men living today have inherited their Y chromosomes from a man who lived 140,000–500,000 years ago, probably in Africa. He has been named Y-chromosomal Adam. Based on comparisons of non-sex-specific chromosomes with sex-specific ones, it is now estimated that more men than women participated in the out-of-Africa exodus of early humans.[53]

13.3.1   Mitochondrial DNA

Further information: Human mitochondrial DNA haplogroup

The first lineage to branch off from Mitochondrial Eve is L0. This haplogroup is found in high proportions among the San of Southern Africa, the Sandawe of East Africa. It is also found among the Mbuti people.[54][55]

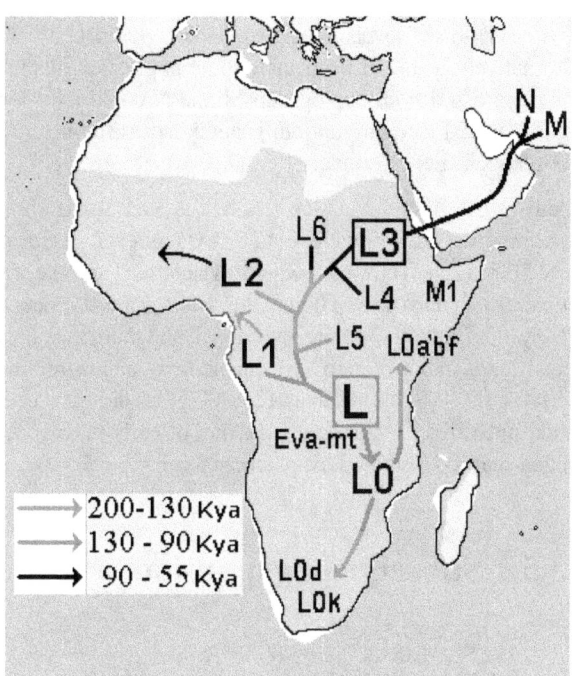

Map of early diversification of modern humans according to mitochondrial population genetics (see: Haplogroup L).

These groups branched off early in human history and have remained relatively genetically isolated since then. Haplogroups L1, L2 and L3 are descendents of L1-6 and are largely confined to Africa. The macro haplogroups M and N, which are the lineages of the rest of the world outside Africa, descend from L3. L3 is about 84,000 years old, and haplogroup M and N are almost identical in age at about 63,000 years old. [56] The relationship between such gene trees and demographic history is still debated when applied to dispersals.[3]

13.3.2 Genomic analysis

Although mitochondrial DNA and Y-chromosomal DNA are particularly useful in deciphering human history, data on the genomes of dozens of population groups have also been studied. In June 2009, an analysis of genome-wide SNP data from the International HapMap Project (Phase II) and CEPH Human Genome Diversity Panel samples was published.[57] Those samples were taken from 1138 unrelated individuals.[57] Before this analysis, population geneticists expected to find dramatic differences among ethnic groups, with derived alleles shared among such groups but uncommon or nonexistent in other groups.[58] Instead the study of 53 populations taken from the HapMap and CEPH data revealed that the population groups studied fell into just three genetic groups: Africans, Eurasians (which includes natives of Europe and the Middle East, and South-

west Asians east to present-day Pakistan), and East Asians, which includes natives of Asia, Japan, Southeast Asia, the Americas, and Oceania.[58] The study determined that most ethnic group differences can be attributed to genetic drift, with modern African populations having greater genetic diversity than the other two genetic groups, and modern Eurasians somewhat more than modern East Asians.[58] The study suggested that natural selection may shape the human genome much more slowly than previously thought, with factors such as migration within and among continents more heavily influencing the distribution of genetic variations.[59] A May 2002 study examined three groups, African, European, and Asian. It found greater genetic diversity among Africans than among Eurasians, and that genetic diversity among Eurasians is largely a subset of that among Africans, supporting the 'out of Africa' model.[60]

13.4 Movement out of Africa

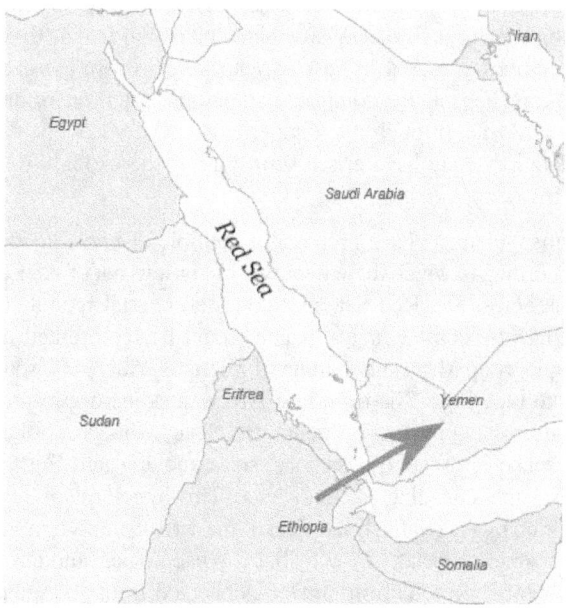

Red Sea crossing

By some 70,000 years ago, a part of the bearers of mitochondrial haplogroup L3 migrated from East Africa into the Near East. The date of this first wave of "out of Africa" migration was called into question in 2011, based on the discovery of stone tools in the United Arab Emirates, indicating the presence of modern humans between 100,000 and 125,000 years ago.[4][61] New research showing slower than previously thought genetic mutations in human DNA published in 2012, indicating a revised dating for the migration of between 90,000 and 130,000 years ago.[62]

Some scientists think that only a few people left Africa

in a single migration that went on to populate the rest of the world,[63] based in the fact that only descendents of L3 are found outside Africa. From that settlement, some others point to the possibility of several waves of expansion. For example, geneticist Spencer Wells says that the early travellers followed the southern coastline of Asia, crossed about 250 kilometres (155 mi) of sea, and colonized Australia by around 50,000 years ago. The Aborigines of Australia, Wells says, are the descendants of the first wave of migrations.[64]

It has been estimated that from a population of 2,000 to 5,000 individuals in Africa,[65] only a small group, possibly as few as 150 to 1,000 people, crossed the Red Sea.[66] Of all the lineages present in Africa only the female descendants of one lineage, mtDNA haplogroup L3, are found outside Africa. If there had been several migrations, one would expect descendants of more than one lineage to be found outside Africa. L3's female descendants, the M and N haplogroup lineages, are found in very low frequencies in Africa (although haplogroup M1 populations are very ancient and diversified in North and Northeast Africa) and appear to be more recent arrivals. A possible explanation is that these mutations occurred in East Africa shortly before the exodus and became the dominant haplogroups after the exodus from Africa through the founder effect. Alternatively, the mutations may have arisen shortly after the exodus from Africa.

Other scientists have proposed a multiple dispersal model according to which there were two migrations out of Africa, one across the Red Sea and along the coastal regions to India (the coastal route), which would be represented by haplogroup M. Another group of migrants with haplogroup N followed the Nile from East Africa, heading northwards and crossing into Asia through the Sinai. This group then branched in several directions, some moving into Europe and others heading east into Asia. This hypothesis is supported by the relatively late date of the arrival of modern humans in Europe as well as by both archaeological and DNA evidence. Results from mtDNA collected from aboriginal Malaysians called Orang Asli and the creation of a phylogentic tree indicate that the hapologroups M and N share characteristics with original African groups from approximately 85,000 years ago and share characteristics with sub-haplogroups among coastal southeast Asian regions, such as Australasia, the Indian Subcontinent, and throughout continental Asia, which had dispersed and separated from its African origins approximately 65,000 years ago. This southern coastal dispersion would have occurred before the original theory of dispersion through the Levant approximately 45,000 years ago.[67] This hypothesis attempts to explain why haplogroup N is predominant in Europe and why haplogroup M is absent in Europe. Evidence of the coastal migration is hypothesized to have been destroyed by the rise in sea levels during the Holocene epoch.[68][69] Alternatively, a small European founder population that initially expressed both haplogroup M and N could have lost haplogroup M through random genetic drift resulting from a bottleneck (i.e. a founder effect).

Today at the Bab-el-Mandeb straits, the Red Sea is about 20 kilometres (12 mi) wide, but 50,000 years ago sea levels were 70 m (230 ft) lower (owing to glaciation) and the water was much narrower. Though the straits were never completely closed, they were narrow enough and there may have been islands in between to have enabled crossing using simple rafts.[70][71] Shell middens 125,000 years old have been found in Eritrea,[72] indicating the diet of early humans included seafood obtained by beachcombing.

13.5   Subsequent expansion

Main article: Early human migrations
From the Near East, these populations spread east to South

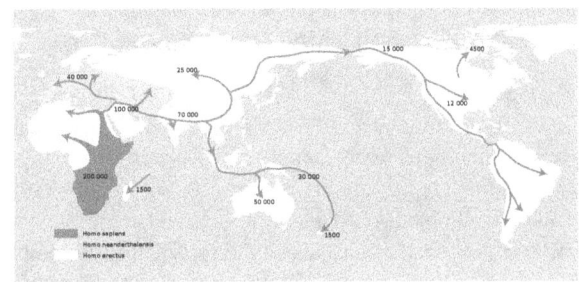

Map of early human migrations[73]
1. Homo sapiens
2. Neanderthals
3. Early Hominids

Asia by 50,000 years ago, and on to Australia by 40,000 years ago, *Homo sapiens* for the first time colonizing territory never reached by *Homo erectus*. Europe was reached by Cro-Magnon some 40,000 years ago. East Asia (Korea, Japan) was reached by 30,000 years ago. It is disputed whether subsequent migration to North America took place around 30,000 years ago, or only considerably later, around 14,000 years ago.[74]

The group that crossed the Red Sea travelled along the coastal route around the coast of Arabia and Persia until reaching India, which appears to be the first major settling point. Haplogroup M is found in high frequencies along the southern coastal regions of Pakistan and India and it has the greatest diversity in India, indicating that it is here where the mutation may have occurred.[75] Sixty percent of the Indian population belong to Haplogroup M.

The indigenous people of the Andaman Islands also belong

to the M lineage. The Andamanese are thought to be off-shoots of some of the earliest inhabitants in Asia because of their long isolation from mainland Asia. They are evidence of the coastal route of early settlers that extends from India along the coasts of Thailand and Indonesia all the way to Papua New Guinea. Since M is found in high frequencies in highlanders from New Guinea as well, and both the Andamanese and New Guineans have dark skin and Afro-textured hair, some scientists think they are all part of the same wave of migrants who departed across the Red Sea ~60,000 years ago in the Great Coastal Migration.

Notably, the findings of Harding et al.[76] show that, at least with regard to dark skin color, the haplotype background of Papua New Guineans at MC1R (one of a number of genes involved in melanin production) is identical to that of Africans (barring a single silent mutation). Thus, although these groups are distinct from Africans at other loci (due to drift, bottlenecks, etc.), it is evident that selection for the dark skin color trait likely continued (at least at MC1R) following the exodus. This would support the hypothesis that suggests that the original migrants from Africa resembled pre-exodus Africans (at least in skin color), and that the present day remnants of this ancient phenotype can be seen among contemporary Africans, Andamanese and New Guineans. Others suggest that their physical resemblance to Africans could be the result of convergent evolution.[77][78]

From Arabia to India the proportion of haplogroup M increases eastwards: in eastern India, M outnumbers N by a ratio of 3:1. However, crossing over into East Asia, Haplogroup N reappears as the dominant lineage. M is predominant in South East Asia but amongst Indigenous Australians N reemerges as the more common lineage. This discontinuous distribution of Haplogroup N from Europe to Australia can be explained by founder effects and population bottlenecks.[79] In addition to genetic analysis, Petraglia et al. also examines the microlithic materials from Indian subcontinent and explains the expansion of population based on the reconstruction of paleoenvironment. He proposed that microlithic industries could be traced back to 35ka in South Asia, and the new technology might be influenced by environmental change and population pressure.[80]

13.6   Competing hypotheses

Main article: Multiregional origin of modern humans

The multiregional hypothesis, initially proposed by Milford Wolpoff, holds that the evolution of humans from *H. erectus* at the beginning of the Pleistocene 1.8 million years BP has been within a single, continuous worldwide population. Proponents of multiregional origin reject the assumption of

an infertility barrier between ancient Eurasian and African populations of *Homo*. Multiregional proponents point to the fossil record and genetic evidence in chromosomal DNA. One study suggested that at least 5% of the human modern gene pool can be attributed to ancient admixture, which in Europe would be from the Neanderthals.[81] But the study also suggests that there may be other reasons why humans and Neanderthals share ancient genetic lineages.[82][83]

13.7   See also

13.8   References

[1] Reid GBR, Hetherington R (2010). *The climate connection: climate change and modern human evolution*. Cambridge, UK: Cambridge University Press. p. 64. ISBN 0-521-14723-9.

[2] Meredith M (2011). *Born in Africa: The Quest for the Origins of Human Life*. New York: PublicAffairs. ISBN 1-58648-663-2.

[3] Groucutt, Huw S.; et al. (2015). "Rethinking the dispersal of Homo sapiens out of Africa". Wiley.

[4] Armitage SJ, Jasim SA, Marks AE, Parker AG, Usik VI, Uerpmann HP; Jasim; Marks; Parker; Usik; Uerpmann (January 2011). "The southern route "out of Africa": evidence for an early expansion of modern humans into Arabia". *Science* **331** (6016): 453–6. Bibcode:2011Sci...331..453A. doi:10.1126/science.1199113. PMID 21273486.

[5] Liu, Wu; Martinón-Torres, María; Cai, Yan-jun; Xing, Song; Tong, Hao-wen; Pei, Shu-wen; Sier, Mark Jan; Wu, Xiao-hong; Edwards, R. Lawrence (14 October 2015). "The earliest unequivocally modern humans in southern China". *Nature*. doi:10.1038/nature15696.

[6] Liu H, Prugnolle F, Manica A, Balloux F; Prugnolle; Manica; Balloux (August 2006). "A geographically explicit genetic model of worldwide human-settlement history". *Am. J. Hum. Genet.* **79** (2): 230–7. doi:10.1086/505436. PMC 1559480. PMID 16826514. Currently available genetic and archaeological evidence is supportive of a recent single origin of modern humans in East Africa. However, this is where the consensus on human settlement history ends, and considerable uncertainty clouds any more detailed aspect of human colonization history.

[7] "This week in Science: Out of Africa Revisited". *Science* **308** (5724): 921. 2005-05-13. doi:10.1126/science.308.5724.921g.

[8] Stringer C (June 2003). "Human evolution: Out of Ethiopia". *Nature* **423** (6941): 692–3, 695. Bibcode:2003Natur.423..692S. doi:10.1038/423692a. PMID 12802315.

[9] Johanson D. "Origins of Modern Humans: Multiregional or Out of Africa?". *ActionBioscience*. American Institute of Biological Sciences.

[10] "Modern Humans – Single Origin (Out of Africa) vs Multiregional".

[11] Balter M (January 2011). "Was North Africa the launch pad for modern human migrations?" (PDF). *Science* **331** (6013): 20–3. Bibcode:2011Sci...331...20B. doi:10.1126/science.331.6013.20. PMID 21212332.

[12] Scerri, Eleanor M. L.; Drake, Nick A.; Jennings, Richard; Groucutt, Huw S. (2014-10-01). "Earliest evidence for the structure of Homo sapiens populations in Africa". *Quaternary Science Reviews* **101**: 207–216. doi:10.1016/j.quascirev.2014.07.019.

[13] Cruciani F, Trombetta B, Massaia A, Destro-Bisol G, Sellitto D, Scozzari R; Trombetta; Massaia; Destro-Bisol; Sellitto; Scozzari (June 2011). "A revised root for the human Y chromosomal phylogenetic tree: the origin of patrilineal diversity in Africa". *Am. J. Hum. Genet.* **88** (6): 814–8. doi:10.1016/j.ajhg.2011.05.002. PMC 3113241. PMID 21601174.

[14] Smith TM, Tafforeau P, Reid DJ, Grün R, Eggins S, Boutakiout M, Hublin JJ; Tafforeau; Reid; Grün; Eggins; Boutakiout; Hublin (April 2007). "Earliest evidence of modern human life history in North African early Homo sapiens". *Proc. Natl. Acad. Sci. U.S.A.* **104** (15): 6128–33. Bibcode:2007PNAS..104.6128S. doi:10.1073/pnas.0700747104. PMC 1828706. PMID 17372199.

[15] Robert Jurmain; Lynn Kilgore; Wenda Trevathan (20 March 2008). *Essentials of Physical Anthropology*. Cengage Learning. pp. 266–. ISBN 978-0-495-50939-4. Retrieved 14 June 2011.

[16] Wolpoff MH, Hawks J, Caspari R; Hawks; Caspari (May 2000). "Multiregional, not multiple origins". *Am. J. Phys. Anthropol.* **112** (1): 129–36. doi:10.1002/(SICI)1096-8644(200005)112:1<129::AID-AJPA11>3.0.CO;2-K. PMID 10766948.

[17]

[18] http://www.theguardian.com/science/2013/feb/04/neanderthals-modern-humans-research

[19]

[20] The Human Stew

[21] http://www.plosgenetics.org/article/info%3Adoi%2F10.1371%2Fjournal.pgen.1002837

[22] Peter Lafreniere (22 September 2010). *Adaptive Origins: Evolution and Human Development*. Taylor & Francis. p. 90. ISBN 978-0-8058-6012-2. Retrieved 14 June 2011.

[23] Robinson D, Ash PM (2010). *The Emergence of Humans: An Exploration of the Evolutionary Timeline*. New York: Wiley. ISBN 0-470-01315-X.

[24] Palmer D (2006). *Prehistoric Past Revealed: The Four Billion Year History of Life on Earth*. Berkeley: University of California Press. p. 43. ISBN 0-520-24827-9.

[25] Regal B (2004). *Human evolution: a guide to the debates*. Santa Barbara, Calif: ABC-CLIO. pp. 73–75. ISBN 1-85109-418-0.

[26] Bowler 2003, p. 213

[27] "The descent of man Chapter 6 – On the Affinities and Genealogy of Man". Darwin-online.org.uk. Retrieved 2011-01-11.

[28] Jackson JP Jr (2001). ""In Ways Unacademical": The Reception of Carleton S. Coon's The Origin of Races" (pdf). *Journal of the History of Biology* **34** (2): 247–285. doi:10.1023/A:1010366015968.

[29] "Allan Wilson: Revolutionary Evolutionist". *New Zealanders Heroes*.

[30] Adcock GJ, Dennis ES, Easteal S, Huttley GA, Jermiin LS, Peacock WJ, Thorne A; Dennis; Easteal; Huttley; Jermiin; Peacock; Thorne (January 2001). "Mitochondrial DNA sequences in ancient Australians: Implications for modern human origins". *Proc. Natl. Acad. Sci. U.S.A.* **98** (2): 537–42. Bibcode:2001PNAS...98..537A. doi:10.1073/pnas.98.2.537. PMC 14622. PMID 11209053.

[31] Australia Challenges Out-of-Africa Theory ABC News January 9, 2003

[32] Cooper A, Rambaut A, Macaulay V, Willerslev E, Hansen AJ, Stringer C; Rambaut; MacAulay; Willerslev; Hansen; Stringer (June 2001). "Human origins and ancient human DNA". *Science* **292** (5522): 1655–6. doi:10.1126/science.292.5522.1655. PMID 11388352.

[33] Smith CI, Chamberlain AT, Riley MS, Stringer C, Collins MJ; Chamberlain; Riley; Stringer; Collins (September 2003). "The thermal history of human fossils and the likelihood of successful DNA amplification" (PDF). *J. Hum. Evol.* **45** (3): 203–17. doi:10.1016/S0047-2484(03)00106-4. PMID 14580590.

[34] Schiller J (2010). *Human Evolution: Neanderthals & Homo sapiens*. CreateSpace. p. 66. ISBN 978-1-4515-4608-8. Retrieved 29 November 2011.

[35] Pugach I, Delfin F, Gunnarsdóttir E, Kayser M, Stoneking M; Delfin; Gunnarsdóttir; Kayser; Stoneking (2013). "Genome-wide data substantiate Holocene gene flow from India to Australia". *Proc Natl Acad Sci U S A* **110** (5): 1803–1808. Bibcode:2013PNAS..110.1803P. doi:10.1073/pnas.1211927110. PMC 3562786. PMID 23319617.

[36] Eric Delson; Ian Tattersall; John A. Van Couvering (2000). *Encyclopedia of human evolution and prehistory*. Taylor & Francis. p. 677–. ISBN 978-0-8153-1696-1. Retrieved 14 June 2011.

[37] White TD, Asfaw B, DeGusta D, Gilbert H, Richards GD, Suwa G, Howell FC; Asfaw; Degusta; Gilbert; Richards; Suwa; Howell (June 2003). "Pleistocene Homo sapiens from Middle Awash, Ethiopia". *Nature* **423** (6941): 742–7. Bibcode:2003Natur.423..742W. doi:10.1038/nature01669. PMID 12802332.

[38] "160,000-year-old fossilized skulls uncovered in Ethiopia are oldest anatomically modern humans". University of California, Berkeley. 2003.

[39] Clive Finlayson (11 October 2009). *The humans who went extinct: why Neanderthals died out and we survived*. Oxford University Press US. p. 68. ISBN 978-0-19-923918-4. Retrieved 14 June 2011.

[40] Liu H, Prugnolle F, Manica A, Balloux F; Prugnolle; Manica; Balloux (August 2006). "A geographically explicit genetic model of worldwide human-settlement history". *Am. J. Hum. Genet.* **79** (2): 230–7. doi:10.1086/505436. PMC 1559480. PMID 16826514.

[41] Hershkovitz, Israel; Marder, Ofer; Ayalon, Avner; Bar-Matthews, Miryam; Yasur, Gal; Boaretto, Elisabetta; Caracuta, Valentina; Alex, Bridget; et al. (2015). "Levantine cranium from Manot Cave (Israel) foreshadows the first European modern humans". *Nature*. Epub ahead of print. doi:10.1038/nature14134. PMID 25629628.

[42] "55,000-Year-Old Skull Fossil Sheds New Light on Human Migration out of Africa". *Science News*. Retrieved 2 February 2015.

[43] Bowler JM, Johnston H, Olley JM, Prescott JR, Roberts RG, Shawcross W, Spooner NA.; Johnston; Olley; Prescott; Roberts; Shawcross; Spooner (2003). "New ages for human occupation and climatic change at Lake Mungo, Australia". *Nature* **421** (6925): 837–40. Bibcode:2003Natur.421..837B. doi:10.1038/nature01383. PMID 1259451.

[44] Olleya JM, Roberts RG, Yoshida H and Bowler JM (2006). "Single-grain optical dating of grave-infill associated with human burials at Lake Mungo, Australia". *Quaternary Science Reviews* **25** (19–20): 2469–2474. Bibcode:2006QSRv...25.2469O. doi:10.1016/j.quascirev.2005.07.022.

[45] Hu Y, Shang H, Tong H, Nehlich O, Liu W, Zhao C, Yu J, Wang C, Trinkaus E, Richards MP; Shang; Tong; Nehlich; Liu; Zhao; Yu; Wang; Trinkaus; Richards (July 2009). "Stable isotope dietary analysis of the Tianyuan 1 early modern human". *Proc. Natl. Acad. Sci. U.S.A.* **106** (27): 10971–4. Bibcode:2009PNAS..10610971H. doi:10.1073/pnas.0904826106. PMC 2706269. PMID 19581579.

[46] Brown P (August 1992). "Recent human evolution in East Asia and Australasia". *Philos. Trans. R. Soc. Lond., B, Biol. Sci.* **337** (1280): 235–42. doi:10.1098/rstb.1992.0101. PMID 1357698.

[47] Shen G, Wang W, Wang Q, Zhao J, Collerson K, Zhou C, Tobias PV; Wang; Wang; Zhao; Collerson; Zhou; Tobias (December 2002). "U-Series dating of Liujiang hominid site in Guangxi, Southern China". *J. Hum. Evol.* **43** (6): 817–29. doi:10.1006/jhev.2002.0601. PMID 12473485.

[48] "Ancestral tools". Handprint.com. 1999-08-01. Retrieved 2011-01-11.

[49] "Middle to upper paleolithic transition". Wsu.edu. Retrieved 2011-01-11.

[50] Jones, Marie; John Savino (2007). *Supervolcano: The Catastrophic Event That Changed the Course of Human History (Could Yellowstone be Next?)*. Franklin Lakes, NJ: New Page Books. ISBN 1-56414-953-6.

[51] Cann RL, Stoneking M, Wilson AC; Stoneking; Wilson (1987). "Mitochondrial DNA and human evolution". *Nature* **325** (6099): 31–6. Bibcode:1987Natur.325...31C. doi:10.1038/325031a0. PMID 3025745.

[52] Vigilant L, Stoneking M, Harpending H, Hawkes K, Wilson AC; Stoneking; Harpending; Hawkes; Wilson (September 1991). "African populations and the evolution of human mitochondrial DNA". *Science* **253** (5027): 1503–7. Bibcode:1991Sci...253.1503V. doi:10.1126/science.1840702. PMID 1840702.

[53] Keinan A, Mullikin JC, Patterson N, Reich D; Mullikin; Patterson; Reich (January 2009). "Accelerated genetic drift on chromosome X during the human dispersal out of Africa". *Nat. Genet.* **41** (1): 66–70. doi:10.1038/ng.303. PMC 2612098. PMID 19098910.

[54] Gonder MK, Mortensen HM, Reed FA, de Sousa A, Tishkoff SA; Mortensen; Reed; De Sousa; Tishkoff (March 2007). "Whole-mtDNA genome sequence analysis of ancient African lineages". *Mol. Biol. Evol.* **24** (3): 757–68. doi:10.1093/molbev/msl209. PMID 17194802.

[55] Chen YS, Olckers A, Schurr TG, Kogelnik AM, Huoponen K, Wallace DC; Olckers; Schurr; Kogelnik; Huoponen; Wallace (April 2000). "mtDNA variation in the South African Kung and Khwe-and their genetic relationships to other African populations". *Am. J. Hum. Genet.* **66** (4): 1362–83. doi:10.1086/302848. PMC 1288201. PMID 10739760.

[56] Macaulay, V; Hill, C; Achilli, A; Rengo, C; Clarke, D; Meehan, W; Blackburn, J; Semino, O; Scozzari, R; et al. (2005). "Single, Rapid Coastal Settlement of Asia Revealed by Analysis of Complete Mitochondrial Genomes". *Science* **308** (5724): 1034–6. doi:10.1126/science.1109792. PMID 15890885.

[57] Coop G, Pickrell, Novembre, Kudaravalli, Li, Absher, Myers, Cavalli-Sforza, Feldman, Pritchard (June 2009). Schierup MH, ed. "The role of geography in human adaptation". *PLoS Genet.* **5** (6): e1000500. doi:10.1371/journal.pgen.1000500. PMC 2685456. PMID 19503611.

[58] Brown, David (June 22, 2009). "Among Many Peoples, Little Genomic Variety". The Washington Post. Retrieved 2009-06-25.

[59] "Geography And History Shape Genetic Differences In Humans". Science Daily. June 7, 2009. Retrieved 2009-06-25.

[60] Yu, Ning; et al. (May 2002). "Larger Genetic Differences Within Africans Than Between Africans and Eurasians". *Genetics* (Genetics Society of America). Retrieved 7 April 2013.

[61] "Humans may have left Africa earlier than thought". Apnews.myway.com. Retrieved 2011-06-14.

[62] Catherine Brahic (24 Nov 2012). "Our True Dawn". *New Scientist* (Reed Business Information) (2892): 34–7. ISSN 0262-4079.

[63] "Both Australian Aborigines and Europeans Rooted in Africa". News.softpedia.com. Retrieved 2011-01-11.

[64] Rincon, Paul (April 24, 2008). "Human line 'nearly split in two'". BBC News. Retrieved 2009-12-31.

[65] Zhivotovsky; Rosenberg, NA; Feldman, MW; et al. (2003). "Features of Evolution and Expansion of Modern Humans, Inferred from Genomewide Microsatellite Markers". *American Journal of Human Genetics* **72** (5): 1171–86. doi:10.1086/375120. PMC 1180270. PMID 12690579.

[66] Stix, Gary (2008). "The Migration History of Humans: DNA Study Traces Human Origins Across the Continents". Retrieved 2011-06-14.

[67] Macaulay V, Hill C, Achilli A, Rengo C, Clarke D, Meehan W, Blackburn J, Semino O, Scozzari R, Cruciani F, Taha A, Shaari NK, Raja JM, Ismail P, Zainuddin Z, Goodwin W, Bulbeck D, Bandelt H-J, Oppenheimer S, Torroni A and Richards M. (2005). Single, Rapid Coastal Settlement of Asia Revealed by Analysis of Complete Mitochondrial Genomes. doi:10.1126/science.1109792

[68] Searching for traces of the Southern Dispersal, by Dr. Marta Mirazón Lahr, et al.

[69] "A single origin, several dispersal hypothesis". Biomedcentral.com. 2004-10-29. Retrieved 2011-01-11.

[70] Fernandes et. al (June 2006). "Absence of post-Miocene Red Sea land bridges: biogeographic implications". *Journal of Biogeography* **33** (6): 961–966. doi:10.1111/j.1365-2699.2006.01478.x.

[71] Beyin, Amanuel (February 2011). "Upper Pleistocene Human Dispersals out of Africa: A Review of the Current State of the Debate". *International Journal of Evolutionary Biology* **2011** (615094): 1–17. doi:10.4061/2011/615094.

[72] Walter RC, Buffler RT, Bruggemann JH, Guillaume MM, Berhe SM, Negassi B, Libsekal Y, Cheng H, Edwards RL, von Cosel R, Néraudeau D, Gagnon M (May 2000). "Early human occupation of the Red Sea coast of Eritrea during the last interglacial". *Nature* **405** (6782): 65–9. doi:10.1038/35011048. PMID 10811218.

[73] Literature: Göran Burenhult: Die ersten Menschen, Weltbild Verlag, 2000. ISBN 3-8289-0741-5

[74] Goebel, Ted; Waters, Michael R.; O'Rourke, Dennis H. (2008). "The Late Pleistocene dispersal of modern humans in the Americas" (PDF). *Science* **319** (5869): 1497–1502. doi:10.1126/science.1153569. PMID 18339930. Retrieved 2010-02-05.

[75] Metspalu M, Kivisild T, Metspalu E, Parik J, Hudjashov G, Kaldma K, Serk P, Karmin M, Behar DM, Gilbert MT, Endicott P, Mastana S, Papiha SS, Skorecki K, Torroni A, Villems R (August 2004). "Most of the extant mtDNA boundaries in south and southwest Asia were likely shaped during the initial settlement of Eurasia by anatomically modern humans". *BMC Genet.* **5**: 26. doi:10.1186/1471-2156-5-26. PMC 516768. PMID 15339343.

[76] Harding, et al. 2000, p. 1355

[77] "Evolution of Human Languages". Ehl.santafe.edu. Retrieved 2011-01-11.

[78] Endicott P, Gilbert MT, Stringer C, Lalueza-Fox C, Willerslev E, Hansen AJ, Cooper A (January 2003). "The genetic origins of the Andaman Islanders". *Am. J. Hum. Genet.* **72** (1): 178–84. doi:10.1086/345487. PMC 378623. PMID 12478481.

[79] Ingman M, Gyllensten U (July 2003). "Mitochondrial genome variation and evolutionary history of Australian and New Guinean aborigines". *Genome Res.* **13** (7): 1600–6. doi:10.1101/gr.686603. PMC 403733. PMID 12840039.

[80] Petraglia, M.; Clarkson, C.; Boivin, N.; Haslam, M.; Korisettar, R.; Chaubey, G.; Arnold, L. (2009). "Population increase and environmental deterioration correspond with microlithic innovations in South Asia ca. 35,000 years ago". *Proceedings of the National Academy of Sciences* **106** (30): 12261–12266. doi:10.1073/pnas.0810842106.

[81] Plagnol V, Wall JD (July 2006). "Possible ancestral structure in human populations". *PLoS Genet.* **2** (7): e105. doi:10.1371/journal.pgen.0020105. PMC 1523253. PMID 16895447. ..strong evidence for ancient admixture in both a European and a West African population (p ≈ 10–7), with contributions to the modern gene pool of at least 5%. While Neanderthals form an obvious archaic source population candidate in Europe..

[82] Green RE, Krause J, Briggs AW, et al. (May 2010). "A draft sequence of the Neandertal genome". *Science* **328** (5979): 710–22. Bibcode:2010Sci...328..710G. doi:10.1126/science.1188021. PMID 20448178. Lay summary – *BBC News*.

[83] Blum MG, Jakobsson M (October 2010). "Deep divergences of human gene trees and models of human origins". *Mol. Biol. Evol.* **28** (2): 889–98. doi:10.1093/molbev/msq265. PMID 20930054.

13.9 Further reading

- Mellars, Paul (2006). "Going East: New Genetic and Archaeological Perspectives on the Modern Human Colonization of Eurasia" **313**. Science. pp. 769–800.

- Bowler, Peter J. (2003). *Evolution: The History of an Idea* (3rd ed.). Berkeley: University of California Press. ISBN 0-520-23693-9.

- Darwin, Charles (1871). "The Descent of Man, and Selection in Relation to Sex" (1st ed.). London: John Murray. Retrieved 2009-09-05.

- Gibbons A (May 2001). "Human anthropology. Modern men trace ancestry to African migrants". *Science* **292** (5519): 1051–2. doi:10.1126/science.292.5519.1051b. PMID 11352048.

- Underhill PA, Passarino G, Lin AA, Shen P, Mirazón Lahr M, Foley RA, Oefner PJ, Cavalli-Sforza LL (January 2001). "The phylogeography of Y chromosome binary haplotypes and the origins of modern human populations" (PDF). *Ann. Hum. Genet.* **65** (Pt 1): 43–62. doi:10.1046/j.1469-1809.2001.6510043.x. PMID 11415522.

- Neanderthals 'mated with modern humans', *BBC News*, 21 April 1999

- New analysis shows three human migrations out of Africa – Replacement theory 'demolished', *Washington University in St. Louis*, 2 February 2006

- Harding RM, Healy E, Ray AJ, Ellis NS, Flanagan N, Todd C, Dixon C, Sajantila A, Jackson IJ, Birch-Machin MA, Rees JL (April 2000). "Evidence for variable selective pressures at MC1R". *Am. J. Hum. Genet.* **66** (4): 1351–61. doi:10.1086/302863. PMC 1288200. PMID 10733465.

- Long JC, Kittles RA (August 2003). "Human genetic diversity and the nonexistence of biological races" (PDF). *Hum. Biol.* **75** (4): 449–71. doi:10.1353/hub.2003.0058. PMID 14655871.

- Risch, N., Burchard, E., Ziv, E. and Tang, H. (2002). "Categorization of humans in biomedical research: genes, race and disease". *Genome Biology* **3** (7): comment2007.2001 – comment2007.2012. doi:10.1186/gb-2002-3-7-comment2007. PMC 139378. PMID 12184798.

- Tishkoff SA, Kidd KK (November 2004). "Implications of biogeography of human populations for 'race' and medicine". *Nat. Genet.* **36** (11 Suppl): S21–7. doi:10.1038/ng1438. PMID 15507999.

- Cavalli-Sforza F, Cavalli-Sforza LL (1995). *The great human diasporas: the history of diversity and evolution*. Boston: Addison-Wesley. ISBN 0-201-44231-0.

- Crow TJ, ed. (2004). *The Speciation of Modern Homo sapiens (Proceedings of the British Academy)*. London: British Academy. ISBN 0-19-726311-9.

- Foley R (1995). *Humans before humanity: an evolutionary perspective*. Oxford: Blackwell. ISBN 0-631-20528-4.

- Olson S (2003). *Mapping human history: genes, race, and our common origins*. Boston: Houghton Mifflin. ISBN 0-618-35210-4.

- Oppenheimer, Stephen (2003). *The Real Eve: Modern Man's Journey Out of Africa*. New York, NY: Carroll & Graf. ISBN 0-7867-1192-2.

- McKie R, Stringer C (1997). *African exodus: the origins of modern humanity*. London: Pimlico. ISBN 0-7126-7307-5.

- Sykes, Bryan (2004). *The Seven Daughters of Eve: The Science That Reveals Our Genetic Ancestry*. Corgi Adult. ISBN 0-552-15218-8.

- Wade N (2006). *Before the Dawn : Recovering the Lost History of Our Ancestors*. Penguin Press HC, The. ISBN 1-59420-079-3.

- Wells S (2004). *Journey of Man: Genetic Odyssey*. Random House. ISBN 0-8129-7146-9.

- Wells, Spencer (2006). *Deep ancestry: inside the Genographic Project*. Washington, D.C: National Geographic. ISBN 0-7922-6215-8.

- Manica A, Amos W, Balloux F, Hanihara T; Amos; Balloux; Hanihara (July 2007). "The effect of ancient population bottlenecks on human phenotypic variation". *Nature* **448** (7151): 346–8. Bibcode:2007Natur.448..346M. doi:10.1038/nature05951. PMC 1978547. PMID 17637668. Lay summary – *Science Daily*.

- Scholz CA, Johnson TC, Cohen AS, King JW, Peck JA, Overpeck JT, Talbot MR, Brown ET, Kalindekafe L, Amoako PY, Lyons RP, Shanahan TM, Castañeda IS, Heil CW, Forman SL, McHargue LR, Beuning KR, Gomez J, Pierson J; Johnson; Cohen; King; Peck; Overpeck; Talbot; Brown; Kalindekafe; Amoako; Lyons; Shanahan; Castañeda; Heil; Forman; McHargue; Beuning; Gomez; Pierson (October 2007). "East African megadroughts between 135 and 75 thousand years ago and bearing on early-modern human origins". *Proc. Natl. Acad. Sci. U.S.A.* **104** (42): 16416–21. Bibcode:2007PNAS..10416416S. doi:10.1073/pnas.0703874104. PMC 1964544. PMID 17785420. Lay summary – *Science Daily*.

- Cohen AS, Stone JR, Beuning KR, Park LE, Reinthal PN, Dettman D, Scholz CA, Johnson TC, King JW, Talbot MR, Brown ET, Ivory SJ; Stone; Beuning; Park; Reinthal; Dettman; Scholz; Johnson; King; Talbot; Brown; Ivory (October 2007). "Ecological consequences of early Late Pleistocene megadroughts in tropical Africa". *Proc. Natl. Acad. Sci. U.S.A.* **104** (42): 16422–7. Bibcode:2007PNAS..10416422C. doi:10.1073/pnas.0703873104. PMC 2034256. PMID 17925446. Lay summary – *Science Daily*.

- Smith L (2007-10-09). "Climate change led mankind out of Africa". Times Online.

- Russell S (2008-02-22). "DNA studies trace human migration from Africa". San Francisco Chronicle.

- Serre D, Langaney A, Chech M, Teschler-Nicola M, Paunovic M, Mennecier P, Hofreiter M, Possnert G, Pääbo S (March 2004). "No evidence of Neandertal mtDNA contribution to early modern humans". *PLoS Biol.* **2** (3): E57. doi:10.1371/journal.pbio.0020057. PMC 368159. PMID 15024415.

- Stringer, Chris (2011). *The Origin of Our Species*. London: Allen Lane. ISBN 978-1-84614-140-9.

13.10   External links

- The Human Family Tree – by Spencer Wells – National Geographic

- An mtDNA view of the peopling of the world by Homo sapiens (archived version)

- National Geographic: Atlas of the Human Journey

- Bradshaw Foundation: The Journey of Mankind

- *Human Evolution*. (2011). In Encyclopædia Britannica. Retrieved from http://www.britannica.com/EBchecked/topic/275670/human-evolution

Documentaries

- DNA Mysteries – The Search for Adam – by Spencer Wells – National Geographic, 2008

- The Real Eve: Modern Man's Journey Out of Africa – by Stephen Oppenheimer – Discovery Channel, 2002

- Journey of Man: A Genetic Odyssey (movie) by Spencer Wells – PBS and National Geographic Channel, 2003

Chapter 14

Upper Paleolithic

Venus of Laussel, an Upper Paleolithic (Aurignacian) carving.

The **Upper Paleolithic** (or **Upper Palaeolithic**, *Late Stone Age*) is the third and last subdivision of the Paleolithic or Old Stone Age as it is understood in Europe, Africa and Asia. Very broadly, it dates to between 50,000 and 10,000 years ago, roughly coinciding with the appearance of behavioral modernity and before the advent of agriculture.

14.1 Overview

See also: Recent African origin of modern humans and Behavioral modernity

Modern humans (*i.e. Homo sapiens*) are believed to have emerged about 195,000 years ago in Africa.[1][2] Though these humans were modern in anatomy, their lifestyle changed very little from their contemporaries, such as *Homo erectus* and the Neanderthals. They used the same crude stone tools. Archaeologist Richard G. Klein, who has worked extensively on ancient stone tools, describes the stone tool kit of archaic hominids as impossible to categorize. It was as if the Neanderthals made stone tools, and were not much concerned about their final forms. He argues that almost everywhere, whether Asia, Africa or Europe, before 50,000 years ago all the stone tools are much alike and unsophisticated.

About 50,000 years ago, there was a marked increase in the diversity of artifacts. For the first time in Africa, bone artifacts and the first art appear in the archeological record. The first evidence of human fishing is also noted, from artifacts in places such as Blombos cave in South Africa. Firstly among the artifacts of Africa, archeologists found they could differentiate and classify those of less than 50,000 years into many different categories, such as projectile points, engraving tools, knife blades, and drilling and piercing tools. These new stone-tool types have been described as being distinctly differentiated from each other, as if each tool had a specific purpose. Between 45,000 and 43,000 years ago, this new tool technology spread with human migration to Europe. The new technology generated a population explosion of modern humans which is believed to have led to the extinction of the Neanderthals. The invaders, commonly referred to as the Cro-Magnons, left many sophisticated stone tools, carved and engraved pieces on bone, ivory and antler, cave paintings and Venus figurines.[3][4][5]

This shift from the Middle to Upper Paleolithic is called

119

the Upper Paleolithic Revolution. The Neanderthals continued to use Mousterian stone tool technology and possibly Chatelperronian technology. These tools disappeared from the archeological record at around the same time the Neanderthals themselves disappeared from the fossil record, about 40,000 years ago.[6] The Upper Paleolithic has the earliest known evidence of organized settlements, in the form of campsites, some with storage pits. These were often located in narrow valley bottoms, possibly associated with hunting of passing herds of animals. Some sites may have been occupied year round, though more commonly they appear to have been used seasonally; peoples moved between the sites to exploit different food sources at different times of the year. Hunting was important, and caribou/wild reindeer "may well be the species of single greatest importance in the entire anthropological literature on hunting."[7]

Technological advances included significant developments in flint tool manufacturing, with industries based on fine blades rather than simpler and shorter flakes. Burins and racloirs were used to work bone, antler and hides. Advanced darts and harpoons also appear in this period, along with the fish hook, the oil lamp, rope, and the eyed needle.

Artistic work blossomed, with Venus figurines, cave painting, carvings and engravings on bone or ivory (such as the Swimming Reindeer), petroglyphs, and exotic raw materials found far from their sources, suggesting emerging trading links. More complex social groupings emerged, supported by more varied and reliable food sources and specialized tool types. This probably contributed to increasing group identification or ethnicity.[8] These group identities produced distinctive symbols and rituals which are an important part of modern human behavior.

The changes in human behavior have been attributed to the changes in climate during the period, which encompasses a number of global temperature drops. This meant a worsening of the already bitter climate of what is popularly (but incorrectly) called the last ice age. Such changes may have reduced the supply of usable timber and forced people to look at other materials. In addition, flint becomes brittle at low temperatures and may not have functioned as a tool.

Some scholars have argued that the appearance of complex or abstract language made these behavior changes possible. The complexity of the new human capabilities hints that humans were less capable of planning or foresight before 40,000 years, while the emergence of cooperative and coherent communication marked a new era of cultural development.[9] This theory is not widely accepted, since human phylogenetic separation dates to the Middle Palaeolithic (see Pre-language). While the latter view is better supported by phylogenetic inference, the material "evidence" is ambiguous.

14.2   Changes in climate and geography

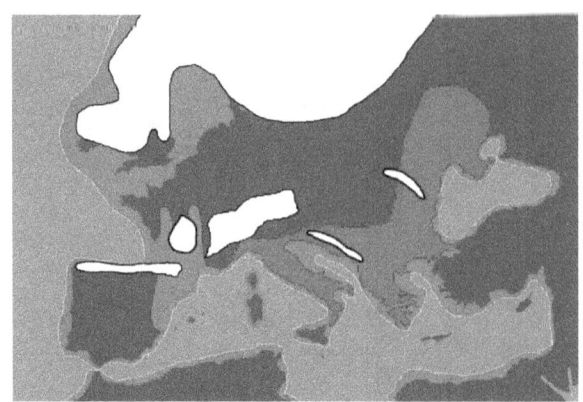

European LGM refuges, 18,000 BC.
Solutrean and Proto Solutrean Cultures
Epi Gravettian Culture

The climate of the period in Europe saw dramatic changes, and included the Last Glacial Maximum, the coldest phase of the last glacial period, which lasted from about 24,500 to 18,000–17,000 BC, being coldest at the end, before a relatively rapid warming (all dates vary somewhat for different areas, and in different studies). During the Maximum, most of Northern Europe was covered by an ice-sheet, forcing human populations into the areas known as Last Glacial Maximum refugia, including modern Italy and the Balkans, parts of the Iberian Peninsula and areas around the Black Sea. This period saw cultures such as the Solutrean in France and Spain. Human life may have continued on top of the ice sheet, but we know next to nothing about it, and very little about the human life that preceded the European glaciers. In the early part of the period, up to about 30,000 BC, the Mousterian Pluvial made northern Africa, including the Sahara, well-watered and with lower temperatures than today; after the end of the Pluvial the Sahara became arid.

The Last Glacial Maximum was followed by the Allerød oscillation, a warm and moist global interstadial that occurred around 11,500 to 10,800 BC. Then there was a very rapid onset, perhaps within as little as a decade, of the cold and dry Younger Dryas climate period, giving sub-arctic conditions to much of northern Europe. The Pre-Boreal rise in temperatures also began sharply around 9600 BC, and by its end around 8501 BC had brought temperatures nearly to present day levels, though the climate was wetter. This period saw the Upper Paleolithic give way to the start of the following Mesolithic cultural period.

As the glaciers receded sea levels rose; the English Channel,

Irish Sea and North Sea were land at this time, and the Black Sea a fresh-water lake. In particular the Atlantic coastline was initially far out to sea in modern terms in most areas, though the Mediterranean coastline has retreated far less, except in the north of the Adriatic and the Aegean. The rise in sea levels continued until at least 5,500 BC, so evidence of human activity along Europe's coasts in the Upper Paleolithic is mostly lost, though some traces are recovered by fishing boats and marine archaeology, especially from Doggerland, the lost area beneath the North Sea.

14.3 Timeline

Map of findings of Upper Paleolithic art in Europe.

14.3.1 50,000 BC

50,000 BC

- start of the Mousterian Pluvial in North Africa

45,000—43,000

- Earliest evidence of modern humans found in Europe, in Southern Italy.[10]

43,000—41,000

- At Ksar Akil in Lebanon, ornaments and skeletal remains of modern humans are dated to this period
- Denisova hominins live in the Altai Mountains

14.3.2 40,000 BC

40,000—35,000 BC

The Venus of Brassempouy is preserved in the Musée d'Archéologie Nationale at Saint-Germain-en-Laye, near Paris.

- Early cultural center in the Swabian Alps, earliest figurative art (Venus of Schelklingen), beginning of the Aurignacian
- The first flutes appear in Germany
- Lion-Human created from Hohlenstein-Stadel. It is now in Ulmer Museum, Ulm, Germany.

39,000 BC

- Most of the giant vertebrates and megafauna in Australia became extinct, around the time of the arrival of humans[11]

38,000 BC

- Examples of cave art in Spain are dated to around 38,000 BC, making them the oldest examples of art yet discovered in Europe. Scientists theorize that the paintings may have been made by Neanderthals, rather than by *homo sapiens.* (BBC) (*Science*)

38,000 BC—29,000 BC

- Wall painting with horses, rhinoceroses and aurochs, Chauvet Cave, Vallon-Pont-d'Arc, Ardéche gorge, France, is made. Discovered in December 1994.

35,000 BC

- Zar, Yataghyeri, Damjili and Taghlar caves in Azerbaijan
- First evidence of people inhabiting Japan [12]

32,000 BC

- Human populations around Europe figure out how to harden clay figures by firing them in an oven at high temperatures

30,000 BC

- First ground stone tools appear in Japan[13]
- Invention of the bow and arrow[14]
- End of the Mousterian Pluvial in North Africa

14.3.3   30,000 BC

29,000—25,000 BC

- Venus of Dolní Věstonice. It is the oldest known ceramic in the world.
- The Red Lady of Paviland lived around 29,000–26,000 years ago. Recent evidence has come to light that he was a tribal chief.

24,000 BC

- Start of the second Mousterian Pluvial in North Africa.

23,000 BC

- Venus of Petřkovice (*Petřkovická venuše* in Czech) from Petřkovice in Ostrava, Czech Republic, was made. It is now in Archeological Institute, Brno.

22,000 BC

- Last Glacial Maximum: Venus of Brassempouy, Grotte du Pape, Brassempouy, Landes, France, was made. It is now at Musée des Antiquités Nationales, St.-Germain-en-Laye.

- Venus of Willendorf, Austria, was made. It is now at Naturhistorisches Museum, Vienna.

20,000 BC

- End of the second Mousterian Pluvial in North Africa.

14.3.4   20,000 BC

Lascaux, a UNESCO World Heritage Site.

- Last Glacial Maximum. Mean Sea Levels are believed to be 110 to 120 meters (361 to 394 ft) *lower than present*,[15] with the direct implication that many coastal and lower riverine valley archaeological sites of interest are today under water.

18,000 BC

- Spotted Horses, Pech Merle cave, Dordogne, France are painted. Discovered in December 1994.

18,000 BC—11,000 BC

- Ibex-headed spear thrower, from Le Mas d'Azil, Ariège, France, is made. It is now at Musée de la Préhistoire, Le Mas d'Azil.

18,000 BC—12,000 BC

- Mammoth-bone village in Mezhirich, Ukraine is inhabited.

17,000 BC

- Spotted human hands, Pech Merle cave, Dordogne, France are painted. Discovered in December 1994.

17,000 BC—15,000 BC

- Hall of Bulls, Lascaux caves, is painted. Discovered in 1940. Closed to the public in 1963.

- Bird-Headed man with bison and Rhinoceros, Lascaux caves, is painted.

- Lamp with ibex design, from La Mouthe cave, Dordogne, France, is made. It is now at Musée des Antiquités Nationales, St.-Germain-en-Laye.

16,500 BC

- Paintings in Cosquer cave, where the cave mouth is now under water at Cap Margiou, France were made.

15,000 BC

- Bison, Le Tuc d'Audoubert, Ariège, France.

14.3.5   16,000 BC

15,000 BC–12,000 BC

- Paleo-Indians move across North America, then southward through Central America.

- Pregnant woman and deer (?), from Laugerie-Basse, France was made. It is now at Musée des Antiquités Nationales, St.-Germain-en-Laye.

14,000 BC

- Paleo-Indians searched for big game near what is now the Hovenweep National Monument.

- Bison, on the ceiling of a cave at Altamira, Spain, is painted. Discovered in 1879. Accepted as authentic in 1902.

- Domestication of Reindeer.[16]

13,000 BC

- Beginning of the Holocene extinction.

- earliest evidence of warfare (found in the Americas)

14.3.6   12,000 BC

11,500 BC—10,000 BC

- Wooden buildings in South America (Chile), first pottery vessels (Japan).

11,000 BC

- First evidence of human settlement in Argentina.

- The Arlington Springs Man dies on the island of Santa Rosa, off the coast of California.

- Human remains deposited in caves which are now located off the coast of Yucatán.[17]

10,500 BC

- Stone Age Creswellian culture settlement on Hengistbury Head dates from around this year

14.4   Cultures

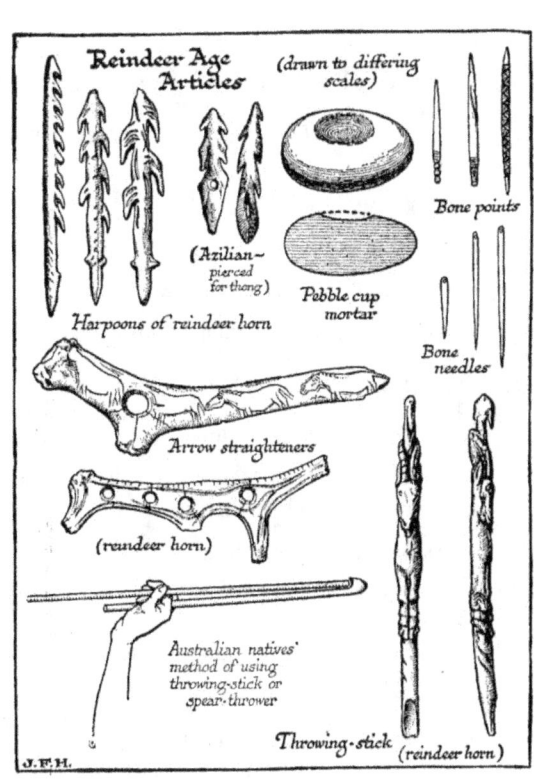

Reindeer Age articles

The Upper Paleolithic in the Franco-Cantabrian region:

- The Châtelperronian culture was located around central and south western France, and northern Spain. It appears to be derived from the Mousterian culture, and represents the period of overlap between Neanderthals and *Homo sapiens*. This culture lasted from approximately 45,000 BP to 40,000 BP.[6]

- The Aurignacian culture was located in Europe and south west Asia, and flourished between 32,000 BC and 21,000 BC. It may have been contemporary with the Périgordian (a contested grouping of the earlier Châtelperronian and later Gravettian cultures).

- The Gravettian culture was located across Europe. Gravettian sites generally date between 26,000 BC to 20,000 BC.

- The Solutrean culture was located in eastern France, Spain, and England. Solutrean artifacts have been dated to around 19000 BC before mysteriously disappearing around 15,000 BC.

- The Magdalenian culture left evidence from Portugal to Poland during the period from 16,000 BC to 8000 BC.

From the Synoptic table of the principal old world prehistoric cultures:

- central and east Europe:
    - 32,000 BC, Gravettian culture in southern Ukraine.[18]
    - 30,000 BC, Szeletian culture
    - 20,000 BC, Pavlovian, Aurignacian cultures
    - 11,000 BC, Ahrensburg culture
    - 10,000 BC, Epigravettian culture

- north and west Africa, and Sahara:
    - 30,000 BC, Aterian culture
    - 10,000 BC, Ibero-Maurusian (a.k.a. Oranian, Ouchtatian), and Sebilian cultures
    - 8000 BC, Capsian culture

- central, south, and east Africa:
    - 50,000 BC, Fauresmithian culture
    - 30,000 BC, Stillbayan culture
    - 10,000 BC, Lupembian culture
    - 9000 BC, Magosian culture
    - 7000 BC, Wiltonian culture
    - 3000 BC, beginning of hunter-gatherer art in southern Africa

- West Asia (including Middle East):
    - 50,000 BC, Jabroudian culture
    - 40,000 BC, Amoudian culture
    - 30,000 BC, Emirian culture
    - 20,000 BC, Aurignacian culture
    - 10,000 BC, Kebarian, Athlitian cultures

- south, central and northern Asia:
    - 30,000 BC, Angara culture
    - 9000 BC, Khandivili culture

- east and southeast Asia:
    - 80,000 BC, Ordosian culture
    - 50,000 BC, Ngandong culture
    - 30,000 BC, Sen-Doki culture
    - c. 14,000 BC, Jōmon period starts in Ancient Japan.
    - 10,000 BC, pre-Jōmon ceramic culture
    - 8000 BC, Hoabinhian culture
    - 7000 BC, Jōmon culture

14.5   See also

- Late Glacial Maximum
- Neolithic
- Neolithic Europe
- Behavioral modernity
- Cro-Magnon 1
- Sungir
- Cultural universal

14.6   References

- Gilman, Antonio (1996). "Explaining the Upper Palaeolithic Revolution". Pp. 220–239 (Chap. 8) in *Contemporary Archaeology in Theory: A Reader*. Cambridge, MA: Blackwell.

[1] Pleistocene Homo sapiens from Middle Awash, Ethiopia **Nature 423, 742-747** (12 June 2003) http://www.nature.com/nature/journal/v423/n6941/pdf/nature01669.pdf

[2] Out of Africa: modern human origins special feature: middle and later Pleistocene hominins in Africa and Southwest Asia **Proc Natl Acad Sci U S A**. 22 September 2009;106(38):16046-50. http://www.ncbi.nlm.nih.gov/pmc/articles/PMC2752549/pdf/zpq16046.pdf

[3] Biological origins of modern human behavior part3

[4] Biological origins of modern human behavior part 1

[5] "'Modern' Behavior Began 40,000 Years Ago In Africa", *Science Daily*, July 1998

[6] http://www.nature.com/nature/journal/v512/n7514/full/nature13621.html

[7] "In North America and Eurasia the species has long been an important resource—in many areas *the* most important resource—for peoples' inhabiting the northern boreal forest and tundra regions. Known human dependence on caribou/wild reindeer has a long history, beginning in the Middle Pleistocene (Banfield 1961:170; Kurtén 1968:170) and continuing to the present....The caribou/wild reindeer is thus an animal that has been a major resource for humans throughout a tremendous geographic area and across a time span of tens of thousands of years." Ernest S. Burch, Jr. "The Caribou/Wild Reindeer as a Human Resource", *American Antiquity*, Vol. 37, No. 3 (July 1972), pp. 339–368.

[8] Gilman, Antonio. 1996. Explaining the Upper Palaeolithic Revolution. Pp. 220-239 (Chap. 8) in Contemporary Archaeology in Theory: A Reader. Cambridge, MA: Blackwell

[9] "No Last Word on Language Origins", Bellarmine University

[10] http://www.nytimes.com/2011/11/03/science/fossil-teeth-put-humans-in-europe-earlier-than-thought.html?scp=1&sq=kents%20cavern&st=cse&_r=0

[11] "Humans killed off Australia's giant beasts". BBC News. 24 March 2012.

[12] Prehistoric Archaeological Periods in Japan, Charles T. Keally

[13] "Prehistoric Japan, New perspectives on insular East Asia", Keiji Imamura, University of Hawaii Press, Honolulu, ISBN 0-8248-1853-9

[14] McClellan, pg 11

[15] Sea level data from *main article:* Cosquer cave

[16] Lloyd, J. & Mitchinson, J.: *The Book of General Ignorance*. Faber & Faber, 2006.

[17] http://www.msnbc.msn.com/id/5955043/

[18] Carpenter, Jennifer (20 June 2011). "Early human fossils unearthed in Ukraine". BBC. Retrieved 21 June 2011.

14.7 External links

- The Upper Paleolithic Revolution

- Picture Gallery of the Paleolithic (reconstructional palaeoethnology), Libor Balák at the Czech Academy of Sciences, the Institute of Archaeology in Brno, The Center for Paleolithic and Paleoethnological Research

Chapter 15

Later Stone Age

The **Later Stone Age** (or **LSA**) is a period in African prehistory which follows the Early Stone Age and Middle Stone Age. The Later Stone Age along with the Early Stone Age and Middle Stone Age are often confused with the Lower Paleolithic, Middle Paleolithic, and Upper Paleolithic. In the 1920s, it became clear to archaeologists that the existing chronological system of Upper, Middle and Lower Paleolithic were not a suitable correlate to the prehistoric past in Africa. The terms Early, Middle, and Later Stone Age were developed to address this issue. Some scholars, however, still view these two chronologies as parallel, arguing that they both represent the development of behavioral modernity.[1] The Later Stone Age is associated with the advent of modern human behavior in Africa, although definitions of this concept and means of studying it are up for debate. The transition from the Middle Stone Age to the Later Stone Age is thought to have occurred first in eastern Africa between 50,000 and 39,000 years ago. It is also thought that Later Stone Age peoples and/or their technologies spread out of Africa over the next several thousand years.[2]

15.1 Origins

Originally, the Later Stone Age was defined as several stone industries and/or cultures which included other evidence of human activity, such as ostrich eggshell beads and worked bone implements, and lacked Middle Stone Age stone tools other than those recycled and reworked. LSA peoples were directly linked with biologically and behaviorally modern populations of hunter/gatherers, some being directly identified as San "Bushmen." This definition has changed since its creation with the discovery of ostrich eggshell beads and bone harpoons in contexts which predate the LSA by tens of thousands of years.[3] The Later Stone Age was also long distinguished from the earlier Middle Stone Age as the time in which modern human behavior developed in Africa. This definition has become more tenuous as evidence for such modern human behaviors is found in sites which predate the LSA significantly.

15.2 Transition from Middle Stone Age

The LSA follows the Middle Stone Age and begins about 50,000 years ago. The LSA is characterized by a wider variety in stone artifacts than in the previous MSA period. These artifacts vary with time and location, unlike Middle Stone Age technology which appeared to have been relatively unchanged for several hundreds of thousands of years. LSA technology is also characterized by the use of bone tools. The LSA was associated with modern human behavior,[4] but this view was modified after discoveries in MSA sites such as Blombos Cave and Pinnacle Point.

LSA sites also greatly outnumber MSA sites in Africa, a trend that could indicate an increase in population numbers. The greater number of LSA sites could also result from bias towards better preservation of younger sites which have had fewer chances to be destroyed.[5]

15.3 Lithic Technology

Differences in stone tool technologies are often used to distinguish between the Middle Stone Age and the Later Stone Age. The larger prepared platform flake-based stone tool industries of the Middle Stone Age, such as Levallois were increasingly replaced with industries that focused on producing blades and bladelets on cores with simple platforms.[6] African stone tool technologies are divided into modes as proposed by Grahame Clark in 1969 and outlined by Lawrence Barham and Peter Mitchell as follows:[7]

- Mode 1: Oldowan tool industries, also known as pebble tool industries

- Mode 2: Tools made through bifacial reduction produced from large flakes or cores

- Mode 3: Flake tools from prepared cores

- Mode 4: Punch-struck blades that are adapted into a variety of different tools

- Mode 5: Microlith portions of composite tools that may include wood or bone, often abruptly retouched or backed

The lithic technologies of the Later Stone Age often fall into Modes 4 and 5. They have been further broken into four stages within the LSA.[3]

1. Microlithic industries dated to between ca. 40,000 and ca. 19,000 B.P. labeled early LSA (ELSA), or as late MSA, or as MSA/LSA transitions or interfaces

2. Nonmicrolithic, bladelet-poor industries with dates between ca. 40,000 B.P. and ca. 19,000 B.P.

3. Microlithic industries with bladelets dated between ca. 18,000 and ca. 12,000 B.P.

4. Nonmicrolithic, bladelet-poor industries dating between 12,000 and 8000 B.P.

15.4   Potential problems

The end of the Later Stone Age took place when groups adopted technologies such as metallurgy to replace the use of stone tools. This process happened at different rates across the continent, and it is worth noting that the term "LSA" is typically used by archaeologists today to refer primarily to stone tool-using hunter/gatherer populations in southern Africa. The model of the LSA "*human revolution*" is no longer favored by many archaeologists working in Africa due to the increasing evidence for development of modern human behavior earlier than 40,000-50,000 years ago.

15.5   See also

- Upper Paleolithic

- Middle Stone Age

- Enkapune Ya Muto

- Mumba Cave

- Mumbwa Cave

15.6   Footnotes

[1] Henshilwood, Christopher S.; Marean, Curtis W. (December 2003). "The Origin of Modern Human Behavior". *Current Anthropology* **44** (5): 627–651. doi:10.1086/377665.

[2] Ambrose, Stanley H. (1998). "Chronology of the Later Stone Age and Food Production in East Africa". *Journal of Archaeological Science* **25**: 377–392. doi:10.1006/jasc.1997.0277.

[3] Wadley, Lyn (September 1993). "The Pleistocene Later Stone Age South of the Limpopo River". *Journal of World Prehistory* **7** (3): 243–296. doi:10.1007/bf00974721.

[4] Klein, Richard (2003). "Body before behavior". *The Dawn of Human Culture*. New York, N.Y.: Wiley. ISBN 0-471-25252-2.

[5] McBrearty S., Brooks, A. S. (2000). The revolution that wasn't: a new interpretation of the origin of modern human behavior. J Hum Evol. 39(5):453–563. doi:10.1006/jhev.2000.0435 PMID 11102266

[6] Ambrose, Stanley H. (1998). "Chronology of the Later Stone Age and Food Production in East Africa". *Journal of Archaeological Science* **25**: 377–392. doi:10.1006/jasc.1997.0277.

[7] Barham, Lawrence; Mitchell, Peter (2009). *The First Africans: African Archaeology From the Earliest Toolmakers to Most Recent Foragers*. New York: Cambridge University Press.

15.7   Further reading

- Deacon, Hilary (1999). "Learning about the past". *Human Beginnings in South Africa*. Cape Town: D. Phillips. ISBN 0-86486-417-5.

- The Stone Age of southern Tanzania

Chapter 16

Behavioral modernity

Behavioral modernity is a suite of behavioral and cognitive traits that distinguishes current *Homo sapiens* from anatomically modern humans, hominins, and other primates. Although often debated, most scholars agree that modern human behavior can be characterized by abstract thinking, planning depth, symbolic behavior (e.g. art, ornamentation, music), exploitation of large game, blade technology, among others.[1][2] Underlying these behaviors and technological innovations are cognitive and cultural foundations that have been documented experimentally and ethnographically. Some of these human universal patterns are cumulative cultural adaptation, social norms, language, cooperative breeding, and extensive help and cooperation beyond close kin.[3] These traits have been viewed as largely responsible for the human replacement of Neanderthals in Western Europe, along with the climatic conditions of the Last Glacial Maximum, and the peopling of the rest of the world.[2][4]

Arising from differences in the archaeological record, a debate continues as to whether anatomically modern humans were behaviorally modern as well. There are many theories on the evolution of behavioral modernity. These generally fall into two camps: gradualist and cognitive approaches. The Later Upper Paleolithic Model refers to the idea that modern human behavior arose through cognitive, genetic changes abruptly around 40–50,000 years ago.[5] Other models focus on how modern human behavior may have arisen through gradual steps; the archaeological signatures of such behavior only appearing through demographic or subsistence-based changes.[1][2][6][7][8]

16.1 Definition

In order to classify what traits should be included in modern human behavior, it is necessary to define behaviors that are universal among living human groups. Examples of these human universals are abstract thought, planning, trade, cooperative labor, body decoration, control and use of fire, among others. Along with these traits, humans possess a heavy reliance on social learning.[9][10] This cumulative cultural change or cultural "ratchet" separates human culture from social learning in animals. As well, a reliance on social learning may be responsible in part for humans' rapid adaptation to many environments outside of Africa.

There is also an important distinction to be made between when humans developed the ability to *invent*, in contrast to developing the ability to *adopt*, modern human behavior. As a modern analogy, there is no shortage of musicians in the world trying to compose new and original music, but only a handful every year that successfully manage to compose lasting worldwide hit songs; yet essentially all of the other aspiring composer musicians can almost trivially learn to play those hit songs once they've heard them (with analogous undertakings in literature, art, science and technology etc.). A dramatic and sudden increase in complexity of human behavior is thus fully plausible even if significantly less than 1% of humanity developed the genetic ability to "invent", provided that the remaining 99% had no significant problems with "adopting" those inventions. There is potentially an evolutionary abyss between *inventing* and *adopting*; for instance, *Homo erectus* and *Homo ergaster* produced with little advancement essentially the same sharpened stone tools for over a million years, but there is no scientific evidence at hand that could prove that they were incapable of producing composite stone tools, such as spears, if shown how to do so.

It is thus not established if the early *Homo sapiens* had the genetic requirements to be able to adopt modern human behavior, such as religious beliefs, through cultural interaction. If indeed the early *Homo sapiens* had the ability to learn modern human behavior, once invented by other groups, there is no geographic restriction where modern behavior originated. However, if the early *Homo sapiens* hypothetically were genetically inhibited from adopting modern human behaviors, since cultural universals are found in all cultures including some of the most isolated indigenous groups, these traits must have evolved or have been invented in Africa prior to the exodus.[11][12][13][14]

Archaeologically a number of empirical traits have been used as indicators of modern human behavior. While these are often debated[15] a few are generally agreed upon. Archaeological evidence of behavioral modernity are:[2][5]

- burial

- fishing

- figurative art (cave paintings, petroglyphs, figurines)

- systematic use of pigment (such as ochre) and jewelry for decoration or self-ornamentation

- Using bone material for tools

- Transport of resources long distances

- Blade technology

- Diversity, standardization, and regionally distinct artifacts

- Hearths

- Composite tools

16.1.1 Critiques

Several critiques have been placed against the traditional concept of behavioral modernity, both methodologically and philosophically.[2][15] Shea (2011) outlines a variety of problems with this concept, arguing instead for "behavioral variability", which, according to the author, better describes the archaeological record. The use of trait lists, according to Shea (2011), runs the risk of taphonomic bias, where some sites may yield more artifacts than others despite similar populations; as well, trait lists can be ambiguous in how behaviors may be empirically recognized in the archaeological record.[15] Shea (2011) in particular cautions that population pressure, cultural change, or optimality models, like those in human behavioral ecology, might better predict changes in tool types or subsistence strategies than a change from "archaic" to "modern" behavior.[15] Some researchers argue that a greater emphasis should be placed on identifying only those artifacts which are unquestionably, or purely, symbolic as a metric for modern human behavior.[2]

16.2 Theories and Models

16.2.1 Late Upper Paleolithic Model or "Revolution"

The Late Upper Paleolithic Model, or Upper Paleolithic Revolution, refers to the idea that, though anatomically modern humans first appear around 150,000 years ago, they were not cognitively or behaviorally "modern" until around 50,000 years ago, leading to their expansion into Europe and Asia.[5][16][17] These authors note that traits used as a metric for behavioral modernity do not appear as a package until around 40–50,000. Klein (1995) specifically describes evidence of fishing, bone shaped as a tool, hearths, significant artifact diversity, and elaborate graves are all absent before this point.[5] Although assemblages before 50,000 years ago show some diversity the only distinctly modern tool assemblages appear in Europe at 48,000.[16] According to these authors, art only becomes common beyond this switching point, signifying a change from archaic to modern humans.[5] Most researchers argue that a neurological or genetic change, perhaps one enabling complex language such as FOXP2, caused this revolutionary change in our species.[5][17]

16.2.2 Alternative Models

Contrasted with this view of a spontaneous leap in cognition among ancient humans, some authors, primarily working in African archaeology, point to the gradual accumulation of "modern" behaviors, starting well before the 50,000 year benchmark of the Upper Paleolithic Revolution models.[1][2][18] Howiesons Poort, Blombos, and other South African archaeological sites, for example, show evidence of marine resource acquisition, trade, and abstract ornamentation at least by 80,000 years ago.[1][6] Given evidence from Africa and the Middle East, a variety of hypotheses have been put forth to describe an earlier, gradual transition from simple to more complex human behavior. Some authors have pushed back the appearance of fully modern behavior to around 80,000 years ago in order to incorporate the South African data.[18]

Others focus on the slow accumulation of different technologies and behaviors across time. These researchers[1][2] describe how anatomically modern humans could have been cognitively the same and what we define as behavioral modernity is just the result of thousands of years of cultural adaptation and learning. D'Errico and others have looked at Neanderthal culture rather than early human behavior for clues into behavioral modernity.[4] Noting that Neanderthal assemblages often portray similar traits as those listed for modern human behavior, researchers stress that the foundations for behavioral modernity may in fact lay deeper in our hominin ancestors.[19] If both modern humans and Neanderthals express abstract art and complex tools then "modern human behavior" cannot be a derived trait for our species.

Cultural evolutionary models may also shed light on why although evidence of behavioral modernity exists before

50,000 years ago it is not expressed consistently until that point. With small population sizes, human groups would have been affected by demographic and cultural evolutionary forces that may not have allowed for complex cultural traits.[7][8][9][10] According to some authors[7] until population density became significantly high, complex traits could not have been maintained effectively. It is worth noting that some genetic evidence supports a dramatic increase in population size before human migration out of Africa.[17] High local extinction rates within a population also can significantly decrease the amount of diversity in neutral cultural traits, regardless of cognitive ability.[8]

16.3 Archaeological Evidence

16.3.1 Africa

Before the "Out of Africa" theory was generally accepted, there was no consensus on where our species evolved and, consequently, where modern human behavior arose. Now, however, African archaeology has become extremely important in discovering where our species began. Since human expansion into Europe around 48,000 years ago is generally accepted as already "modern",[16] the question becomes whether behavioral modernity appeared in Africa well before 50,000 years ago as a late Upper Paleolithic "revolution" which prompted migration out of Africa, or arose outside Africa and diffused back.

A variety of evidence of abstract imagery, widened subsistence strategies, and other "modern" behaviors have been discovered in Africa, especially South Africa. The Blombos Cave site in South Africa, for example, is famous for rectangular slabs of ochre engraved with geometric designs. Using multiple dating techniques, the site was confirmed to be around 77,000 years old.[20] Beads and other personal ornamentation have been found from Morocco which might be as old as 130,000 years old; as well, the Cave of Hearths in South Africa has yielded a number of beads significantly before 50,000 years ago.[1]

Expanding subsistence strategies beyond big-game hunting and the consequential diversity in tool types has been noted as signs of behavioral modernity. A number of South African sites have shown an early reliance on aquatic resources from fish to shellfish. Pinnacle Point, in particular, shows exploitation of marine resources as early as 120,000 years ago, perhaps in response to more arid conditions inland.[6] Establishing a reliance on predictable shellfish deposits, for example, could reduce mobility and facilitate complex social systems and symbolic behavior. Blombos Cave and Site 440 in Sudan both show evidence of fishing as well. Taphonomic change in fish skeletons from

Blombos Cave have been interpreted as capture of live fish, clearly an intentional human behavior.[1]

16.3.2 Europe

While traditionally described as evidence for the later Upper Paleolithic Model,[5] European archaeology has shown that the issue is more complex. A variety of stone tool technologies are present at the time of human expansion into Europe and show evidence of modern behavior. Despite the problems of conflating specific tools with cultural groups, the Aurignacian tool complex, for example, is generally taken as a purely modern human signature.[21][22] The discovery of "transitional" complexes, like "proto-Aurignacian", have been taken as evidence of human groups progressing through "steps of innovation".[21] If, as this might suggest, human groups were already migrating into eastern Europe around 40,000 years and only afterward show evidence of behavioral modernity, then either the cognitive change must have diffused back into Africa or was already present before migration.

In light of a growing body of evidence of Neanderthal culture and tool complexes some researchers have put forth a "multiple species model" for behavioral modernity.[4][19][23] Neanderthals were often cited as being an evolutionary dead-end, apish cousins who were less advanced than their human contemporaries. Personal ornaments were relegated as trinkets or poor imitations compared the cave art produced by *H. sapiens*. Despite this, European evidence has shown a variety of personal ornaments and artistic artifacts produced by Neanderthals; for example, the Neanderthal site of Grotte du Renne has produced grooved bear, wolf, and fox incisors, ochre and other symbolic artifacts.[23] Though burials are few and controversial, there have been circumstantial evidence of Neanderthal ritual burials.[19] There are two options to describe this symbolic behavior among Neanderthals: they copied cultural traits from arriving modern humans or they had their own cultural traditions comparative with behavioral modernity. If they just copied cultural traditions, which is debated by several authors,[4][19] they still possessed the capacity for complex culture described by behavioral modernity. As discussed above, if Neanderthals also were "behaviorally modern" then it cannot be a species-specific derived trait.

16.3.3 Asia

Most debates surrounding behavioral modernity have been focused on Africa or Europe but an increasing amount of focus has been placed on East Asia. This region offers a unique opportunity to test hypotheses of multi-regionalism, replacement, and demographic effects.[24] Unlike Europe, where initial migration occurred around 50,000 years ago, human remains have been dated in China to around 100,000 years ago.[25] This early evidence of human expansion calls into question behavioral modernity as an impetus for migration.

Stone tool technology is particularly of interest in East Asia. Following Homo erectus migrations out of Africa, Acheulean technology never seems to appear beyond present-day India and into China. Analogously, Mode 3, or Levallois technology, is not apparent in China following later hominin dispersals.[26] This lack of more advanced technology has been explained by serial founder effects and low population densities out of Africa.[27] Though tool complexes comparative to Europe are missing or fragmentary, other archaeological evidence shows behavioral modernity. For example, the peopling of the Japanese archipelago offers an opportunity to investigate the early use of watercraft. Though one site, Kanedori in Honshu, does suggest the use of watercraft as early as 84,000 years ago, there is no other evidence of hominins in Japan until 50,000 years ago.[24]

The Zhoukoudian cave system near Beijing has been excavated since the 1930s and has yielded precious data on early human behavior in East Asia. Though disputed, there is evidence of possible human burials and interred remains in the cave dated to around 34-20,000 years ago.[24] These remains have associated personal ornaments in the form of beads and worked shell, suggesting symbolic behavior. Along with possible burials, numerous other symbolic objects like punctured animal teeth and beads, some dyed in red ochre, have all been found at Zhoukoudian.[24] Though fragmentary, the archaeological record of eastern Asia shows evidence of behavioral modernity before 50,000 years ago but, like the African record, it is not fully apparent until that time.

16.4 See also

16.5 References

[1] McBrearty, Sally; Brooks, Allison (2000). "The revolution that wasn't: a new interpretation of the origin of modern human behavior". *Journal of Human Evolution* **39**: 453–563.

[2] Henshilwood, Christopher; Marean, Curtis (2003). "The Origin of Modern Human Behavior: Critique of the Models and Their Test Implications". *Current Anthropology* **44** (5): 627–651.

[3] Hill, Kim; et al. (2009). "The Emergence of Human Uniqueness: Characters Underlying Behavioral Modernity". *Evolutionary Anthropology* **18**: 187–200.

[4] D'Errico, F; et al. (1998). "Neanderthal Acculturation in Western Europe? A Critical Review of the Evidence and Its Interpretation". *Current Anthropology* **39** (S1): S1-S44.

[5] Klein, Richard (1995). "Anatomy, behavior, and modern human origins". *Journal of World Prehistory* **9**: 167–198.

[6] Marean, Curtis; et al. (2007). "Early human use of marine resources and pigment in South Africa during the Middle Pleistocene". *Nature* **449**.

[7] Powell, Adam; et al. (2009). "Late Pleistocene Demography and the Appearance of Modern Human Behavior". *Science* **324**: 1298–1301.

[8] Premo, Luke; Kuhn, Steve (2010). "Modeling Effects of Local Extinctions on Culture Change and Diversity in the Paleolithic". *PLoS One* **5** (12).

[9] Boyd, Robert; Richerson, Peter (1988). *Culture and the Evolutionary Process* (2 ed.). University of Chicago Press. ISBN 9780226069333.

[10] Nakahashi, Wataru (2013). "Evolution of improvement and cumulative culture". *Theoretical Population Biology* **83**: 30–38.

[11] Wade, Nicholas (2003-07-15). "leap to language". New York Times. Retrieved 2009-09-10.

[12] Buller, David (2005). *Adapting Minds: Evolutionary Psychology and the Persistent Quest for Human Nature*. PMIT Press. p. 468. ISBN 0-262-02579-5.

[13] "80,000-year-old Beads Shed Light on Early Culture". Livescience.com. 2007-06-18. Retrieved 2009-09-10.

[14] "three distinct human populations". Accessexcellence.org. Retrieved 2009-09-10.

[15] Shea, John (2011). "Homo sapiens Is as Homo sapiens Was". *Current Anthropology* **52** (1): 1–35.

[16] Hoffecker, John (2009). "The spread of modern humans in Europe". *PNAS* **106** (38): 16040–16045.

[17] Tattersall, Ian (2009). "Human origins: Out of Africa". *PNAS* **106** (38): 16018–16021.

[18] Foley, Robert; Lahr, Marta (1997). "Mode 3 Technologies and the Evolution of Modern Humans". *Cambridge Archaeological Journal* **7** (1): 3–36.

[19] D'Errico, Francesco (2003). "The Invisible Frontier A Multiple Species Model for the Origin of Behavioral Modernity". *Evolutionary Anthropology* **12**.

[20] Henshilwood, Christopher; et al. (2002). "Emergence of Modern Human Behavior: Middle Stone Age Engravings from South Africa". *Science* **295** (5558): 1278–1280.

[21] Joris, Olaf; Street, Martin (2008). "At the end of the 14C time scaledthe Middle to Upper Paleolithic record of western Eurasia". *Journal of Human Evolution* **55**: 782–802.

[22] Anikovich, M.; et al. (2007). "Early Upper Paleolithic in Eastern Europe and Implications for the Dispersal of Modern Humans". *Science* **315** (5809): 223–226.

[23] Abadia, Oscar Moro; Gonzalez Morales, Manuel R. (2010). "REDEFINING NEANDERTHALS AND ART: AN ALTERNATIVE INTERPRETATION OF THE MULTIPLE SPECIES MODEL FOR THE ORIGIN OF BEHAVIOURAL MODERNITY". *Oxford Journal of Archaeology* **29** (3): 229–243.

[24] Norton, Christopher; Jin, Jennie (2009). "The Evolution of Modern Human Behavior in East Asia: Current Perspectives". *Evolutionary Anthropology* **18**: 247–260.

[25] Liu, Wu; et al. (2010). "Human remains from Zhirendong, South China, and modern human emergence in East Asia". *PNAS* **107** (45): 19201–19206.

[26] Norton, Christopher; Bae, K. (2008). "The Movius Line sensu lato (Norton et al. 2006) further assessed and defined". *Journal of Human Evolution* **55**: 1148–1150.

[27] Lycett, Stephen; Norton, Christopher (2010). "A demographic model for Palaeolithic technological evolution: The case of East Asia and the Movius Line". *Quaternary International* **211**: 55–65.

16.6   External links

- Steven Mithen (1999), *The Prehistory of the Mind: The Cognitive Origins of Art, Religion and Science*, Thames & Hudson, ISBN 978-0-500-28100-0.

- Artifacts in Africa Suggest An Earlier Modern Human

- Tools point to African origin for human behaviour

- Key Human Traits Tied to Shellfish Remains, nytimes 2007/10/18

- "Python Cave" Reveals Oldest Human Ritual, Scientists Suggest

Chapter 17

Origin of the domestic dog

This article is about the origin of the domestic dog. For dog breeding, see Dog breeding.

The **origin of the domestic dog** (*Canis lupus familiaris* or *Canis familiaris*) is not clear. Whole genome sequencing indicates that the dog, the gray wolf, and the extinct Taymyr wolf diverged at around the same time 27,000–40,000 years before present.[1] These dates imply that the earliest dogs arose in the time of human hunter-gatherers and not agriculturists.[2] Modern dogs are more closely related to ancient wolf fossils that have been found in Europe than they are to modern gray wolves,[3] with nearly all genetic commonalities with the gray wolf due to admixture,[2] but several Arctic dog breeds have commonalities with the Taymyr wolf of North Asia due to admixture.[1]

The dog diverged immediately prior to the Last Glacial Maximum when much of Eurasia was a steppe/tundra biome.

17.1 Dog evolution

Dog evolution (from the Latin *evolutio*: "unrolling")[4] is the biological descent with modification[5] that led to the domestic dog. This process encompasses small-scale evolution (changes in gene frequency in a population from one generation to the next) and large-scale evolution (the descent of different species from a common ancestor over many generations).[6]

17.1.1 Paleoecology

The dog came into being towards the last peak of the last Ice Age, in very cold and dry climatic conditions.

During the peak of the last Ice Age - known as the last glacial maximum - a vast mammoth steppe stretched from Spain across Eurasia and over the Bering land bridge into Alaska and the Yukon. The continent of Europe was much colder and drier than it is today, with polar desert in the north and the remainder steppe or tundra. Forest and woodland was almost non-existent, except for isolated pockets in the mountain ranges of southern Europe.[7] The Late Pleistocene was characterized by a series of severe and rapid climate oscillations with regional temperature changes of up to 16 °C, which has been correlated with Pleistocene megafaunal extinctions. There is no evidence of megafaunal extinctions at the height of the LGM, indicating that increasing cold and glaciation were not factors. Multiple events appear to also involve the rapid replacement of one species by one within the same genus, or one population by another within the same species, across a broad area. As some species became extinct, so too did their predators.[8] Modern humans' ancestors first reached Europe with their remains dated 43,000-45,000 years BP discovered in Italy[9] and in Britain.[10]

Into this environment came the dog.

See further: Paleoecology at this time

17.1.2 Wolf-like lineage

Early DNA studies indicated that the dog is descended from a wolf-like lineage.

In 1868, Charles Darwin proposed that domestic dogs were phenotypically so diverse that they likely had originated from two or more wild canis species.[11]:16 All species within the *Canis* genus, the wolf-like canids, are phylogenetically closely related with 78 chromosomes and can potentially interbreed.[12] Later, others thought that "the wolf is the most probable ancestor and closest relative of the domestic dog."[13]:54

The development of molecular biology allows the inference of evolutionary relationships about species and to represent them in a phylogenetic tree; however, these are not without their limitations. DNA studies may give unresolvable results due to the specimens selected, the genome technology used, and the assumptions made by the researchers.[14] A panel of genetic markers can be chosen, for example from mitochondrial Cytochrome b. The techniques used to extract, locate and compare sequences can be applied using advances in genome technology to observe longer lengths of base pairs that give better phylogenetic resolution.[15] These techniques can be applied to the maternal mitochondrial control region, mitochondrial D-loop, mitochondrial genome, paternal Y chromosomes, microsatellites, single nucleotide polymorphisms, nuclear DNA and the whole genome.

In June 1993, a study of 736 base pairs of the mitochondrial Cytochrome b gene of the wolf-like canids found that there was a close kinship between domestic dogs, gray wolves, coyotes and Simien jackals, but with a distance from the African wild dog and from the golden, side-striped and black-backed jackals. The domestic dog was "an extremely close relative of the gray wolf, differing from it by at most 0.2% of Mitochondrial DNA [mDNA] sequence. In comparison, the gray wolf differs from its closest wild relative, the coyote, by about 4% of mDNA sequence." Therefore, the study concluded that the molecular genetic evidence did not support theories that dogs arose from jackal ancestors. The study warned of "the need for caution in the interpretation of phylogenies based on mDNA; such gene trees are not necessarily species trees and may not accurately reflect the phylogenetic affiliations or divergence time." The study proposed the hypothesis that because of the diversity of dog remains found in archaeological sites, dogs may be derived from several different ancestral gray wolf populations.[16]

In the same year 1993, the domestic dog *Canis familiaris* was reclassified as *Canis lupus familiaris*, a sub-species of *Canis lupus*, in *Mammal Species of the World*.[17][18] However, *Canis familiaris* is also accepted due to a nomenclature debate over the naming of wild and domestic subspecies.[19]

In 1995, a researcher found that the archaeological record did not show any distinctive dog breeds until 3,000–4,000 years ago and that most breeds had been developed over the past seven hundred years.[20]

In 1997, a study was conducted of 261 base pairs on the control region of 140 dogs across 67 breeds and 162 wolves across 27 populations. The control region of the dogs and wolves was highly polymorphic, with dogs revealing 26 haplotypes, and wolves revealing 27 haplotypes, of which four had a widespread distribution. Dog haplotypes could not be partitioned into breeds: for example, eight German shepherds revealed five distinct sequences, and six golden retrievers revealed four sequences. No dog sequence differed from any wolf sequence by more than 12 substitutions but differed from coyote sequences by at least 20 substitutions and 2 insertions, clearly supporting a wolf ancestry for dogs. The 26 dog haplogroups formed four distinct clades; Clade 1 included 19 haplotypes, Clade 2 included one haplotype, Clade 3 included three haplotypes, and Clade 4 included three haplotypes which were identical or very similar to a wolf haplotype found in Romania and western Russia and which suggests recent hybridization. A further analysis of 1,030 base pairs on the control region of 24 dogs provided greater phylogenetic resolution and supported four monophyletic clades. The wolf and coyote ancestral lines diverged one million years ago based on the fossil record, and comparing their sequence divergence with those of the dog implies that dogs diverged 135,000 years ago. The estimate may be inflated as it is based on some assumptions, but it implies that the origin of dogs is more ancient than 14,000 YBP as suggested by the archaeological record. The study proposed the hypothesis that because dog haplotypes form four monophyletic clades, early domestic dogs may not have been morphologically distinct from their wild relatives until the change 10,000-15,000 years ago when humans moved from nomadic hunter-gatherer societies into agricultural societies and imposed selective regimes that resulted in marked phenotopic divergence from wolves. After the origin of dogs from a wolf ancestor, dogs and wolves may have continued to exchange genes.[21]

In 1999, a review of the scientific literature regarding the genetic origin of the dog proposed a number of hypotheses. The molecular data indicated that dogs have protein alleles in common with wolves, share highly polymorphic microsatellites, and have mitochondrial DNA sequences similar or identical to those found in gray wolves. The mitochondrial control region DNA sequences show an average divergence between dogs and wolves of 1.5%, compared to dogs and coyotes, their next-closest relative, at 7.5%. Therefore, this indicated that the origin of the dog was from wolves. The archaeological record suggests that dogs were in Europe and the Middle East approximately 14,000 years ago, but the genetic record shows 135,000 YBP, which indicates that the morphological change was associated with artificial selection as humans shifted from hunter-gatherer to agrarian societies. Alternately, dogs may have had a more recent origin but are descended from a now extinct species of canid whose closest living relative was the gray wolf. The DNA sequences of dogs form four distinct clades, each with a separate ancestry from wolves that indicates four separate domestication events. One of the clades shows a wolf sequence that is identical to a dog sequence, suggesting a very recent interbreeding or domestication event. Once dogs were domesticated and spread over a wide area, occasional interbreeding would have transferred wolf mDNA to them.[12]

See further: Hybrid

17.1.3 Probable ancestor

During the LGM, there were two types of wolf. A large, heavily-built megafaunal wolf spanned the cold north of the Holarctic that specialised in preying on megafauna. Another more gracile form lived in the warmer south in refuges from the glaciation. When the planet warmed and the LGM came to an end, whole species of megafauna became extinct along with their predators, leaving the more gracile form to dominate the Holarctic. This wolf we know today as the modern gray wolf, which is the dog's sister but not its ancestor - the dog shows a closer genetic relationship with the extinct megafaunal wolf.

Within the species *Canis lupus*, phylogenetic analysis strongly supports the hypothesis that dogs and gray wolves are reciprocally monophylic taxa that form two sister clades.[21]:1687[2]:4

In 2010, a study compared the mDNA haplotypes of 947 modern gray wolves from across Europe with the published sequences of 24 Pleistocene wolves from western Europe dated between 1,200-44,000 years BP. The study found that phylogenetically the haplotypes represented two haplogroups and referred to these as haplogroup 1 and 2. The 947 European wolves revealed 27 different haplotypes with haplogroup 1 forming a monophyletic clade, and all other haplotypes forming haplogroup 2. Comparison with gray wolves from other regions revealed that haplogroups 1 and 2 could be found spread across Eurasia, but only haplogroup 1 could be found in North America. The Pleistocene wolf samples from western Europe all belonged to haplogroup 2, which suggested a long-term predominance in this region. A comparison of current and past frequencies indicated that in Europe haplogroup 2 became outnumbered by haplogroup 1, but in North America haplogroup 2 became extinct and was replaced by haplogroup 1 after the Last Glacial Maximum.[24] Access into North America was available between 20,000-11,000 years ago, after the

Wisconsin glaciation had retreated but before the Bering land bridge became inundated by the sea.[25] Therefore, haplogroup 1 was able to enter into North America during this period.

Analysis of stable isotopes, which offer conclusions about the diet and therefore the ecology of the extinct wolf populations, suggest that the Pleistocene wolves from haplogroup 2 mainly preyed on Pleistocene megafaunal species,[26][27] which became rare at the beginning of the Holocene 12,000 years ago.[28]:2 "Thus, Pleistocene wolves across Northern Eurasia and America may actually have represented a continuous and almost panmictic population that was genetically and probably also ecologically distinct from the wolves living in this area today."[29]:R610 "The Pleistocene Eurasian wolves are morphologically and genetically comparable to the Pleistocene eastern-Beringian wolves."[30]:791 Some of the ancient European and Beringian wolves shared a common haplotype (a17).[24]:8 The specialized Pleistocene wolves, thus, did not contribute to the genetic diversity of modern wolves. Rather, modern wolf populations across the Holarctic are likely be the descendants of wolves from populations that came from more southern refuges as suggested previously[31] for the North American wolves.[29]:R611

These two haplogroups exclude the older-lineage Himalayan wolf and the Indian gray wolf.

See also: Beringian wolf
See also: Megafaunal wolf

The fossil remains of the direct ancestor of the dog have yet to be found, and so the probable ancestor is not yet confirmed.

17.1.4 First divergence

Time of divergence

The ancestral dog and the ancestral modern gray wolf diverged from a common ancestor at least 27,000 years ago.

In May 2015, a study was conducted on a partial rib-bone of a wolf (named Taimyr-1) found near the Bolshaya Balakhnaya River in the Taymyr Peninsula, North Asia, that was AMS radiocarbon dated to 34,900 years BP. The sample provided the first draft of the entire nuclear genome of a Pleistocene carnivore, and the sequence was deposited in the European Nucleotide Archive and classified as *Canis lupus* because the genome sequence was found to be substantially closer to modern gray wolves than it was to modern coyotes. The data was compared to the genotypes of 532

DNA evidence indicates that the dog, the modern gray wolf (above) and the now-extinct Taimyr wolf triverged from an extinct wolf-like canid that lived in Europe.

dogs from 48 breeds and 15 gray wolves from Europe, the Middle East, China, and North America.[1]

Using the Taimyr-1 specimen's radiocarbon date, its genome sequence and that of a modern wolf, a direct estimate of the genome-wide mutation rate in dogs/wolves could be made to calculate the time of divergence. The data showed that the Taimyr-1 lineage was separate to modern wolves and dogs and indicated that the Taimyr-1 genotype, gray wolves and dogs triverged from a now-extinct common ancestor before the peak of the Last Glacial Maximum 27,000-40,000 years ago. The separation of the dog and wolf did not have to coincide with selective breeding by humans.[1]:page3[32]

This derived mutation rate was much slower than that assumed in previous studies. Such an early divergence is consistent with several paleontological reports of dog-like canids up to 36,000 years old, as well as evidence that domesticated dogs most likely accompanied early colonizers into the Americas.[1]

The ancestral fossils have not yet been found.

Place of divergence

> *Modern dogs show a closer genetic association with ancient, extinct canids from Europe[3] and Arctic north-east Siberia.[33]*

In 2002, a study looked at 582 base pairs on the mitochondrial DNA of 654 dogs and 38 Eurasian wolves. The study found that 96% of the dogs formed three major clades, which were composed of 71 haplotypes. There was a total number of 44 haplotypes found in East Asia, of which 30 were unique to the region. The total number of haplotypes found in Europe was 20. The study assumed that the number of haplotypes would be higher in the ancestral

population, and therefore the dog originated from one domestication event in East Asia and spread from there across Eurasia and into South-West Asia and Europe.[34]

In August 2009, a study looked at 680 base pairs on the mitochondrial D-loop, 300 SNPs and 89 microsatellite markers of 318 African village dogs. The village dogs were found to be more genetically distinct when compared to non-native dogs, and with a similar haplotype diversity as had been found in East Asian village dogs. This finding called to question an East Asian origin of the domestic dog in an earlier 2002 study which appeared to have included many East Asian village dogs but few from other regions.[14]

In September 2009, a study looked of 16,159 base pairs on the mitochondrial genomes of 169 dogs and 582 base-pairs of the mtDNA control regions of 1,543 dogs and eight wolves to reveal three haplogroups composed of ten sub-haplogroups. All ten sub-haplogroups could be found only in south-eastern Asia south of the Yangtze River and with diversity decreasing across Eurasia, with seven in Central China, five in North China and south-western Asia, until only four sub-haplogroups could be found in Europe. Therefore, this study concluded that dogs originated in a single domestication event from this region less than 16,300 YBP.[15]

In 2010, a study looked at 48,000 SNPs from 912 dogs from 85 breeds and 225 grey wolves from 11 populations. The dog breeds shared a higher proportion of multi-locus haplotypes unique to grey wolves from the Middle East. Therefore, this study concluded that Middle Eastern grey wolves were the dominant source of dog diversity, European wolves to some European breeds, and Asian wolves to some Asian breeds, rather than wolves from Asia for all breeds as was suggested by an earlier 2009 mDNA study. Alternatively, there may have been significant admixture between some regional breeds and regional wolves.[35]

In 2011, a study looked at 14,437 base pairs of Y-chromosome DNA sequences from 151 dogs sampled world-wide and eight wolves to reveal that the dogs exhibited five haplogroups composed of 28 haplotypes. Two of the haplogroups could be found world-wide. One could be found primarily in East Asia (including Siberia), North America, but at low frequencies in south-western Asia, Scandinavia, and Britain but not in continental Europe. One haplogroup could be found East Asia and in low frequencies in south-western Asia. One haplogroup consisted of only four individuals, with one in eastern Siberia and 3 in Africa. The south-western part of south-eastern Asia that is south of the Yangtze River (comprising South-East Asia and the Chinese provinces of Yunnan and Guangxi) provided 16 dogs representing 11 haplotypes which showed the highest diversity. Therefore, based on both mDNA markers from the 2009 study and Y-chromosome markers from this study that are in agreement, 50% of the dog gene pool is shared universally, but based on the wider diversity found in the south-western part of South-eastern Asia dogs have originated from this region.[36]

In 2015, a study looked at 85,805 genetic markers of autosomal, maternal mitochondrial genome and paternal Y chromosome diversity in 4,676 purebred dogs from 161 breeds and 549 village dogs from 38 countries. Some dog populations in the Neotropics and the South Pacific are almost completely derived from European stock, and other regions show clear admixture between indigenous and European dogs. The indigenous dog populations of Vietnam, India, and Egypt show minimal evidence of European admixture, and exhibit indicators consistent with a Central Asian domestication origin.[37]

Genetic studies comparing the dog with extant gray wolves did not result in agreement among researchers. In 1868, Charles Darwin wrote that some authors at the time proposed an unknown or extinct species was the ancestor of the dog.[11] In 1934, an eminent paleontologist indicated that the ancestor of the dog lineage may have been the extinct *Canis lupus variabilis*.[38] The advent of rapid and inexpensive DNA sequencing technology has made it possible to significantly increase the resolving power of genetic data taken from both modern and ancient domestic dog genomes. Attention was now turned to ancient DNA.[39]

The 14,500-year-old upper-right jaw found in Kesslerloch Cave, Switzerland, is the sister to 2/3 of modern dogs. (courtesy Hannes Napierala)

Europe In November 2013, a study analysed the complete and partial mitochondrial genome sequences of 18 fossil canids dating from 1,000 to 36,000 YBP from the Old and New Worlds, and compared these with the complete mitochondrial genome sequences from modern wolves and dogs. The data indicate that 22% of the dogs sampled are sister to modern wolves from Sweden and the Ukraine with a most recent common ancestor 9,200 years ago (else admixture with wolves as dogs were clearly domesticated by

this time), and that 78% are sister to one or more ancient canids from Europe. Some 64% of the dogs are sister to a 14,500 YBP wolf sequence with a most recent common ancestor 32,100 YBP. This group of dogs matches three fossil pre-Columbian New World dogs between 1,000 and 8,500 YBP, which supports the hypothesis that pre-Columbian dogs in the New World share ancestry with modern dogs and that they likely arrived with the first humans to the New World. The data from this study indicates a European origin of dogs 18,800–32,100 years ago which supports the hypothesis that dog domestication preceded the emergence of agriculture and was initiated close to the Last Glacial Maximum when hunter-gatherers preyed on megafauna.[3]

Arctic North-East Siberia In 2015, a study looked at the mitogenome contol region sequences of 13 ancient canid remains and one modern wolf from five sites across Arctic north-east Siberia. The 14 canids revealed nine haplotypes, three of which were on record and the others unique. Four of the Siberian canids dated 28,000 YBP, and one *Canis c.f. variabilis* dated 360,000 YBP, were as divergent as the ancient European specimens found in an earlier study, and the European origin of domestic dogs may not be conclusive. The phylogenetic relationship of the extracted sequences showed that the haplotype from specimen S805 (28,000 YBP) was one step away from another haplotype S902 (8,000 YBP) that represents the domestic dog lineages. Several ancient haplotypes were oriented around S805, including *Canis c.f. variabilis* (360,000 YBP), Belgium (36,000 YBP - the "Goyet dog") and Belgium (30,000 YBP), and Konsteki, Russia (22,000 YBP). Given the position of the S805 haplotype, it may potentially represent a direct link from the putative progenitor (including *Canis c.f. variabilis*) to the domestic dog and modern wolf lineages.[33]

See further: Hybrid speciation and Introgression

17.1.5   Ancestral dog

In 1978, a researcher proposed that due to their similar behavior patterns that the dog and wolf shared a common ancestor prior to the dog's domestication, and that the "dog was the dog before it was domesticated".[40] In 1983, a study proposed that the ancestor of *Canis familiaris* was a wild *Canis familiaris*.[41] In 1999, a study emphasized that while molecular genetic data seem to support the origin of dogs from wolves, dogs may have descended from a now extinct species of canid whose closest living relative was the wolf.[12]

Goyet dog – 36,000 BP

Genus Canis, species indeterminate

In 2009, a study looked at 117 skulls of recent and fossil large canids. None of the 10 canid skulls from the Belgian caves of Goyet, Trou du Frontel, Trou de Nutons, and Trou de Chaleux could be classified, so the team took as their basic assumption that all of these canid samples were wolves.[42] The DNA sequence of seven of the skulls indicated seven unique haplotypes that represented ancient wolf lineages lost until now. The osteometric analysis of the skulls showed that one large canid fossil from Goyet was clearly different from recent wolves, resembling most closely the Eliseevichi-1 dogs (15,000 years YBP) and so was identified as a Paleolithic dog (see below).[27][43]

In November 2013, a DNA study sequenced three haplotypes from the ancient Belgium canids (the Goyet dog - Belgium 36,000 YBP cataloged as *Canis* species, and with Belgium 30,000 YBP and 26,000 years YBP cataloged as *Canis lupus*) and found they formed the most diverging group. Although the cranial morphology of the Goyet dog has been interpreted as dog-like, its mitochondrial DNA relation to other canids places it as an ancient sister-group to all modern dogs and wolves rather than a direct ancestor. Belgium 26,000 YBP has been found to be uniquely large but was found not to be related to the Beringian wolf. This Belgium canid clade may represent a phenotypically distinct and not previously recognized population of gray wolf, or the Goyet dog may represent an aborted domestication episode.[3]

Altai dog – 33,000 BP

Genus Canis, species indeterminate

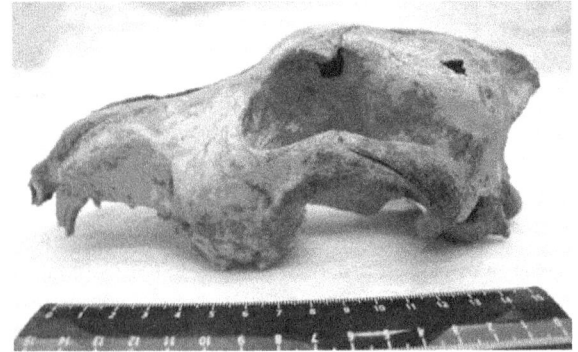

33,000-year-old skull of a dog-like canid found in the Altai Mountains. It has no direct descendants today.

In 2011, a study looked at the well-preserved 33,000-year-old skull and left mandible of a dog-like canid that was excavated from Razboinichya Cave in the Altai Mountains

of southern Siberia (Central Asia). The morphology was compared to the skulls and mandibles of large Pleistocene wolves from Predmosti, Czech Republic, dated 31,000 YBP, modern wolves from Europe and North America, and prehistoric Greenland dogs from the Thule period (1,000 YBP or later) to represent large-sized but unimproved fully domestic dogs. "The Razboinichya Cave cranium is virtually identical in size and shape to prehistoric Greenland dogs" and not the ancient nor modern wolves. However, the lower carnassial tooth fell within the lower range of values for prehistoric wolves and was only slightly smaller than modern European wolves, and the upper carnassial tooth fell within the range of modern wolves. "We conclude, therefore, that this specimen may represent a dog in the very early stages of domestication, i.e. an incipient dog, rather than an aberrant wolf... The Razboinichya Cave specimen appears to be an incipient dog...and probably represents wolf domestication disrupted by the climatic and cultural changes associated with the Last Glacial Maximum".[44]

In March 2013, a DNA study of the Altai dog deposited the sequence in GenBank with a classification of *Canis lupus familiaris* (dog). "The analyses revealed that the unique haplotype of the Altai dog is more closely related to modern dogs and prehistoric New World canids than it is to contemporary wolves... This preliminary analysis affirms the conclusion that the Altai specimen is likely an ancient dog with shallow divergence from ancient wolves. These results suggest a more ancient history of the dog outside of the Middle East or East Asia." The haplotype groups closest to the Altai dog included such diverse breeds as the Tibetan mastiff, Newfoundland, Chinese crested, cocker spaniel and Siberian husky.[45]

In November 2013, a study looked at 18 fossil canids and compared these with the complete mitochondrial genome sequences from 49 modern wolves and 77 modern dogs. A more comprehensive analysis of the complete mtDNA found that the phylogenetic position of the Altai dog as being either dog or wolf was inconclusive and cataloged its sequence as *Canis species*. The sequence strongly suggests a position at the root of a clade uniting two ancient wolf genomes, two modern wolves, as well as two dogs of Scandinavian origin. However, the study does not support its recent common ancestry with the great majority of modern dogs. The study suggests that it may represent an aborted domestication episode.[3]

Paleolithic dog – 27,000 BP

Detailed DNA analysis yet to be conducted

In 2002, a study looked at the fossil skulls from two large canids dated at 13,905 YBP that had been found buried within metres of what was once a mammoth-bone hut at the Upper Paleolithic site of Eliseevichi-1 in the Brayansk region of central Russia, and using an accepted morphologically-based definition of domestication declared them to be "Ice Age dogs".[46] In 2013, a study recalibrated the age of the Eliseevichi-1 specimens to 15,000 YBP and classified them as *Canis lupus familiaris*(dog).[3] In 2009, a study looked at these two early dog skulls in comparison to other much earlier but morphologically similar fossil skulls that had been found across Europe and concluded that the earlier specimens were "Paleolithic dogs", which were morphologically and genetically distinct from Pleistocene wolves that lived in Europe at that time. The study proposed, based on the genetic evidence of the timeline and European location, the archaeological evidence of the Paleolithic dog remains being found at known European hunting camp-sites, and based on morphology and collagen analysis that showed their diet had been restricted compared to wolves, that the Paleolithic dog was domesticated. The study hypothesized that the Paleolithic dogs may have provided the stock from which early dogs came, or alternatively that they are a type of wolf that is not known to science.[27]

See also Paleolithic dog.

17.1.6 Second divergence

Gray wolf admixture

There was admixture between the ancestral dog, the ancestral modern gray wolf, and the golden jackal.

The ancestral dog triverged into the dingo, Basenji and boxer lineages, and the ancestral modern gray wolf split into today's gray wolves.

In January 2014, a study analysed the whole-genome sequences of three wolves (*Canis lupus*) to represent the regions of Eurasia where domestication has been hypothesized to have taken place – Croatia (Europe), Israel (Middle East), and China (East/South-East Asia), plus an Australian dingo and a Basenji, being divergent lineages to the reference boxer genome, and so maximize the odds to capture distinct alleles present in the earliest dogs. These lineages are also geographically distinct, with modern Basenjis tracing their ancestry to hunting dogs of western Africa, while dingoes are free-living semi-feral dogs of Australia that arrived there at least 3,500 years ago. The natural range of wolves has never extended this far south, and due to geographic isolation they are less likely to have overlapped and admixed with wolves in the recent past. For some analyses,

Dog breeds like this Tamaskan Dog look like wolves due to admixture.

data were leveraged from a companion study of 12 additional dog breeds.[2]

The data provided significant evidence of admixture between the Israeli wolf and the Basenji, the Israeli wolf and the boxer, and between the Chinese wolf and dingo. The Chinese wolf with dingo likely represents ancient admixture in Eastern Eurasia, and the Israeli wolf with Basenji and boxer likely represents ancient admixture in Western Eurasia. The fact that these lineages have been geographically isolated from wolves in the recent past suggests that this gene flow was ancestral and has likely affected most dog lineages. There was significant gene flow between the golden jackal and the Israeli wolf, as well as the population ancestral to the dog and wolf samples.[2]

One test indicated that dogs and modern wolves form sister clades, meaning that the dog is a sister to the modern wolf and they share a common ancestor. Supporting this, another test indicated that none of the sampled wolf populations is more closely related to dogs than any of the others, and dogs diverged from wolves at about the same time as wolves diverged from each other. This implies that the wolf population(s) from which dogs originated has gone extinct and the current wolf diversity from each region represents novel, younger wolf lineages.[2]

The data indicate that the golden jackal and the ancestor of the wolf/dog diverged 400,000 years ago. Dogs and wolves then diverged into the ancestral dog and the ancestral modern gray wolf. The ancestral modern gray wolf population triverged into the three populations studied. Not long after, the ancestral dog populations diverged into the dingo lineage, the basenji lineage and the reference boxer lineage.

There was a 16-fold population bottleneck for dogs since this divergence.[2]

There was a three-fold population decline for the three wolf samples since divergence, and it appears to have occurred well in advance of direct extermination campaigns by humans and within the timeframe of environmental and biotic changes associated with the ending of the Pleistocene era, namely changes in climate and prey, including megafaunal extinctions. This indicates that before the divergence of dogs from wolves there was much more wolf diversity. The results support a recent divergence between dogs and wolves followed by a dramatic reduction in population size.[2]

AMY2B (Alpha-Amylase 2B) is a gene that codes a protein that assists with the first step in the digestion of dietary starch and glycogen. An expansion of this gene in dogs would enable early dogs to exploit a starch-rich diet as they fed on refuse from agriculture. Data indicated that the wolves and dingo had just two copies of the gene and the Siberian Husky that is associated with hunter-gatherers had just three or four copies, whereas the Saluki that is associated with the Fertile Crescent where agriculture originated had 29 copies. The results show that on average, modern dogs have a high copy number of the gene, whereas wolves and dingoes do not. The high copy number of AMY2B variants likely already existed as a standing variation in early domestic dogs, but expanded more recently with the development of large agriculturally based civilizations. This suggests that at the beginning of the domestication process, dogs may have been characterized by a more carnivorous diet than their modern-day counterparts, a diet held in common with early hunter-gatherers.[2]

The Greenland dog carries 3.5% shared genetic material with the 35,000 years BP Taymyr wolf specimen.

Taimyr wolf admixture

There was admixture between Taimyr-1 and those breeds associated with high latitudes.

In May 2015, a study compared the ancestry of the Taimyr-1 wolf lineage to that of dogs and gray wolves.

Comparison to the gray wolf lineage indicated that Taimyr-1 was basal to gray wolves from the Middle East, China, Europe and North America but shared a substantial amount of history with the present-day gray wolves after their divergence from the coyote. This implies that the ancestry of the majority of gray wolf populations today stems from an ancestral population that lived less than 35,000 years ago but before the inundation of the Bering Land Bridge with the subsequent isolation of Eurasian and North American wolves.[1]:21

A comparison of the ancestry of the Taimyr-1 lineage to the dog lineage indicated that some modern dog breeds have a closer association with either the gray wolf or Taimyr-1 due to admixture. The Saarloos wolfdog showed more association with the gray wolf, which is in agreement with the documented historical crossbreeding with gray wolves in this breed. Taimyr-1 shared more alleles (i.e. gene expressions) with those breeds that are associated with high latitudes - the Siberian husky and Greenland dog that are also associated with arctic human populations, and to a lesser extent the Shar Pei and Finnish spitz. An admixture graph of the Greenland dog indicates a best-fit of 3.5% shared material, although an ancestry proportion ranging between 1.4% and 27.3% is consistent with the data. This indicates admixture between the Taimyr-1 population and the ancestral dog population of these four high-latitude breeds. These results can be explained either by a very early presence of dogs in northern Eurasia or by the genetic legacy of Taimyr-1 being preserved in northern wolf populations until the arrival of dogs at high latitudes. This introgression could have provided early dogs living in high latitudes with phenotypic variation beneficial for adaption to a new and challenging environment. It also indicates that the ancestry of present-day dog breeds descends from more than one region.[1]:3–4

An attempt to explore admixture between Taimyr-1 and gray wolves produced unreliable results.[1]:23

17.2 Dog domestication

Dog domestication (from the Latin *domesticus*: "belonging to the house")[47] is a process by which the dog has become adapted to man and the captive environment by some combination of genetic changes occurring over generations.[48] The process of dog domestication is unknown; the two main theories are self-domestication and human domestication.[49] As a result of this process there is also evidence of convergent evolution having occurred between dogs and humans.[50]

17.2.1 Archaeological evidence

Archaeological evidence locates the earliest dog remains along with human remains 15,000 years ago at the Eliseevich-I Upper Paleolithic site, Russian Plain, Europe. Domesticated dogs are more clearly identified when they are associated with human occupation, and those interred side-by-side with human remains provide the most conclusive evidence.[51]

17.2.2 Domestication of the dog

Definition

Domestication is a process by which a population of animals becomes adapted to man and the captive environment by some combination of genetic changes occurring over generations.[48] Domestication is an evolutionary process in which one population of a species is reproductively isolated from another intentionally by humans.[64] This reproductive isolation leads to divergent adaptation and results in a specialization process. As a result of the changes in the selection pressures on the given species, the process of domestication produces evolutionary changes in certain aspects of the characteristic behavior of the domesticated species just as it affects the anatomy and morphology of the certain species as well. For example, such behavioral change in dogs is the decreased level of aggression, which shows in morphological changes such as teeth size.[65]

Theory

As a result of the archaeological evidence, humans formed a theory to help explain the observations. The current theory of dog domestication is based on comparisons between the dog and the extant (i.e. living today) gray wolf. However, in 2002 a study highlighted a number of inconsistencies with this comparison and proposed that the ancestor of the dog appears more likely to have been a generalist canid and not the specialized gray wolf.[53] Recently two DNA studies indicated that the ancestor of the dog was not the extant gray wolf and that the fossil remains of their common ancestor have yet to be found.[1][3] In 1983, a study proposed that the ancestor of *Canis familiaris* was a wild *Canis familiaris*.[41] In 2015, a study found that the dog, the gray wolf and the now-extinct Taimyr wolf all triverged from a common ancestor 40,000 years ago i.e. without human intervention.[1]

The process of domestication is unknown, but the two main hypotheses are self-domestication and human domestication.[49]:210

Self-domestication The first of the two main hypotheses of dog domestication is self-domestication by wolves. Some wolves moved into a mutually beneficial relationship with prehistoric humans. They scavenged on the remains of the prey animals left by the prehistoric people at the human settlements or the kill sites. Those wolves that were less anxious and aggressive thrived, continued to follow the prehistoric humans, and colonized the human-dominated environments, generation after generation. Gradually, the first primitive dogs emerged from this group.[66][67][68]

Cooperation Two recent DNA studies indicate a dog-wolf divergence time of greater than 15,000 years ago.[1][3] An evolutionary scenario consistent with these results is that dog domestication was initiated close to the Last Glacial Maximum when hunter-gathers preyed on megafauna. Conceivably, proto-dogs might have taken advantage of carcasses left on site by early hunters, assisted in the capture of prey, or provided defense from large competing predators at kills.[3]

A leading evolutionary biologist stated, "But if domestication occurred in association with hunter-gatherers, one can imagine wolves first taking advantage of the carcasses that humans left behind – a natural role for any large carnivore – and then over time moving more closely into the human niche through a co-evolutionary process. The idea of wolves following hunter-gatherers also helps to explain the eventual genetic divergence that led to the appearance of dogs. Wolves following the migratory patterns of these early human groups would have given up their territoriality and would have been less likely to reproduce with resident territorial wolves. We have an analog of this process today, in the only migratory population of wolves known existing in the tundra and boreal forest of North America. This population follows the barren-ground caribou during their thousand-kilometer migration. When these wolves return from the tundra to the boreal forest during the winter, they do not reproduce with resident wolves there that never migrate. We feel this is a model for domestication and the reproductive divergence of the earliest dogs from wild wolves. We know also that there were distinct wolf populations existing ten of thousands of years ago. One such wolf, which we call the megafaunal wolf, preyed on large game such as horses, bison and perhaps very young mammoths. Isotope data show that they ate these species, and the dog may have been derived from a wolf similar to these ancient wolves in the late Pleistocene of Europe."[69]

See further: Reproductive isolation

See further: Megafaunal wolf

Natural selection without humans Dogs can infer the name of an object and have been shown to learn the names of over 1,000 objects. Dogs can follow the human pointing gesture; even nine-week-old puppies can follow a basic human pointing gesture without being taught. New Guinea singing dogs, a half-wild proto-dog endemic to the remote alpine regions of New Guinea, as well as dingoes in the remote Outback of Australia are also capable of this. These examples demonstrate an ability to read human gestures that arose early in domestication and did not require human selection. "Humans did not develop dogs, we only fine-tuned them down the road."[50]:92

A dog's cranium is 15% smaller than an equally heavy wolf's, and the dog is less aggressive and more playful. Other species pairs show similar differences. Bonobos, like chimpanzees, are a close genetic cousin to humans, but unlike the chimpanzees, bonobos are not aggressive and do not participate in lethal inter-group aggression or kill within their own group. The most distinctive features of a bonobo are its cranium, which is 15% smaller than a chimpanzee's, and its less aggressive and more playful behavior. In other examples, the guinea pig's cranium is 13% smaller than its wild cousin the cavy, and domestic fowl show a similar reduction to their wild cousins. Possession of a smaller cranium for holding a smaller brain is a telltale sign of domestication. Bonobos appear to have domesticated themselves.[50]:104

In the "farm fox" experiment, humans selectively bred foxes against aggression, causing a domestication syndrome. The foxes were not selectively bred for smaller craniums and teeth, floppy ears, or skills at using human gestures, but these traits were demonstrated in the friendly foxes. Natural selection favours those that are the most successful at reproducing, not the most aggressive. Selection against aggression made possible the ability to cooperate and communicate among foxes, dogs and bonobos. Perhaps it did the same thing for humans.[50]:114[70]

Human domestication The second of the two main hypotheses of dog domestication is domestication by humans. Paleolithic people actively selected wolf pups for several reasons: they could be used as pets, they could be kept

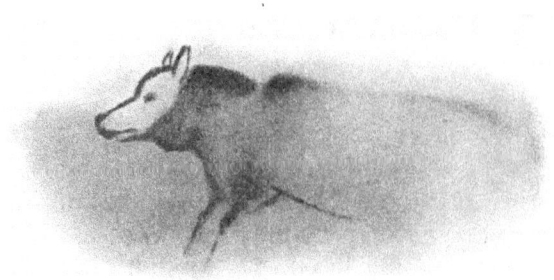

Polychrome cave painting of a wolf-like canid 17,000 years ago, Font-de-Gaume, France

for utilitarian, ceremonial and symbolic uses, as social storage, or combat and/or as living tools.[68][71][72][73][74] The most docile or interesting animals could have been permitted to reproduce.[20][72][73] After several generations of unconscious and later of conscious selection of human-defined behavioral traits, the first dogs emerged.[75]

Orphaned wolf-pups Studies have shown that some wolf pups taken at an early age and reared by humans are easily tamed and socialized,[13]:140 and one study has demonstrated that adult wolves can be socialized.[13]:141 Some researchers propose that humans adopted orphaned wolf pups and breastfed them alongside human babies.[76][77] In Alaska and other northern areas where people still live close to wolves, wolf pups are sometimes captured and some become acceptable as pets or sled dogs. These breedings over generations would become more dog-like.[13]:pages55-56

Against this proposition, at the time of domestication near the Last Glacial Maximum humans were already the top predator and had no need for wolves that would grow to eat five kilograms of meat per wolf per day at a time when food was very scarce. Starvation would have been a real threat to many carnivores in the Ice Age, and competition for food would have been fierce.[50]:29 Other researchers attempting to socialize wolf pups after they reached 21 days of age found it very time-consuming and seldom practical or reliable in achieving success.[78]

See further: Socialization - dogs and wolves

Human selection The "farm fox" experiment attempted to reenact how domestication may have occurred.[79] Researchers worked with farmed silver foxes selectively bred over 35 generations for tamability, i.e. becoming less fearful and less aggressive. The "domestic" foxes were tamer to humans than others, but they also showed new physical traits even though the physical traits were not originally selected for. These include spotted or black-and-white coats, floppy ears, tails that curl over their backs, barking vocalization, and earlier sexual maturity. One researcher found that the migration of certain melanocytes (which determine colour) was delayed, resulting in a black and white 'star' pattern.

One criticism of this experiment was later made by the author, who stated that the living conditions of the foxes in the farm would have been very different to those of wolf puppies in Paleolithic camps.[80] A further criticism based on information obtained after the experiment's publication was that the definition of "tame" was changed at least once during the experiment, and that some of the foxes that were classified as neither tame nor aggressive also exhibited these changes, indicating that some factor other than human selection for tameness may have been at work during domestication.[81] When humans restrict dogs' breeding diversity, another variable also comes into play that may have contributed to the change – inbreeding.[76]:30

See further: Reproductive isolation

In 2014, a whole genome study of the DNA differences between wolves and dogs found that dogs did not show a reduced fear response but did show greater synaptic plasticity. Synaptic plasticity is widely believed to be the cellular correlate of learning and memory, and this change may have altered the learning and memory abilities of dogs in comparison to wolves.[82]

See further: Dog learning by inference

In August 2015, a study looked at over 100 pig genome sequences to ascertain their process of domestication. The process of domestication was assumed to have been initiated by humans, involved few individuals and relied on reproductive isolation between wild and domestic forms. The study found that the assumption of reproductive isolation with population bottlenecks were not supported. The study indicated that pigs were domesticated separately in Western Asia and China, with Western Asian pigs introduced into Europe where they crossed with wild boar. The study also found that despite back-crossing with wild pigs, the genomes of domestic pigs have strong signatures of selection at DNA loci that affect behavior and morphology. The study concluded that human selection for domestic traits likely counteracted the homogenizing effect of gene flow from wild boars and created 'islands of domestication' in the genome. The same process may also apply to other domesticated animals, including dogs.[83][84]

17.2.3 Convergent evolution between dogs and humans

Behavioral evidence

Convergent evolution is when distantly related species inde-

pendently evolve similar solutions to the same problem. For example, fish, penguins and dolphins have each separately evolved flippers as a solution to the problem of moving through the water. What has been found between dogs and humans is something less frequently demonstrated: psychological convergence. Dogs have independently evolved to be cognitively more similar to humans than we are to our closest genetic relatives.[50]:60 Dogs have evolved specialized skills for reading human social and communicative behavior. These skills seem more flexible – and possibly more human-like – than those of other animals more closely related to humans phylogenetically, such as chimpanzees, bonobos and other great apes. This raises the possibility that convergent evolution has occurred: both *Canis familiaris* and *Homo sapiens* might have evolved some similar (although obviously not identical) social-communicative skills – in both cases adapted for certain kinds of social and communicative interactions with human beings.[70]

The pointing gesture is a human-specific signal, is referential in its nature, and is a foundation building-block of human communication. Human infants acquire it weeks before the first spoken word.[85] In 2009, a study compared the responses to a range of pointing gestures by dogs and human infants. The study showed little difference in the performance of 2-year-old children and dogs, while 3-year-old children's performance was higher. The results also showed that all subjects were able to generalize from their previous experience to respond to relatively novel pointing gestures. These findings suggest that dogs demonstrating a similar level of performance as 2-year-old children can be explained as a joint outcome of their evolutionary history as well as their socialization in a human environment.[86]

Later studies support coevolution in that dogs can discriminate the emotional expressions of human faces,[87] and that most people can tell from a bark whether a dog is alone, being approached by a stranger, playing, or being aggressive,[88] and can tell from a growl how big the dog is.[89]

Biological evidence

In 2013, a DNA sequencing study indicated that parallel evolution in humans and dogs is most apparent in the genes for digestion and metabolism, neurological processes, and cancer, likely as a result of shared selection pressures.[90][91]

In 2014, a study compared the hemoglobin levels of village dogs and people on the Chinese lowlands with those on the Tibetan Plateau. It found the hemoglobin levels higher for both people and dogs in Tibet, suggesting that Tibetan dogs might share similar adaptive strategies as the Tibetan people. A population genetic analysis then showed a significant convergence between humans and dogs in Tibet.[92]

In 2015, a study found that when dogs and their owners interact, extended eye contact (mutual gaze) increases oxytocin levels in both the dog and its owner. As oxytocin is known for its role in maternal bonding, it is considered likely that this effect has supported the coevolution of human-dog bonding.[93] One observer has stated, "The dog could have arisen only from animals predisposed to human society by lack of fear, attentiveness, curiosity, necessity, and recognition of advantage gained through collaboration....the humans and wolves involved in the conversion were sentient, observant beings constantly making decisions about how they lived and what they did, based on the perceived ability to obtain at a given time and place what they needed to survive and thrive. They were social animals willing, even eager, to join forces with another animal to merge their sense of group with the others' sense and create an expanded super-group that was beneficial to both in multiple ways. They were individual animals and people involved, from our perspective, in a biological and cultural process that involved linking not only their lives but the evolutionary fate of their heirs in ways, we must assume, they could never have imagined. Powerful emotions were in play that many observers today refer to as love – boundless, unquestioning love."[81]:40

Lupification of humans

Isn't it strange that, our being such an intelligent primate, we didn't domesticate chimpanzees as companions instead? Why did we choose wolves even though they are strong enough to maim or kill us?[94]

Bison surrounded by gray wolf pack

In 2002, a study proposed that immediate human ancestors and wolves may have domesticated each other through a strategic alliance that would change both respectively into humans and dogs. The effects of human psychology, hunting practices, territoriality and social behavior would have been profound.[95]

Marking of territory with signs such as pecked cupules, hand stencils and prints, abraded grooves, and finger impressions in once-soft mud are enduring signs used to mark occupation. They also became the first symbolic objects i.e. art. Wolves mark their territory with urine, but humans do not have the keen sense of smell as wolves and would have needed to use something more easily recognizable and enduring to mark their territory. Humans may have learned to mark their territory after watching wolves and dogs.[95]

Hunting large animals in packs is a distinctive wolf behavioral trait. There is no evidence of big game hunting in pre-*sapiens* groups, but big-game hunting is very typical of *homo sapiens* that, in addition to climate change, may have contributed to the extinction of many large mammals. Early humans moved from scavenging and small-game hunting to big-game hunting by living in larger, socially more-complex groups, learning to hunt in packs, and developing powers of cooperation and negotiation in complex situations. As these are characteristics of wolves, dogs and humans, it can be argued that these behaviors were enhanced once wolves and humans began to cohabit. Communal hunting led to communal defense. Wolves actively patrol and defend their scent-marked territory, and perhaps humans had their sense of territoriality enhanced by living with wolves.[95]

New forms of bonding might assist in living in large, complex and varied social groups. One of the keys to recent human survival has been the negotiation of situations by forming partnerships. Strong bonds exist between same-sex wolves, dogs and humans – bonds less fickle than exist between other same-sex animal pairs. Today, the most widespread form of inter-species bonding occurs between humans and dogs. The concept of friendship has ancient origins, but it may have been enhanced through the interspecies relationship to give a survival advantage.[95]

In 2003, a study compared the behavior and ethics of chimpanzees, wolves and humans. Humans' genetically closest relative appears to be a frightful caricature of human egoism, and even in their maternal behavior, warmth and affection are reduced to nursing and the occasional comforting hug. Cooperation among group members is limited to occasional hunting episodes or the persecution of a competitor, always aimed for one's own advantage. The closest approximation to human morality that can be found in nature is that of the gray wolf, *Canis lupus*. Wolves' ability to cooperate in well-coordinated drives to hunt prey, carry items too heavy for an individual, provisioning not only their own young but also the other pack members, babysitting etc. are rivaled only by that of human societies. Similar forms of cooperation are observed in two closely related canids, the African Cape hunting dog and the Asian dhole, therefore it is reasonable to assume that canid sociality and cooperation are old traits that in terms of evolution predate human sociality and cooperation. Today's wolves may even be less social than their ancestors, as they have lost access to big herds of ungulates and now tend more toward a lifestyle similar to coyotes, jackals, and even foxes.[94]

Reindeer moved in large herds across the mammoth steppe and were preyed upon by carnivores.

The mammoth steppe was the Eurasian tundra and grass steppe ecosystem which once stretched from Spain to the far east of Siberia, and at times continued into North America. On this steppe the wolves' ability to hunt in packs, to share risk fairly among pack members, and to cooperate moved them to the top of the food pyramid above lions, hyenas and bears. Some, but not all, wolves followed the great reindeer herds, eliminating the unfit, the weaklings, the sick and the aged, and therefore improved the herd. These wolves had become the first pastoralists hundreds of thousands of years before humans also took to this role. The wolves' advantage over their competitors was that they were able to keep pace with the herds, move fast and enduringly, and make the most efficient use of their kill by their ability to "wolf down" a large part of their quarry before other predators had detected the kill. The authors of the study propose that during the last ice age, some of our ancestors teamed up with those pastoralist wolves. Many of our ancestors remained gatherers and scavengers, or specialized as fish-hunters, hunter-gatherers, and hunter-gardeners. However, some ancestors adopted the pastoralist wolves' lifestyle as herd followers and herders of reindeer, horses, and other hoofed animals. They harvested the best stock for themselves while the wolves kept the herd strong. These pastoralists later became herders. From a biologist's vantage point, the interwining process of hominization and canization makes sense only if viewed in terms of coevolution.[94]

17.3 References

[1] Skoglund, P. (2015). "Ancient wolf genome reveals an early divergence of domestic dog ancestors and admixture into high-latitude breeds". *Current Biology* **25** (11): 1515–9. doi:10.1016/j.cub.2015.04.019. PMID 26004765.

[2] Freedman, A. (2014). "Genome sequencing highlights the dynamic early history of dogs". *PLoS genetics* **10** (1): e1004016. doi:10.1371/journal.pgen.1004016. PMC 3894170. PMID 24453982.

[3] Thalmann, O. (2013). "Complete mitochondrial genomes of ancient canids suggest a European origin of domestic dogs". *American Journal of Physical Anthropology* **145** (4): 653–7. doi:10.1002/ajpa.21526. PMC 3133791. PMID 21541929.

[4] "Evolution". *Oxford Dictionaries*. Oxford University Press. 2014.

[5] Darwin, 1859. *On the Origin of Species*, Chapter XIII

[6] University of California Museum of Paleontology. "Understanding Evolution". Retrieved 2015.

[7] Jonathan Adams. "Europe During the Last 150,000 Years". Oak Ridge National Laboratory, Oak Ridge, USA.

[8] Cooper, A. (2015). "Abrupt warming events drove Late Pleistocene Holarctic megafaunal turnover". *Science* **349** (6248): 602–6. doi:10.1126/science.aac4315. PMID 26250679.

[9] Benazzi, S. (2011). "Early dispersal of modern humans in Europe and implications for Neanderthal behaviour". *Nature* **479** (7374): 525–8. doi:10.1038/nature10617. PMID 22048311.

[10] Higham, T. (2011). "The earliest evidence for anatomically modern humans in northwestern Europe". *Nature* **479** (7374): 521–4. doi:10.1038/nature10484. PMID 22048314.

[11] Darwin, Charles (1868). "Chapter 1: Domestic Dogs and Cats". *The Variation of Animals and Plants under Domestication*. Vol. 1. John Murray, London.

[12] Wayne, R. (1999). "Origin, genetic diversity, and genome structure of the domestic dog". *BioEssays* **21** (3): 247–57. doi:10.1002/(SICI)1521-1878(199903)21:3<247::AID-BIES9>3.0.CO;2-Z. PMID 10333734.

[13] Scott, J. (1965). *Genetics and the social behavior of the dog: The classic study*. University of Chicago Press. ISBN 978-0-226-74338-7.

[14] Boyko, A. (2009). "Complex population structure in African village dogs and its implications for inferring dog domestication history". *Proceedings of the National Academy of Sciences* **106** (33): 13903. doi:10.1073/pnas.0902129106.

[15] Pang, J. (2009). "mtDNA data indicate a single origin for dogs south of Yangtze River, less than 16,300 years ago, from numerous wolves". *Molecular Biology and Evolution* **26** (12): 2849–64. doi:10.1093/molbev/msp195. PMC 2775109. PMID 19723671.

[16] Wayne, R. (1993). "Molecular evolution of the dog family". *Trends in Genetics* **9** (6): 218–24. doi:10.1016/0168-9525(93)90122-X. PMID 8337763.

[17] Wozencraft, W. Christopher (1993). "Order Carnivora". In Wilson, D.E.; Reeder, D.M. *Animal Species of the World:A Taxonomic and Geographic Reference* (2 ed.). Washington, D.C.: Smithsonian Institution Press. pp. 280–281. ISBN 1560982179. Page 281- "COMMENTS: *Canis familiaris* has page priority over *Canis lupus*, but both were published simultaneously in Linnaeus (1758), and *Canis lupus* has been universally used for this species."

[18] Smithsonian - Animal Species of the World database. "Canis lupus".

[19] Gentry A. (2004). "The naming of wild animal species and their domestic derivatives". *Journal of Archaeological Science* **31** (5): 645. doi:10.1016/j.jas.2003.10.006.

[20] Clutton-Brock, J. (1995). "Chapter 1". In James Serpell. *The Domestic Dog: Its Evolution, Behaviour and Interactions with People.* Cambridge University Press Press. pp. 7–20.

[21] Vila, C. (1997). "Multiple and ancient origins of the domestic dog". *Science* **276** (5319): 1687–9. doi:10.1126/science.276.5319.1687. PMID 9180076.

[22] Koepfli, K P (2015). "Genome-wide Evidence Reveals that African and Eurasian Golden Jackals Are Distinct Species". doi:10.1016/j.cub.2015.06.060.

[23] Aggarwal, R. K. (2007). "Mitochondrial DNA coding region sequences support the phylogenetic distinction of two Indian wolf species". doi:10.1111/j.1439-0469.2006.00400.x.

[24] Pilot, M.; et al. (2010). "Phylogeographic history of grey wolves in Europe". *BMC Evolutionary Biology* **10**: 104. doi:10.1186/1471-2148-10-104. PMC 2873414. PMID 20409299.

[25] Tamm, E. (2007). "Beringian standstill and spread of Native American founders". *PLoS ONE* **2** (9): e829. doi:10.1371/journal.pone.0000829. PMC 1952074. PMID 17786201.

[26] Leonard, J. (2007). "Megafaunal extinctions and the disappearance of a specialized wolf ecomorph". *Current Biology* **17** (13): 1146–50. doi:10.1016/j.cub.2007.05.072. PMID 17583509.

[27] Germonpre, M. (2009). "Fossil dogs and wolves from Palaeolithic sites in Belgium, the Ukraine and Russia: Osteometry, ancient DNA and stable isotopes". *Journal of Archaeological Science* **36** (2): 473. doi:10.1016/j.jas.2008.09.033.

[28] Hofreiter, M. (2010). "Diversity lost: Are all Holarctic large mammal species just relic populations?". *BMC Biology* **8**: 46. doi:10.1186/1741-7007-8-46. PMC 2858106. PMID 20409351.

[29] Hofreiter, M. (2007). "Pleistocene extinctions: Haunting the survivors". *Current Biology* **17** (15): R609–11. doi:10.1016/j.cub.2007.06.031. PMID 17686436.

[30] Germonpre, M. (2012). "Palaeolithic dogs and the early domestication of the wolf: A reply to the comments of Crockford and Kuzmin". *Journal of Archaeological Science* **40**: 786. doi:10.1016/j.jas.2012.06.016.

[31] Leonard, J. (2005). "Legacy lost: Genetic variability and population size of extirpated US grey wolves (*Canis lupus*)". *Molecular Ecology* **14** (1): 9–17. doi:10.1111/j.1365-294X.2004.02389.x. PMID 15643947.

[32] "Ancient wolf genome pushes back dawn of the dog". *Nature*. 2015.

[33] Lee, E. (2015). "Ancient DNA analysis of the oldest canid species from the Siberian Arctic and genetic contribution to the domestic dog". *PLoS ONE* **10** (5): e0125759. doi:10.1371/journal.pone.0125759. PMC 4446326. PMID 26018528.

[34] Savolainen, P. (2002). "Genetic evidence for an East Asian origin of domestic dogs". *Science* **298** (5598): 1610–3. doi:10.1126/science.1073906. PMID 12446907.

[35] vonHoldt, B. (2010). "Genome-wide SNP and haplotype analyses reveal a rich history underlying dog domestication". *Nature* **464** (7290): 898–902. doi:10.1038/nature08837. PMC 3494089. PMID 20237475.

[36] Ding, Z. (2011). "Origins of domestic dog in Southern East Asia is supported by analysis of Y-chromosome DNA". *Heredity* **108** (5): 507–14. doi:10.1038/hdy.2011.114. PMC 3330686. PMID 22108628.

[37] Shannon, L (2015). "Genetic structure in village dogs reveals a Central Asian domestication origin". doi:10.1073/pnas.1516215112.

[38] Pei, W. (1934). "The carnivora from locality 1 of Choukoutien". *Palaeontologia Sinica, Series C, vol. 8, Fascicle 1*. Geological Survey of China, Beijing. pp. 1–45.

[39] Larson, G (2012). "Rethinking dog domestication by integrating genetics, archeology, and biogeography". doi:10.1073/pnas.1203005109.

[40] Fox, M W (1978). "11". *The dog:its domestication and behavior.* Garland STPM Press, New York. p. 248.

[41] Manwell, C. & C. M. A. Baker. 1983. "Origin of the dog: From wolf or wild Canis familiaris?" *Speculations in Science and Technology* 6 (3): 213–224.

[42] Shipman, P. (2011). *The Animal Connection: A New Perspective on What Makes Us Human.* W W Norton & Co New York. p. 218.

[43] Royal Belgium Institute of Natural Sciences. "Goyet skull photo".

[44] Ovodov, N. (2011). "A 33,000-year-old incipient dog from the Altai Mountains of Siberia: Evidence of the earliest domestication disrupted by the Last Glacial Maximum". *PLoS ONE* **6** (7): e22821. doi:10.1371/journal.pone.0022821. PMC 3145761. PMID 21829526.

[45] Druzhkova, A. (2013). "Ancient DNA analysis affirms the canid from Altai as a primitive dog". *PLoS ONE* **8** (3): e57754. doi:10.1371/journal.pone.0057754. PMC 3590291. PMID 23483925.

[46] Sablin, M. (2002). "The earliest Ice Age dogs: Evidence from Eliseevichi I". *Current Anthropology* **43** (5): 795. doi:10.1086/344372.

[47] "Domesticate". *Oxford Dictionaries*. Oxford University Press. 2014.

[48] Grandin, T. (1998). *Genetics and the Behavior of Domestic Animals*. Academic Press, San Diego, CA. ISBN 978-0-12-394586-0.

[49] Germonpre, M. (2014). "Palaeolithic dogs and Pleistocene wolves revisited: A reply to Morey". *Journal of Archaeological Science* **54**: 210. doi:10.1016/j.jas.2014.11.035.

[50] Hare, B. (2013). *The Genius of Dogs*. Penguin Publishing Group.

[51] Boudadi-Maligne, M. (2014). "A biometric re-evaluation of recent claims for Early Upper Palaeolithic wolf domestication in Eurasia". *Journal of Archaeological Science* **45**: 80. doi:10.1016/j.jas.2014.02.006.

[52] Lawrence, B. (1967). "Early domestic dogs". *Zeitschrift für Säugetierkunde* **32**: 44–59.

[53] Koler-Matznick, J. (2002). "The origin of the dog revisited". *Anthrozoos: A Multidisciplinary Journal of the Interactions of People & Animals* **15** (2): 98. doi:10.2752/089279302786992595.

[54] Garcia, M. (2005). "Ichnologie générale de la grotte Chauvet". *Bulletin de la Société préhistorique française* **102**: 103. doi:10.3406/bspf.2005.13341.

[55] Morey, D., ed. (2010). *Dogs: Domestication and the Development of a Social Bond*. Cambridge University Press. p. 24. ISBN 9780521757430.

[56] Leisowska, A (2015). "Autopsy carried out in Far East on world's oldest dog mummified by ice". Retrieved October 19, 2015.

[57] Davis, F. (1978). "Evidence for domestication of the dog 12,000 years ago in the Natufian of Palestine". *Nature* **276** (5688): 608. doi:10.1038/276608a0.

[58] Tito, R. (2011). "Brief communication: DNA from early Holocene American dog". *American Journal of Physical Anthropology* **145** (4): 653–7. doi:10.1002/ajpa.21526. PMC 3133791. PMID 21541929.

[59] Henriksen, B. (1976). *Værdborg I: Excavations 1943-44: A Settlement of the Maglemose Culture*. Copenhagen: Akademisk forlag.

[60] Susan J. Crockford, *A Practical Guide to In Situ Dog Remains for the Field Archaeologist*, 2009

[61] Losey, R. (2011). "Canids as persons: Early Neolithic dog and wolf burials, Cis-Baikal, Siberia". *Journal of Anthropological Archaeology* **30** (2): 174. doi:10.1016/j.jaa.2011.01.001.

[62] Oestigaard, F., ed. (2008). "The materiality of death bodies, burials, beliefs". *BAR International Series 17682008*.

[63] Witt, K. (2014). "DNA analysis of ancient dogs of the Americas: Identifying possible founding haplotypes and reconstructing population histories". *Journal of Human Evolution* **79**: 105–18. doi:10.1016/j.jhevol.2014.10.012. PMID 25532803.

[64] Kretchmer, K. (1975). "Effects of domestication on animal behaviour". *Veterinary Record* **96** (5): 102–8. doi:10.1136/vr.96.5.102. PMID 1090069.

[65] Lakatos, G. (2011). "Evolutionary approach to communication between humans and dogs". doi:10.4415/ANN_11_04_08 (inactive 2015-09-30).

[66] Crockford, S. (2000). Crockford, S., ed. *A commentary on dog evolution: Regional variation, breed development and hybridization with wolves*. Archaeopress BAR International Series 889. pp. 11–20. ISBN 978-1841710891.

[67] Coppinger, R. (2001). *Dogs: A Startling New Understanding of Canine Origin, Behavior & Evolution*. ISBN 0684855305.

[68] Russell, N. (2012). *Social Zooarchaeology: Humans and Animals in Prehistory*. Cambridge University Press. ISBN 978-0-521-14311-0.

[69] Wolpert, S. (2013), "Dogs likely originated in Europe more than 18,000 years ago, UCLA biologists report", *UCLA News Room*, retrieved December 10, 2014

[70] Hare B. (2005). "Human-like social skills in dogs?". *Trends in Cognitive Sciences* **9** (9): 439–44. doi:10.1016/j.tics.2005.07.003. PMID 16061417.

[71] Crabtree, P. (1987). "A new model for the domestication of the dog". *Mus. Appl. Sci. Cent. Archaeol. J* **4**: 98–102.

[72] Germonpre, M. (2010). Shipman, P., ed. "The animal connection and human evolution". *Current Anthropology* **51** (4): 519. doi:10.1086/653816.

[73] Serpell, J. (1989). "Pet-keeping and animal domestication: A reappraisal". In Clutton-Brock, J. *The Walking Larder: Patterns of Domestication, Pastoralism and Predation*. Unwin-Hyman, London. pp. 10–20.

[74] Shipman, P. (2010). "The animal connection and human evolution". *Current Anthropology* **51** (4): 519. doi:10.1086/653816.

[75] Trut, L. (2009). "Animal evolution during domestication: The domesticated fox as a model". *BioEssays* **31** (3): 349–60. doi:10.1002/bies.200800070. PMC 2763232. PMID 19260016.

[76] Derr, M. (2004). *Dog's Best Friend: Annals of the Dog-Human Relationship*. University of Chicago Press. ISBN 0-226-14280-9.

[77] Grandin, T. (2005). *Animals in Translation*. Scribner, New York, New York. ISBN 0-7432-4769-8.

[78] Klinghammer, E. (1987). "Chapter 2: Socialization and management of wolves in captivity". In Frank, H. *Man and Wolf: Advances, Issues, and Problems in Captive Wolf Research*. Dr W. Junk Publishers. pp. 31–61. ISBN 90-6193-614-4.

[79] Trut, L. (1999). "Early canid domestication: The farm-fox experiment: Foxes bred for tamability in a 40-year experiment exhibit remarkable transformation that suggest an interplay between behavioral genetics and development". *American Scientist* **87** (2): 160. doi:10.1511/1999.2.160.

[80] Trut, L. (2004). "An experiment on fox domestication and debatable issues of evolution of the dog". *Russian Journal of Genetics* **40** (6): 644. doi:10.1023/B:RUGE.0000033312.92773.c1.

[81] Derr, M. (2011). *How the Dog Became the Dog: From Wolves to Our Best Friends*. Penguin Group USA. ISBN 1-59020-700-9.

[82] Li, Y. (2014). "Domestication of the dog from the wolf was promoted by enhanced excitatory synaptic plasticity: A hypothesis". *Genome Biology and Evolution* **6** (11): 3115–21. doi:10.1093/gbe/evu245. PMC 4255776. PMID 25377939.

[83] Frantz, L. (2015). "Evidence of long-term gene flow and selection during domestication from analyses of Eurasian wild and domestic pig genomes". doi:10.1038/ng.3394.

[84] Pennisi, E. (2015). "The taming of the pig took some wild turns". doi:10.1126/science.aad1692.

[85] Butterworth, G. (2003). "Pointing is the royal road to language for babies".

[86] Lakatos, G. (2009). "A comparative approach to dogs' (*Canis familiaris*) and human infants' comprehension of various forms of pointing gestures". *Animal Cognition* **12** (4): 621–31. doi:10.1007/s10071-009-0221-4. PMID 19343382.

[87] Muller, C. (2015). "Dogs can discriminate the emotional expressions of human faces". *Current Biology* **25** (5): 601–5. doi:10.1016/j.cub.2014.12.055. PMID 25683806.

[88] Hare, B. (2013). "What Are Dogs Saying When They Bark?". *Scientific American*. Retrieved 17 March 2015

[89] Sanderson, K. (2008). "Humans can judge a dog by its growl". *Nature*. doi:10.1038/news.2008.852.

[90] Wang, G. (2013). "The genomics of selection in dogs and the parallel evolution between dogs and humans". *Nature Communications* **4**: 1860. doi:10.1038/ncomms2814. PMID 23673645.

[91] Cossins, D. (2003), *Dogs and Human Evolving Together*, retrieved January 12, 2014

[92] Wang, G. (2014). "Genetic convergence in the adaptation of dogs and humans to the high-altitude environment of the Tibetan Plateau". *Genome Biology and Evolution* **6** (8): 2122–8. doi:10.1093/gbe/evu162. PMC 4231634. PMID 25091388.

[93] Nagasawa, M. (2015). "Oxytocin-gaze positive loop and the coevolution of human-dog bonds". doi:10.1126/science.1261022.

[94] Schleidt, W. (2003). "Co-evolution of humans and canids: An alternative view of dog domestication: Homo homini lupus?" (PDF). *Evolution and Cognition* **9** (1): 57–72.

[95] Paul Taçon (2002). "Dogs make us human". *Nature Australia* (Australian Museum) **27** (4): 52–61. Journal no longer published

Chapter 18

Spear-thrower

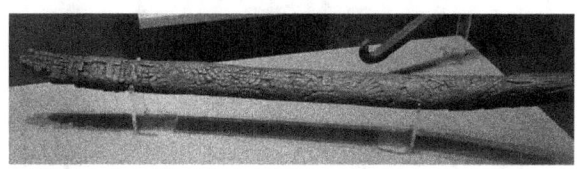

Carved Aztec atlatl at the National Museum of Anthropology and History in Mexico City

Depiction of an atlatl

A **spear-thrower** or **atlatl** (/ˈɑːt.lɑːtəl/[1] /ˈæt.lætəl/; Nahuatl: *ahtlatl* Nahuatl pronunciation: [ˈaʔt͡ɬat͡ɬ]) is a tool that uses leverage to achieve greater velocity in dart-throwing, and includes a bearing surface which allows the user to store energy during the throw.

It may consist of a shaft with a cup or a spur at the end that supports and propels the butt of the dart. The spear-thrower is held in one hand, gripped near the end farthest from the cup. The dart is thrown by the action of the upper arm and wrist. The throwing arm together with the atlatl acts as a lever. The spear-thrower is a low-mass, fast-moving extension of the throwing arm, increasing the length of the lever. This extra length allows the thrower to impart force to the dart over a longer distance, thus imparting more energy and ultimately higher speeds.[2]

Common modern ball throwers (molded plastic shafts used for throwing tennis balls for dogs to fetch) use the same principle.

A spear-thrower is a long-range weapon and can readily impart to a projectile speeds of over 150 km/h (93 mph).[3]

Spear-throwers appear very early in human history in several parts of the world, and have survived in use in traditional societies until the present day, as well as being revived in recent years for sporting purposes. In the United States the Nahuatl word *atlatl* is often used for revived uses of spear-throwers, and in Australia the Aboriginal word *woomera*.

The ancient Greeks and Romans used a leather thong or loop, known as an *ankule* or *amentum*, as a spear-throwing device.[4]

18.1 Design

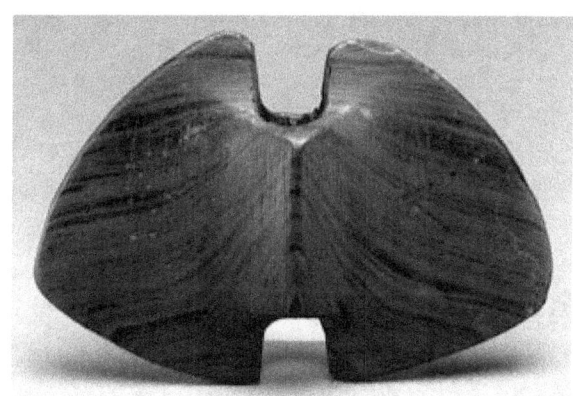

Bannerstone atlatl weight, ca. 2000 BC. Archaic peoples; Ohio.

Spear-thrower designs may include improvements such as thong loops to fit the fingers, the use of flexible shafts, stone balance weights, and thinner, highly flexible darts for added power and range. Darts resemble large arrows or thin spears and are typically from 1.2 to 2.7 m (3.9 to 8.9 ft) in length and 9 to 16 mm (3/8" to 5/8") in diameter.

Another important improvement to the spear-thrower's design was the introduction of a small weight (between 60 and 80 grams) strapped to its midsection. Some atlatlists maintain that stone weights add mass to the shaft of the device, causing resistance to acceleration when swung and resulting in a more forceful and accurate launch of the dart. Others claim that spear-thrower weights add only stability to a cast, resulting in greater accuracy.

Based on previous work done by William S. Webb, William R. Perkins[5] claims that spear-thrower weights, commonly called "bannerstones," and characterized by a centered hole in a symmetrically shaped carved or ground stone, shaped wide and flat with a drilled hole and thus a little like a large wingnut, are a rather ingenious improvement to the design that created a silencing effect when swung. The use of the device would reduce the telltale "zip" of a swung atlatl to a more subtle "woof" sound that did not travel as far and was less likely to alert prey or other humans. Robert Berg's theory is that the bannerstone was carried by hunters as a spindle weight to produce string from natural fibers gathered while hunting, for the purpose of tying on fletching and hafting stone or bone points.

18.1.1 Woomera

The Woomera design is distinctly different to most other spear-throwers, in that it has a curved, hollow shape, which allows for it to be used for other purposes.

18.1.2 Artistic designs

Several Stone Age spear-throwers (usually now incomplete) are decorated with carvings of animals: the British Museum has a mammoth, and there is a hyena in France. Many pieces of decorated bone may have belonged to *batons de commandement*.

The Aztec atlatl was often decorated with snake designs and feathers.[6]

18.2   History

Wooden darts were known at least since the Middle Paleolithic (Schöningen, Torralba, Clacton-on-Sea and Kalambo Falls). While the spear-thrower is capable of casting a dart well over one hundred meters, it is most accurately used at distances of twenty meters or less. Seven spears were found in the Schöningen 13 II-4 layer, dating from about 400,000 years ago and thought to represent activities of *Homo heidelbergensis*.[7] The spearthrower is believed to have been in use by *Homo sapiens* since the Upper Paleolithic (around 30,000 years ago).[8] Most stratified European finds come from the Magdalenian (late upper Palaeolithic). In this period, elaborate pieces, often in the form of animals, are common. The earliest secure data concerning atlatls has come from several caves in France dating to the Upper Paleolithic, about 21,000 to 17,000 years ago. The earliest known example is a 17,500-year-old Solutrean atlatl made of reindeer antler and found at Combe Saunière (Dordogne), France.[9]

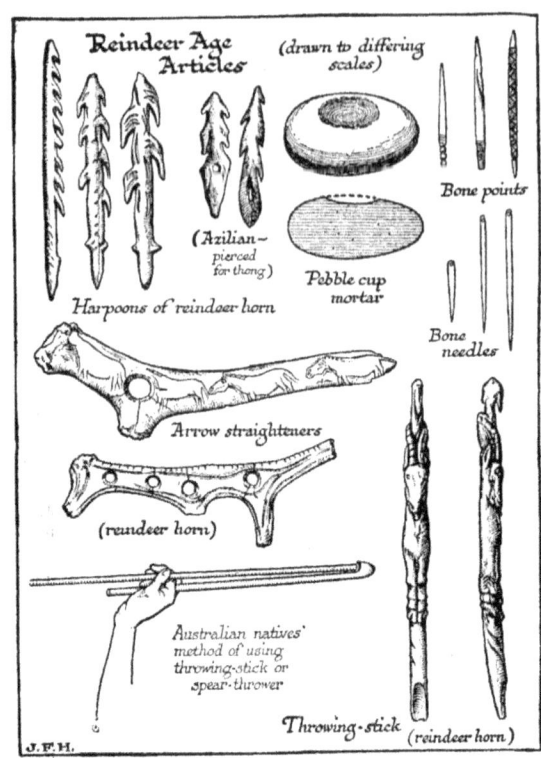

Reindeer Age articles

In Europe, the spear-thrower was supplemented by the bow and arrow, in the Epi-Paleolithic. Along with improved ease-of-use, the bow offered the advantage that the bulk of elastic energy is stored in the throwing device, rather than the projectile; arrow shafts can therefore be much smaller, and have looser tolerances for spring constant and weight distribution than atlatl darts. This allowed for more forgiving flint knapping: dart heads designed for a particular spear thrower tend to differ in mass by only a few percent. By the Iron Age, the amentum, a strap attached to the shaft, was the standard European mechanism for throwing lighter javelins. The amentum gives not only range, but also spin to the projectile.[10]

The spear-thrower was used by early Americans as well. It seems to have been introduced to America during the immigration across the Bering Land Bridge, and despite the later introduction of the bow and arrow, atlatl use was widespread at the time of first European contact. Complete wooden spear-throwers have been found on dry sites in the western USA, and in waterlogged environments in Florida and Washington. Several Amazonian Indian tribes also used the atlatl, for fishing and hunting. Some even preferred this weapon over the bow and arrow, and used it not only in combat but also in sports competitions. Such was the case with the Tarairiu, a Tapuya tribe of migratory foragers and raiders inhabiting the forested mountains and

Ceremonial atlatl, Peru 0-300 A.D., Lombards Museum

in) long and 3 to 4.5 cm (1 $\frac{1}{4}$ to 1 $\frac{3}{4}$ in) wide, this spear thrower was a tapering piece of wood carved of brown hard-wood. Well-polished, it was shaped with a semi-circular outer half and had a deep groove hollowed out to receive the end of the javelin, which could be engaged by a horizontal wooden peg or spur lashed with a cotton thread to the proximal and narrower end of the throwing board, where a few scarlet parrot feathers were tied for decoration. [Their] darts or javelins… were probably made of a two-meter long wooden cane with a stone or long and serrated hard-wood point, sometimes tipped with poison. Equipped with their uniquely grooved atlatl, they could hurl their long darts from a great distance with accuracy, speed, and such deadly force that these easily pierced through the protective armor of the Portuguese or any other enemy.".[11]

Among the Tlingit of Southeast Alaska approximately one dozen very old elaborately carved specimens they call "shee áan" (sitting on a branch) remain in museum collections [12] and private collections, one having sold at auction for more than $100,000.

The people of New Guinea and Australian Aborigines also use spear-throwers. In the mid Holocene,[13] Australians developed spear-throwers, known as *woomeras*.[14][15]

As well as its practical use as a hunting weapon, it may also have had social effects. John Whittaker, an anthropologist at Grinnell College, Iowa, suggests the device was a social equalizer in that it requires skill rather than muscle power alone. Thus women and children would have been able to participate in hunting,[3] although in recent Australian societies, spear-throwers are restricted by custom to male use.

Whittaker said the stone-tipped projectiles from the Aztec atlatl were not powerful enough to penetrate Spanish steel plate armor, but they were strong enough to penetrate the mail, leather and cotton armor that most Spanish soldiers wore.[6] Whittaker said the Aztecs started their battles with atlatl darts followed with melee combat using the macuahuitl.[6]

18.3 *Bâtons de commandement*

Another type of Stone Age artefact that is sometimes excavated is the *bâton de commandement*. These are shorter, normally less than one foot long, and made of antler, with a hole drilled through them. When first found in the nineteenth century, they were interpreted by French archaeologists to be symbols of authority, like a modern Field Mar-

highland savannahs of Rio Grande do Norte in mid-17th-century Brazil. Anthropologist Harald Prins offers the following description:

> "The atlatl, as used by these Tarairiu warriors, was unique in shape. About 88 cm (35

shal's baton, and so named *bâtons de commandement* ("batons of command"). Though debate over their function continues, tests with replicas have found them, when used with a cord, very effective aids to spear or dart throwing.[16] Another theory is that they were "arrow-straighteners", and the examples in the 1920 illustration at right are so labelled.

18.4   Modern times

The thrower is checking to see that the dart has been correctly located on the spur of the spear-thrower; next she will turn her head in the other direction, aim and throw

In modern times, some people have resurrected the dart thrower for sports, often using the term atlatl, throwing either for distance and/or for accuracy. Throws of almost 260 m (850 ft) have been recorded.[17] Colleges reported to field teams in this event include Grinnell College in Iowa, Franklin Pierce University in New Hampshire, Alfred University in New York, and the University of Vermont.[18] There are numerous tournaments, with spears and spear-throwers built with both ancient and with modern materials. Similar devices are available to throw tennis balls for dogs to chase, and in the sport of jai alai.

Atlatl are sometimes used in modern times for hunting. There are meetings and events where people can throw darts.[19] A few examples of the locations of such competitions are in Oregon,[20] Rhode Island and in Lexington, Kentucky[21] held yearly. In the U.S., the Pennsylvania Game Commission has given preliminary approval for the legalization of the atlatl for hunting certain animals.[22] The animals that would be allowed to atlatl hunters have yet to be determined, but attention is focused on deer. Currently, Alabama allows the atlatl for deer hunting, while a handful of other states list the device as legal for rough fish (those not sought for sport or food), some game birds and non-game mammals.[23] Starting in 2007, Missouri al-

lowed use of the atlatl for hunting wildlife (excluding deer and turkey), and starting in 2010, also allowed deer hunting during the firearms portion of the deer season (except the muzzleloader portion).[24][25] Starting in 2012, Missouri allowed the use of atlatls during the Fall archery deer and turkey hunting seasons and starting in 2014 allowed the use of atlatl during the Spring turkey hunting season as well.[26] Missouri also allows the use of the atlatl for fishing, with some restrictions (similar to the restrictions for spearfishing and bowfishing).[27] The Nebraska Game and Parks Commission allows the use of atlatls for the taking of deer as of 2013.[28]

The woomera is still used today by some Australian Aborigines for hunting in remote parts of Australia. Yup'ik Eskimo hunters still use the atlatl, known locally as "nuqaq" (nook-ak), in villages near the mouth of the Yukon River for seal hunting.

18.5   Competitions

Chimney Point state historic site in Addison, Vermont hosts the annual Northeast Open Atlatl Championship. In 2009, the Fourteenth Annual Open Atlatl Championship was held on Saturday and Sunday, September 19 and 20. On the Friday before the Championship, a workshop was open to teach modern and traditional techniques of atlatl and dart construction, flint knapping, hafting stone points, and cordage making.[29]

The World Atlatl Association stages an annual event of spear-throwing at Valley of Fire State Park in Nevada.[3]

There was an atlatl competition at the Ohio Pawpaw Festival each year.[30] Another annual atlatl competition is at Bois D' Arc Primitive Skills Gathering and Knap-in, held every September in southern Missouri.[31]

Atlatl associations around the world[32] host a number of local atlatl competitions.

In the sixth episode of the fourth season of the television competition, *Top Shot*, the elimination round consisted of two contestants using the atlatl at ranges of 30, 45, and 60 feet.

18.6   See also

- Aztec warfare

- Hunter-gatherer societies

- Woomera (spear-thrower)

- Amentum

- Kestros

- Swiss arrow

18.7 References

[1] "atlatl". Dictionary.com. Retrieved October 12, 2006.

[2] "Atlatl History and Physics". Tasigh.org. Retrieved 2013-09-21.

[3] "Girls on top". The Economist. April 12, 2008. Retrieved December 2010.

[4] Howard L. Blackmore. (2000) *Hunting Weapons: From the Middle Ages to the Twentieth Century*, page 103 "... the air'.31 A device which enabled all but the heaviest of spears to be cast a respectable distance was the spear thrower. ... It was known to the Greeks as the ankuli and to Romans as the amentum.3 The spear was rested in the hand and ..."

[5] "Atlatls & Primitive Technology". Retrieved 2012-08-19.

[6] "Mexicolore". Mexicolore. 2012-12-21. Retrieved 2013-09-21.

[7] Terberger, Thomas (2006). "From the First Humans to the Mesolithic Hunters in the Northern German Lowlands, Current Results and Trends". In Hansen, Keld Møller; Pedersen, Kristoffer Buck. *Across the western Baltic* (PDF). Sydsjællands Museums Publikationer Vol. 1. ISBN 87-983097-5-7.

[8] McClellan, James Edward; Dorn, Harold (2006). *Science and technology in world history: an introduction*. JHU Press. p. 11. ISBN 978-0-8018-8360-6. Retrieved December 2010.

[9] Peregrine, Peter N.; Ember, Melvin (2001). "Europe". *Encyclopedia of Prehistory* 4. Springer. ISBN 978-0-306-46258-0.

[10] Gardiner, E. Norman (1907). "Throwing the Javelin". *The Journal of Hellenic Studies* **27**: 249–273. doi:10.2307/624444.

[11] Prins, Harald E.L. (2010). "The Atlatl as Combat Weapon in 17th-Century Amazonia: Tapuya Indian Warriors in Dutch Colonial Brazil" (PDF). *The Atlatl* **23** (2): 1–3.

[12] "Fenimore Art Museum | Atlatl". Collections.fenimoreartmuseum.org. 1992-12-08. Retrieved 2013-09-21.

[13] Laet, Sigfried J. de & International Commission for the New Edition of the History of the Scientific and Cultural Development of Mankind & International Commission for a History of the Scientific and Cultural Development of Mankind. History of mankind (1994). History of humanity. Routledge; Paris : Unesco, London; New York, p.1064

[14] Palter, John L. "Design and construction of Australian spearthrower projectiles and hand-thrown spears." Archaeology & Physical Anthropology in Oceania (1977): 161-172. APA

[15] Cundy, B. J. (1989). Formal variation in Australian spear and spearthrower technology (Vol. 546). British Archaeological Reports Ltd.

[16] *Primitive Technology: A Book of Earth Skills - Google Books*. Books.google.co.uk. Retrieved 2013-09-21.

[17] "The Atlatl". Flight-toys.com. Retrieved 2009-06-08.

[18] "Our Towns: A College Team Takes Aim". Parade.com. 2010-05-30. Retrieved 2013-09-21.

[19] "Thunderbird Atlatl". Thunderbird Atlatl. Retrieved 2013-09-21.

[20] "Echoes in Time, Workshops in Early Living Skills". *Echoes in Time*. 26 July 2014. Retrieved 13 November 2014.

[21] "Expired Website". Atlatl.bravehost.com. Retrieved 2013-09-21.

[22] "Pennsylvania May Let Hunters Use Prehistoric Weapon". RedOrbit.com. 2005-11-13. Retrieved 2009-06-08.

[23] "Spear near in Pennsylvania? from espn.com". Sports.espn.go.com. 2006-06-21. Retrieved 2009-06-08.

[24] "2013 Firearms Deer Hunting | Missouri Department of Conservation". Mdc.mo.gov. 2013-09-15. Retrieved 2013-09-21.

[25] Wang, Regina (2010-09-28). "Atlatl makes debut for Missouri deer season". Columbia Missourian. Retrieved 2013-09-21.

[26] Conservation Action: Meeting of the May 2011 Conservation Commission | Missouri Department of Conservation at the Wayback Machine (archived November 16, 2012)

[27] "Missouri Department of Conservation, Wildlife Code: Sport Fishing: Seasons, Methods, Limits" (PDF). Sos.mo.gov. Retrieved 2013-09-21.

[28] "Whitetail and Mule Deer Hunting". State of Nebraska. Retrieved December 21, 2014.

[29] Archived September 5, 2009 at the Wayback Machine

[30] "2013 Ohio Pawpaw Festival". Ohiopawpawfest.com. Retrieved 2013-09-21.

[31] "Bois D' Arc Primitive Skills Gathering and Knap-in". Retrieved 2013-10-06.

[32] "Atlatl Associations". Thunderbird Atlatl. Retrieved 2013-09-21.

- Garrod,D. (1955) Palaeolithic spear throwers. *Proc. Prehist. Soc.* 21, pp. 21–35.

- Hunter, W. (1992) "Reconstructing a Generic Basket Maker Atlatl", *Bulletin of Primitive Technology*, No. 4.

- Knecht, H. (1997) Projectile technology, New York, Plenum Press, 408 p. ISBN 0-306-45716-4

- Nuttall, Zelia (1891). *The atlatl or spear-thrower of the ancient Mexicans*. Cambridge, Massachusetts: Peabody Museum of American Archaeology and Ethnology. OCLC 3536622.

- Perkins,W. (1993) "Atlatl Weights, Function and Classification", *Bulletin of Primitive Technology*, No. 5.

- Prins, Harald E.L. (2010). The Atlatl as Combat Weapon in 17th-Century Amazonia: Tapuya Indian Warriors in Dutch Colonial Brazil. *The Atlatl,* Vol.23, No.2, pp. 1–3. http: //waa.basketmakeratlatl.com/wp-content/uploads/ 2013/02/Tapuya-Atlatl-Article-by-Harald-Prins-25 May 2010.pdf

- Stodiek, U. (1993) Zur Technik der jungpaläolithischen Speerschleuder (Tübingen).

18.8   External links

- World Atlatl Association Web Site

- Atlatl reference page

- Graphic of a spear thrower in use.

Chapter 19

Mesolithic

Mesolithic microliths

In archaeology, **mesolithic** (Greek: μεσος, *mesos* "middle"; λιθος, *lithos* "stone") is the culture between paleolithic and neolithic. The term "Epipaleolithic" is often used for areas outside northern Europe, but was also the preferred synonym used by French archaeologists until the 1960s.

Mesolithic has different time spans in different parts of Eurasia. It was originally post-Pleistocene, pre-agricultural material in northwest Europe about 10,000 to 5,000 BC, but material from the Levant (about 20,000 to 9,500 BC) is also labelled mesolithic.

19.1 Information

The term "Mesolithic" is in competition with another term, "Epipaleolithic", which means the "final Upper Palaeolithic industries occurring at the end of the final glaciation which appear to merge technologically into the Mesolithic".[1]

In the archaeology of northern Europe, for example for archaeological sites in Great Britain, Germany, Scandinavia, Ukraine, and Russia, the term "Mesolithic" is almost always used. In the archaeology of other areas, the term "Epipaleolithic" may be preferred by most authors, or there may be divergences between authors over which term to use or what meaning to assign to each. In the New World, neither term is used (except provisionally in the Arctic).

- Some authors use the term "Epipaleolithic" for those cultures that are late developments of hunter-gatherer traditions but not in transition toward agriculture, reserving the term "Mesolithic" for those cultures, like the Natufian culture, that are transitional between hunter-gatherer and agricultural practices.

- Other authors use the term Mesolithic for a variety of Late Paleolithic cultures subsequent to the end of the last glacial period whether they are transitional towards agriculture or not.

A Spanish scholar, Alfonso Moure, says in this regard:[2]

> In the terminology of prehistoric archeology, the most widespread trend is to use the term "Epipaleolithic" for the industrial complexes of post-glacial hunter-gatherer groups. Conversely, those that are in course of transition toward artificial food production are assigned to the "Mesolithic".

In the archaeology of sub-Saharan Africa, Lower Paleolithic is replaced by "Early Stone Age", Middle Paleolithic is replaced by "Middle Stone Age" and Upper Paleolithic by "Later Stone Age" according to the terminology introduced by John Hilary Goodman and Clarence van Riet Lowe of South Africa in the early 20th century. Therefore, care must be taken in translating "Mesolithic" as "Middle Stone Age", as the latter term has an unrelated technical meaning in the context of African archaeology.

19.2 History of the concept

The three -lithics are subdivisions of the Stone Age in the three-age system developed since classical times and given a modern archaeological meaning by Christian Jürgensen Thomsen, a Danish archaeologist, in the early 19th century. Subdivisions of "earlier" and "later" were added to the Stone Age by Thomsen and especially his junior colleague and employee Jens Jacob Asmussen Worsaae. John Lubbock kept these divisions in his work *Pre-historic Times* in 1865 and introduced the terms Paleolithic ("Old Stone Age") and Neolithic ("New Stone Age") for them. He saw no need for an intermediate category.

When Hodder Westropp introduced the Mesolithic in 1866, as a technology intermediate between Paleolithic and Neolithic, a storm of controversy immediately arose around it. A British school led by John Evans denied any need for an intermediate. The ages blended together like the colors of a rainbow, he said. A European school led by Louis Laurent Gabriel de Mortillet asserted that there was a gap between the earlier and later. Edouard Piette claimed to have filled the gap with his discovery of the Azilian Culture. Knut Sterjna offered an alternative in the Epipaleolithic, a continuation of the use of Paleolithic technology. By the time of Vere Gordon Childe's work, *The Dawn of Europe* (1947), which affirms the Mesolithic, sufficient data had been collected to determine that the Mesolithic was in fact necessary and was indeed a transition and intermediary between the Paleolithic and the Neolithic.[3]

19.3   Characteristics

The start and end dates of the Mesolithic vary by geographical region. Childe's view prevails that the term generally covers the period between the end of the Pleistocene and the start of the Neolithic. The times of these events vary greatly; moreover, the various Mesolithics within the span might be as short as roughly a thousand years or as long as roughly 15,000 years depending on the circumstances. If the Mesolithic is more similar to the Paleolithic it is called the Epipaleolithic.

The Paleolithic was an age of purely hunting and gathering while in the Neolithic domestication of plants and animals had occurred. Some Mesolithic peoples continued with intensive hunting. Others were practising the initial stages of domestication. Some Mesolithic settlements were villages of huts. Others were walled cities.

The type of tool remains the diagnostic factor: The Mesolithic featured composite devices manufactured with Mode V chipped stone tools (microliths). The Paleolithic had utilized Modes I-IV and the Neolithic mainly abandoned the chipped microliths in favor of polished, not chipped, stone tools.

19.3.1   The Levant

Mesolithic 1

The first period, known as **Mesolithic 1** (Kebarian culture; 20–18,000 BC to 12,150 BC), followed the Aurignacian or Levantine Upper Paleolithic periods throughout the Levant. By the end of the Aurignacian, gradual changes took place in stone industries. Small stone tools called microliths and retouched bladelets can be found for the first time. The microliths of this culture period differ greatly from the Aurignacian artifacts. This period is more properly called Epipaleolithic.

By 20,000 to 18,000 BC the climate and environment had changed, starting a period of transition. The Levant became more arid and the forest vegetation retreated, to be replaced by steppe. The cool and dry period ended at the beginning of Mesolithic 1. The hunter-gatherers of the Aurignacian would have had to modify their way of living and their pattern of settlement to adapt to the changing conditions. The crystallization of these new patterns resulted in Mesolithic 1. New types of settlements and new stone industries developed.

The inhabitants of a small Mesolithic 1 site in the Levant left little more than their chipped stone tools behind. The industry was of small tools made of bladelets struck off single-platform cores. Besides bladelets, burins and end-scrapers were found. A few bone tools and some ground stone have also been found. These so-called Mesolithic sites of Asia are far less numerous than those of the Neolithic and the archeological remains are very poor.

Mesolithic 2

Main article: Natufian culture

The second period, **Mesolithic 2**, is also called the Natufian culture. The change from Mesolithic 1 to Natufian culture can be dated more closely. The latest date from a Mesolithic 1 site in the Levant is 12,150 BC. The earliest date from a Natufian site is 11,140 BC. This period is characterized by the early rise of agriculture that would later emerge into the Neolithic period. Radiocarbon dating places the Natufian culture between 12,500 and 9500 BC, just before the end of the Pleistocene.[4] This period is characterised by the beginning of agriculture.[5] The earliest known battle occurred during the Mesolithic period at a site in Egypt known as Cemetery 117.

Natufian culture is commonly split into two subperiods: Early Natufian (12,500–10,800 BC) (Christopher Delage gives c. 13,000–11,500 BP uncalibrated, equivalent to c. 13,700 to 11,500 BC)[6] and Late Natufian (10,800–9500

BC). The Late Natufian most likely occurred in tandem with the Younger Dryas.

19.3.2 Europe

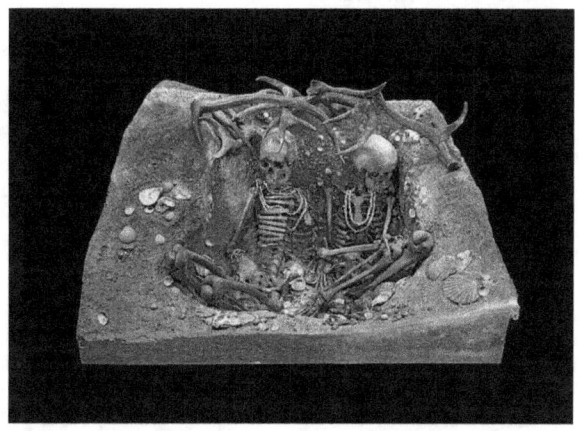

Two skeletons of women aged between 25 and 35 years, dated between 6740 and 5680 BP, both of whom died a violent death. Found at Téviec, France in 1938.

The Mesolithic began with the Holocene warm period around 11,660 BP and ended with the introduction of farming, the date of which varied in each geographical region. Regions that experienced greater environmental effects as the last glacial period ended have a much more apparent Mesolithic era, lasting millennia. In northern Europe, for example, societies were able to live well on rich food supplies from the marshlands created by the warmer climate. Such conditions produced distinctive human behaviors that are preserved in the material record, such as the Maglemosian and Azilian cultures. Such conditions also delayed the coming of the Neolithic until as late as 5000–4000 BC in northern Europe.

As the "Neolithic package" (including farming, herding, polished stone axes, timber longhouses and pottery) spread into Europe, the Mesolithic way of life was marginalized and eventually disappeared. Mesolithic adaptations such as sedentism, population size and use of plant foods are cited as evidence of the transition to agriculture.[7] In one sample from the Blätterhöhle in Hagen, it seems that the descendants of Mesolithic people maintained a foraging lifestyle for more than 2000 years after the arrival of farming societies in the area.[8] In north-Eastern Europe, the hunting and fishing lifestyle continued into the Medieval period in regions less suited to agriculture.

19.3.3 Ceramic Mesolithic

In North-Eastern Europe, Siberia, and certain southern European and North African sites, a "ceramic Mesolithic" can be distinguished between 7000-3850 BC. Russian archaeologists prefer to describe such pottery-making cultures as Neolithic, even though farming is absent. This pottery-making Mesolithic culture can be found peripheral to the sedentary Neolithic cultures. It created a distinctive type of pottery, with point or knob base and flared rims, manufactured by methods not used by the Neolithic farmers. Though each area of Mesolithic ceramic developed an individual style, common features suggest a single point of origin.[9] The earliest manifestation of this type of pottery may be in the region around Lake Baikal in Siberia. It appears in the Elshan or Yelshanka or Samara culture on the Volga in Russia c. 7000 BC,[10] and from there spread via the Dnieper-Donets culture to the Narva culture of the Eastern Baltic. Spreading westward along the coastline it is found in the Ertebølle culture of Denmark and Ellerbek of Northern Germany, and the related Swifterbant culture of the Low Countries.[11]

19.4 Mesolithic cultures

Periodization: The Levant: 20,000 to 9500 BC; Europe: 9660 to 5000 BC; Elsewhere: 10,000 to 400 BC

Some notable Mesolithic cultures:

- Azilian culture
- Balkan mesolithic cultures
- Capsian culture
- Fosna-Hensbacka culture
- Harifian culture
- Kebaran culture
- Jōmon cultures
- Jeulmun culture
- Komsa culture
- Kongemose culture
- Kunda culture
- Lepenski Vir culture
- Maglemosian culture
- Natufian culture

- Neman culture

- Nøstvet and Lihult cultures

- Sauveterrian culture

- Tardenoisian culture

- Zarzian culture

19.5   List of Mesolithic sites

Some notable Mesolithic sites:

- Lepenski Vir, Serbia: 7000 BC

- Star Carr, England: 8700 BC

- Pulli settlement, Estonia: 9000 BC

- Franchthi cave, Greece: 20,000–3000 BC

- Cramond, Scotland: 8500 BC

- Mount Sandel, Ireland: 7010 BC

- Howick house, England: 7000 BC

- Newbury, England

- Swifterbant culture, The Netherlands

- Aveline's Hole, Somerset, England: 8000 BC

- Shigir Idol, Russia: 9500 BC

19.6   See also

- 10th millennium BC

- 9th millennium BC

- 8th millennium BC

- 7th millennium BC

- Holocene

- Jōmon period

- Archaic period in North America

- List of Stone Age art

19.7   Notes

[1] Bahn, Paul, *The Penguin Archaeology Guide*, Penguin, London, pp. 141. ISBN 0-14-051448-1.

[2] A. Mouré. *El Origen del Hombre*, 1999. ISBN 84-7679-127-5.

[3] Linder, F., 1997. *Social differentiering i mesolitiska jägarsamlarsamhällen.* Institutionen för arkeologi och antik historia, Uppsala universitet. Uppsala.

[4] Munro, Natalie D. (2003). "Small game, the Younger Dryas, and the transition to agriculture in the southern Levant" (PDF). *Mitteilungen der Gesellschaft für Urgeschichte* **12**: 47–71.

[5] Bar-Yosef, Ofer (1998). "The Natufian Culture in the Levant, Threshold to the Origins of Agriculture" (PDF). *Evolutionary Anthropology* **6** (5): 159–177. doi:10.1002/(SICI)1520-6505(1998)6:5<159::AID-EVAN4>3.0.CO;2-7.

[6] Delage, Christopher, *The Last Hunter-gatherers in the Near East*, British Archaeological Reports (1 June 2004), ISBN 978-1-84171-389-2

[7] T.Douglas Price, *Europe's first farmers* (Cambridge University Press 2000), page 5.

[8] 2000 Years of Parallel Societies in Stone Age Central Europe. Ruth Bollongino, Olaf Nehlich, Michael P. Richards, Jörg Orschiedt, Mark G. Thomas, Christian Sell, Zuzana Fajkošová, Adam Powell, Joachim Burger. Science. Published Online October 10, 2013. DOI: 10.1126/science.1245049 http://www.sciencemag.org/content/early/2013/10/09/science.1245049

[9] De Roevers, p.162-163

[10] D. W. Anthony, Pontic-Caspian Mesolithic and Early Neolithic societies at the time of the Black Sea Flood: a small audience and small effects, in V. Yanko-Hombach, A.A. Gilbert, N. Panin and P. M. Dolukhanov (eds.), *The Black Sea Flood Question: changes in coastline, climate and human settlement* (2007), pp. 245-370 (361); D. W. Anthony, *The Horse, The Wheel and Language* (2007), pp.148-9, p. 480, note 19.

[11] Detlef Gronenborn, Beyond the models: Neolithisation in Central Europe, *Proceedings of the British Academy*, vol. 144 (2007), pp. 73-98 (87).

19.8   Further reading

- Dragoslav Srejovic *Europe's First Monumental Sculpture: New Discoveries at Lepenski Vir.* (1972) ISBN 0-500-39009-6

19.9 External links

- Official Lepenski Vir Site in Serbian

- Mesolithic Miscellany — Newsletter and Information on the European Mesolithic

- 20th Century Mesolithic Sites in Mandla (Madhya Pradesh), India, discovered by Dr. Babul Roy: , , and

- Picture Gallery of the Paleolithic (reconstructional palaeoethnology), Libor Balák at the Czech Academy of Sciences, the Institute of Archaeology in Brno, The Center for Paleolithic and Paleoethnological Research

- Gazetteer of Mesolithic sites in England and Wales with a gazetteer of Upper Palaeolithic sites in England and Wales. Wymer JJ and CJ Bonsall, 1977 Council for British Archaeology Research Report No 20

- UNESCO World Heritaga Site in India - Rock Shelters of Bhimbetka

Chapter 20

Microlith

For the catalytic reactor, see Microlith (catalytic reactor).
For other uses, see Microlith (disambiguation).

A **microlith** is a small stone tool usually made of flint or

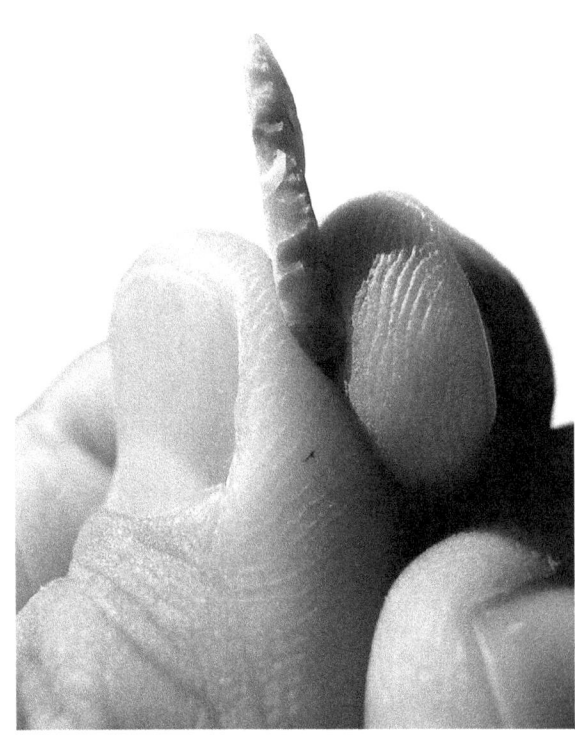

Backed edge bladelet

chert and typically a centimetre or so in length and half a centimetre wide. It is produced from either a small blade (microblade) or a larger blade-like piece of flint by abrupt or truncated retouching, which leaves a very typical piece of waste, called a microburin. The microliths themselves are sufficiently worked so as to be distinguishable from workshop waste or accidents.

Two families of microliths are usually defined: laminar and geometric. An assemblage of microliths can be used to date an archeological site. Laminar microliths are associated with the end of the Upper Paleolithic and the beginning of the Epipaleolithic era; geometric microliths are characteristic of the Mesolithic and the Neolithic. Geometric microliths may be triangular, trapezoid or lunate. Microlith production generally declined following the introduction of agriculture (8000 BCE) but continued later in cultures with a deeply rooted hunting tradition.

Regardless of type, microliths were used to form the points of hunting weapons, such as spears and (in later periods) arrows, and other artifacts and are found throughout Africa, Asia and Europe. They were utilised with wood, bone, resin and fiber to form a composite tool or weapon, and traces of wood to which microliths were attached have been found in Sweden, Denmark and England. An average of between six and eighteen microliths may often have been used in one spear or harpoon, but only one or two in an arrow.

20.1 Types of microlith

20.1.1 Laminar and non-geometric microliths

Laminar microliths date from at least the Gravettian culture or possibly the start of the Upper Paleolithic era, and they are found all through the Mesolithic and Neolithic eras. Noilles Burins and Microgravettes indicate that the production of microliths had already started in the Gravettian culture.[1] This style of flint working flourished during the Magdalenian period and persisted in numerous Epipaleolithic traditions all around the Mediterranean basin. These microliths are slightly larger than the geometric microliths that followed and were made from the flakes of flint obtained *ad hoc* from a small nucleus or from a depleted nucleus of flint. They were produced either by percussion or by the application of a variable pressure (although pressure is the best option, this method of producing microliths is complicated and was not the most commonly used technique).[2]

Truncated blade

There are three basic types of laminar microlith. The truncated blade type can be divided into a number of sub-types depending on the position of the truncation (for example, oblique, square or double) and according to its form, for example, concave or convex. "Raclette scrapers" are notable for their particular form, being blades or flakes whose edges have been sharply retouched until they are semicircular or even shapeless. Raclettes are indefinite cultural indicators, as they appear from the Upper Paleolithic through to the Neolithic.

- Flint blade

    - Truncated bladelet

    - Backed edge bladelet

    - Dufour bladelet

Backed edge blades

Backed edge blades have one of the edges, generally a side one, rounded or chamfered by abrupt retouching. There are fewer types of these blades, and may be divided into those where the entire edge is rounded and those where only a part is rounded, or even straight. They are fundamental in the blade-forming processes, and from them, innumerable other types were developed.[3] Dufour bladelets are up to three centimeters in length, finely shaped with a curved profile whose retouches are semi-abrupt and which characterize a particular phase of the Aurignacian period. Solutrean backed edge blades display pronounced and abrupt retouching, so that they are long and narrow and, although rare, characterize certain phases of the Solutrean period. Ouchtata bladelets are similar to the others, except that the retouched back is not uniform but irregular; this type of microlith characterizes certain periods of the Epipaleolithic saharans. The Ibero-Maurusian and the Montbani bladelet, with a partial and irregular lateral retouching, is characteristic of the Italian Tardenoisian.[4]

Micro points

These are very sharp bladelets formed by abrupt retouching. There are a huge numbers of regional varieties of these microliths, nearly all of which are very hard to distinguish (especially those from the western area) without knowing the archaeological context in which they appear. The following is a small selection. Omitted are the foliaceous tips (also called leafed tips), which are characterized by a covering retouch and which constitute a group apart.[5]

- The *Châtelperrón point* is not a true microlith, although it is close to the required dimensions. Its antiquity and its short, curved blade edge make it the antecedent of many laminar microliths.

- The *Micro-gravette* or *Gravette micro point* is a microlith version of the *Gravette point* and is a narrow bladelet with an abrupt retouch, which gives it a characteristically sharp edge when compared to other types.

- The *Azilian point* links the Magdalenian microlith points with those from the western Epipaleolithic. They can be identified by a rough and invasive retouching.

- The *Ahrensburgian point* is also a *peripheral paleolithic* or western Epipaleolithic piece, but with a more specific morphology, as it is formed on a blade (not on a bladelet), is obliquely truncated and has a small tongue that possibly served as a haft on a spear point.

The next group contains a number of points from the Middle East characterized as cultural markers.

- The Emireh point from the Upper Paleolithic is almost the same as one found in Châtelperrón, which is likely to be contemporary, although they are slightly shorter and also appear to be fashioned from a blade and not a bladelet.

- The *El-Wad point* is from the end of the Upper Paleolithic from the same area, made from a very long, thin bladelet.

- The *El-Khiam point* has been identified by the Spanish archeologist González Echegaray in Protoneolithic sites in Jordan. They are little known but easy to identify by two basal notches, doubtless used as a haft.[6]

    - Châtelperrón points

    - Micro-gravette

    - Azilian point

    - Ahrensburguien point

    - Emireh point

    - El-Wad point

    - El-Khiam point

    - Adelaide point

The *Adelaide point* is found in Australia. Its construction, based on truncations on a blade, has a nearly trapezoidal form. The Adelaide point emphasizes the range of variation in both time and culture of the laminar microliths; it also shows their technological differences, but sometimes morphological similarities, with geometric microliths. Laminar microliths can also sometimes be described as trapezoidal, triangular or lunate.[7] However, as we will see below, they are distinct from the geometric microliths because of the strokes used in the manufacture of geometric microliths, which mainly involved the microburin technique.

20.1.2 Geometric microliths

While geometric microliths first make their appearance in the Magdalenian,[8] initially as elongated triangles and later as trapezoids (although the microburin technique is seen from the Perigordian), they are mostly seen during the Epipaleolithic and the Neolithic. They remained in existence even into the Copper Age and Bronze Age, competing with "leafed" and then metallic arrowheads.

Geometric microliths are a clearly defined type of stone tool, at least in their basic forms. They can be divided into trapezoid, triangular and lunate (half-moon) forms, although there are many subdivisions of each of these types. A microburin is included among the illustrations below because, although it is not a geometrical microlith (or even a tool),[9] it is now seen as a characteristic waste product from the manufacture of these geometric microliths:

- Microburin
- Trapeze
- Triangle
- Lunate

Microburin technique

All the currently known geometric microliths share the same fundamental characteristics – only their shapes vary. They were all made from blades or from microblades (nearly always of flint), using the microburin technique (which implies that it is not possible to conserve the remains of the heel or the conchoidal flakes from the blank). The pieces were then finished by a percussive retouching of the edges (generally leaving one side with the natural edge of the blank), giving the piece its definitive polygonal form. For example, in order to make a triangle, two adjacent notches were retouched, leaving free the third edge or *base*[3] (using the terminology of Fortea). They generally have one long axis and concave or convex edges, and

it is possible for them to have a gibbosity (hump) or indentations. Triangular microliths may be isosceles, scalene or equilateral. In the case of trapezoid geometric microliths, on the other hand, the notches are not retouched, leaving a portion of the natural edge between them. Trapezoids can be further subdivided into symmetrical, asymmetrical and those with concave edges. Lunate microliths have the least diversity of all and may be either semicircular or segmental.

Archeological findings and the analysis of wear marks, or use-wear analysis, has shown that, predictably, the tips of spears, harpoons and other light projectiles of varying size received the most wear. Microliths were also used from the Neolithic on arrows, although a decline in this use coincided with the appearance of bifacial or "leafed" arrowheads that became widespread in the Chalcolithic period, or Copper Age (that is, stone arrowheads were increasingly made by a different technique during this later period).

20.2 Weapons and tools

Not all the different types of laminar microliths had functions that are clearly understood. It is likely that they contributed to the points of spears or light projectiles, and their small size suggests that they were fixed in some way to a shaft or handle.[10][11]

Backed edge bladelets are particularly abundant at a site in France that preserves habitation from the late Magdalenian – the Pincevent. In the remains of some of the hearths at this location, bladelets are found in groups of three, perhaps indicating that they were mounted in threes on their handles. A javelin tip made of horn has been found at this site with grooves made for flint bladelets that could have been secured using a resinous substance. Signs of much wear and tear have been found on some of these finds.[12]

Specialists have carried out lithic or microwear analysis on artefacts, but it has sometimes proved difficult to distinguish those fractures made during the process of fashioning the flint implement from those made during its use. Microliths found at Hengistbury Head in Dorset, England, show features that can be confused with chisel marks, but which might also have been produced when the tip hit a hard object and splintered.[13] Microliths from other locations have presented the same problems of interpretation.[14]

An exceptional piece of evidence for the use of microliths has been found in the excavations of the cave at Lascaux in the French Dordogne. Twenty backed edge bladelets were found with the remains of a resinous substance and the imprint of a circular handle (a horn). It appears that the bladelets might have been fixed in groups like the teeth of a harpoon or similar weapon.

In all these locations, the microliths found have been backed edge blades, tips and crude flakes. Despite the great number of geometric microliths that have been found in Western Europe, few examples show any clear evidence of their use, and all the examples are from the Mesolithic or Neolithic periods. Despite this, there is unanimity amongst researchers that these items were used to increase the penetrating potential of light projectiles such as harpoons, assegais, javelins and arrows.

20.3 Discoveries

20.3.1 France

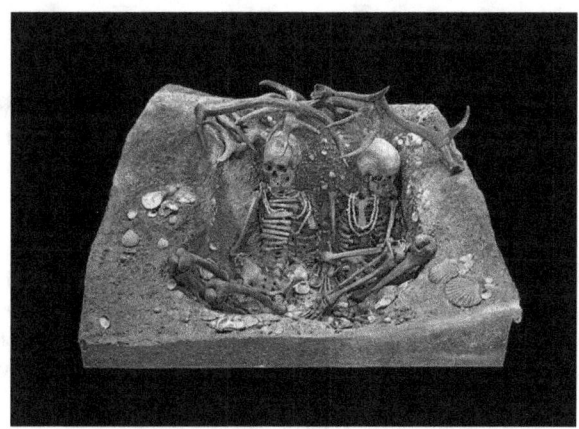

Two skeletons in the Tomb of Téviec

In France, one unusual find stands out: in the Mesolithic cemetery of Téviec, in Morbihan, one of the skeletons that has been found has a geometric microlith lodged in one of its vertebra. All indications suggest that the person died because of this projectile; whether by intention or by accident is unknown. It is widely agreed that geometric microliths were mainly used in hunting and fishing, but they may also have been used as weapons.[1]

20.3.2 Scandinavia

Well-preserved examples of arrows with microliths in Scandinavia have been found at Loshult, at Osby in Sweden, and Tværmose, at Vinderup in Denmark. These finds, which have been preserved practically intact due to the special conditions of the peat bogs, have included wooden arrows with microliths attached to the tip by resinous substances and cords.

20.3.3 England

There are many examples of possible tools from Mesolithic deposits in England. Possibly the best known is a microlith from Star Carr in Yorkshire that retains residues of resin, probably used to fix it to the tip of a projectile. Recent excavations have found other examples. Archeologists at the Risby Warren V site in Lincolnshire have uncovered a row of eight triangular microliths that are equidistantly aligned along a dark stain indicating organic remains (possibly the wood from an arrow shaft). Another clear indication is from the Readycon Dene site in West Yorkshire, where 35 microliths appear to be associated with a single projectile. In Urra Moor, North Yorkshire, 25 microliths give the appearance of being related to one another, due to the extreme regularity and symmetry of their arrangement in the ground.[15]

The study of English and European artifacts in general has revealed that projectiles were made with a widely variable number of microliths: in *Tværmose* there was only one, in *Loshult* there were two (one for the tip and the other as a fin),[16] in White Hassocks, in West Yorkshire, more than 40 have been found together; the average is between 6 and 18 pieces for each projectile.[15]

20.3.4 India

Early research regard the microlithic industry in India as a Holocene phenomenon, however a new research provides solid data to put the South Asia microliths industry up to 35 ka across whole South Asia subcontinent. This new research also synthesizes the data from genetic, paleoenvironmental and archaeological research, and proposes that the emergence of microlith in India subcontinent could reflect the increase of population and adaptation of environmental deterioration.[17]

20.4 Dating

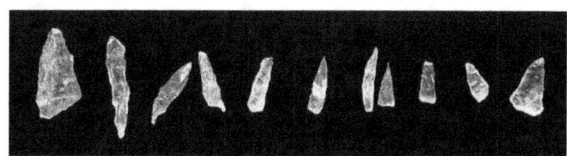

Crystal spear tips, ca. 8000–7000 BCE, on display at Sion History museum

Laminar microliths are common artifacts from the Upper Paleolithic and the Epipaleolithic, to such a degree that numerous studies have used them as markers to date different phases of prehistoric cultures.

During the Epipaleolithic and the Mesolithic, the presence of laminar or geometric microliths serves to date the deposits of different cultural traditions. For instance, in the Atlas Mountains of northwest Africa, the end of the Upper Paleolithic period coincides with the end of the Aterian tradition of producing laminar microliths, and deposits can be dated by the presence or absence of these artifacts. In the Near East, the laminar microliths of the Kebarian culture were superseded by the geometric microliths of the Natufian tradition a little more than 11,000 years ago. This pattern is repeated throughout the Mediterranean basin and across Europe in general.[3][18]

A similar thing is found in England, where the preponderance of elongated microliths, as opposed to other frequently occurring forms, has permitted the Mesolithic to be separated into two phases: the Earlier Mesolithic of about 8300–6700 BCE, or the ancient and laminar Mesolithic, and the Later Mesolithic, or the recent and geometric Mesolithic. Deposits can be thus dated based upon the assemblage of artifacts found.[19]

20.5 References

[1] Piel-Desruisseaux, Jean-Luccite (1986). *Outils préhistoriques. Forme. Fabrication. Utilisation.* Masson, Paris. ISBN 2-225-80847-3. (pages 147–9)

[2] Pelegrin, Jacques (1988). "Débitage expérimental par pression. Du plus petit au plus grand". Journée d'études technologiques en Préhistoire (Notes et monographies techniques, nº 25). Technologie préhistorique. ISBN 2-222-04235-6. (pages 37–53)

[3] Fortea Pérez, Francisco Javier (1973). *Los complejos microlaminares y geométricos del Epipaleolítico mediterráneo español.* Universidad de Salamanca. ISBN 84-600-5678-3.

[4] Brézillon, Michel (1971). *La dénomination des objets de pierre taillée.* París: Editions du CNRS. pages 263–7.

[5] Brézillon, Michel (1971). *La dénomination des objets de pierre taillée.* París: Editions du CNRS. pages 292–340.

[6] González Echegaray, J. (1964). *Excavaciones en la terraza de El Khiam (Jordania).* Bibliotheca Praehistorica Hispana.

[7] Geometric shapes, as we have seen, are present in many laminar microliths: for example the *Dufour bladelet* is an elongated lunate shape, the *El-Emireh point* is a triangle and the *Adelaide point* is a trapeze, the *El-Wad point* is spindle shaped; and there are many other examples.

[8] Bordes, F. y Fitte, P. (1964). "Microlithes du Magdalénien supérieur de la Gare de Gouze (Dordogne)". *Miscelánea en homenaje al Abate Henri Breuil. Vol. I.* Barcelona. page 264.

[9] Some of the earlier researchers, such as Octobon Octobon, E. (1920). "La question tardenoisienne. Montbani". *Revue Anthropologique.* page 107.), Peyrony and Noone (Peyrony, D. y Noone H. V. V. (1938). "Usage possible des microburins" **2** (numéro 3). Bulletin de la Société Préhistorique Française., believed that these microburins had a useful function. Currently it has been demonstrated that these microburins did not have a function, at least not intentionally, although it cannot be ruled out that they were not reused at some point.

[10] Laming-Emperaire, Annette (1980). "Los cazadores depredadores del posglacial y del Mesolítico". *La Prehistoria.* Editorial Labor, Barcelona. ISBN 84-335-9309-9. (page 68)

[11] Piel-Desruisseaux, Jean-Luc (1986). *Outils préhistoriques. Forme. Fabrication. Utilisation.* Masson, Paris. ISBN 2-225-80847-3. (pages 123-127)

[12] Personal communication: Excavation Director André Leroi-Gourhan

[13] Barton, R. N. E. y Bergman, C. A. (1982). "Hunters at Hengistbury: some evidence from experimental archaeology" **14** (Number 2). World Archaeology. ISSN 0043-8243.

[14] M. Lenoir has found knapping similar to that used in chiseled bladelets from Gironde, but considered this to be a coincidence and attributed the marks to the fact that the microliths were mounted on the tip of a projectile. A similar line of enquiry has also been followed by Lawrence H. Keeley, who has studied a wide range of bladelets from the French site at Buisson Campin, in Verberie, Oise.

[15] Myers, Andrew (1989). "Reliable and mantainable technological strategies in the Mesolithic of mainland Britain". *Time, energy and stone tools: New directions in Archaeology (edited by Robin Torrence).* Cambridge: Cambridge University Press. pp. 78–91. ISBN 0-521-25350-0.

[16] Petersson, M. (1951). *Microlithen als Pfeilspitzen. Ein Fund aus dem Lilla-Loshult Moor: Ksp. Loshult, Skane.* Meddelanden fram Lunds Universitets (Historika Museum). (Pagies 123–37).

[17] Petraglia; et al. (2009). "Population increase and environmental deterioration correspond with microlithic innovations in South Asia ca. 35,000 years ago". *Proceedings of the National Academy of Sciences* **106** (30): 12261–12266. doi:10.1073/pnas.0810842106.

[18] Professor Fortea has been able to distinguish two traditions in the Epipaleolithic period based in the Spanish Mediterranean , the "Microlaminar Complex" (with three separate phases: that of Sant Grégori de Falset, that based on the Cova de Les Mallaetes in Valencia and that of the *Epigravettian*) and the "Geometric Complex" (with two phases: the Filador and the Cocina, which receive their names from caves located on the eastern coast of Spain).

[19] Myers, Andrew (1989). "Reliable and mantainable technological strategies in the Mesolithic of mainland Britain". *Time, energy and stone tools: New directions in Archaeology (edited by Robin Torrence)*. Cambridge: Cambridge University Press. p. 78. ISBN 0-521-25350-0. The same author has suggested that the geometric microliths may replace one or two rows of teeth in the bone harpoons commonly found in the Upper Paleolithic at the end of the Upper Magdalanian (page 84).

20.6 External links

- Media related to Microliths at Wikimedia Commons

Chapter 21

Bow and arrow

For the consultancy firm, see Bow & Arrow.

The **bow and arrow** is a projectile weapon system (a bow with arrows) that predates recorded history and is common to most cultures. Archery is the art, practice, or skill of applying it.

21.1 Description

A bow is a flexible arc which shoots aerodynamic projectiles called arrows. A string joins the two ends of the bow and when the string is drawn back, the ends of the bow are flexed. When the string is released, the potential energy of the flexed stick is transformed into the velocity of the arrow.[1] Archery is the art or sport of shooting arrow from bows.[2]

Today, bows and arrows are used primarily for hunting and for the sport of archery. Though they are still occasionally used as weapons of war, the development of gunpowder and muskets, and the growing size of armies, led to their replacement in warfare several centuries ago in much of the world.

Someone who makes bows is known as a bowyer,[3] and one who makes arrows is a fletcher[4] —or in the case of the manufacture of metal arrow heads, an arrow smith.[5]

21.2 History

Main article: History of archery

The bow and arrow appears around the transition from the Upper Paleolithic to the Mesolithic. After the end of the last glacial period, use of the bow seems to have spread to every continent, including the New World, except for Australia.[6]

The oldest extant bows in one piece are the elm Holmegaard bows from Denmark which were dated to 9,000 BCE. High performance wooden bows are currently made following the Holmegaard design. The Stellmoor bow fragments from northern Germany were dated to about 8,000 BCE, but they were destroyed in Hamburg during the Second World War, before carbon 14 dating was available; their age is attributed by archaeological association.[7] Microliths discovered on the south coast of Africa suggest that arrows may be at least 70,000 years old. [8]

The bow was an important weapon for both hunting and warfare from prehistoric times until the widespread use of gunpowder in the 16th century. Organised warfare with bows ended in the mid 17th century in Europe, but it persisted into the early 19th century in Eastern cultures and in tribal warfare in the New World. It has recently been used as a weapon of tribal warfare in some parts of Sub-Saharan Africa; an example was documented in 2009 in Kenya when Kisii people and Kalenjin people clashed resulting in four deaths.[9][10]

The British upper class led a revival of archery from the late 18th century.[11] Sir Ashton Lever, an antiquarian and collector, formed the Toxophilite Society in London in 1781, with the patronage of George, the Prince of Wales.

21.3 Construction

21.3.1 Parts of the bow

The basic elements of a bow are a pair of curved elastic limbs, traditionally made from wood, joined by a riser. Both ends of the limbs are connected by a string known as the bow string.[1] By pulling the string backwards the archer exerts compressive force on the string-facing section, or belly, of the limbs as well as placing the outer section, or back, under tension. While the string is held, this stores the energy later released in putting the arrow to flight. The force required to hold the string stationary at full draw is often used to express the power of a bow, and is known as its draw weight, or weight.[12][13] Other things being equal, a higher draw weight means a more powerful bow, which is able to project heavier arrows at the same velocity or the same arrow at a greater velocity.

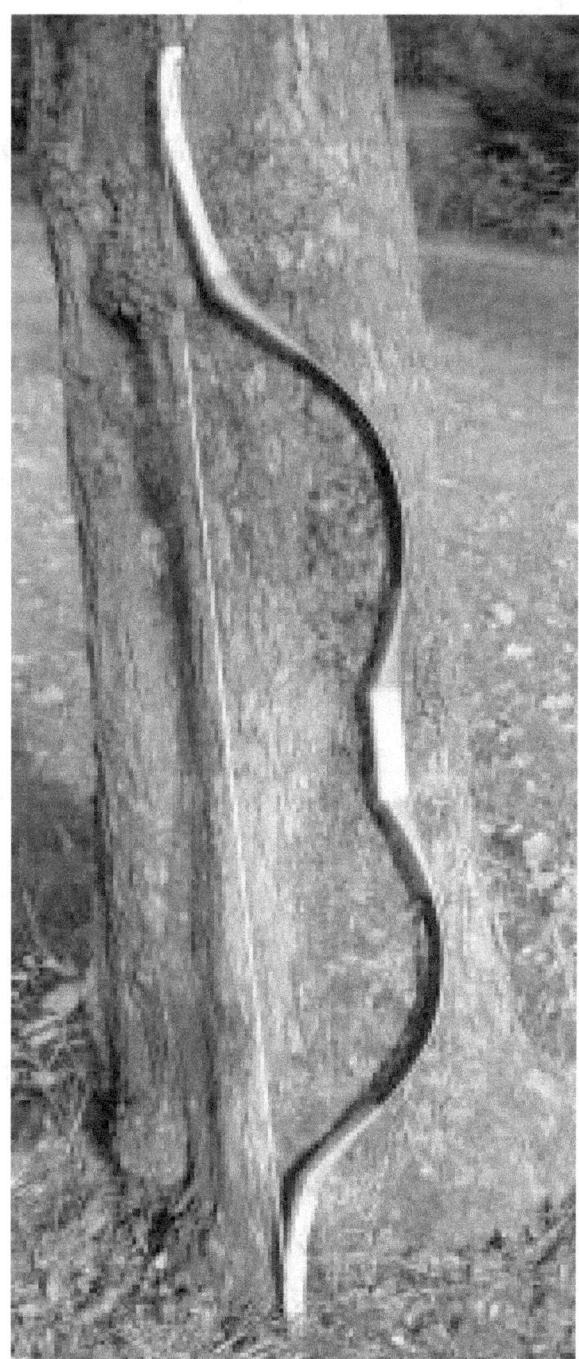

A modern reconstruction, in fiberglass and wood, of a historical composite bow

Scythians shooting with bows, Panticapeum (modern Kertch), 4th century BCE.

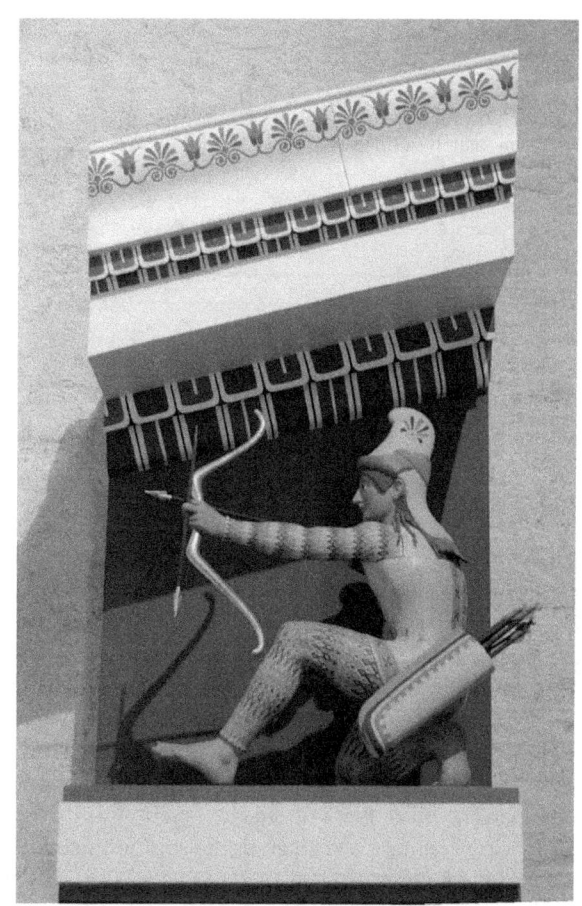

Polychrome small-scale model of the archer XI of the west pediment of the Temple of Aphaea, ca. 505–500 BCE.

The various parts of the bow can be subdivided into further sections. The topmost limb is known as the upper limb, while the bottom limb is the lower limb. At the tip of each limb is a nock, which is used to attach the bowstring to the limbs. The riser is usually divided into the grip, which is held by the archer, as well as the arrow rest and the bow window. The arrow rest is a small ledge or extension above the grip which the arrow rests upon while being aimed. The bow window is that part of the riser above the grip, which contains the arrow rest.[1]

In bows drawn and held by hand, the maximum draw weight

is determined by the strength of the archer.[13] The maximum distance the string could be displaced and thus the longest arrow that could be loosed from it, a bow's draw length, is determined by the size of the archer.[14]

A composite bow uses a combination of materials to create the limbs, allowing the use of materials specialized for the different functions of a bow limb. The classic composite bow uses wood for lightness and dimensional stability in the core, horn to store energy in compression, and sinew for its ability to store energy in tension. Such bows, typically Asian, would often use a stiff end on the limb end, having the effect of a recurve.[15] In this type of bow, this is known by the Arabic name 'siyah'.[16]

Modern construction materials for bows include laminated wood, fiberglass, metals,[17] and carbon fiber components.

21.3.2   Arrows

Main article: Arrow

An arrow usually consists of a shaft with an arrowhead at-

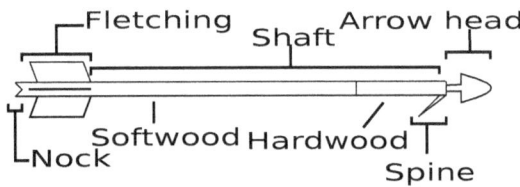

Schematic of an arrow showing its parts.

tached to the front end, with fletchings and a nock at the other.[18] Modern arrows are usually made from carbon fibre, aluminum, fiberglass, and wood shafts. Carbon shafts have the advantage that they do not bend or warp, but they can often be too light weight to shoot from some bows and are expensive. Aluminum shafts are less expensive than carbon shafts, but they can bend and warp from use. Wood shafts are the least expensive option but often will not be identical in weight and size to each other and break more often than the other types of shafts.[19] Arrow sizes vary greatly across cultures and range from very short ones that require the use of special equipment to be shot to ones in use in the Amazon River jungles that are 8.5 feet (2.6 metres) long. Most modern arrows are 22 inches (56 cm) to 30 inches (76 cm) in length.[18]

Arrows come in many types, among which are breasted, bob-tailed, barrelled, clout, and target.[18] A breasted arrow is thickest at the area right behind the fletchings, and tapers towards the nock and head.[20] A bob-tailed arrow is thickest right behind the head, and tapers to the nock.[21] A barrelled arrow is thickest in the centre of the arrow.[22] Target

arrows are those arrows used for target shooting rather than warfare or hunting, and usually have simple arrowheads.[23]

For safety reasons, a bow should never be fired without an arrow nocked; without an arrow, the energy that is normally transferred into the projectile is instead directed back into the bow itself, which will cause damage to the bow's limbs.

21.3.3   Arrowheads

Main article: Arrowhead

The end of the arrow that is designed to hit the target is called the arrowhead. Usually, these are separate items that are attached to the arrow shaft by either tangs or sockets. Materials used in the past for arrowheads include flint, bone, horn, or metal. Most modern arrowheads are made of steel, but wood and other traditional materials are still used occasionally. A number of different types of arrowheads are known, with the most common being bodkins, broadheads, and piles.[24] Bodkin heads are simple spikes made of metal of various shapes, designed to pierce armour.[21] A broadhead arrowhead is usually triangular or leaf-shaped and has a sharpened edge or edges. Broadheads are commonly used for hunting.[25] A pile arrowhead is a simple metal cone, either sharpened to a point or somewhat blunt, that is used mainly for target shooting. A pile head is the same diameter as the arrow shaft and is usually just fitted over the tip of the arrow.[26] Other heads are known, including the blunt head, which is flat at the end and is used for hunting small game or birds, and is designed to not pierce the target nor embed itself in trees or other objects and make recovery difficult.[21] Another type of arrowhead is a barbed head, usually used in warfare or hunting.[18]

21.3.4   Bowstrings

Main article: Bowstring

Bowstrings may have a nocking point marked on them, which serves to mark where the arrow is fitted to the bowstring before firing.[27] The area around the nocking point is usually bound with thread to protect the area around the nocking point from wear by the archer's hands. This section is called the serving.[28] At one end of the bowstring a loop is formed, which is permanent. The other end of the bowstring also has a loop, but this is not permanently formed into the bowstring but is constructed by tying a knot into the string to form a loop. Traditionally this knot is known as the archer's knot, but is a form of the timber hitch. The knot can be adjusted to lengthen or shorten the bowstring. The adjustable loop is known as the "tail".[29] The string is often twisted (this being called the "flemish twist").

Bowstrings have been constructed of many materials throughout history, including fibres such as flax, silk, and hemp.[30] Other materials used were animal guts, animal sinews, and rawhide. Modern fibres such as Dacron or Kevlar are now used in commercial bowstring construction, as well as steel wires in some compound bows.[31] Compound bows have a mechanical system of pulley cams over which the bowstring is wound.[28] Nylon is useful only in emergency situations, as it stretches too much.[32]

21.4 Types of bow

Bow and arrow in heraldry

There is no one accepted system of classification of bows.[33] Bows may be described by various characteristics including the materials used, the length of the draw that they permit, the shape of the bow in sideways view, and the shape of the limb in cross-section.[34]

Common types of bow include

- Recurve bow: a bow with the tips curving away from the archer. The curves straighten out as the bow is drawn and the return of the tip to its curved state after release of the arrow adds extra velocity to the arrow.[35]

- Reflex bow: a bow whose entire limbs curve away from the archer when unstrung. The curves are opposite to the direction in which the bow flexes while drawn.[35]

- Self bow: a bow made from one piece of wood.[28]

- Longbow: a self bow with limbs rounded in cross-section, about the same height as the archer so as to allow a full draw, usually over 5 feet (1.5 metres) long. The traditional European longbow was usually made of yew wood, but other woods are also used.[36]

- Flatbow: the limbs are approximately rectangular in cross-section. This was traditional in many Native American societies and was found to be the most efficient shape for bow limbs by American engineers in the 20th century.

- Composite bow: a bow made of more than one material.[34]

- Takedown bow: a bow that can be demounted for transportation, usually consisting of 3 parts: 2 limbs and a Riser.

- Compound bow: a bow with mechanical aids to help with drawing the bowstring. Usually, these aids are pulleys at the tips of the limbs.[15]

21.5 Crossbow

Main article: Crossbow

In a crossbow, the limbs of the bow, called a **prod**, are attached at right angles to a crosspiece or stock in order to allow for mechanical pulling and holding of the string. The mechanism that holds the drawn string has a release or trigger that allows the string to be released.[37] A crossbow shoots a "bolt" or "quarrel", rather than an arrow.[38]

21.6 Citations

- Collins, Desmond (1973). *Background to archaeology: Britain in its European setting* (Revised ed.). Cambridge University Press. ISBN 0-521-20155-1.

- Elmer, R. P. (1946). *Target Archery: With a History of the Sport in America*. New York: A. A. Knopf. OCLC 1482628.

- Heath, E. G. (1978). *Archery: The Modern Approach*. London: Faber and Faber. ISBN 0-571-04957-5.

- Paterson, W. F. (1984). *Encyclopaedia of Archery*. New York: St. Martin's Press. ISBN 0-312-24585-8.

- Sorrells, Brian J. (2004). *Beginner's Guide to Traditional Archery*. Mechanicsburg, PA: Stackpole Books. ISBN 978-0-8117-3133-1.

- Stone, George Cameron (1999) [1934]. *A Glossary of the Construction, Decoration, and Use of Arms and Armor in All Countries and in All Times* (Reprint ed.). Mineola: Dover Publications. ISBN 0-486-40726-8.

21.7   References

[1] Paterson *Encyclopaedia of Archery* pp. 27-28

[2] Paterson *Encyclopaedia of Archery* p. 17

[3] Paterson *Encyclopaedia of Archery* p. 31

[4] Paterson *Encyclopaedia of Archery* p. 56

[5] Paterson *Encyclopaedia of Archery* p. 20

[6] M. H. Monroe, *Aboriginal Weapons and Tools* "The favoured weapon of the Aborigines was the spear and spear thrower. The fact that they never adopted the bow and arrow has been debated for a long time. During post-glacial times the bow and arrow were being used in every inhabited part of the world except Australia. A number of reasons for this have been put forward [...] Captain Cook saw the bow and arrow being used on an island close to the mainland at Cape York, as it was in the Torres Strait islands and New Guinea. But the Aborigines preferred the spear. "

[7] Collins *Background to Archaeology*

[8] http://www.nature.com/nature/journal/v491/n7425/full/ nature11660.html Kyle S. Brown, Curtis W. Marean, et al. *An early and enduring advanced technology originating 71,000 years ago in South Africa* Nature 491, 590–593 (22 November 2012) doi:10.1038/nature11660

[9] http://www.dengedenge.com/2009/10/ traditional-war-for-land-in-africa/ Bow and arrow-warfare in todays Africa

[10] http://www.time.com/time/photogallery/0,29307, 1722198,00.html

[11] Johnes, Martin. "Archery—Romance-and-Elite-Culture-in-England-and-Wales—c-1780-1840 Martin Johnes. Archery, Romance and Elite Culture in England and Wales, c. 1780–1840". Swansea.academia.edu. Retrieved 2013-03-26.

[12] Paterson *Encyclopaedia of Archery* p. 111

[13] Sorrells *Beginner's Guide* pp. 20-21

[14] Sorrells *Beginner's Guide* pp. 19-20

[15] Paterson *Encyclopaedia of Archery* p. 38

[16] Elmer *Target Archery*

[17] Heath *Archery* pp. 15-18

[18] Paterson *Encyclopaedia of Archery* pp. 18-19

[19] Sorrells *Beginner's Guide* pp. 21-22

[20] Paterson *Encyclopaedia of Archery* p. 32

[21] Paterson *Encyclopaedia of Archery* pp. 25-26

[22] Paterson *Encyclopaedia of Archery* p. 24

[23] Paterson *Encyclopaedia of Archery* p. 103

[24] Paterson *Encyclopaedia of Archery* p. 19

[25] Paterson *Encyclopaedia of Archery* p. 33

[26] Paterson *Encyclopaedia of Archery* p. 85

[27] Paterson *Encyclopaedia of Archery* p. 80

[28] Paterson *Encyclopaedia of Archery* pp. 93-94

[29] Heath *Archery* pp. 27-28

[30] Grow your own bowstring

[31] Paterson *Encyclopaedia of Archery* pp. 28-29

[32] Paracord 550 used as bow string

[33] Paterson *Encyclopaedia of Archery* p. 37

[34] Heath *Archery* pp. 14-16

[35] Paterson *Encyclopaedia of Archery* pp. 90-91

[36] Paterson *Encyclopaedia of Archery* pp. 73-75

[37] Paterson *Encyclopaedia of Archery* p. 41

[38] Paterson *Encyclopaedia of Archery* p. 26

21.8   Further reading

- Gad Rausing, "The Bow", Lund University Acta Archaeologica Lundensia Serie in 8° No 6, 1967

- *The Traditional Bowyers Bible Volume 1*. 1992 The Lyons Press. ISBN 1-58574-085-3

- *The Traditional Bowyers Bible Volume 2*. 1992 The Lyons Press. ISBN 1-58574-086-1

- *The Traditional Bowyers Bible Volume 3*. 1994 The Lyons Press. ISBN 1-58574-087-X

- *The Traditional Bowyers Bible Volume 4*. 2008 The Lyons Press. ISBN 978-0-9645741-6-8

- U. Stodiek/H. Paulsen, "Mit dem Pfeil, dem Bogen..." Techniken der steinzeitlichen Jagd. (Oldenburg 1996).

- Gray, David, "Bows of the World". The Lyons Press, 2002. ISBN 1-58574-478-6.

- Comstock, Paul. "The Bent Stick"

21.9 External links

- The Asian Traditional Archery Research Network

- Simon Archery Collection From The Manchester Museum, The University of Manchester

- An Approach to the Study of Ancient Archery using Mathematical Modeling

Chapter 22

Canoe

For other uses, see Canoe (disambiguation).

A **canoe** is a lightweight narrow boat, typically pointed at

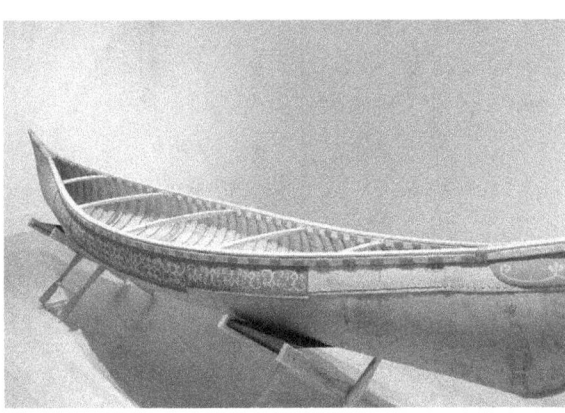

Birchbark canoe at Abbe Museum, a division of the Smithsonian Institution which specializes in Native America culture and is located in Bar Harbor, Maine

A B.N. Morris Canoe Company wood-and-canvas canoe built approximately 1912

both ends and open on top, propelled by one or more seated or kneeling paddlers facing the direction of travel using a single-bladed paddle.[1] In some European countries, like the United Kingdom, the term canoe is often used for both canoes and kayaks,[2] with canoes called *Canadian canoes* then. This is seen in the International Canoe Federation nomenclature.

Canoes are used for racing, whitewater canoeing, touring and camping, freestyle, and general recreation. The intended use of the canoe dictates its hull shape and length and construction material.

Historically, canoes were dugouts or made of bark on a wood frame,[3] but construction materials evolved to canvas on a wood frame, then to aluminum. Most modern canoes are made of molded plastic or composites such as fiberglass. Until the mid-1800s the canoe was an important means of transport for exploration and trade, but then transitioned to recreational or sporting use. Canoeing has been part of the Olympics since 1936. In places where the canoe played a key role in history, such as the northern United States, Canada, and New Zealand, the canoe remains an important theme in popular culture.

Canoes can be adapted to many purposes, for example with the addition of sails, outboard motors, and outriggers.

Voyageur canoe shooting the rapids

These antique dug out canoes are in the courtyard of the Old Military Hospital in the Historic Center of Quito.

22.1 History

The word canoe comes from the Carib kenu (dugout), via the Spanish canoa.[4]

Constructed between 8200 and 7600 BC, and found in the Netherlands, the Pesse canoe may be the oldest known canoe. Excavations in Denmark reveal the use of dugouts and paddles during the Ertebølle period, (ca 5300 BC – 3950 BC).[5]

Australian Aboriginal people made canoes using a variety of materials, including bark and hollowed out tree trunks.[6] The indigenous people of the Amazon commonly used Hymenaea trees.

Many indigenous peoples of the Americas built bark canoes. They were usually skinned with birch bark over a light wooden frame, but other types could be used if birch was scarce. At a typical length of 4.3 m (14 ft) and weight of 23 kg (50 lb), the canoes were light enough to be portaged, yet could carry a lot of cargo, even in shallow water. Although susceptible to damage from rocks, they are easily repaired.[7] Their performance qualities were soon recognized by early European immigrants, and canoes played a key role in the exploration of North America,[8] with Samuel de Champlain canoeing as far as the Georgian Bay in 1615. René de Bréhant de Galinée a French missionary who explored the Great Lakes in 1669 declared: "The convenience of these canoes is great in these waters, full of cataracts or waterfalls, and rapids through which it is impossible to take any boat. When you reach them you load canoe and baggage upon your shoulders and go overland until the navigation is good; and then you put your canoe back into the water, and embark again.[9] American painter, author and traveler George Catlin wrote that the bark canoe was "the most beautiful and light model of all the water crafts that ever were invented."[10]

Native American groups of the north Pacific coast made dugout canoes in a number of styles for differ-

ent purposes, from western red-cedar (*Thuja plicata*) or yellow-cedar (*Chamaecyparis nootkatensis*), depending on availability.[11] Different styles were required for ocean-going vessels versus river boats, and for whale-hunting versus seal-hunting versus salmon-fishing. The Quinault of Washington State built shovel-nose canoes, with double bows, for river travel that could slide over a logjam without portaging. The Kootenai of British Columbia province made sturgeon-nosed canoes from pine bark, designed to be stable in windy conditions on Kootenay Lake.[12]

The first explorer to cross the North American continent, Alexander Mackenzie, used canoes extensively, as did David Thompson and the Lewis and Clark Expedition.

In the North American fur trade the Hudson's Bay Company's voyageurs used three types of canoe:[13]

- The rabaska or *canot du maître* was designed for the long haul from the St. Lawrence River to western Lake Superior. Its dimensions were: length approximately 11 m (35 ft), beam 1.2 to 1.8 m (4 to 6 ft), and height about 76 cm (30 in). It could carry 60 packs weighing 41 kg (90 lb), and 910 kg (2,000 lb) of provisions. With a crew of eight or ten (paddling or rowing), they could make three knots over calm waters. Four to six men could portage it, bottom up. Henry Schoolcraft declared it "altogether one of the most eligible modes of conveyance that can be employed upon the lakes." Archibald McDonald of the Hudson's Bay Company wrote: "I never heard of such a canoe being wrecked, or upset, or swamped ... they swam like ducks."[14]

- The canot du nord (French: "canoe of the north"), a craft specially made and adapted for speedy travel, was the workhorse of the fur trade transportation system. About one-half the size of the Montreal canoe, it could carry about 35 packs weighing 41 kg (90 lb) and was manned by four to eight men. It could be carried by two men and was portaged in the upright position.[14]

- The express canoe or canot léger, was about 4.6 m (15 ft) long and were used to carry people, reports, and news.

The birch bark canoe was used in a 6,500 kilometres (4,000 mi) supply route from Montreal to the Pacific Ocean and the Mackenzie River, and continued to be used up to the end of the 19th century.[15]

Also popular for hauling freight on inland waterways in 19th Century North America were the York boat and the batteau.

In 19th-century North America, the birch-on-frame construction technique evolved into the wood-and-canvas canoes made by fastening an external waterproofed canvas shell to planks and ribs by boat builders Old Town Canoe, E.M. White Canoe, Peterborough Canoe Company and at the Chestnut Canoe Company[16] in New Brunswick.

Although canoes were once primarily a means of transport, with industrialization they became popular as recreational or sporting watercraft. John MacGregor popularized canoeing through his books, and in 1866 founded the Royal Canoe Club in London and in 1880 the American Canoe Association. The Canadian Canoe Association was founded in 1900, and the British Canoe Union in 1936.

Sprint canoe was a demonstration sport at the 1924 Paris Olympics and became an Olympic discipline at the 1936 Berlin Olympics.[17] The International Canoe Federation was formed in 1946 and is the umbrella organization of all national canoe organizations worldwide.

In recent years First Nations in British Columbia, Washington State have been revitalizing the ocean-going canoe tradition. Beginning in the 1980s, the Heiltsuk and Haida were early leaders in this movement. The paddle to Expo '86 in Vancouver by the Heiltsuk, and the 1989 Paddle to Seattle were early instances of this. In 1993 a large number of Canoes paddled from up and down the coast to Bella Bella in its first canoe festival - 'Qatuwas.[18] The revitalization continued - an Tribal Journeys began with trips to various communities held most years.

22.2 Hull design

Hull design must meet different, often conflicting, requirements for speed, carrying capacity, maneuverability, and stability[19] The canoe's hull speed can be calculated using the principles of ship resistance and propulsion.

- Length: this is often stated by manufacturers as the

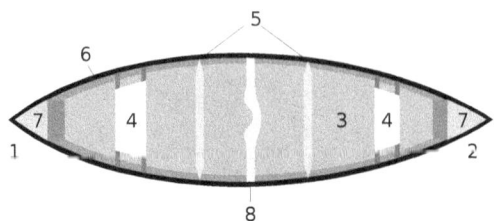

1 Bow, 2 Stern, 3 Hull, 4 Seat, 5 Thwart, 6 Gunwale, 7 Deck, 8 Yoke

Prospector canoe showing rocker at the stern

overall length of the boat, but what counts in performance terms is the length of the waterline, and more specifically its value relative to the displacement of the canoe. Displacement is the amount of water displaced by the boat. It is equal to the total weight of the boat and its contents, since a floating body displaces its own weight in water. When a canoe is paddled through water, it takes an effort to push all of the displaced water out of the way. Canoes are displacement hulls: the longer the waterline relative to its displacement, the faster it can be paddled.[20] Among general canoeists, 5.2 m (17 ft) is the most popular length, providing a good compromise between maneuverability and speed.[21]

- Width (beam): a wider boat provides more stability at the expense of speed. A canoe cuts through the water like a wedge, and a shorter boat needs a narrower beam to reduce the angle of the wedge cutting through the water.[21]

- Depth: a higher-sided boat stays drier in rough water. The cost of high sides is extra weight and extra wind resistance.[21]

- Stability and bottom shape: the hull can be optimized for initial stability (the boat feels steady when it sits flat on the water) or final stability (resistance to capsizing).

A flat-bottomed hull has high initial stability, while a rounded or V-shaped hull has high final stability.[22] The fastest flat water canoes have sharp V-bottoms to cut through the water. But they are difficult to turn and have a deeper draft which makes them less suitable for shallows. Flat-bottomed canoes are most popular among recreational canoeists. At the cost of speed, they have shallow draft, turn better, and more cargo space. The reason a flat bottom canoe has lower final stability is that it the hull must wrap a sharper angle between the bottom and the sides, compared to a more round-bottomed boat.[21]

- Keel: an external keel makes a canoe track (hold its course) better, and can stiffen a floppy bottom, but it can get stuck on rocks and decrease stability in rapids.[22]

- Profile, the shape of the canoe's sides. Sides which flare out above the waterline deflect water but require the paddler to reach out over the side of the canoe. If the gunwale width is less than the waterline width or the maximum width the canoe is said to have tumble-home. This makes paddling easier.[23]

- Rocker: viewed from the side of the canoe, rocker is the amount of curve in the hull, much like the curve of a banana. A straight keeled canoe, with no rocker, is meant for covering long distances in a straight line. The full length of the hull is in the water, so it tracks well and has good speed. As the rocker increases, so does the ease of turning, at the cost of tracking.[24] Native American birch bark canoes were often characterized by extreme rocker.[21]

- Hull symmetry: viewed from above, a symmetrical hull has its widest point at the center of the hull and both ends are identical. An asymmetrical hull typically has the widest section aft of center line, creating a longer bow and improving speed.[24]

22.3 Materials and construction

22.3.1 Modern

- Plastic: Royalex is a composite material, comprising an outer layer of vinyl and hard acrylonitrile butadiene styrene plastic (ABS) and an inner layer of ABS foam, bonded by heat treatment.[25] As a canoe material, Royalex lighter, more resistant to UV damage, is more rigid, and has greater structural memory than non-composite plastics such as polyethylene. Royalex canoes are, however, more expensive than aluminium canoes or canoes made from traditionally molded or

Aluminum canoe

roto-molded polyethylene hulls.[25] It is heavier, and less suited for high-performance paddling than fiber-reinforced composites, such as fiberglass, kevlar, or graphite. Roto-molded polyethylene is a cheaper alternative to Royalex.

- Fiber reinforced composites: Fiberglass is the most common material used in manufacturing canoes.[26] Fiberglass is not expensive, can be molded to any shape, and is easy to repair.[21] Kevlar is popular with paddlers looking for a light boat that will not be taken in whitewater. Fiberglass and Kevlar are strong but lack rigidity. Boats are built by draping the cloth on a mold, then impregnating it with a liquid resin. A gel coat on the outside gives a smoother appearance.[21]

- Polycarbonate: Lexan is used in transparent canoes.

- Aluminum: Before the invention of fiberglass, this was the standard choice for whitewater canoeing. It is good value and very strong by weight.[21] This material was once more popular but is being replaced by modern lighter materials. "It is tough, durable, and will take being dragged over the bottom very well", as it has no gel or polymer outer coating which would make it subject to abrasion. The hull does not degrade from long term exposure to sunlight, and "extremes of hot and cold do not [affect] the material". It can dent, is difficult to repair, is noisy, can get stuck on underwater objects, and requires buoyancy chambers to assist in keeping the canoe afloat in a capsize.[27]

- Folding canoes usually consist of a PVC skin around an aluminum frame.

- Inflatable: These contain no rigid frame members and can be deflated, folded and stored in a bag. The more durable types consist of an abrasion-resistant nylon or rubber outer shell, with separate PVC air chambers for the two side tubes and the floor.[28]

22.3.2   Traditional

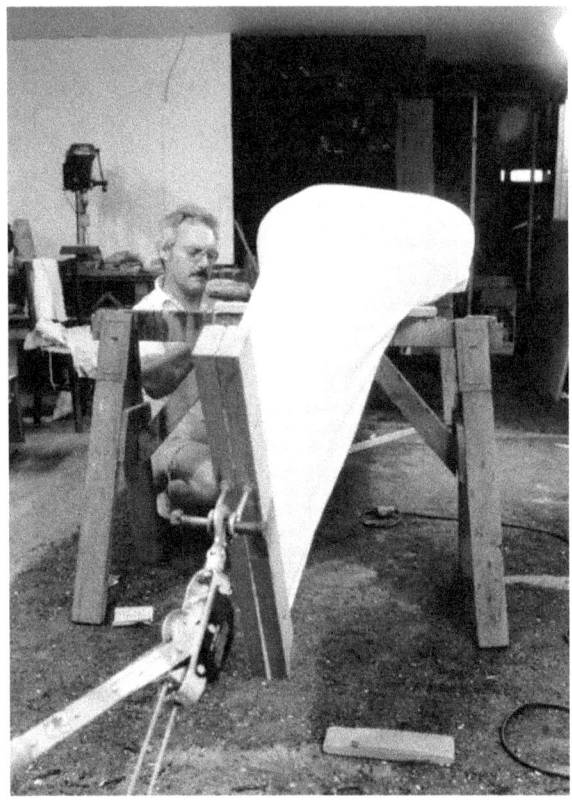

Stretching canvas on a canoe

These materials and techniques are used by artisans and produce boats that some consider more attractive, but which are more fragile than those made with modern methods.[29]

- Bark: the indigenous peoples of eastern Canada and the northeast United States made canoes using the bark of the paper birch, which was harvested in early spring by stripping off the bark in one piece, using wooden wedges. Next, the two ends (stem and stern) were sewn together and made watertight with the pitch of balsam fir. The ribs of the canoe, called verons in Canadian French, were made of white cedar, and the hull, ribs, and thwarts were fastened using watap, a binding usually made from the roots of various species of conifers, such as the white spruce, black spruce, or cedar, and caulked with pitch.[30][31]

- Dugout: Many indigenous groups from around the world made dugout canoes, by carving out a single piece of wood; either a whole trunk, or a slab of trunk from particularly large trees.[11][32]

- Reed: Some peoples, with less access to suitable trees, made canoes from bundled reeds. Papyrus was used in Egypt, Totora in South America, and Tule in California.

- Canvas on wood frame: while similar to bark canoes in the use of ribs, and a waterproof covering, the construction method is different, being built by bending ribs over a solid mold. Once removed from the mold, the decks, thwarts and seats are installed, and canvas is stretched tightly over the hull. The canvas is then treated with a combination of varnishes and paints to render it more durable and watertight.[33]

- Wood strips: these are built by securing narrow, flexible strips of wood, usually cedar, edge-to-edge over a building jig that defines the shape of the hull. Once the strips are glued together, a transparent fiberglass and epoxy coating is applied to the canoe inside and out.

- Clinker, lapstrake, or carvel: a wooden construction using longitudinal planks to form the hull. Traditionally planking is nailed together with copper tacks. Once the planking is completed, steam-bent ribs are inserted into the hull and fastened with nails or rivets.

- Stitch and glue: plywood panels are stitched together to form a hull shape, and the seams are reinforced with fiberglass tape and thickened epoxy.

22.4   Symbolism

La Chasse-galerie by Henri Julien

In Canada, the canoe has been a theme in history and folklore, and is a symbol of Canadian identity.[34] From 1935 to 1986 the Canadian silver dollar depicted a voyageur and an aboriginal paddling a canoe with the Northern Lights in the background.

The Chasse-galerie is a French-Canadian tale of voyageurs who, after a night of heavy drinking on New Year's Eve at a remote timber camp want to visit their sweethearts some 100 leagues (about 400 km) away. Since they have to be back in time for work the next morning they make a pact with the devil. Their canoe will fly through the air, on condition that they not mention God's name or touch the cross of any church steeple as they fly by in the canoe. One version of this fable ends with the coup de grâce when, still high in the sky, the voyageurs complete the hazardous journey but the canoe overturns, so the devil can honour the pact to deliver the voyageurs and still claim their souls.

In John Steinbeck's novella *The Pearl* set in Mexico, the main character's canoe is a means of making a living that has been passed down for generations and represents a link to cultural tradition.[35]

The Māori, indigenous Polynesian people arrived in New Zealand in several waves of canoe voyage. Canoe traditions are important to the identity of Māori. Whakapapa (genealogical links) back to the crew of founding canoes served to establish the origins of tribes, and defined tribal boundaries and relationships.[36]

22.5 Types

Modern canoe types are usually categorized by the intended use. Many modern canoe designs are hybrids (a combination of two or more designs, meant for multiple uses). The purpose of the canoe will also often determine the materials used. Most canoes are designed for either one person (solo) or two people (tandem), but some are designed for more than two people.

Women's C2

22.5.1 Sprint

Sprint canoe is also known as flatwater racing. The paddler kneels on one knee, and uses a single-blade paddle.[37] Canoes have no rudder, so the boat must be steered by the athlete's paddle using a j-stroke. Canoes may be entirely open or be partly covered. The minimum length of the opening on a C1 is 280 cm (110 in). Boats are long and streamlined with a narrow beam, which makes them very unstable. A C4 can be up to 9 m (30 ft) long and weigh 30 kg (66 lb).[38] ICF classes include C1 (solo), C2 (crew of two), and C4 (crew of four). Race distances at the 2012 Olympic Games were 200 and 1000 meters.

22.5.2 Slalom and wildwater

Whitewater slalom canoe

In ICF whitewater slalom paddlers negotiate their way down a 300 m (980 ft) of whitewater rapids, through a series of up to 25 gates (pairs of hanging poles.) The colour of the poles indicates the direction in which the paddlers must pass through; time penalties are assessed for striking poles or missing gates. Categories are C1 (solo) and C2 (tandem), the latter for two men, and C2M (mixed) for one woman and one man.[39] C1 boats must have a minimum weight and width of 10 kg (22 lb) and 0.65 m (2 ft 2 in) and be not more than 3.5 m (11 ft) long. C2s must have a minimum weight and width of 15 kg (33 lb) and 0.75 m (2 ft 6 in), and be not more that 4.1 m (13 ft). Rudders are prohibited. Canoes are decked and propelled by single-bladed paddles, and the competitor must kneel.[40]

In ICF wildwater canoeing athletes paddle a course of class III to IV whitewater (using the international scale of river difficulty), passing over waves, holes and rocks of a natural riverbed in events lasting either 20–30 minutes ("Classic" races) or 2–3 minutes ("Sprint" races). Categories are C1 and C2, for both women and men. C1s must have a minimum weight and width of 12 kg (26 lb) and 0.7 m (2 ft 4 in), and a maximum length of 4.3 m (14 ft). C2s must have a minimum weight and width of 18 kg (40 lb) and 0.8 metres (2 ft 7 in), and a maximum length of 5 metres (16 ft). Rudders are prohibited. The canoes are decked boats which must be propelled by single bladed paddles and inside which the paddler kneels.[41]

22.5.3   Marathon

Marathons are long-distance races which may include portages. Under ICF rules minimum canoe weight is 10 and 14 kg (22 and 31 lb) for C1 and C2 respectively. Other rules can vary by race, for example in the Classique Internationale de Canots de la Mauricie athletes race in C2s, with a maximum length of 5.6 m (18 ft 6 in), minimum width of 69 cm (27 in) at 8 cm (3 in) from the bottom of the centre of the craft, minimum height of 38 cm (15 in) at the bow and 25 cm (10 in) at the centre and stern.[42] The Texas Water Safari, at 422 km (262 mi), includes an open class, the only rule being the vessel must be human-powered, and although novel setups have been tried, the fastest so far has been the six-man canoe.[43]

22.5.4   General recreation

A square-stern canoe is an asymmetrical canoe with a squared-off stern for the mounting of an outboard motor, and is meant for lake travel or fishing. (In practice, use of a side bracket on a double-ended canoe often is more comfortable for the operator, with little or no loss of performance.) Since mounting a rudder on the square stern is very easy, such canoes often are adapted for sailing.

22.5.5   Touring and camping

In North America, a "touring canoe" is a good-tracking boat, good for wind-blown lakes and large rivers. A "tripping canoe" is a touring canoe with larger capacity for wilderness travel and is often designed with more rocker for better maneuverability on whitewater rivers but requiring some skill on the part of the canoeist in open windy waters when lightly loaded. Touring canoes are often made of lighter materials and built for comfort and cargo space. Tripping canoes such as the Chestnut Prospector derivates, and the Old Town Trippers, are typically made of heavier and tougher materials, and are usually a more traditional design. A Prospector canoe is a generic name for copies of the Chestnut model, a popular type of tripping canoe marked by a symmetrical hull and a relatively large amount of rocker, giving a nice balance for wilderness tripping. This model

Touring canoe

also has the ability to carry large amounts of gear while being maneuverable enough for rapids. This makes it a superb large capacity wilderness boat, but requires skill on windy, broad waters when lightly loaded. It is made in a variety of materials. For home construction, 4 mm ($^3/_{16}$ in) plywood is commonly used, mainly marine ply, using the "stitch and glue" technique. Commercially built canoes are commonly built of fibreglass, HDPE, Kevlar, Carbon Fiber, and Royalex which although relatively heavy, is very durable.

A touring canoe is sometimes covered with a greatly extended deck, forming a "cockpit" for the paddlers. A cockpit has the advantage that the gunwales can be made lower and narrower so the paddler can reach the water more easily, and the sides of the boat can be higher, keeping the boat dryer.

22.5.6   Freestyle

Playboating decked canoe

A canoe specialized for whitewater play and tricks. Most are identical to short, flat-bottomed kayak playboats except for internal outfitting. The paddler kneels and uses a single-blade canoe paddle. Playboating is a discipline of white-water canoeing where the paddler performs various technical moves in one place (a playspot), as opposed to down-river where the objective is to travel the length of a section of river (although whitewater canoeists will often stop and play en-route). Specialized canoes known as playboats can be used.

22.6 Image gallery

- Wood-and-canvas canoe by Joe Seliga

- Wood-and-canvas canoe in the Boundary Waters Canoe Area Wilderness

- *Spearing Salmon By Torchlight*, an oil painting by Paul Kane

- *Canoe Manned by Voyageurs Passing a Waterfall (Ontario)*, oil painting by Frances Anne Hopkins

- Ojibwe women in canoe on Leech Lake

- A dugout canoe of pirogue type in the Solomon Islands

- Canoe in Kerala, India

- Canoe in Vietnam in the Mekong delta

22.7 See also

- Kennebec Boat and Canoe Company

- B.N. Morris Canoe Company

- E.H. Gerrish Canoe Company

- E.M. White Canoe Company

- Thompson Brothers Boat Manufacturing Company

- Old Town Canoe Company

- Peterborough Canoe Company

- Chestnut Canoe Company

- Carleton Canoe Company

22.8 References

[1] "Canoe". Merriam-Webster Dictionary. Retrieved 20 October 2012.

[2] "Buying a canoe or kayak". gocanoeing.org. Retrieved 8 September 2014.

[3] "Dugout Canoe". The Canadian Encyclopedia. Retrieved 30 January 2013.

[4] "The history of the canoe". canoe.ca. Retrieved 27 September 2012.

[5] "Dugouts and paddles". Retrieved 8 October 2012.

[6] "Carved wooden canoe, National Museum of Australia". Nma.gov.au. Retrieved 2013-04-25.

[7] "Bark canoes". Canadian Museum of Civilization. Retrieved 8 October 2012.

[8] "Our Canoeing Heritage". The Canadian Canoe Museum. Retrieved 8 October 2012.

[9] Kellogg, Louise Phelps (1917). *Early Narratives of the Northwest. 1634–1699*. New York. pp. 172–173.

[10] Catlin, George (1989). *Letters and Notes on the Manners. Customs, and Conditions of the North American Indians* (reprint ed.). New York. p. 415.

[11] Pojar and MacKinnon (1994). *Plants of the Pacific Northwest Coast*. Vancouver, British Columbia: Lone Pine Publishing. ISBN 1-55105-040-4.

[12] Nisbet, Jack (1994). *Sources of the River*. Seattle, Washington: Sasquatch Books. ISBN 1-57061-522-5.

[13] "The Canoe". The Hudson's Bay Company. Retrieved 6 October 2012.

[14] "Portage Trails in Minnesota, 1630s-1870s" (PDF). United States Department of the Interior National Park Service. Retrieved 20 November 2012.

[15] "Canoeing". The Canadian Encyclopedia. Retrieved 8 October 2012.

[16] "A Venerable Chestnut". Canada Science and Technology Museum. Retrieved 8 October 2012.

[17] "Canoe / kayak sprint equipment and history". olympic.org. Retrieved 29 September 2012.

[18] Neel, David The Great Canoes: Reviving a Northwest Coast Tradition. Douglas & McIntyre. 1995. ISBN 1-55054-185-4

[19] *Canoeing : outdoor adventures*. Champaign, IL: Human Kinetics. 2008. ISBN 0-7360-6715-9.

[20] Winters, John. "Speaking Good Boat: Part 1". Retrieved 18 October 2012.

[21] Davidson, James & John Rugge (1985). *The Complete Wilderness Paddler*. Vintage. pp. 38–39. ISBN 0-394-71153-X.

[22] "How to Choose a Canoe: A Primer on Modern Canoe Design". GORP. Retrieved 7 October 2012.

[23] "Canoe Design". Retrieved 8 October 2012.

[24] "The Hull Truth". Mad River Canoe. Retrieved 7 October 2012.

[25] "Royalex (RX)". Retrieved 20 November 2010.

[26] "Canoe Materials". Frontenac Outfittesr. Retrieved 7 October 2012.

[27] "Buying The Right Canoe". Retrieved 6 October 2012.

[28] James Weir, *Discover Canoeing: A Complete Introduction to Open Canoeing*, p.17, Pesda Press, 2010, ISBN 1906095124

[29] "Buying The Right Canoe - Materials". Retrieved 6 October 2012.

[30] Margry, Pierre (1876–1886). *Decouvertes et etablissements des francais dans l'ouest et dans le sud de l'Amerique Septentrionale (1614–1754). 6 vols*. Paris.

[31] Tom Vennum, Charles Weber, Earl Nyholm (Director) (1999). *Earl's Canoe: A Traditional Ojibwe Craft*. Smithsonian Center for Folklife Programs and Cultural Studies. Retrieved 2012-12-03.

[32] Olympic Peninsula Intertribal Cultural Advisory Committee (2002). *Native Peoples of the Olympic Peninsula*. Norman, Oklahoma: University of Oklahoma Press. ISBN 0-8061-3552-2.

[33] "The Wood and Canvas Canoe". Wooden Canoe Heritage Association. Retrieved 26 October 2012.

[34] "The Canoe". McGill University. Retrieved 16 October 2012.

[35] "The Pearl: Themes, Motifs, & Symbols". Spark Notes. Retrieved 16 October 2012.

[36] "Story: Canoe traditions". The Encyclopedia of New Zealand. Retrieved 16 October 2012.

[37] "Canoe sprint". International Canoe Federation. Retrieved 22 November 2012.

[38] "Canoe Sprint Overview". International Canoe Federation. Retrieved 22 November 2012.

[39] "About Canoe Slalom". International Canoe Federation. Retrieved 22 November 2012.

[40] "Rules for Canoe Slalom" (PDF). International Canoe Federation. Retrieved 22 November 2012.

[41] "Wildwater Competition rules 2011" (PDF). International Canoe Federation. Retrieved 22 November 2012.

[42] "La Classique Internationale de Canots de la Mauricie: Rules and Regulations". Retrieved 30 November 2012.

[43] "Texas Water Safari: History". Retrieved 30 November 2012.

22.9 External links

- Media related to Canoes at Wikimedia Commons

Chapter 23

Natufian culture

Ain Mallaha

Ein Gev

Shuqba cave

Tell Abu Hureyra

Wadi an-Natuf

Mureybet

Nahal Oren

Beidha

Map of Israel showing important sites that were occupied by the Natufian culture (clickable map)

The **Natufian culture** /nəˈtjuːfiən/ was an Epipaleolithic culture that existed from 12,500 to 9,500 BC in the Levant, a region in the Eastern Mediterranean. It was unusual in that it was sedentary, or semi-sedentary, before the introduction of agriculture. The Natufian communities are possibly the ancestors of the builders of the first Neolithic settlements of the region, which may have been the earliest in the world. There is some evidence for the deliberate cultivation of cereals, specifically rye, by the Natufian culture, at Tell Abu Hureyra, the site of earliest evidence of agriculture in the world.[1] Generally, though, Natufians made use of wild cereals. Animals hunted included gazelles.[2] According to Christy G. Turner II, there is an archaeological and physical anthropological reason for a relationship between the modern Semitic-speaking populations of the Levant and the Natufians.[3]

The term "Natufian" was coined by Dorothy Garrod who studied the Shuqba cave in Wadi an-Natuf, in the western Judean Mountains, about halfway between Tel Aviv and Ramallah.[4]

23.1 Dating

Radiocarbon dating places this culture from the terminal Pleistocene to the very beginning of the Holocene, from 12,500 to 9,500 BC.[5]

The period is commonly split into two subperiods: Early Natufian (12,500–10,800 BC) and Late Natufian (10,800–9,500 BC). The Late Natufian most likely occurred in tandem with the Younger Dryas (10,800 to 9,500 BC). In the Levant, there are more than a hundred kinds of cereals, fruits, nuts, and other edible parts of plants, and the flora of the Levant during the Natufian period was not the dry, barren, and thorny landscape of today, but rather woodland.[6]

23.2   Precursors and associated cultures

The Natufian developed in the same region as the earlier Kebaran complex, and is generally seen as a successor which developed from at least elements within that earlier culture. There were also other cultures in the region, such as the Mushabian culture of the Negev and Sinai, which are sometimes distinguished from the Kebaran, and sometimes also seen as having played a role in the development of the Natufian.

More generally there has been discussion of the similarities of these cultures with those found in coastal North Africa. Graeme Barker notes there are: "similarities in the respective archaeological records of the Natufian culture of the Levant and of contemporary foragers in coastal North Africa across the late Pleistocene and early Holocene boundary".[7]

Ofer Bar-Yosef has argued that there are signs of influences coming from North Africa to the Levant, citing the microburin technique and "microlithic forms such as arched backed bladelets and La Mouillah points."[8] But recent research has shown that the presence of arched backed bladelets, La Mouillah points, and the use of the microburin technique was already apparent in the Nebekian industry of the Eastern Levant.[9] And Maher et al. state that, "Many technological nuances that have often been always highlighted as significant during the Natufian were already present during the Early and Middle EP [Epipalaeolithic] and do not, in most cases, represent a radical departure in knowledge, tradition, or behavior."[10]

Authors such as Christopher Ehret have built upon the little evidence available to develop scenarios of intensive usage of plants having built up first in North Africa, as a precursor to the development of true farming in the Fertile Crescent, but such suggestions are considered highly speculative until more North African archaeological evidence can be gathered.[11][12] In fact, Weiss et al. have shown that the earliest known intensive usage of plants was in the Levant 23,000 years ago at the Ohalo II site.[13][14] Anthropologist C. Loring Brace in a recent study on cranial metric traits however, was also able to identify a "clear link" to Sub-Saharan African populations for early Natufians based on his observation of gross anatomical similarity with extant populations found mostly in the Sahara.[15] Brace believes that these populations later became assimilated into the broader continuum of Southwest Asian populations.

According to Bar-Yosef and Belfer-Cohen, "It seems that certain preadaptive traits, developed already by the Kebaran and Geometric Kebaran populations within the Mediterranean park forest, played an important role in the emergence of the new socioeconomic system known as the Natufian culture."[16]

23.3   Settlements

Remains of a wall of a Natufian house

Settlements occur in the woodland belt where oak and *Pistacia* species dominated. The underbrush of this open woodland was grass with high frequencies of grain. The high mountains of Lebanon and the Anti-Lebanon, the steppe areas of the Negev desert in Israel and Sinai, and the Syro-Arabian desert in the east were much less favoured for Natufian settlement, presumably due to both their lower carrying capacity and the company of other groups of foragers who exploited this region.[17]

The habitations of the Natufian are semi-subterranean, often with a dry-stone foundation. The superstructure was probably made of brushwood. No traces of mudbrick have been found, which became common in the following Pre-Pottery Neolithic A (PPNA). The round houses have a diameter between three and six meters, and they contain a central round or subrectangular fireplace. In Ain Mallaha traces of postholes have been identified. "Villages" can cover over 1,000 square meters. Smaller settlements have been interpreted by some researchers as camps. Traces of rebuilding in almost all excavated settlements seem to point to a frequent relocation, indicating a temporary abandonment of the settlement. Settlements have been estimated to house 100–150 people, but there are three categories: small, median, and large, ranging from 15 sq. m to 1,000 sq. m. There are no definite indications of storage facilities.

23.4 Lithics

The Natufian had a microlithic industry, based on short blades and bladelets. The microburin technique was used. Geometric microliths include lunates, trapezes and triangles. There are backed blades as well. A special type of retouch (Helwan retouch) is characteristic for the early Natufian. In the late Natufian, the Harif-point, a typical arrowhead made from a regular blade, became common in the Negev. Some scholars use it to define a separate culture, the Harifian.

Sickle blades appear for the first time. The characteristic sickle-gloss shows that they have been used to cut the silica-rich stems of cereals and form an indirect proof for incipient agriculture. Shaft straighteners made of ground stone indicate the practice of archery. There are heavy ground-stone bowl mortars as well.

23.5 Other finds

There was a rich bone industry, including harpoons and fish hooks. Stone and bone were worked into pendants and other ornaments. There are a few human figurines made of limestone (El-Wad, Ain Mallaha, Ain Sakhri), but the favourite subject of representative art seems to have been animals. Ostrich-shell containers have been found in the Negev.

23.6 Subsistence

The Natufian people lived by hunting and gathering. The preservation of plant remains is poor because of the soil conditions, but wild cereals, legumes, almonds, acorns and pistachios may have been collected. Animal bones show that gazelle (*Gazella gazella* and *Gazella subgutturosa*) were the main prey. Additionally deer, aurochs and wild boar were hunted in the steppe zone, as well as onagers and caprids (ibex). Water fowl and freshwater fish formed part of the diet in the Jordan River valley. Animal bones from Salibiya I (12,300 – 10,800 BP) have been interpreted as evidence for communal hunts with nets.

23.7 Development of agriculture

According to one theory,[18] it was a sudden change in climate, the Younger Dryas event (ca. 10,800 to 9500 BCE), that inspired the development of agriculture. The Younger Dryas was a 1,000-year-long interruption in the higher temperatures prevailing since the Last Glacial Maximum, which produced a sudden drought in the Levant. This would have endangered the wild cereals, which could no longer compete with dryland scrub, but upon which the population had become dependent to sustain a relatively large sedentary population. By artificially clearing scrub and planting seeds obtained from elsewhere, they began to practice agriculture. However, this theory of the origin of agriculture is controversial in the scientific community.[19]

23.8 Domesticated dog

See also: Origin of the domestic dog § Archaeological evidence

It is at Natufian sites that some of the earliest archaeological evidence for the domestication of the dog is found. At the Natufian site of Ain Mallaha in Israel, dated to 12,000 BCE, the remains of an elderly human and a four-to-five-month-old puppy were found buried together.[20] At another Natufian site at the cave of Hayonim, humans were found buried with two canids.[20]

23.9 Art

Main article: Ain Sakhri lovers

The *Ain Sakhri lovers*, a carved stone object held at the British Museum, is the oldest known depiction of a couple having sex. It was found in the Ain Sakhri cave in the Judean desert.[21]

23.10 Burials

Burials made of shell, teeth (of red deer), bones, and stone. There are pendants, bracelets, necklaces, earrings, and belt-ornaments as well.

In 2008, the grave of a Natufian 'priestess' was discovered (in most media reports referred to as a shaman[18] or witch doctor).[22] The burial contained complete shells of 50 tortoises, which are thought to have been brought to the site and eaten during the funeral feast.[23]

23.11 Long distance exchange

At Ain Mallaha (in Northern Israel), Anatolian obsidian and shellfish from the Nile valley have been found. The source of malachite beads is still unknown.

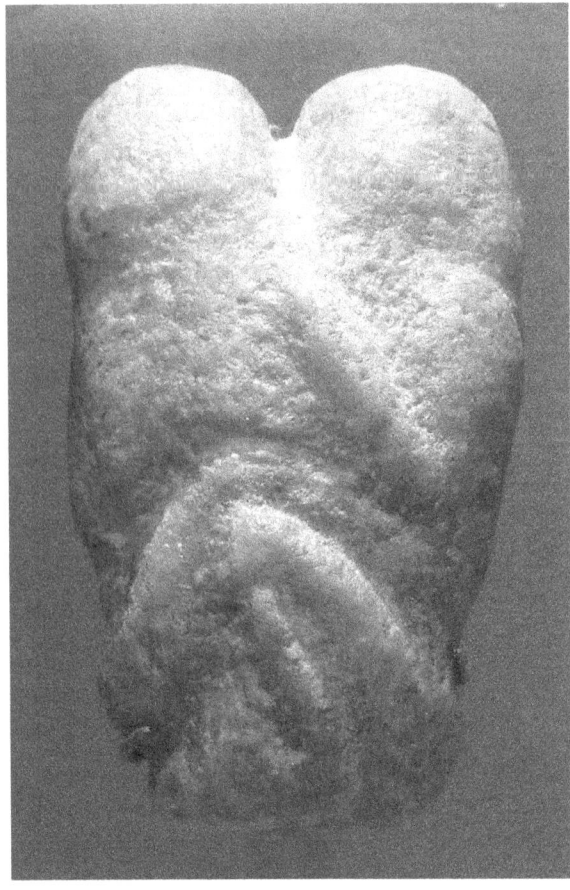

The Ain Sakhri lovers. British Museum: 1958,1007.1

23.12   Language

While the period involved makes it difficult to speculate on any language associated with the Natufian culture, linguists who believe it is possible to speculate this far back in time have written on this subject. As with other Natufian subjects, opinions tend to either emphasize North African connections or Eurasian connections. The view that the Natufians spoke the Afro-Asiatic language is accepted by Vitaly Shevoroshkin.[24] Alexander Militarev and others have argued that the Natufian may represent the culture which spoke Proto-Afroasiatic,[25] which he in turn believes has a Eurasian origin associated with the concept of Nostratic languages. The possibility of Natufians speaking the proto-Afro-Asiatic, and that the language was introduced into Africa from the Levant, is approved by Colin Renfrew with caution, as a possible hypothesis for proto-Afro-Asiatic dispersal.[26]

Some scholars, for example Christopher Ehret, Roger Blench and others, contend that the Afroasiatic Urheimat is to be found in North or North East Africa, probably in the area of Egypt, the Sahara, Horn of Africa or Sudan.[27][28][29][30][31] Within this group, Ehret, who like Militarev believes Afroasiatic may already have been in existence in the Natufian period, would associate Natufians only with the Near Eastern pre-Proto-Semitic branch of Afroasiatic.

23.13   Craniofacial research

The Epipalaeolithic Natufian of Israel from whom the Neolithic realm was assumed to arise is described as having a clear link to Sub-Saharan Africa. The Sub-Saharan element in the remains is also said to be of almost equal importance to that of the Eurasian element. The authors, however, remain cautious because of the small sample size. The authors further speculate that the admixture process between Neolithic people and in situ foragers diluted any discoverable trace of Sub-Saharan ancestry that may have been present.[32]

23.14   Sites

Natufian sites include:

- Syria: Tell Abu Hureyra, Mureybet, Yabrud III

- Israel/Palestine: Ain Mallaha (Eynan), El-Wad, Ein Gev, Hayonim cave, Nahal Oren, Salibiya I, Kfar Ha-Horesh, Jericho, Shuqba cave

- Jordan: Beidha

- Lebanon: Jeita III, Borj Barajne, Jabal es Saaïdé, Aamiq II

- Egypt: Mushabian culture

- Turkey: Göbekli Tepe, Nevalı Çori

23.15   See also

- Synoptic table of the principal old world prehistoric cultures

23.16   References

[1] Moore, Andrew M. T.; Hillman, Gordon C.; Legge, Anthony J. (2000), *Village on the Euphrates: From Foraging to Farming at Abu Hureyra*, Oxford: Oxford University Press, ISBN 0-19-510806-X

[2] Kottak, Conrad P. (2005), *Window on Humanity: A Concise Introduction to Anthropology*, Boston: McGraw-Hill, pp. 155–156, ISBN 0-07-289028-2

[3] John D. Bengtson. *In Hot Pursuit of Language in Prehistory: Essays in the four fields of anthropology. In honor of Harold Crane Fleming*. p. 22.

[4] New fieldwork at Shuqba Cave and in Wadi en-Natuf, Western Judea

[5] Munro, Natalie D. (2003), "Small game, the Younger Dryas, and the transition to agriculture in the southern Levant" (PDF), *Mitteilungen der Gesellschaft für Urgeschichte* **12**: 47–71

[6] Bar-Yosef, Ofer (1998), "The Natufian Culture in the Levant, Threshold to the Origins of Agriculture" (PDF), *Evolutionary Anthropology* **6** (5): 159–177, doi:10.1002/(SICI)1520-6505(1998)6:5<159::AID-EVAN4>3.0.CO;2-7

[7] Barker G (2002) Transitions to farming and pastoralism in North Africa, in Bellwood P, Renfrew C (2002), *Examining the Farming/Language Dispersal Hypothesis*, pp 151–161.

[8] Bar-Yosef O (1987) Pleistocene connections between Africa and SouthWest Asia: an archaeological perspective. *The African Archaeological Review*; Chapter 5, pg 29-38

[9] Richter, Tobias. Interaction before Agriculture: Exchanging Material and Sharing Knowledge in the Final Pleistocene Levant (2011)doi:10.1017/S0959774311000060

[10] Maher, Lisa A. Richter, Tobias. Stock, Jay T. The Pre-Natufian Epipaleolithic: Long-Term Behavioral Trends in the Levant. Evolutionary Anthropology 21:69–81 (2012).

[11] Ehret (2002) The Civilizations of Africa: A History to 1800. Charlottesville: University Press of Virginia

[12] Bellwood P (2005) Blackwell, Oxford. Page 97

[13] Weiss E, Kislev ME, Simchoni O, Nadel D, and Tschauner H. 2008. Plant-food preparation area on an Upper Paleolithic brush hut floor at Ohalo II, Israel. Journal of Archaeological Science 35(8):2400-2414.

[14] Nadel D, Piperno DR, Holst I, Snir A, and Weiss E. 2012. New evidence for the processing of wild cereal grains at Ohalo II, a 23 000-year-old campsite on the shore of the Sea of Galilee, Israel. Antiquity 86(334):990-1003.

[15] Brace, C.L., Seguchi, N., Quintyn, C.B., Fox, S.C., Nelson, A.R., Manolis, S.K., and Qifeng, P. (2006). The questionable contribution of the Neolithic and the Bronze Age to European craniofacial form. Proc. Natl. Acad. Sci. USA 103, 242–24

[16] Ofer Bar-Yosef and Anna Belfer-Cohen. The Origins of Sedentism and Farming Communities in the Levant. Journal of World Prehistory, Vol. 3, No. 4 (December 1989), pp. 447-498

[17] Ofer Bar-Yosef, The Natufian culture and the Early Neolithic: Social and economic trends in Southwestern Asia, chapter 10 in Peter Bellwood and Colin Renfrew (eds.), *Examining the Farming/Language Dispersal Hypothesis* (2002), p.114.

[18] "Oldest Shaman Grave Found". National Geographic 04-Nov-2008

[19] Balter, Michael (2010), "Archaeology: The Tangled Roots of Agriculture", *Science* **327** (5964): 404–406, doi:10.1126/science.327.5964.404, PMID 20093449, retrieved 4 February 2010

[20] Clutton-Brock, Juliet (1995), "Origins of the dog: domestication and early history", in Serpell, James, *The domestic dog: its evolution, behaviour and interactions with people*, Cambridge: Cambridge University Press, ISBN 0-521-41529-2

[21] BBC. A History of the World. Ain Sakhri Lovers

[22] "Archaeologists discover 12,000 year-old grave of witch doctor". Daily Mail 04-Nov-2008

[23] "Hebrew U. unearths 12,000-year-old skeleton of 'petite' Natufian priestess". By Bradley Burston. Haaretz, 05-Nov-2008

[24] Winfried Nöth (1994). *Origins of Semiosis: Sign Evolution in Nature and Culture*. p. 293.

[25] Roger Blench,Matthew Spriggs (2003). *Archaeology and Language IV: Language Change and Cultural Transformation*. p. 70.

[26] John A. Hall, I. C. Jarvie (2005). *Transition to Modernity: Essays on Power, Wealth and Belief*. p. 27.

[27] Blench R (2006) Archaeology, Language, and the African Past, Rowman Altamira, ISBN 0-7591-0466-2, ISBN 978-0-7591-0466-2, http://books.google.com/books?doi= esFy3Po57A8C

[28] Ehret C, Keita SOY, Newman P (2004) The Origins of Afroasiatic a response to Diamond and Bellwood (2003) in the Letters of SCIENCE 306, no. 5702, p. 1680 10.1126/science.306.5702.1680c

[29] Bernal M (1987) Black Athena: the Afroasiatic roots of classical civilization, Rutgers University Press, ISBN 0-8135-3655-3, ISBN 978-0-8135-3655-2. http://books.google.com/books?id=yFLm_M_OdK4C

[30] Bender ML (1997), Upside Down Afrasian, Afrikanistische Arbeitspapiere 50, pp. 19-34

[31] Militarev A (2005) Once more about glottochronology and comparative method: the Omotic-Afrasian case, Аспекты компаративистики - 1 (Aspects of comparative linguistics - 1). FS S. Starostin. Orientalia et Classica II (Moscow), p. 339-408. http://starling.rinet.ru/Texts/fleming.pdf

[32] Loring, C. Brace. "The questionable contribution of the Neolithic and the Bronze Age to European craniofacial form".

23.17 Further reading

- Balter, Michael (2005), *The Goddess and the Bull*, New York: Free Press, ISBN 0-7432-4360-9

- Bar-Yosef, Ofer (1998), "The Natufian Culture in the Levant, Threshold to the Origins of Agriculture" (PDF), *Evolutionary Anthropology* **6** (5): 159–177, doi:10.1002/(SICI)1520-6505(1998)6:5<159::AID-EVAN4>3.0.CO;2-7

- Bar-Yosef, Ofer; Belfer-Cohen, Anna (1999), "Encoding information: unique Natufian objects from Hayonim Cave, Western Galilee, Israel", *Antiquity* **73**: 402–409

- Bar-Yosef, Ofer (1992), Valla, Francois R., ed., *The Natufian Culture in the Levant*, Ann Arbor: International Monographs in Prehistory, ISBN 1-879621-03-7

- Campana, Douglas V.; Crabtree, Pam J. (1990), "Communal Hunting in the Natufian of the Southern Levant: The Social and Economic Implications", *Journal of Mediterranean Archaeology* **3** (2): 223–243

- Clutton-Brock, Juliet (1999), *A Natural History of Domesticated Mammals* (2nd ed.), Cambridge: Cambridge University Press, ISBN 0-521-63247-1

- Dubreuil, Laure (2004), "Long-term trends in Natufian subsistence: a use-wear analysis of ground stone tools", *Journal of Archaeological Science* **31** (11): 1613–1629, doi:10.1016/j.jas.2004.04.003

- Munro, Natalie D. (August–October 2004), "Zooarchaeological measures of hunting pressure and occupation intensity in the Natufian: Implications for agricultural origins" (PDF), *Current Anthropology* **45**: S5, doi:10.1086/422084 S6-S33.

- Simmons, Alan H. (2007), *The Neolithic Revolution in the Near East: Transforming the Human Landscape*, University of Arizona Press, ISBN 978-0816529667

23.18 External links

- Epi-Palaeolithic (European Mesolithic) Natufian Culture of Israel (The History of the Ancient Near East)

- *Cultural Complexity (Hierarchical Societies [Socio-Economic-Political Inequalities]) in Mesopotamia: An Outline*

Chapter 24

Khiamian

El Khiam

Map of Israel showing the typesite El Khiam that was occupied in the Khiamian (clickable map)

The **Khiamian** (also referred to as El Khiam or El-Khiam)

A shepherd with sheep on a mountainside. Sheep were among the first animals to be domesticated by humankind; the domestication date is estimated to fall between nine and eleven thousand years ago in Mesopotamia.[1][2][3][4]

is a period of the Near-Eastern Neolithic, marking the transition between the Natufian and the Pre-Pottery Neolithic A. Some sources date it from about 10,000 to 9,500 BCE.[5] It currently dates between 10,200 and 8800 BC according to the ASPRO chronology.

The Khiamian owes its name to the site of El Khiam, situated on banks of the Dead Sea, where researchers have recovered the oldest chert arrows heads, with lateral notchs, the so-called "El Khiam points".[5] They have served to identify sites of this period, which are found in Israel, as well as in Jordan (Azraq), Sinai (Abu Madi), and to the north as far as the Middle Euphrates (Mureybet).

Aside from the appearance of El Khiam arrow heads, the Khiamian is placed in the continuity of the Natufian, without any major technical innovations. However, for the first time houses were built on the ground level itself, and not half below ground as was previously done. Otherwise, the bearers of the El Khiam culture were still hunter-gatherers, and agriculture at that time was then still rather primitive, based on what has been reported on sites of this period.[6] Newer discoveries show that in the Middle East and Anatolia some experiments with agriculture were being made by 10,900 BC.[7] and that there may already have been experimenting with wild grain processing by around 19,000 BC at Ohalo II.[8]

The Khiamien also sees a change occur in the symbolic aspects of culture, as evidenced by the appearance of small female statuettes, as well as by the burying of aurochs skulls. According to Jacques Cauvin, it is the beginning of the worship of the Woman and the Bull, as evidenced in the following periods of the Near-Eastern Neolithic.[9]

24.1 References

[1] Ensminger

[2] Weaver

[3] Simmons & Ekarius

187

[4] Krebs, Robert E. & Carolyn A. (2003). *Groundbreaking Scientific Experiments, Inventions & Discoveries of the Ancient World.* Westport, CT: Greenwood Press. ISBN 0-313-31342-3.

[5] C. Calvet. 2007. *Zivilisationen – wie die Kultur nach Sumer kam.* Munich. p. 126.

[6] K. Schmidt. 2008. *Sie bauten die ersten Tempel. Das rätselhafte Heiligtum der Steinzeitjäger.* Munich. pp.283.

[7] Turneya, C.S.M. and H. Brown. 2007. "Catastrophic early Holocene sea level rise, human migration and the Neolithic transition in Europe." Quaternary Science Reviews 26: 2036–2041.

[8] Research pushes back history of crop development 10,000 years

[9] J. Cauvin. 2000. *The birth of the gods and the origins of agriculture.* Cambridge. p. 25. Cf. C. Calvet. 2007. *Zivilisationen – wie die Kultur nach Sumer kam.* Munich. p. 127.

24.2   References

- C. Calvet. 2007. *Zivilisationen – wie die Kultur nach Sumer kam.* Munich.

- J. Cauvin. 2000. *The birth of the gods and the origins of agriculture.* Cambridge.

- Klaus Schmidt. 2008. *Sie bauten die ersten Tempel. Das rätselhafte Heiligtum der Steinzeitjäger.* Munich. pp. 283.

- C.S.M. Turneya and H. Brown. 2007. "Catastrophic early Holocene sea level rise, human migration and the Neolithic transition in Europe." Quaternary Science Reviews 26: 2036–2041.

Chapter 25

Tahunian

The **Tahunian** is variously referred to as an archaeological culture, flint industry and period of the Palestinian Stone Age around Wadi Tahuna near Bethlehem. It was discovered and termed by Denis Buzy during excavations in 1928.[1][2]

Due to the early date and problems with the stratigraphy of the excavations at Wadi Tahuna, a great deal of debate has been put forward regarding the definition and position of the Tahunian within the sequences of Mesolithic, Epipaleolithic, Natufian, Khiamian, Heavy Neolithic, Pre-Pottery Neolithic A, Pre-Pottery Neolithic B and Neolithic and its relation to other Neolithic cultures such as the Qaraoun culture. In the search for naming conventions for the culture that started the Neolithic Revolution, this has reduced Avi Gopher to calling it a *"Tahunian Pandora's box"*, resulting in offshoots in terminology such as **Proto-Tahunian**.[3] It is no longer widely used but would appear to be an early PPNB culture of the Levantine corridor of around 8800 BC according to the ASPRO chronology.

25.1 References

[1] Buzy, Denis., Une Industrie Mesolithique en Palestine, In : Revue biblique, ISSN 0035-0907, vol.37 1 4, pp. 558–578, Planches XXVII-XXXI, 1928.

[2] Moore, A.M.T. (1978). *The Neolithic of the Levant, Neolithic Palestinian Tahunian.* Oxford University, Unpublished Ph.D. Thesis. pp. Selected Excerpt on the Tahunian Period.

[3] Avi Gopher (November 1994). *Arrowheads of the neolithic Levant: a seriation analysis.* Eisenbrauns. pp. 10–. ISBN 978-0-931464-76-8. Retrieved 11 January 2012.

Chapter 26

Heavy Neolithic

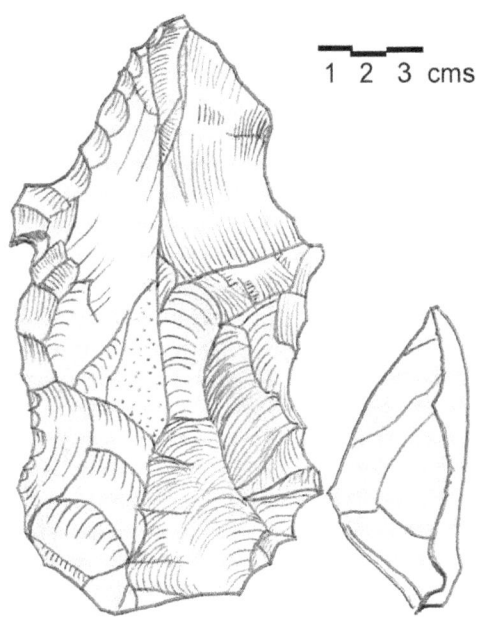

Heavy Neolithic tools of the Qaraoun culture found at Mtaileb I - Massive nosed scraper on a flake with irregular jagged edges, notches and "noses". Light grey and streaky silicious limestone.

Heavy Neolithic (alternatively, **Gigantolithic**) is a style of large stone and flint tools (or industry) associated primarily with the Qaraoun culture in the Beqaa Valley, Lebanon, dating to the Epipaleolithic or early Pre-pottery Neolithic at the end of the Stone Age.[1] The type site for the Qaraoun culture is Qaraoun II.[2]

26.1 Naming

The term "Heavy Neolithic" was translated by Lorraine Copeland and Peter J. Wescombe from Henri Fleisch's term *"gros Neolithique"*, suggested by Dorothy Garrod (in a letter dated February 1965) for adoption to describe the particular flint industry that was identified at sites near Qaraoun in the Beqaa Valley.[3] The industry was also termed "Gigantolithic" and confirmed as Neolithic by Alfred Rust and Dorothy Garrod.

26.2 Characteristics

Gigantolithic was initially mistaken for Acheulean or Levalloisian by some scholars. Diana Kirkbride and Henri de Contenson suggested that it existed over a wide area of the fertile crescent. Heavy Neolithic industry occurred before the invention of pottery and is characterized by huge, coarse, heavy tools such as axes, picks and adzes including bifaces. There is no evidence of polishing at the Qaraoun sites or indeed of any arrowheads, burins or millstones. Henri Fleisch noted that the culture that produced this industry may well have led a forest way of life before the dawn of agriculture.[4] Jacques Cauvin proposed that some of the sites discovered may have been factories or workshops as many artifacts recovered were rough outs.[5] James Mellaart suggested the industry dated to a period before the Pottery Neolithic at Byblos (10,600 to 6900 BC according to the ASPRO chronology) and noted *"Aceramic cultures have not yet been found in excavations but they must have existed here as it is clear from Ras Shamra and from the fact that the Pre-Pottery B complex of Palestine originated in this area, just as the following Pottery Neolithic cultures can be traced back to the Lebanon."*[6] Maya Haidar Boustani has called for discussion on the chronological problem when reliable data on the flint workshops becomes available.[7] She looked towards the work of Ron Barkai and H. Taute as being of possible use in this research.[8][9]

A notable stratified excavation of Heavy Neolithic material took place at Adloun II (Bezez Cave), conducted by Diana Kirkbride and Dorothy Garrod. Materials extracted from the upper layers were however disturbed.[7] The morphology of the tools has noted similarities to the Campignian industry in France.[10] Due to the disturbance of the upper layers and lack of radiocarbon dating or the materials at the time of this excavation, the placement of the Qaroun culture into the chronology of the ancient Near East remains undetermined from these excavations.[11]

The industry has been found at surface stations in the Beqaa Valley and on the seaward side of the mountains. Heavy Neolithic sites were found near sources of flint and were thought to be factories or workshops where large, coarse flint tools were roughed out to work and chop timber. Chisels, flake scrapers and picks were also found with little, if any sign of arrowheads, sickles (except for Orange slices) or pottery. Finds of waste and debris at the sites were usually plentiful, normally consisting of Orange slices, thick and crested blades, discoid, cylindrical, pyramidal or Levallois cores.[12] Andrew Moore suggested that many of the sites were used as flint factories that complimented settlements in the surrounding hills.[13]

The identification of Heavy Neolithic sites in Lebanon was complicated by the fact that the assemblages found at these sites included tools made with *all* techniques used during earlier periods. Bifaces are found both with and without a cortex, along with grattoir de cote, triangular flakes, tortoise cores, discoid cores and steep scrapers. This presented particular problems with sites where Heavy Neolithic material was mixed with that from the Lower Paleolithic and Middle Paleolithic, such as at Mejdel Anjar I and Dakoue. Although tools similar to Heavy Neolithic ones were found at later Neolithic surfaces sites, little relationship could be established between those found at the later Neolithic tells, where flints were often sparse, especially at those of later dates. The relationship and dividing line between the related Shepherd Neolithic zone of the north Bekaa Valley could also not be clearly defined but was suggested to be in the area around Douris and Qalaat Tannour. Not enough exploration has been carried out yet to conclude whether the bands of Neolithic surface sites continues north into the areas around Zahle and Rayak.[14]

26.3 Sites

Apart from the type site, Qaraoun II, other sites with Heavy Neolithic finds include Qaraoun I, Adloun II, Akbiyeh, Beit Mery II, Dikwene II, Hadeth South, Jbaa, Jebel Aabeby, Jdeideh I, Jdeideh III, Mtaileb I (Rabiya), Ourrouar II, Sin el Fil, Sarafand, Tell Mureibit near Kasimiyeh, Fadaous Sud, Baidar ech Chamout, Kfar Tebnit, Wadi Koura, Wadi Yaroun and other suggested sites at Flaoui, Sidon III, the Akkar plain foothills and the Plain of Zghorta. Others found in the Beqaa Valley include Ard Saouda, Nabi Zair, Tell Khardane, Mejdel Anjar I, Dakoue, Kefraya, Tell Zenoub, Kamid al lawz I, Bustan Birke, Joub Jannine III, Amlaq Qatih, Tayibe, Taire II, Khallet Michte I, Khallet Michte II, Khallet el Hamra, Douwara, Douris and Moukhtara with other possible sites at Tell Ain el

Meten and El Biré.[6][14][15] The Heavy Neolithic industry has also been identified at the Palestinian archaeological sites around Wadi al-Far'a; (Wadi Farah, Shemouniyeh and Wadi Sallah (occupational) excavated by Francis Turville-Petre.[10][13][14]

26.4 Gallery

- Double ended pick, triangular section with narrowing, jagged edges at both ends.

- Mini blade core on a split cobble.

- Thick and heavy biface, retouched all over with jagged and irregular edges.

26.5 References

[1] Lorraine Copeland; P. Wescombe (1965). *Inventory of Stone-Age sites in Lebanon, p. 43*. Imprimerie Catholique. Retrieved 21 July 2011.

[2] Cauvin, Jacques. and Cauvin, Marie-Claire., Des ateliers "campigniens" au Liban. pp. 103-116 in M. Maziéres (ed.) La préhistoire probléme et tendances. Hommabge á Raymond Vaufrey. Éditions CNRS, Paris, 1968.

[3] Fleisch, Henri, Nouvelles stations préhistoriques au Liban, BSPF, vol. 51, pp. 564-565, 1954.

[4] Fleisch, Henri, Les industries lithiques récentes de la Békaa, République Libanaise, Acts of the 6th C.I.S.E.A., vol. XI, no. 1, Paris, 1960.

[5] Cauvin, Jacques., Le néolithique de Mouchtara (Liban-Sud), L'Anthropologie, vol. 67, 5-6, p. 509, 1963.

[6] Mellaart, James, Earliest Civilizations in the Near East, p. 46, Thames and Hudson, London, 1965.

[7] E. J. Peltenburg; Alexander Wasse; Council for British Research in the Levant (2004). *Maya Haïdar Boustani, Flint workshops of the Southern Beqa' valley (Lebanon): preliminary results from Qar'oun* in Neolithic revolution: new perspectives on southwest Asia in light of recent discoveries on Cyprus. Oxbow Books. ISBN 978-1-84217-132-5. Retrieved 18 January 2012.

[8] Barkai, Ron., Make my axe: flint axe production and resharpening at EPPNB Nahal Lavan 109. pp. 73-92 in I. Canneva, C. Lemorini, D. Zampetti and P. Biagi (eds.) Beyond tools Proceedings of the Third Workshop on PPN chipped lithic industries. Department of Classical and Near Eastern Studies Ca'Foscari University of Venice, 1-4 November 1998. Studies in Early Near Eastern Production, Subsistence and Environment 9, Ex Oriente : Berlin, 2001

[9] Taute, W., The Pre-Pottery Neolithic flint mining and work-shop activities southwest of the Dead Sea, Israel (Ramat Tamar and Mesad Mazzal). pp. 495-509 in H.G. Gebel and S.K. Kozlowski (eds.) Neolithic chipped stone industries of the Fertile Crescent. Proceedings of the First Workshop on PPN chipped lithic industries. Free University of Berlin, 29 March-2 April 1993. Studies in Early Near Eastern Production, Subsistence and Environment 1. Ex Oriente : Berlin, 1994.

[10] Francis Adrian Joseph Turville-Petre; Dorothea M. A. Bate; Sir Arthur Keith; British School of Archaeology in Jerusalem (1927). *Researches in prehistoric Galilee, 1925-1926, p. 108*. The Council of the School. Retrieved 22 July 2011.

[11] Derek Arthur Roe; L. Copeland (1983). *Adlun in the Stone Age: the excavations of D.A.E. Garrod in the Lebanon, 1958-1963*. B.A.R. ISBN 978-0-86054-203-2. Retrieved 23 August 2012.

[12] Moore, A.M.T. (1978). *The Neolithic of the Levant*. Oxford University, Unpublished Ph.D. Thesis. p. 443.

[13] Moore, A.M.T. (1978). *The Neolithic of the Levant*. Oxford University, Unpublished Ph.D. Thesis. pp. 465–469.

[14] L. Copeland; P. Wescombe (1966). *Inventory of Stone-Age Sites in Lebanon: North, South and East-Central Lebanon,*. Impr. Catholique. Retrieved 1 January 2012.

[15] Moore, A.M.T. (1978). *The Neolithic of the Levant*. Oxford University, Unpublished Ph.D. Thesis. pp. 444–446.

Chapter 27

Shepherd Neolithic

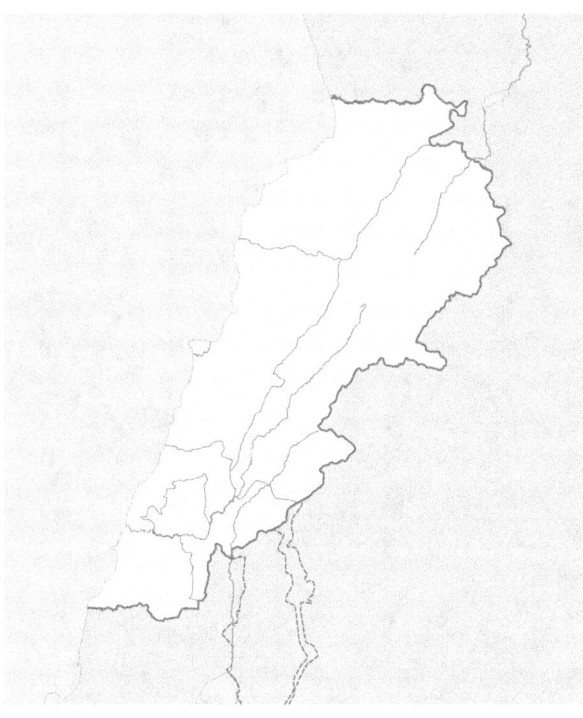

- Qaa
- Maqne
- Douris
- Hermel
- Kamouh el Hermel
- Qalaat Tannour
- Rayak North
- Riha Station

Map of Lebanon showing important sites that were occupied in the Shepherd Neolithic (clickable map)

Shepherd Neolithic is a name given by archaeologists to

A shepherd with sheep on a mountainside. Sheep were among the first animals to be domesticated by humankind; the domestication date is estimated to fall between nine and eleven thousand years ago in Mesopotamia.[1][2][3][4] Henri Fleisch suggested that the Shepherd Neolithic industry could have been used by nomadic shepherds.[5][6]

a style (or industry) of small flint tools from the Hermel plains in the north Beqaa Valley, Lebanon.[7] The Shepherd Neolithic industry has been insufficiently studied and was provisionally named based on a limited typology collected by Jesuit archaeologist "Père" Henri Fleisch.[8] Lorraine Copeland and Peter J. Wescombe suggested it was possibly *"of quite late date"*.[8]

27.1 Characteristics

Shepherd Neolithic material can be found dispersed over a wide area of the north Beqaa Valley in low concentrations. M. Billaux and Henri Fleisch suggested that the flints were of a higher quality than the brittle flint in the nearby conglomerates indicating that they had been imported from somewhere else. Three groups of flint could be determined; light brown, red-brown and that varied but was usually grey-chocolate that was distinguished with a radiant "desert shine". Characteristics of the industry include smallness in size, commonly between 2.5 cm and 4 cm and frequently being quite thick, unlike geometric microliths. The small number of tools within the assemblage is another distinguishable characteristic, including short denticulated or notched blades, end scrapers, transverse racloirs on thin flakes and borers with strong points. They also display a lack of recognizable typology although Levallois technique was occasionally observed to have been used. They also show signs of having been heavily worked with cores being re-used and turned into scrapers. Fleisch suggested the industry was Epipaleolithic as it is evidently not Paleolithic, Mesolithic or even Pottery Neolithic. He further suggested that the industry could have been used by nomadic shepherds.[5][6]

The relationship and dividing line between the related Heavy Neolithic zone of the south Beqaa Valley could also not be clearly defined but was suggested to be in the area around Douris and Qalaat Tannour. Not enough exploration had been carried out to conclude whether the bands of Neolithic surface sites continues south into the areas around Zahle and Rayak.[5]

27.2   Sites

The type sites of the Shepherd Neolithic are at Qaa and Maqne I, with other sites with Shepherd Neolithic finds include Douris, Hermel II, Hermel III, Kamouh el Hermel, Qalaat Tannour, Wadi Boura I and possibly at Rayak North, Riha Station and Serain.

27.3   References

[1] Ensminger

[2] Weaver

[3] Simmons & Ekarius

[4] Krebs, Robert E. & Carolyn A. (2003). *Groundbreaking Scientific Experiments, Inventions & Discoveries of the Ancient World*. Westport, CT: Greenwood Press. ISBN 0-313-31342-3.

[5] L. Copeland; P. Wescombe (1966). *Inventory of Stone-Age Sites in Lebanon: North, South and East-Central Lebanon,*. Impr. Catholique. Retrieved 1 January 2012.

[6] Fleisch, Henri., Notes de Préhistoire Libanaise : 1) Ard es Saoude. 2) La Bekaa Nord. 3) Un polissoir en plein air. BSPF, vol. 63.

[7] Fleisch, Henri., Les industries lithiques récentes de la Békaa, République Libanaise, Acts of the 6th C.I.S.E.A., vol. XI, no. 1. Paris, 1960.

[8] Lorraine Copeland; P. Wescombe (1965). *Inventory of Stone-Age sites in Lebanon, p. 43*. Imprimerie Catholique. Retrieved 21 July 2011.

Chapter 28

Trihedral Neolithic

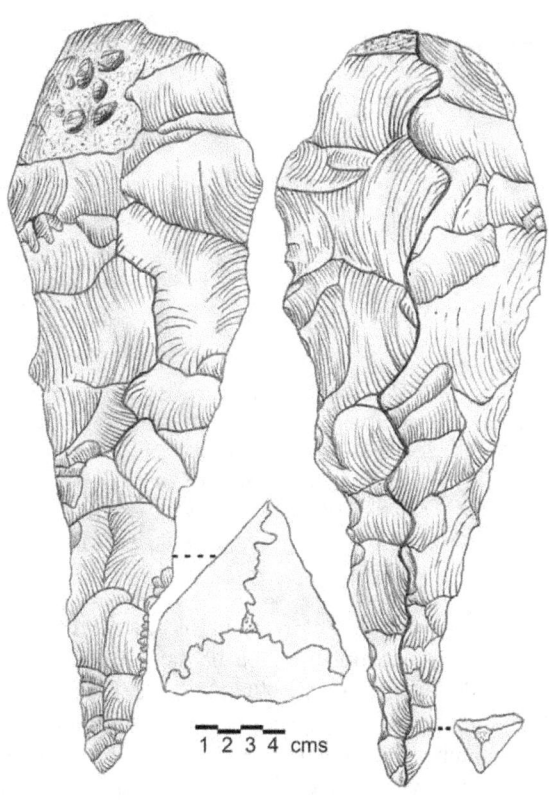

Trihedral Neolithic axe or pick from Joub Jannine II, Lebanon. Cream flint patinated to brown. In the collection of the Museum of Lebanese Prehistory at the Saint Joseph University, Beirut, Lebanon.

28.1   References

[1] Fleisch, Henri., Les industries lithiques récentes de la Békaa, République Libanaise, Acts of the 6th C.I.S.E.A., vol. XI, no. 1. Paris, 1960.

[2] Lorraine Copeland; P. Wescombe (1965). *Inventory of Stone-Age sites in Lebanon, p. 43*. Imprimerie Catholique. Retrieved 21 July 2011.

Trihedral Neolithic is a name given by archaeologists to a style (or industry) of striking spheroid and trihedral flint tools from the archaeological site of Joub Jannine II in the Beqaa Valley, Lebanon.[1] The style appears to represent a highly specialized Neolithic industry. Little comment has been made of this industry.[2]

Chapter 29

Pre-Pottery Neolithic

●
Jericho

Map of Palestine showing important sites that were occupied in the Pre-Pottery Neolithic (clickable map)

The **Pre-Pottery Neolithic** (**PPN**, around 8500-5500

Map of the world showing approximate centers of origin of agriculture and its spread in prehistory: the Fertile Crescent (11,000 BP), the Yangtze and Yellow River basins (9,000 BP) and the New Guinea Highlands (9,000–6,000 BP), Central Mexico (5,000–4,000 BP), Northern South America (5,000–4,000 BP), sub-Saharan Africa (5,000–4,000 BP, exact location unknown), eastern North America (4,000–3,000 BP).[1]

BCE)[2] represents the early Neolithic in the Levantine and upper Mesopotamian region of the Fertile Crescent. It succeeds the Natufian culture of the Epipaleolithic (Mesolithic) as the domestication of plants and animals was in its beginnings, possibly triggered by the Younger Dryas. The Pre-Pottery Neolithic culture came to an end around the time of the 8.2 kiloyear event, a cool spell lasting several hundred years centred on 6200 BCE.

29.1 Pre-Pottery Neolithic A

Main article: Pre-Pottery Neolithic A

The Pre-Pottery Neolithic is divided into Pre-Pottery Neolithic A (**PPNA** 8500 BCE - 7600 BCE) and the following Pre-Pottery Neolithic B (**PPNB** 7600 BCE - 6000 BCE).[3] These were originally defined by Kathleen Kenyon in the type site of Jericho (Palestine). The Pre-Pottery Neolithic precedes the ceramic Neolithic (Yarmukian). At 'Ain Ghazal in Jordan the culture continued a few more centuries as the so-called Pre-Pottery Neolithic C culture.

Around 8000 BCE during the Pre-Pottery Neolithic A (PPNA) the world's first town Jericho appeared in the Levant.

29.2 Pre-Pottery Neolithic B

Main article: Pre-Pottery Neolithic B

PPNB differed from PPNA in showing greater use of domesticated animals, a different set of tools, and new architectural styles.

29.3 Pre-Pottery Neolithic C

Work at the site of 'Ain Ghazal in Jordan has indicated a later Pre-Pottery Neolithic C period. Juris Zarins has proposed that a Circum Arabian Nomadic Pastoral Complex developed in the period from the climatic crisis of 6200 BCE, partly as a result of an increasing emphasis in PPNB cultures upon domesticated animals, and a fusion with Harifian hunter gatherers in the Southern Levant, with affiliate connections with the cultures of Fayyum and the Eastern Desert of Egypt. Cultures practicing this lifestyle spread down the Red Sea shoreline and moved east from Syria into southern Iraq.[4]

29.4 See also

- Pre-history of the Southern Levant

- History of pottery in the Southern Levant

29.5 References

[1] Diamond, J.; Bellwood, P. (2003). "Farmers and Their Languages: The First Expansions". *Science* **300** (5619): 597–603. Bibcode:2003Sci...300..597D. doi:10.1126/science.1078208. PMID 12714734.

[2] Richard, Suzanne *Near Eastern archaeology* Eisenbrauns; illustrated edition (1 Aug 2004) ISBN 978-1-57506-083-5 p.244

[3] Richard, Suzanne *Near Eastern archaeology* Eisenbrauns; illustrated edition (1 Aug 2004) ISBN 978-1-57506-083-5 p.244

[4] Zarins, Juris (1992) "Pastoral Nomadism in Arabia: Ethnoarchaeology and the Archaeological Record," in Ofer Bar-Yosef and A. Khazanov, eds. "Pastoralism in the Levant"

29.6 Further reading

- Ofer Bar-Yosef, The PPNA in the Levant – an overview. Paléorient 15/1, 1989, 57-63.

- J. Cauvin, Naissance des divinités, Naissance de l'agriculture. La révolution des symboles au Néolithique (CNRS 1994). Translation (T. Watkins) The birth of the gods and the origins of agriculture (Cambridge 2000).

Chapter 30

Neolithic

An array of Neolithic artifacts, including bracelets, axe heads, chisels, and polishing tools. Neolithic stone artifacts are by definition polished and, except for specialty items, not chipped.

The **Neolithic** 🔊/ˌniːəˈlɪθɪk/[1] **Age**, **Era**, or **Period**, from νέος (néos, "new") and λίθος (líthos, "stone"), or **New Stone Age**, was a period in the development of human technology, beginning about 10,200 BC, according to the ASPRO chronology, in some parts of the Middle East, and later in other parts of the world[2] and ending between 4,500 and 2,000 BC.

Traditionally considered the last part of the Stone Age, the Neolithic followed the terminal Holocene *Epipaleolithic* period and commenced with the beginning of farming, which produced the "Neolithic Revolution". It ended when metal tools became widespread (in the Copper Age or Bronze Age; or, in some geographical regions, in the Iron Age). The Neolithic is a progression of behavioral and cultural characteristics and changes, including the use of wild and domestic crops and of domesticated animals.[3]

The beginning of the Neolithic culture is considered to be in the Levant (Jericho, modern-day West Bank) about 10,200–8,800 BC. It developed directly from the Epipaleolithic Natufian culture in the region, whose people pioneered the use of wild cereals, which then evolved into true farming. The Natufian period was between 12,000 and 10,200 BC, and the so-called "proto-neolithic" is now included in the Pre-Pottery Neolithic (PPNA) between 10,200 and 8,800 BC. As the Natufians had become dependent on wild cereals in their diet, and a sedentary way of life had begun among them, the climatic changes associated with the Younger Dryas are thought to have forced people to develop farming. By 10,200–8,800 BC, farming communities arose in the Levant and spread to Asia Minor, North Africa and North Mesopotamia. Early Neolithic farming was limited to a narrow range of plants, both wild and domesticated, which included einkorn wheat, millet and spelt, and the keeping of dogs, sheep and goats. By about 6,900–6,400 BC, it included domesticated cattle and pigs, the establishment of permanently or seasonally inhabited settlements, and the use of pottery.[4]

Not all of these cultural elements characteristic of the Neolithic appeared everywhere in the same order: the earliest farming societies in the Near East did not use pottery. In other parts of the world, such as Africa, South Asia and Southeast Asia, independent domestication events led to their own regionally distinctive Neolithic cultures that arose completely independent of those in Europe and Southwest Asia. Early Japanese societies and other East Asian cultures used pottery *before* developing agriculture.[5][6]

Unlike the Paleolithic, when more than one human species existed, only one human species (*Homo sapiens sapiens*) reached the Neolithic.[7] *Homo floresiensis* may have survived right up to the very dawn of the Neolithic, about 12,200 years ago.[8]

The term *Neolithic* derives from the Greek νεολιθικός, *neolithikos*, from νέος *neos*, "new" + λίθος *lithos*, "stone", literally meaning "New Stone Age". The term was invented by Sir John Lubbock in 1865 as a refinement of the three-age system.

30.1 Periods by pottery phase

In the Middle East, cultures identified as Neolithic began appearing by in the 10th millennium BCE.[2] Early development occurred in the Levant (e.g., Pre-Pottery Neolithic A and Pre-Pottery Neolithic B) and from there spread eastwards and westwards. Neolithic cultures are also attested in southeastern Anatolia and northern Mesopotamia by c. 8,000 BCE.

The prehistoric Beifudi site near Yixian in Hebei Province, China, contains relics of a culture contemporaneous with the Cishan and Xinglongwa cultures of about 5,000–6,000 BCE, neolithic cultures east of the Taihang Mountains, filling in an archaeological gap between the two Northern Chinese cultures. The total excavated area is more than 1,200 square yards (1,000 m^2; 0.10 ha), and the collection of neolithic findings at the site encompasses two phases.[9]

30.1.1   Neolithic 1 – Pre-Pottery Neolithic A (PPNA)

The Neolithic 1 (PPNA) period began roughly 10,000 years ago in the Levant.[2] A temple area in southeastern Turkey at Göbekli Tepe dated around 9,500 BC may be regarded as the beginning of the period. This site was developed by nomadic hunter-gatherer tribes, evidenced by the lack of permanent housing in the vicinity and may be the oldest known human-made place of worship.[10] At least seven stone circles, covering 25 acres (10 ha), contain limestone pillars carved with animals, insects, and birds. Stone tools were used by perhaps as many as hundreds of people to create the pillars, which might have supported roofs. Other early PPNA sites dating to around 9,500 to 9,000 BCE have been found in Jericho, Israel (notably Ain Mallaha, Nahal Oren, and Kfar HaHoresh), Gilgal in the Jordan Valley, and Byblos, Lebanon. The start of Neolithic 1 overlaps the Tahunian and Heavy Neolithic periods to some degree.

The major advance of Neolithic 1 was true farming. In the proto-Neolithic Natufian cultures, wild cereals were harvested, and perhaps early seed selection and re-seeding occurred. The grain was ground into flour. Emmer wheat was domesticated, and animals were herded and domesticated (animal husbandry and selective breeding).

In the 21st century, remains of figs were discovered in a house in Jericho dated to 9,400 BCE. The figs are of a mutant variety that cannot be pollinated by insects, and therefore the trees can only reproduce from cuttings. This evidence suggests that figs were the first cultivated crop and mark the invention of the technology of farming. This occurred centuries before the first cultivation of grains.[11]

Settlements became more permanent with circular houses, much like those of the Natufians, with single rooms. However, these houses were for the first time made of mudbrick. The settlement had a surrounding stone wall and perhaps a stone tower (as in Jericho). The wall served as protection from nearby groups, as protection from floods, or to keep animals penned. Some of the enclosures also suggest grain and meat storage.

30.1.2   Neolithic 2 – Pre-Pottery Neolithic B (PPNB)

The Neolithic 2 (PPNB) began around 8,800 BCE according to the ASPRO chronology in the Levant (Jericho, Israel).[2] As with the PPNA dates, there are two versions from the same laboratories noted above. This system of terminology, however, is not convenient for southeast Anatolia and settlements of the middle Anatolia basin. This era was before the Mesolithic era. A settlement of 3,000 inhabitants was found in the outskirts of Amman, Jordan. Considered to be one of the largest prehistoric settlements in the Near East, called 'Ain Ghazal, it was continuously inhabited from approximately 7,250 – 5,000 B.[12]

Settlements have rectangular mud-brick houses where the family lived together in single or multiple rooms. Burial findings suggest an ancestor cult where people preserved skulls of the dead, which were plastered with mud to make facial features. The rest of the corpse could have been left outside the settlement to decay until only the bones were left, then the bones were buried inside the settlement underneath the floor or between houses.

30.1.3   Neolithic 3 – Pottery Neolithic (PN)

The Neolithic 3 (PN) began around 6,400 BCE in the Fertile Crescent.[2] By then distinctive cultures emerged, with pottery like the Halafian (Turkey, Syria, Northern Mesopotamia) and Ubaid (Southern Mesopotamia). This period has been further divided into **PNA** (Pottery Neolithic A) and **PNB** (Pottery Neolithic B) at some sites.

The Chalcolithic period began about 4500 BCE, then the Bronze Age began about 3500 BCE, replacing the Neolithic cultures.

30.2   Periods by region

30.2.1   Fertile Crescent

Around 10,200 BC the first fully developed Neolithic cultures belonging to the phase Pre-Pottery Neolithic A (PPNA) appeared in the fertile crescent.[2] Around 10,700 to 9,400 BC a settlement was established in Tell Qaramel, 10 miles north of Aleppo. The settlement included 2 temples dating back to 9,650.[13] Around 9000 BC during the PPNA, one of the world's first towns, Jericho, appeared in the Levant. It was surrounded by a stone and marble wall and contained a population of 2000–3000 people and a massive stone tower.[14] Around 6,400 BC the Halaf culture appeared in Lebanon, Israel and Palestine, Syria, Anatolia, and Northern Mesopotamia and subsisted on dryland agriculture.

In 1981 a team of researchers from the Maison de l'Orient et de la Méditerranée, including Jacques Cauvin and Oliver Aurenche divided Near East neolithic chronology into ten periods (0 to 9) based on social, economic and cultural characteristics.[15] In 2002 Danielle Stordeur and Frédéric Abbès advanced this system with a division into five periods. Natufian (1) between 12,000 and 10,200 BC, Khiamian (2) between 10,200-8,800 BC, PPNA: Sultanian (Jericho), Mureybetian, early PPNB (*PPNB ancien*) (3) between 8,800-7,600 BC, middle PPNB (*PPNB moyen*) 7,600-6,900 BC, late PPNB (*PPNB récent*) (4) between 7,500 and 7,000 BC and a PPNB (sometimes called PPNC) transitional stage (*PPNB final*) (5) where Halaf and dark faced burnished ware begin to emerge between 6,900-6,400 BC.[16] They also advanced the idea of a transitional stage between the PPNA and PPNB between 8,800 and 8,600 BC at sites like Jerf el Ahmar and Tell Aswad.[17]

30.2.2   Southern Mesopotamia

Alluvial plains (Sumer/Elam). Little rainfall makes irrigation systems necessary. Ubaid culture from 6,900 BC.

30.2.3   North Africa

Algerian cave paintings depicting hunting scenes

Domestication of sheep and goats reached Egypt from the Near East possibly as early as 6,000 BC.[18][19][20] Graeme Barker states "The first indisputable evidence for domestic plants and animals in the Nile valley is not until the early

fifth millennium bc in northern Egypt and a thousand years later further south, in both cases as part of strategies that still relied heavily on fishing, hunting, and the gathering of wild plants" and suggests that these subsistence changes were not due to farmers migrating from the Near East but was an indigenous development, with cereals either indigenous or obtained through exchange.[21] Other scholars argue that the primary stimulus for agriculture and domesticated animals (as well as mud-brick architecture and other Neolithic cultural features) in Egypt was from the Middle East.[22][23][24]

30.2.4   Europe

Main article: Neolithic Europe

In southeast Europe agrarian societies first appeared in the

Female figure from Tumba Madžari, Republic of Macedonia

7th millennium BC, attested by one of the earliest farming sites of Europe, discovered in Vashtëmi, southeastern Albania and dating back to 6,500 BC.[25][26] Anthropomorphic figurines have been found in the Balkans from 6000 BC,[27] and in Central Europe by c. 5800 BC (La Hoguette). Among the earliest cultural complexes of this area are the Sesklo culture in Thessaly, which later expanded in the Balkans giving rise to Starčevo-Körös (Cris), Linearbandkeramik, and Vinča. Through a combination of

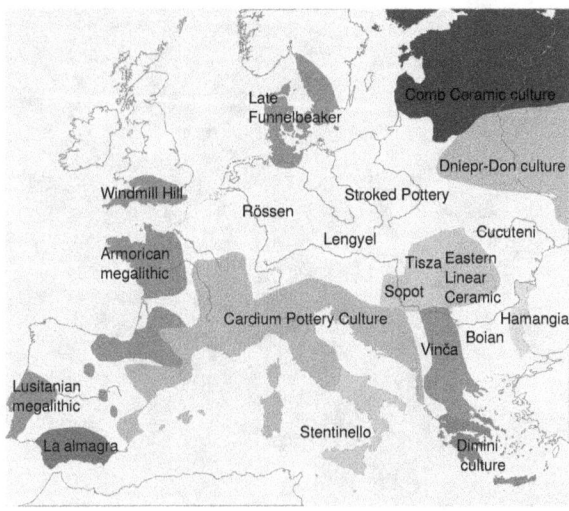

Map showing distribution of some of the main culture complexes in Neolithic Europe, c.3,500 BC

Skara Brae, Scotland. Evidence of home furnishings (shelves)

Dolmen of Cava dei Servi, Sicily

cultural diffusion and migration of peoples, the Neolithic traditions spread west and northwards to reach northwest-

ern Europe by around 4500 BC. The Vinča culture may have created the earliest system of writing, the Vinča signs, though archaeologist Shan Winn believes they most likely represented pictograms and ideograms rather than a truly developed form of writing.[28] The Cucuteni-Trypillian culture built enormous settlements in Romania, Moldova and Ukraine from 5300 to 2300 BC. The megalithic temple complexes of Ġgantija on the Mediterranean island of Gozo (in the Maltese archipelago) and of Mnajdra (Malta) are notable for their gigantic Neolithic structures, the oldest of which date back to c. 3600 BC. The Hypogeum of Ħal-Saflieni, Paola, Malta, is a subterranean structure excavated c. 2500 BC; originally a sanctuary, it became a necropolis, the only prehistoric underground temple in the world, and showing a degree of artistry in stone sculpture unique in prehistory to the Maltese islands. After 2500 BC, the Maltese Islands were depopulated for several decades until the arrival of a new influx of Bronze Age immigrants, a culture that cremated its dead and introduced smaller megalithic structures called dolmens to Malta.[29] In most cases there are small chambers here, with the cover made of a large slab placed on upright stones. They are claimed to belong to a population certainly different from that which built the previous megalithic temples. It is presumed the population arrived from Sicily because of the similarity of Maltese dolmens to some small constructions found in the largest island of the Mediterranean sea.[30]

30.2.5   South and East Asia

The earliest Neolithic site in South Asia is Mehrgarh, dated to 7500 BC, in the Kachi plain of Baluchistan, Pakistan; the site has evidence of farming (wheat and barley) and herding (cattle, sheep and goats).[31]

In South India, the Neolithic began by 6500 BC and lasted until around 1400 BC when the Megalithic transition period began. South Indian Neolithic is characterized by Ashmounds since 2500 BC in Karnataka region, expanded later to Tamil Nadu.

In East Asia, the earliest sites include Nanzhuangtou culture around 9500 BC to 9000 BC,[32] Pengtoushan culture around 7500 BC to 6100 BC, and Peiligang culture around 7000 BC to 5000 BC.

The 'Neolithic' (defined in this paragraph as using polished stone implements) remains a living tradition in small and extremely remote and inaccessible pockets of West Papua (Indonesian New Guinea). Polished stone adze and axes are used in the present day (as of 2008) in areas where the availability of metal implements is limited. This is likely to cease altogether in the next few years as the older generation die off and steel blades and chainsaws prevail.

In 2012, news was released about a new farming site discovered in Munam-ri, Goseong, Gangwon Province, South Korea, which may be the earliest farmland known to date in east Asia.[33] "No remains of an agricultural field from the Neolithic period have been found in any East Asian country before, the institute said, adding that the discovery reveals that the history of agricultural cultivation at least began during the period on the Korean Peninsula".[34] The farm was dated between 3600 and 3000 B.C. Pottery, stone projectile points, and possible houses were also found. "In 2002, researchers discovered prehistoric earthenware, jade earrings, among other items in the area". The research team will perform accelerator mass spectrometry (AMS) dating to retrieve a more precise date for the site.

30.2.6 America

In Mesoamerica, a similar set of events (i.e., crop domestication and sedentary lifestyles) occurred by around 4500 BC, but possibly as early as 11,000–10,000 BC. These cultures are usually not referred to as belonging to the Neolithic; in America different terms are used such as Formative stage instead of mid-late Neolithic, Archaic Era instead of Early Neolithic and Paleo-Indian for the preceding period.[35] The Formative stage is equivalent to the Neolithic Revolution period in Europe, Asia, and Africa. In the Southwestern United States it occurred from 500 to 1200 C.E. when there was a dramatic increase in population and development of large villages supported by agriculture based on dryland farming of maize, and later, beans, squash, and domesticated turkeys. During this period the bow and arrow and ceramic pottery were also introduced.[36]

30.3   Social organization

During most of the Neolithic age of Eurasia, people lived in small tribes composed of multiple bands or lineages.[37] There is little scientific evidence of developed social stratification in most Neolithic societies; social stratification is more associated with the later Bronze Age.[38] Although some late Eurasian Neolithic societies formed complex stratified chiefdoms or even states, states evolved in Eurasia only with the rise of metallurgy, and most Neolithic societies one the whole were relatively simple and egalitarian.[37] Beyond Eurasia, however, states were formed during the local Neolithic in three areas, namely in the Preceramic Andes with the Norte Chico Civilization,[39][40] Formative Mesoamerica and Ancient Hawai'i.[41] However, most Neolithic societies were noticeably more hierarchical than the Paleolithic cultures that preceded them and hunter-gatherer cultures in general.[42][43]

Anthropomorphic Neolithic figurine

The domestication of large animals (c. 8000 BC) resulted in a dramatic increase in social inequality in most of the areas where it occurred; New Guinea being a notable exception.[44] Possession of livestock allowed competition between households and resulted in inherited inequalities of wealth. Neolithic pastoralists who controlled large herds gradually acquired more livestock, and this made economic inequalities more pronounced.[45] However, evidence of social inequality is still disputed, as settlements such as Catal Huyuk reveal a striking lack of difference in the size of homes and burial sites, suggesting a more egalitarian society with no evidence of the concept of capital, although some homes do appear slightly larger or more elaborately decorated than others.

Families and households were still largely independent economically, and the household was probably the center of life.[46][47] However, excavations in Central Europe have revealed that early Neolithic Linear Ceramic cultures ("*Linearbandkeramik*") were building large arrangements of circular ditches between 4800 BC and 4600 BC. These structures (and their later counterparts such as causewayed enclosures, burial mounds, and henge) required considerable time and labour to construct, which suggests that some influential individuals were able to organise and direct human labour — though non-hierarchical and voluntary work remain possibilities.

There is a large body of evidence for fortified settlements at *Linearbandkeramik* sites along the Rhine, as at least some villages were fortified for some time with a palisade and an outer ditch.[48][49] Settlements with palisades and weapon-traumatized bones have been discovered, such as at the Talheim Death Pit demonstrates "...systematic violence between groups" and warfare was probably much more common during the Neolithic than in the preceding

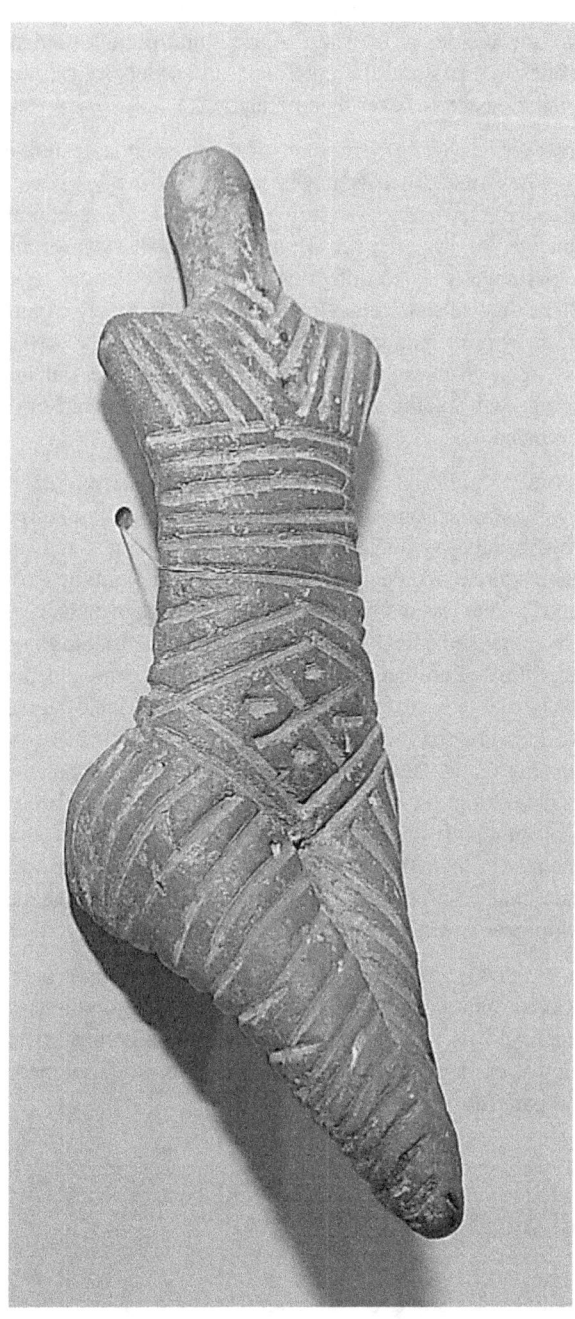

Anthropomorphic Female Neolithic ceramic figurine

case in the chiefdoms of the European Early Bronze Age.[51] Theories to explain the apparent implied egalitarianism of Neolithic (and Paleolithic) societies have arisen, notably the Marxist concept of primitive communism.

30.4   Shelter

Reconstruction of Neolithic house in Tuzla, Bosnia and Herzegovina

The shelter of the early people changed dramatically from the paleolithic to the neolithic era. In the paleolithic, people did not normally live in permanent constructions. In the neolithic, mud brick houses started appearing that were coated with plaster.[52] The growth of agriculture made permanent houses possible. Doorways were made on the roof, with ladders positioned both on the inside and outside of the houses.[52] The roof was supported by beams from the inside. The rough ground was covered by platforms, mats, and skins on which residents slept.[53] Stilt-houses settlements were common in the Alpine and Pianura Padana (Terramare) region.[54] Remains have been found at the Ljubljana Marshes in Slovenia and at the Mondsee and Attersee lakes in Upper Austria, for example.

30.5   Farming

Main article: Neolithic Revolution

A significant and far-reaching shift in human subsistence and lifestyle was to be brought about in areas where crop farming and cultivation were first developed: the previous reliance on an essentially nomadic hunter-gatherer subsistence technique or pastoral transhumance was at first supplemented, and then increasingly replaced by, a reliance upon the foods produced from cultivated lands. These developments are also believed to have greatly encouraged the growth of settlements, since it may be supposed that the

Paleolithic period.[43] This supplanted an earlier view of the Linear Pottery Culture as living a "peaceful, unfortified lifestyle".[50]

Control of labour and inter-group conflict is characteristic of corporate-level or 'tribal' groups, headed by a charismatic individual; whether a 'big man' or a proto-chief, functioning as a lineage-group head. Whether a non-hierarchical system of organization existed is debatable, and there is no evidence that explicitly suggests that Neolithic societies functioned under any dominating class or individual, as was the

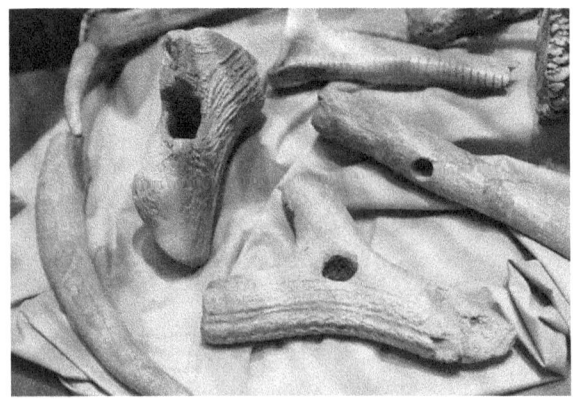

A Cucuteni-Trypillian culture deer antler plough

Food and cooking items retrieved at a European Neolithic site: millstones, charred bread, grains and small apples, a clay cooking pot, and containers made of antlers and wood

other necessities or luxuries. Agricultural life afforded securities that pastoral life could not, and sedentary farming populations grew faster than nomadic.

However, early farmers were also adversely affected in times of famine, such as may be caused by drought or pests. In instances where agriculture had become the predominant way of life, the sensitivity to these shortages could be particularly acute, affecting agrarian populations to an extent that otherwise may not have been routinely experienced by prior hunter-gatherer communities.[45] Nevertheless, agrarian communities generally proved successful, and their growth and the expansion of territory under cultivation continued.

Another significant change undergone by many of these newly agrarian communities was one of diet. Pre-agrarian diets varied by region, season, available local plant and animal resources and degree of pastoralism and hunting. Post-agrarian diet was restricted to a limited package of successfully cultivated cereal grains, plants and to a variable extent domesticated animals and animal products. Supplementation of diet by hunting and gathering was to variable degrees precluded by the increase in population above the carrying capacity of the land and a high sedentary local population concentration. In some cultures, there would have been a significant shift toward increased starch and plant protein. The relative nutritional benefits and drawbacks of these dietary changes and their overall impact on early societal development is still debated.

In addition, increased population density, decreased population mobility, increased continuous proximity to domesticated animals, and continuous occupation of comparatively population-dense sites would have altered sanitation needs and patterns of disease.

30.6 Technology

Main article: Stone tool § Neolithic industries

The identifying characteristic of Neolithic technology is the use of polished or ground stone tools, in contrast to the flaked stone tools used during the Paleolithic era.

Neolithic people were skilled farmers, manufacturing a range of tools necessary for the tending, harvesting and processing of crops (such as sickle blades and grinding stones) and food production (e.g. pottery, bone implements). They were also skilled manufacturers of a range of other types of stone tools and ornaments, including projectile points, beads, and statuettes. But what allowed forest clearance on a large scale was the polished stone axe above all other tools. Together with the adze, fashioning wood for shelter, struc-

increased need to spend more time and labor in tending crop fields required more localized dwellings. This trend would continue into the Bronze Age, eventually giving rise to permanently settled farming towns, and later cities and states whose larger populations could be sustained by the increased productivity from cultivated lands.

The profound differences in human interactions and subsistence methods associated with the onset of early agricultural practices in the Neolithic have been called the *Neolithic Revolution*, a term coined in the 1920s by the Australian archaeologist Vere Gordon Childe.

One potential benefit of the development and increasing sophistication of farming technology was the possibility of producing surplus crop yields, in other words, food supplies in excess of the immediate needs of the community. Surpluses could be stored for later use, or possibly traded for

tures and canoes for example, this enabled them to exploit their newly won farmland.

Neolithic peoples in the Levant, Anatolia, Syria, northern Mesopotamia and Central Asia were also accomplished builders, utilizing mud-brick to construct houses and villages. At Çatal höyük, houses were plastered and painted with elaborate scenes of humans and animals. In Europe, long houses built from wattle and daub were constructed. Elaborate tombs were built for the dead. These tombs are particularly numerous in Ireland, where there are many thousand still in existence. Neolithic people in the British Isles built long barrows and chamber tombs for their dead and causewayed camps, henges, flint mines and cursus monuments. It was also important to figure out ways of preserving food for future months, such as fashioning relatively airtight containers, and using substances like salt as preservatives.

The peoples of the Americas and the Pacific mostly retained the Neolithic level of tool technology until the time of European contact. Exceptions include copper hatchets and spearheads in the Great Lakes region.

30.7 Clothing

Most clothing appears to have been made of animal skins, as indicated by finds of large numbers of bone and antler pins which are ideal for fastening leather. Wool cloth and linen might have become available during the later Neolithic,[55][56] as suggested by finds of perforated stones which (depending on size) may have served as spindle whorls or loom weights.[57][58][59] The clothing worn in the Neolithic Age might be similar to that worn by Ötzi the Iceman, although he was not Neolithic (since he belonged to the later Copper age).

30.8 Early settlements

Neolithic human settlements include:

- Göbekli Tepe in Turkey, c. 11,000–9000 BC

- Tell Qaramel in Syria, 10,700–9400 BC

- Franchthi Cave in Greece, epipalaeolithic (c. 10,000 BC) settlement, reoccupied between 7500 and 6000 BC

- Nanzhuangtou in Hebei, China, 9500-9000 BC

- Byblos in Lebanon believed to have been occupied first between 8800 and 7000 BC,[60]

Reconstruction of a Cucuteni-Trypillian hut, in the Tripillian Museum, Ukraine.

- Jericho in West bank, Neolithic from around 8350 BC, arising from the earlier Epipaleolithic Natufian culture

- Donghulin in China 8000-6200 BC

- Aşıklı Höyük in Central Anatolia, Turkey, an Aceramic Neolithic period settlement, 8200–7400 BC, correlating with the E/MPPNB in the Levant.

- Nevali Cori in Turkey, c. 8000 BC

The Archaeological Site of Çatal Hüyük in the Konya Plain in Turkey

- Pengtoushan culture in China, 7500–6100 BC, rice residues were carbon-14 dated to 8200–7800 BC in type site

- Çatalhöyük in Turkey, 7500 BC

- 'Ain Ghazal in Jordan, 7250–5000 BC

- Chogha Bonut in Iran, 7200 BC

- Jhusi in India, 7100 BC

- Ganj Dareh in Iran, c. 7000 BC

- Lahuradewa in India, 7000 BC[61]

- Jiahu in China, 7000 to 5800 BC

- Mehrgarh in Pakistan, 7000 BC

- Knossus on Crete, c. 7000 BC

- Karanovo in Bulgaria, 6200 BC

- Sesklo in Greece, 6850 BC (with a ±660-year margin of error)

- Dispilio in Greece, c. 5500 BC

- Porodin in Republic of Macedonia, 6500 BC[62]

- Vrshnik (Anzabegovo) in Republic of Macedonia, 6500 BC[62]

- Pizzo di Bodio (Varese), Lombardy in Italy, c. 6320 ±80 BC

- Sammardenchia in Friuli, Italy, c. 6050 ±90 BC,

- Padah-Lin Caves in Burma, c. 6000 BC

- Petnica in Serbia, 6000 BC

- Stara Zagora in Bulgaria, 5500 BC

- Cucuteni-Trypillian culture, 5500–2750 BC, in Ukraine, Moldova and Romania first salt works

- Tell Zeidan in northern Syria, from about 5500 to 4000 BC.

- around 2000 settlements of Trypillian culture, 5400–2800 BC

- Tabon Cave Complex in Quezon, Palawan, Philippines 5000–2000 BC[63][64]

- Hemudu culture in China, 5000–4500 BC, large-scale rice plantation

- The Megalithic Temples of Malta, 3600 BC

- Knap of Howar and Skara Brae, Orkney, Scotland, from 3500 BC and 3100 BC respectively

- Brú na Bóinne in Ireland, c. 3500 BC

- Lough Gur in Ireland from around 3000 BC

- Norte Chico civilization, from 3000 to 1700 BC, 30 Aceramic Neolithic period settlements in Northern Coastal Peru

- Tichit Neolithic village on the Tagant Plateau in central southern Mauritania, 2000–500 BC

- Oaxaca, state in Southwestern Mexico, by 2000 BC Neolithic sedentary villages had been established in the Central Valleys region of this state.

- Lajia in China, 2000 BC

- Mumun pottery period, Neolithic revolution spreads down the Korean Peninsula and permanent settlements are established 1800–1500 BC, Neolithic revolution reaches Japan around 500–300 BC

The world's oldest known engineered roadway, the Sweet Track in England, dates from 3800 BC and the world's oldest free-standing structure is the neolithic temple of Ggantija in Gozo, Malta.

30.9   List of cultures and sites

Excavated dwellings at Skara Brae (Orkney, Scotland), Europe's most complete Neolithic village

Note: Dates are very approximate, and are only given for a rough estimate; consult each culture for specific time periods.

Early Neolithic
Periodization: The Levant: 10,000 to 8500 BC; Europe: 5000 to 4000 BC; Elsewhere: varies greatly, depending on region.

- Beixin culture

- Cishan culture

- Dudeşti culture

- Franchthi Cave people

  - Earliest European Neolithic site: 20th to 3rd millennium BC

- Sesclo village culture

- Starcevo-Criş culture

  - (also known as the Starčevo-Körös-Criş culture)

- Nanzhuangtou

Middle Neolithic
Periodization: The Levant: 8500 to 6500 BC; Europe: 4000 to 3500 BC; Elsewhere: varies greatly, depending on region.

- Baodun culture

  - Jinsha settlement and Sanxingdui mound.

- Catalhoyuk

- Cardium Pottery culture

- Comb Ceramic culture

- Corded Ware culture

- Cortaillod culture

- Cucuteni-Trypillian culture

- Dadiwan culture

- Dawenkou culture

- Daxi culture

  - Chengtoushan settlement

- Grooved ware people

  - Skara Brae, et al.

- Erlitou culture

  - Xia Dynasty

- Ertebølle culture

- Hembury culture

- Hemudu culture

- Hongshan culture

- Houli culture

- Horgen culture

- Liangzhu culture

- Linear Pottery culture

  - Goseck circle, et al.

- Longshan culture

- Majiabang culture

- Majiayao culture

- Peiligang culture

- Pengtoushan culture

- Pfyn culture

- Precucuteni culture

- Qujialing culture

- Shijiahe culture

- Trypillian culture

- Vinča culture

- Windmill Hill culture

  - Stonehenge

- Xinglongwa culture

  - Beifudi site

- Xinle culture

- Yangshao culture

  - Banpo and Xishuipo settlements.

- Zhaobaogou culture

Later Neolithic
Periodization: 6500 to 4500 BC; Europe: 3500 to 3000 BC; Elsewhere: varies greatly, depending on region.

Eneolithic

Main article: Eneolithic

Periodization: Middle East: 4500 to 3300 BC; Europe: 3000 to 1700 BC; Elsewhere: varies greatly, depending on region. In the Americas, the Eneolithic ended as late as the 1800s for some people.

- Beaker culture

- Cucuteni-Trypillian culture

- Funnelbeaker culture

- Gaudo Culture

- Lengyel culture

- Varna culture

30.10   See also

30.11   Footnotes

[1] "Neolithic: definition of Neolithic in Oxford dictionary (British & World English)".

[2] Figure 3.3 from *First Farmers: The Origins of Agricultural Societies* by Peter Bellwood, 2004

[3] Some archaeologists have long advocated replacing "Neolithic" with a more descriptive term, such as "Early Village Communities", but this has not gained wide acceptance.

[4] The potter's wheel was a later refinement that revolutionized the pottery industry.

[5] Habu, Junko (2004). *Ancient Jomon of Japan*. p. 3. ISBN 978-0-521-77670-7.

[6] Xiaohong Wu. "Early Pottery at 20,000 Years Ago in Xianrendong Cave, China". Sciencemag.org. Retrieved 15 January 2015.

[7] "World Museum of Man: Neolithic / Chalcolithic Period". World Museum of Man. Retrieved 21 August 2013.

[8] Lyras; et al. (2008). "The origin of Homo floresiensis and its relation to evolutionary processes under isolation". *Anthropological Science*.

[9] "New Archaeological Discoveries and Researches in 2004 — The Fourth Archaeology Forum of CASS". Institute of Archaeology — Chinese Academy of Social Sciences. Retrieved 2007-09-18.

[10] "The World's First Temple", Archaeology magazine, Nov/Dec 2008 p 23.

[11] "Early Domesticated Fig in the Jordan Valley". *Science*. 2 June 2006. doi:10.1126/science.1125910.

[12] https://www.brown.edu/Departments/Joukowsky_ Institute/courses/architecturebodyperformance/326.html

[13] Yet another sensational discovery by polish archaeologists in Syria. eduskrypt.pl. 21 June 2006

[14] "Jericho", Encyclopædia Britannica

[15] Haïdar Boustani, M., The Neolithic of Lebanon in the context of the Near East: State of knowledge (in French), Annales d'Histoire et d'Archaeologie, Uinversite Saint-Joseph, Beyrouth, Vol. 12–13, 2001–2002. (PDF) . Retrieved on 2011-12-03.

[16] Stordeur, Danielle., Abbès Frédéric., Du PPNA au PPNB : mise en lumière d'une phase de transition à Jerf el Ahmar (Syrie), Bulletin de la Société préhistorique française, Volume 99, Issue 3, pp. 563–595, 2002

[17] PPND – the Platform for Neolithic Radiocarbon Dates – Summary. exoriente. Retrieved on 2011-12-03.

[18] Linseele, V.; et al. (July 2010). "Sites with Holocene dung deposits in the Eastern Desert of Egypt: Visited by herders?" (PDF). *Journal of Arid Environments* **74** (7): 818–828. doi:10.1016/j.jaridenv.2009.04.014.

[19] Hays, Jeffrey (March 2011). "EARLY DOMESTICATED ANIMALS". *Facts and Details*. Retrieved 5 September 2013.

[20] Blench, Roger; MacDonald, Kevin C (1999). *The Origins and Development of African Livestock*. Routledge. ISBN 1-84142-018-2.

[21] Barker, Graeme (25 March 2009). *The Agricultural Revolution in Prehistory: Why Did Foragers Become Farmers?*. Oxford University Press. pp. 292–293. ISBN 978-0-19-955995-4. Retrieved 3 December 2011.

[22] Alexandra Y. Aĭkhenval'd; Robert Malcolm Ward Dixon (2006). *Areal Diffussion and Genetic Inheritance: Problems in Comparative Linguistics*. Oxford University Press, USA. p. 35. ISBN 978-0-19-928308-8.

[23] Fekri A. Hassan (2002). *Droughts, food and culture: ecological change and food security in Africa's later prehistory*. Springer. pp. 164–. ISBN 978-0-306-46755-4. Retrieved 3 December 2011.

[24] Shillington, Kevin (2005). *Encyclopedia of African history: A-G*. CRC Press. pp. 521–. ISBN 978-1-57958-245-6. Retrieved 3 December 2011.

[25] Dawn Fuller (April 16, 2012). "UC research reveals one of the earliest farming sites in Europe". Phys.org. Retrieved April 18, 2012.

[26] "One of Earliest Farming Sites in Europe Discovered". ScienceDaily. April 16, 2012. Retrieved April 18, 2012.

[27] Female figurine, c. 6000 BC, Nea Nikomidia, Macedonia, Veroia, (Archaeological Museum), Greece. Macedonianheritage.gr. Retrieved on 2011-12-03.

[28] Winn, Shan (1981). *Pre-writing in Southeastern Europe: The Sign System of the Vinča Culture ca. 4000 BC*. Calgary: Western Publishers.

[29] Daniel Cilia, "Malta Before Common Era", in *The Megalithic Temples of Malta*. Retrieved 28 January 2007.

[30] S. Piccolo, *op.cit.*, pp. 33-34.

[31] Hirst, K. Kris. 2005. "Mehrgarh". *Guide to Archaeology*

[32] Xiaoyan Yang. "Early millet use in northern China". Pnas.org. Retrieved 15 January 2015.

[33] The Archaeology News Network. 2012. "Neolithic farm field found in South Korea".

[34] The Korea Times. 2012. "East Asia's oldest remains of agricultural field found in Korea".

[35] Gordon R. Willey and Philip Phillips; Philip Phillips (1957). *Method and Theory in American Archaeology*. University of Chicago Press. ISBN 0-226-89888-1.

[36] Kohler TA, M Glaude, JP Bocquet-Appel and Brian M Kemp (2008). "The Neolithic Demographic Transition in the North American Southwest". *American Antiquity 73(4)*: **73** (4): 645–669.

[37] Leonard D. Katz Rigby; S. Stephen Henry Rigby (2000). *Evolutionary Origins of Morality: Cross-disciplinary Perspectives*. United kingdom: Imprint Academic. p. 158. ISBN 0-7190-5612-8.

[38] Langer, Jonas; Killen, Melanie (1998). *Piaget, evolution, and development*. Psychology Press. pp. 258–. ISBN 978-0-8058-2210-6. Retrieved 3 December 2011.

[39] "The Oldest Civilization in the Americas Revealed" (PDF). *CharlesMann*. Science. Retrieved 9 October 2015.

[40] "First Andes Civilization Explored". BBC NEWS. 22 December 2004. Retrieved 9 October 2015.

[41] Hommon, Robert J. (2013). *The ancient Hawaiian state : origins of a political society* (First ed.). New York: Oxford University Press. ISBN 978-0199916122.

[42] "Stone Age," Microsoft Encarta Online Encyclopedia 2007 © 1997–2007 Microsoft Corporation. All Rights Reserved. Contributed by Kathy Schick, B.A., M.A., Ph.D. and Nicholas Toth, B.A., M.A., Ph.D. Archived 2009-11-01.

[43] Russell Dale Guthrie (2005). *The nature of Paleolithic art*. University of Chicago Press. pp. 420–. ISBN 978-0-226-31126-5. Retrieved 3 December 2011.

[44] "Farming Pioneered in Ancient New Guinea". *New Scientist*. New Scientist. Retrieved 9 October 2015.

[45] Bahn, Paul (1996) "The atlas of world archeology" Copyright 2000 The brown Reference Group plc

[46] "Prehistoric Cultures". Museum of Ancient and Modern Art. 2010. Retrieved 5 September 2013.

[47] Hirst, K. Kris. "Çatalhöyük: Urban Life in Neolithic Anatolia". *About.com Archaeology*. About.com. Retrieved 5 September 2013.

[48] Idyllic Theory of Goddess Creates Storm. Holysmoke.org. Retrieved on 2011-12-03.

[49] Krause (1998) under External links, places.

[50] Gimbutas (1991) page 143.

[51] Kuijt, Ian (30 June 2000). *Life in Neolithic farming communities: social organization, identity, and differentiation*. Springer. pp. 317–. ISBN 978-0-306-46122-4. Retrieved 3 December 2011.

[52] Shane, Orrin C. III, and Mine Küçuk. "The World's First City." Archaeology 51.2 (1998): 43–47.

[53] Barber, E. J. W. (1991). *Prehistoric Textiles:The Development of Cloth in the Neolithic and Bronze Ages with Special Reference to the Aegean*. Princeton University Press. ISBN 069100224X.

[54] Alan W. Ertl (15 August 2008). *Toward an Understanding of Europe: A Political Economic Précis of Continental Integration*. Universal-Publishers. p. 308. ISBN 978-1-59942-983-0. Retrieved 28 March 2011.

[55] Harris, Susanna (2009). "Smooth and Cool, or Warm and Soft: Investigatingthe Properties of Cloth in Prehistory". *North European Symposium for Archaeological Textiles X*. Academia.edu. Retrieved 5 September 2013.

[56] "Aspects of Life During the Neolithic Period" (PDF). Teachers' Curriculum Institute. Retrieved 5 September 2013.

[57] Gibbs, Kevin T. (2006). "Pierced clay disks and Late Neolithic textile production". *Proceedings of the 5th International Congress on the Archaeology of the Ancient Near East*. Academia.org. Retrieved 5 September 2013.

[58] Green, Jean M (1993). "Unraveling the Enigma of the Bi: The Spindle Whorl as the Model of the Ritual Disk". *Asian Perspectives* (University of Hawai'i Press) **32** (1): 105–24.

[59] Cook, M (2007). "The clay loom weight, in: Early Neolithic ritual activity, Bronze Age occupation and medieval activity at Pitlethie Road, Leuchars, Fife". *Tayside And Fife Archaeological Journal* **13**: 1–23.

[60] E. J. Peltenburg; Alexander Wasse; Council for British Research in the Levant (2004). *Garfinkel, Yosef., "Néolithique" and "Énéolithique" Byblos in Southern Levantine Context* in Neolithic revolution: new perspectives on southwest Asia in light of recent discoveries on Cyprus*. Oxbow Books. ISBN 978-1-84217-132-5. Retrieved 18 January 2012.

[61] Davis K. Thanjan (12 January 2011). *Pebbles*. Bookstand Publishing. pp. 31–. ISBN 978-1-58909-817-6. Retrieved 4 July 2011.

[62] Developed Neolithic period, 5500 BC. Eliznik.org.uk. Retrieved on 2011-12-03.

[63] "Manunggul Burial Jar". *Virtual Collection of Asian Masterpieces*. Retrieved 5 September 2013.

[64] "Tabon Cave Complex". National Museum of the Philippines. 2011. Retrieved 5 September 2013.

30.12 Bibliography

- Bellwood, Peter (2004). *First Farmers: The Origins of Agricultural Societies*. Wiley-Blackwell. ISBN 0-631-20566-7.

- Pedersen, Hilthart (2008). *Die Jüngere Steinzeit Auf Bornholm*. GRIN Verlag. ISBN 978-3-638-94559-2.

- Piccolo, Salvatore (2013). *Ancient Stones: The Prehistoric Dolmens of Sicily*. Abingdon/GB: Brazen Head Publishing. ISBN 978-0-956-51062-4.

30.13   External links

- Romeo, Nick (Feb. 2015). Embracing Stone Age Couple Found in Greek Cave. "Rare double burials discovered at one of the largest Neolithic burial sites in Europe." *National Geographic Society*

- McNamara, John (2005). "Neolithic Period". World Museum of Man. Retrieved 2008-04-14.

- M.T.C. Affonso and E. Pernicka 2000 'Pre-Pottery Neolithic Clay Figurines from Nevali Çori, Turkey' Internet Archaeology

- Rincon, Paul (11 May 2006). "Brutal lives of Stone Age Britons". BBC News. Retrieved 2008-04-14.

- Current Directions in West African Prehistory – McIntosh & McIntosh (1983)

- "Neolithic". *Encyclopædia Britannica* (11th ed.). 1911.

Chapter 31

Neolithic Revolution

This article is about the introduction of agriculture during the Stone Age. For later historical breakthroughs in agriculture, see agricultural revolution (disambiguation).

The **Neolithic Revolution** or **Neolithic Demographic Transition**, sometimes called the **Agricultural Revolution**, was the wide-scale transition of many human cultures from a lifestyle of hunting and gathering to one of agriculture and settlement, allowing the ability to support an increasingly large population.[1] Archaeological data indicates that the domestication of various types of plants and animals evolved in separate locations worldwide, starting in the geological epoch of the Holocene[2] around 12,500 years ago.[3] It was the world's first historically verifiable revolution in agriculture.

The Neolithic Revolution involved far more than the adoption of a limited set of food-producing techniques. During the next millennia it would transform the small and mobile groups of hunter-gatherers that had hitherto dominated human pre-history into sedentary (here meaning non-nomadic) societies based in built-up villages and towns. These societies radically modified their natural environment by means of specialized food-crop cultivation (e.g., irrigation and deforestation) which allowed extensive surplus food production. These developments provided the basis for densely populated settlements, specialization and division of labour, trading economies, the development of non-portable art and architecture, centralized administrations and political structures, hierarchical ideologies, depersonalized systems of knowledge (e.g., writing), and property ownership. Personal, land and private property ownership led to hierarchical society, class struggle and armies. The first full-blown manifestation of the entire Neolithic complex is seen in the Middle Eastern Sumerian cities (c. 5,500 BP), whose emergence also heralded the beginning of the Bronze Age.

The relationship of the above-mentioned Neolithic characteristics to the onset of agriculture, their sequence of emergence, and empirical relation to each other at various Neolithic sites remains the subject of academic debate, and varies from place to place, rather than being the outcome of universal laws of social evolution.[4][5]

31.1 Agricultural transition

Map of the world showing approximate centers of origin of agriculture and its spread in prehistory: the Fertile Crescent (11,000 BP), the Yangtze and Yellow River basins (9,000 BP) and the New Guinea Highlands (9,000–6,000 BP), Central Mexico (5,000–4,000 BP), Northern South America (5,000–4,000 BP), sub-Saharan Africa (5,000–4,000 BP, exact location unknown), eastern North America (4,000–3,000 BP).[6]

Knap of Howar farmstead on a site occupied from 5,500 to 5,100 BP

The term *Neolithic Revolution* was coined in 1923 by V. Gordon Childe to describe the first in a series of agricultural revolutions in Middle Eastern history. The period is described as a "revolution" to denote its importance, and the great significance and degree of change affecting the communities in which new agricultural practices were gradually adopted and refined.

The beginning of this process in different regions has been dated from 10,000 to 8,000 BC in the Fertile Crescent[3][7] and perhaps 8000 BC in the Kuk Early Agricultural Site of Melanesia[8][9] to 2500 BC in Subsaharan Africa, with some considering the developments of 9000–7000 BC in the Fertile Crescent to be the most important. This transition everywhere seems associated with a change from a largely nomadic hunter-gatherer way of life to a more settled, agrarian-based one, with the inception of the domestication of various plant and animal species—depending on the species locally available, and probably also influenced by local culture. Recent archaeological research suggests that in some regions such as the Southeast Asian peninsula, the transition from hunter-gatherer to agriculturalist was not linear, but region-specific.[10]

There are several competing (but not mutually exclusive) theories as to the factors that drove populations to take up agriculture. The most prominent of these are:

- The **Oasis Theory,** originally proposed by Raphael Pumpelly in 1908, popularized by V. Gordon Childe in 1928 and summarised in Childe's book *Man Makes Himself*.[11] This theory maintains that as the climate got drier due to the Atlantic depressions shifting northward, communities contracted to oases where they were forced into close association with animals, which were then domesticated together with planting of seeds. However, today this theory has little support amongst archaeologists because subsequent climate data suggests that the region was getting wetter rather than drier.[12]

- The **Hilly Flanks** hypothesis, proposed by Robert Braidwood in 1948, suggests that agriculture began in the hilly flanks of the Taurus and Zagros mountains, where the climate was not drier as Childe had believed, and fertile land supported a variety of plants and animals amenable to domestication.[13]

- The **Feasting** model by Brian Hayden[14] suggests that agriculture was driven by ostentatious displays of power, such as giving feasts, to exert dominance. This required assembling large quantities of food, which drove agricultural technology.

- The **Demographic theories** proposed by Carl Sauer[15] and adapted by Lewis Binford[16] and Kent Flannery posit an increasingly sedentary population that expanded up to the carrying capacity of the local environment and required more food than could be gathered. Various social and economic factors helped drive the need for food.

- The **evolutionary/intentionality theory**, developed by David Rindos[17] and others, views agriculture as an evolutionary adaptation of plants and humans. Starting with domestication by protection of wild plants, it led to specialization of location and then full-fledged domestication.

- Peter Richerson, Robert Boyd, and Robert Bettinger[18] make a case for the development of agriculture coinciding with an increasingly stable climate at the beginning of the Holocene. Ronald Wright's book and Massey Lecture Series *A Short History of Progress*[19] popularized this hypothesis.

- The postulated Younger Dryas impact event, claimed to be in part responsible for megafauna extinction and ending the last glacial period, could have provided circumstances that required the evolution of agricultural societies for humanity to survive.[20] The agrarian revolution itself is a reflection of typical overpopulation by certain species following initial events during extinction eras; this overpopulation itself ultimately propagates the extinction event.

- Leonid Grinin argues that whatever plants were cultivated, the independent invention of agriculture always took place in special natural environments (e.g., South-East Asia). It is supposed that the cultivation of cereals started somewhere in the Near East: in the hills of Palestine or Egypt. So Grinin dates the beginning of the agricultural revolution within the interval 12,000 to 9,000 BP, though in some cases the first cultivated plants or domesticated animals' bones are even of a more ancient age of 14–15 thousand years ago.[21]

- Andrew Moore suggested that the Neolithic Revolution originated over long periods of development in the Levant, possibly beginning during the Epipaleolithic. In *"A Reassessment of the Neolithic Revolution"*, Frank Hole further expanded the relationship between plant and animal domestication. He suggested the events could have occurred independently over different periods of time, in as yet unexplored locations. He noted that no transition site had been found documenting the shift from what he termed immediate and delayed return social systems. He noted that the full range of domesticated animals (goats, sheep, cattle and pigs) were not found until the sixth millennium at Tell Ramad. Hole concluded that *"close attention should be paid in future investigations to the western margins of the*

Euphrates basin, perhaps as far south as the Arabian Peninsula, especially where wadis carrying Pleistocene rainfall runoff flowed."[22]

31.2   Domestication of plants

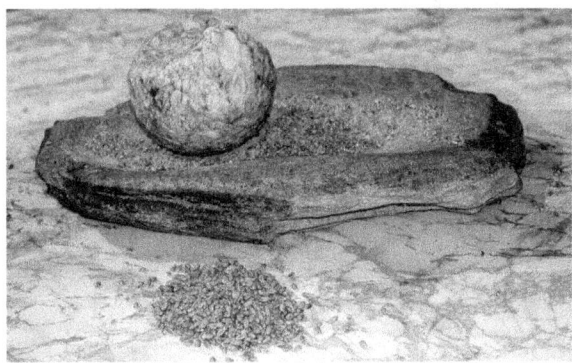

Neolithic grindstone for processing grain

Once agriculture started gaining momentum, human activity resulted in the selective breeding of cereal grasses (beginning with emmer, einkorn and barley), and not simply of those that would favour greater caloric returns through larger seeds. Plants that possessed traits such as small seeds or bitter taste would have been seen as undesirable. Plants that rapidly shed their seeds on maturity tended not to be gathered at harvest, therefore not stored and not seeded the following season; years of harvesting selected for strains that retained their edible seeds longer.

Several plant species, the "pioneer crops" or Neolithic founder crops were named by Daniel Zohary, who highlighted importance of the three cereals, and suggesting domestication of flax, pea, chickpea, bitter vetch and lentil came a little later. Based on analysis of the genes of domesticated plants, he preferred theories of a single, or at most a very small number of domestication events for each taxa that spread in an arc from the Levantine corridor around the fertile crescent and later into Europe.[23][24] Gordon Hillman and Stuart Davies carried out experiments with wild wheat varieties to show that the process of domestication would have happened over a relatively short period of between twenty and two hundred years.[25] Some of these pioneering attempts failed at first and crops were abandoned, sometimes to be taken up again and successfully domesticated thousands of years later: rye, tried and abandoned in Neolithic Anatolia, made its way to Europe as weed seeds and was successfully domesticated in Europe, thousands of years after the earliest agriculture.[26] Wild lentils present a different challenge that needed to be overcome: most of the wild seeds do not germinate in the first year; the first evidence of lentil domestication, breaking dormancy

in their first year, was found in the early Neolithic at Jerf el Ahmar (in modern Syria), and quickly spread south to the Netiv HaGdud site in the Jordan Valley.[26] This process of domestication allowed the founder crops to adapt and eventually become larger, more easily harvested, more dependable in storage and more useful to the human population

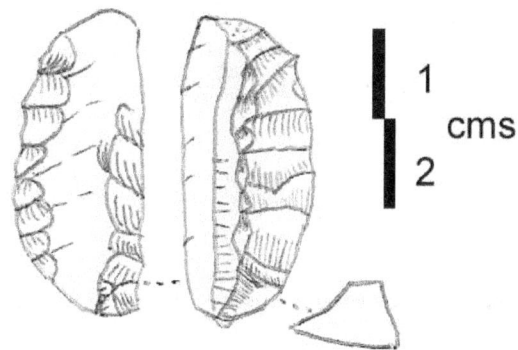

An "Orange slice" sickle blade element with inverse, discontinuous retouch on each side, not denticulated. Found in large quantities at Qaraoun II and often with Heavy Neolithic tools in the flint workshops of the Beqaa Valley in Lebanon. Suggested by James Mellaart to be older than the Pottery Neolithic of Byblos (around 8,400 cal. BP).

Selectively propagated figs, wild barley and wild oats were cultivated at the early Neolithic site of Gilgal I, where in 2006[27] archaeologists found caches of seeds of each in quantities too large to be accounted for even by intensive gathering, at strata datable c. 11,000 years ago. Some of the plants tried and then abandoned during the Neolithic period in the Ancient Near East, at sites like Gilgal, were later successfully domesticated in other parts of the world.

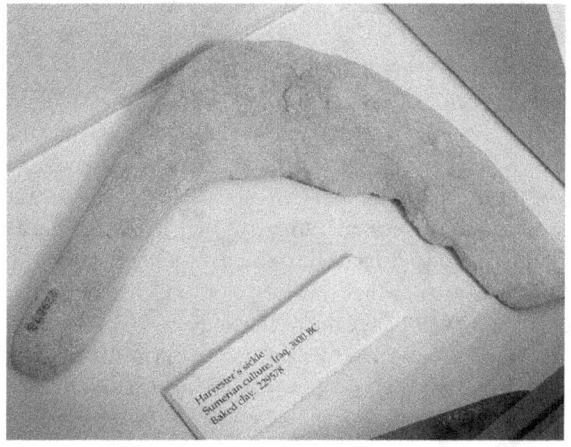

A Sumerian harvester's sickle dated to 5,000 BP

Once early farmers perfected their agricultural techniques

like irrigation, their crops would yield surpluses that needed storage. Most hunter gatherers could not easily store food for long due to their migratory lifestyle, whereas those with a sedentary dwelling could store their surplus grain. Eventually granaries were developed that allowed villages to store their seeds longer. So with more food, the population expanded and communities developed specialized workers and more advanced tools.

The process was not as linear as was once thought, but a more complicated effort, which was undertaken by different human populations in different regions in many different ways.

31.2.1   Agriculture in the Fertile Crescent

Early agriculture is believed to have originated and become widespread in Southwest Asia around 10,000–9,000 BP, though earlier individual sites have been identified. The Fertile Crescent region of Southwest Asia is the centre of domestication for three cereals (einkorn wheat, emmer wheat and barley) four legumes (lentil, pea, bitter vetch and chickpea) and flax.[28] The Mediterranean climate consists of a long dry season with a short period of rain, which may have favored small plants with large seeds, like wheat and barley. The Fertile Crescent also had a large area of varied geographical settings and altitudes and this variety may have made agriculture more profitable for former hunter-gatherers in this region in comparison with other areas with a similar climate .

Finds of large quantities of seeds and a grinding stone at the paleolithic site of Ohalo II in the vicinity of the Sea of Galilee, dated to around 19,400 BP has shown some of the earliest evidence for advanced planning of plant food consumption and suggests that humans at Ohalo II processed the grain before consumption.[29][30] Tell Aswad is oldest site of agriculture with domesticated emmer wheat dated by Willem van Zeist and his assistant Johanna Bakker-Heeres to 8800 BC.[31][32] Soon after came hulled, two-row barley found domesticated earliest at Jericho in the Jordan valley and Iraq ed-Dubb in Jordan.[33] Other sites in the Levantine corridor that show the first evidence of agriculture include Wadi Faynan 16 and Netiv Hagdud.[3] Jacques Cauvin noted that the settlers of Aswad did not domesticate on site, but *"arrived, perhaps from the neighbouring Anti-Lebanon, already equipped with the seed for planting".*[34] The Heavy Neolithic Qaraoun culture has been identified at around fifty sites in Lebanon around the source springs of the River Jordan, however the dating of the culture has never been reliably determined.[35][36]

31.2.2   Agriculture in China

Northern China appears to have been the domestication center for foxtail millet (*Setaria italica*) and broomcorn millet (*Panicum miliaceum*) with evidence of domestication of these species approximately 8,000 years ago.[37] These species were subsequently widely cultivated in the Yellow River basin (7,500 years ago).[37] Rice was domesticated in southern China later on.[37] Soybean was domesticated in northern China 4500 years ago.[38] Orange and peach also originated in China. They were cultivated around 2500 BC.[39][40]

31.2.3   Agriculture in Africa

Nile River Valley, Egypt

On the African continent, three areas have been identified as independently developing agriculture: the Ethiopian highlands, the Sahel and West Africa.[41] By contrast, Agriculture in the Nile River Valley developed the original Neolithic Revolution in the Fertile Crescent. Many grinding stones are found with the early Egyptian Sebilian and Mechian cultures and evidence has been found of a neolithic domesticated crop-based economy dating around 7,000 BP.[42][43] Unlike the Middle East, this evidence appears as a "false dawn" to agriculture, as the sites were later abandoned, and permanent farming then was delayed until 6,500 BP with the Tasian and Badarian cultures and the arrival of crops and animals from the Near East.

Bananas and plantains, which were first domesticated in Southeast Asia, most likely Papua New Guinea, were re-domesticated in Africa possibly as early as 5,000 years ago. Asian yams and taro were also cultivated in Africa.[41]

The most famous crop domesticated in the Ethiopian highlands is coffee. In addition, khat, ensete, noog, teff and finger millet were also domesticated in the Ethiopian highlands. Crops domesticated in the Sahel region include sorghum and pearl millet. The kola nut was first domesticated in West Africa. Other crops domesticated in West Africa include African rice, yams and the oil palm.[41]

Agriculture spread to Central and Southern Africa in the Bantu expansion during the 1st millennium BC to 1st millennium AD.

31.2.4 Agriculture in the Americas

Further information: New World Crops, Ancient Pueblo Peoples, Oasisamerica and Proto-Uto-Aztecan

Maize (corn), beans and squash were among the earliest crops domesticated in Mesoamerica, with maize beginning about 7500 BC, squash, as early as 8000 to 6000 BC and beans by no later than 4000 BC. Potatoes and manioc were domesticated in South America. In what is now the eastern United States, Native Americans domesticated sunflower, sumpweed and goosefoot around 2500 BC. At Guilá Naquitz cave in the Mexican highlands, fragments of maize pollen, bottle gourd and pepo squash were recovered and variously dated between 8000 to 7000 BC. In this area of the world people relied on hunting and gathering for several millennia to come. Sedentary village life based on farming did not develop until the second millennium BC, referred to as the formative period.[3]

31.2.5 Agriculture in Papua New Guinea

Evidence of drainage ditches at Kuk Swamp on the borders of the Western and Southern Highlands of Papua New Guinea shows evidence of the cultivation of taro and a variety of other crops, dating back to 11,000 BP. Two potentially significant economic species, taro (*Colocasia esculenta*) and yam (*Dioscorea* sp.), have been identified dating at least to 10,200 calibrated years before present (cal BP). Further evidence of bananas and sugarcane dates to 6,950 to 6,440 BP. This was at the altitudinal limits of these crops, and it has been suggested that cultivation in more favourable ranges in the lowlands may have been even earlier. CSIRO has found evidence that taro was introduced into the Solomons for human use, from 28,000 years ago, making taro cultivation the earliest crop in the world.[44][45] It seems to have resulted in the spread of the Trans–New Guinea languages from New Guinea east into the Solomon Islands and west into Timor and adjacent areas of Indonesia. This seems to confirm the theories of Carl Sauer who, in "Agricultural Origins and Dispersals", suggested as early as 1952 that this region was a centre of early agriculture.

31.3 Domestication of animals

Further information: Domestication

When hunter-gathering began to be replaced by sedentary food production it became more profitable to keep animals close at hand. Therefore, it became necessary to bring animals permanently to their settlements, although in many cases there was a distinction between relatively sedentary farmers and nomadic herders. The animals' size, temperament, diet, mating patterns, and life span were factors in the desire and success in domesticating animals. Animals that provided milk, such as cows and goats, offered a source of protein that was renewable and therefore quite valuable. The animal's ability as a worker (for example ploughing or towing), as well as a food source, also had to be taken into account. Besides being a direct source of food, certain animals could provide leather, wool, hides, and fertilizer. Some of the earliest domesticated animals included dogs (East Asia, about 15,000 years ago),[46] sheep, goats, cows, and pigs.

31.3.1 Domestication of animals in the Middle East

Dromedary camel caravan in Algeria

The Middle East served as the source for many animals that could be domesticated, such as sheep, goats and pigs. This area was also the first region to domesticate the dromedary camel. Henri Fleisch discovered and termed the Shepherd Neolithic flint industry from the Bekaa Valley in Lebanon and suggested that it could have been used by the earliest nomadic shepherds. He dated this industry to the Epipaleolithic or Pre-Pottery Neolithic as it is evidently not Paleolithic, Mesolithic or even Pottery Neolithic.[36][47] The presence of these animals gave the region a large advantage in cultural and economic development. As the climate in the Middle East changed and became drier, many of the farmers were forced to leave, taking their domesticated animals with them. It was this massive emigration from the Middle East that would later help distribute these animals to the rest of Afrocurasia. This emigration was mainly on an east-west axis of similar climates, as crops usually have a narrow optimal climatic range outside of which they cannot grow for reasons of light or rain changes. For instance, wheat does not normally grow in tropical climates, just like tropical crops such as bananas do not grow in colder climates. Some authors, like Jared Diamond, have postulated that this East-West axis is the main reason why plant and animal domestication spread so quickly from the Fertile Crescent to the rest of Eurasia and North Africa, while it did not reach through the North-South axis of Africa to reach the Mediterranean climates of South Africa, where temperate crops were successfully imported by ships in the last 500 years.[48] Similarly, the African Zebu of central Africa and the domesticated bovines of the fertile-crescent — separated by the dry sahara desert — were not introduced into each other's region.

31.4 Consequence

31.4.1 Social change

It has long been taken for granted that the introduction of agriculture had been an unequivocal progress. This is now questioned in view of findings by archaeologists and paleopathologists showing that nutritional standards of Neolithic populations were generally inferior to that of hunter-gatherers, and that their life expectancy may well have been shorter too, in part due to diseases and harder work - hunter-gatherers must have covered their food needs with about 20 hours' work a week, while agriculture required much more and was at least as uncertain. The hunter-gatherers' diet was more varied and balanced than what agriculture later allowed. Average height went down from 5'10" (178 cm) for men and 5'6" (168 cm) for women to 5'5" (165 cm) and 5'1" (155 cm), respectively, and it took until the twentieth century for average human height to come back to the pre-Neolithic Revolution levels.[49] Agriculturalists had more anaemias and vitamin deficiencies, more spinal deformations and more dental pathologies.[50]

However, the decrease in individual nutrition was accompanied by an increase in population.

The traditional view is that agricultural food production supported a denser population, which in turn supported larger sedentary communities, the accumulation of goods and tools, and specialization in diverse forms of new labor.

The development of larger societies led to the development of different means of decision making and to governmental organization. Food surpluses made possible the development of a social elite who were not otherwise engaged in agriculture, industry or commerce, but dominated their communities by other means and monopolized decision-making.[51] Jared Diamond (in The World Until Yesterday) identifies the availability of milk and/or cereal grains as permitting mothers to raise both an older (e.g. 3 or 4 year old) child and a younger child concurrently, whereas this was not possible previously. The result is that a population can significantly more-rapidly increase its size than would otherwise be the case, resources permitting.

However, recent analyses point out that agriculture also brought about deep social divisions and in particular encouraged inequality between the sexes.[52]

31.4.2 Subsequent revolutions

Domesticated cow being milked in Ancient Egypt.

Andrew Sherratt has argued that following upon the Neolithic Revolution was a second phase of discovery that he refers to as the secondary products revolution. Animals, it appears, were first domesticated purely as a source of meat.[53] The Secondary Products Revolution occurred when it was recognised that animals also provided a number of other useful products. These included:

- hides and skins (from undomesticated animals)

- manure for soil conditioning (from all domesticated animals)

- wool (from sheep, llamas, alpacas, and Angora goats)

- milk (from goats, cattle, yaks, sheep, horses and camels)

- traction (from oxen, onagers, donkeys, horses, camels and dogs)

- guarding and herding assistance (dogs)

Sherratt argues that this phase in agricultural development enabled humans to make use of the energy possibilities of their animals in new ways, and permitted permanent intensive subsistence farming and crop production, and the opening up of heavier soils for farming. It also made possible nomadic pastoralism in semi arid areas, along the margins of deserts, and eventually led to the domestication of both the dromedary and Bactrian camel. Overgrazing of these areas, particularly by herds of goats, greatly extended the areal extent of deserts. Living in one spot would have more easily permitted the accrual of personal possessions and an attachment to certain areas of land. From such a position, it is argued, prehistoric people were able to stockpile food to survive lean times and trade unwanted surpluses with others. Once trade and a secure food supply were established, populations could grow, and society would have diversified into food producers and artisans, who could afford to develop their trade by virtue of the free time they enjoyed because of a surplus of food. The artisans, in turn, were able to develop technology such as metal weapons. Such relative complexity would have required some form of social organisation to work efficiently, so it is likely that populations that had such organisation, perhaps such as that provided by religion, were better prepared and more successful. In addition, the denser populations could form and support legions of professional soldiers. Also, during this time property ownership became increasingly important to all people. Ultimately, Childe argued that this growing social complexity, all rooted in the original decision to settle, led to a second Urban Revolution in which the first cities were built.

31.4.3 Disease

Throughout the development of sedentary societies, disease spread more rapidly than it had during the time in which hunter-gatherer societies existed. Inadequate sanitary practices and the domestication of animals may explain the rise in deaths and sickness following the Neolithic Revolution, as diseases jumped from the animal to the human population. Some examples of diseases spread from animals to humans are influenza, smallpox, and measles.[54] In concordance with a process of natural selection, the humans who first domesticated the big mammals quickly built up immunities to the diseases as within each generation the individuals with better immunities had better chances of survival. In their approximately 10,000 years of shared proximity with animals, such as cows, Eurasians and Africans became more resistant to those diseases compared with the indigenous populations encountered outside Eurasia and Africa.[55] For instance, the population of most Caribbean and several Pacific Islands have been completely wiped out by diseases. According to the Population history of Amer-

Llama overlooking the ruins of the Inca city of Machu Picchu.

ican indigenous peoples, 90% of the population of certain regions of North and South America were wiped out, perhaps by contact with European trappers, before recorded contact with European explorers or colonists. Some cultures like the Inca Empire did have one big mammal domesticated, the Llama, but the Inca did not drink its milk or live in a closed space with their herds, hence limiting the risk of contagion. According to bioarchaeological research, the effects of agriculture on physical and dental health in Southeast Asian rice farming societies from 4000 to 1500 B.P. was not detrimental to the same extent as in other world regions.[56]

The causal link between the type or lack of agricultural development, disease and colonisation is not supported by colonization in other parts of the world. Disease increased after the establishment of British Colonial rule in Africa and India despite the areas having diseases for which Europeans lacked natural immunity. In India, agriculture developed during the Neolithic period with a wide range of animals domesticated. During colonial rule an estimated 23 million people died from cholera between 1865 and 1949, and millions more died from plague, malaria, influenza and tuberculosis. In Africa, European colonization was accompanied by great epidemics, including malaria and sleeping sickness and despite parts of colonized Africa having little or no agriculture, Europeans were more susceptible to disease than the Africans. The increase of disease has been attributed to increased mobility of people, increased population density, urbanisation, environmental deterioration and irrigation schemes that helped to spread malaria rather than the development of agriculture.[57]

31.4.4 Technology

In his book *Guns, Germs, and Steel*, Jared Diamond argues that Europeans and East Asians benefited from an advantageous geographical location that afforded them a head start in the Neolithic Revolution. Both shared the temperate climate ideal for the first agricultural settings, both were near a number of easily domesticable plant and animal species, and both were safer from attacks of other people than civilizations in the middle part of the Eurasian continent. Being among the first to adopt agriculture and sedentary lifestyles, and neighboring other early agricultural societies with whom they could compete and trade, both Europeans and East Asians were also among the first to benefit from technologies such as firearms and steel swords. In addition, they developed resistances to infectious disease, such as smallpox, due to their close relationship with domesticated animals. Groups of people who had not lived in proximity with other large mammals, such as the Australian Aborigines and American indigenous peoples, were more vulnerable to infection and largely wiped out by diseases.

During and after the Age of Discovery, European explorers, such as the Spanish conquistadors, encountered other groups of people who had never or only recently adopted agriculture.

31.5   Archaeogenetics

The dispersal of Neolithic culture from the Middle East has recently been associated with the distribution of human genetic markers. In Europe, the spread of the Neolithic culture has been associated with distribution of the E1b1b lineages and Haplogroup J that are thought to have arrived in Europe from North Africa and the Near East respectively.[58][59] In Africa, the spread of farming, and notably the Bantu expansion, is associated with the dispersal of Y-chromosome haplogroup E1b1a from West Africa.[58]

31.6   Notes and references

[1] Jean-Pierre Bocquet-Appel (July 29, 2011). "When the World's Population Took Off: The Springboard of the Neolithic Demographic Transition". *Science*

333 (6042): 560–561. Bibcode:2011Sci...333..560B. doi:10.1126/science.1208880. PMID 21798934. Retrieved June 10, 2012.

[2] "International Stratigraphic Chart". International Commission on Stratigraphy. Retrieved 2012 12-06.

[3] Graeme Barker (25 March 2009). *The Agricultural Revolution in Prehistory: Why did Foragers become Farmers?*. Oxford University Press. ISBN 978-0-19-955995-4. Retrieved 15 August 2012.

[4] "The Slow Birth of Agriculture", Heather Pringle*

[5] "Wizard Chemi Shanidar", EMuseum, Minnesota State University

[6] Diamond, J.; Bellwood, P. (2003). "Farmers and Their Languages: The First Expansions". *Science* **300** (5619): 597–603. Bibcode:2003Sci...300..597D. doi:10.1126/science.1078208. PMID 12714734.

[7] Thissen, L. 2002. Appendix I, The CANeW 14C databases, Anatolia 10,000-5000 cal. BC. In The Neolithic of Central Anatolia. Internal developments and external relations during the 9th–6th millennia cal BC, Proc. Int. CANeW Round Table, Istanbul 23–24 November 2001, edited by F. Gérard and L. Thissen. Istanbul: Ege Yayınları.

[8] Denham, Tim P.; Haberle, S. G.; et al. (2003). "Origins of Agriculture at Kuk Swamp in the Highlands of New Guinea". *Science* **301** (5630): 189–193. doi:10.1126/science.1085255. PMID 12817084.

[9] The Kuk Early Agricultural Site

[10] Kealhofer, Lisa (2003). "Looking into the gap: land use and the tropical forests of southern Thailand". *Asian Perspectives* **42** (1): 72–95. doi:10.1353/asi.2003.0022.

[11] Gordon Childe (1936). *Man Makes Himself*. Oxford university press.

[12] Scarre, Chris (2005). "The World Transformed: From Foragers and Farmers to States and Empires" in *The Human Past: World Prehistory and the Development of Human Societies* (Ed: Chris Scarre). London: Thames and Hudson. Page 188. ISBN 0-500-28531-4

[13] Charles E. Redman (1978). *Rise of Civilization: From Early Hunters to Urban Society in the Ancient Near East*. San Francisco: Freeman.

[14] Hayden, Brian (1992). "Models of Domestication". In Anne Birgitte Gebauer and T. Douglas Price. *Transitions to Agriculture in Prehistory*. Madison: Prehistory Press. pp. 11–18.

[15] Sauer, Carl O. (1952). *Agricultural origins and dispersals*. Cambridge, MA: MIT Press.

[16] Binford, Lewis R. (1968). "Post-Pleistocene Adaptations". In Sally R. Binford and Lewis R. Binford. *New Perspectives in Archaeology*. Chicago: Aldine Publishing Company. pp. 313–342.

[17] Rindos, David (December 1987). *The Origins of Agriculture: An Evolutionary Perspective*. Academic Press. ISBN 978-0-12-589281-0.

[18] Richerson, Peter J.; Boyd, Robert; et al. (2001). "Was Agriculture Impossible during the Pleistocene but Mandatory during the Holocene?". *American Antiquity* **66** (3): 387–411. doi:10.2307/2694241. JSTOR 2694241.

[19] Wright, Ronald (2004). *A Short History of Progress*. Anansi. ISBN 0-88784-706-4.

[20] Anderson, David G; Albert C. Goodyear; James Kennett; Allen West (2011). "Multiple lines of evidence for possible Human population decline/settlement reorganization during the early Younger Dryas". *Quaternary International* **242** (2): 570–583. doi:10.1016/j.quaint.2011.04.020.

[21] Grinin L.E. Production Revolutions and Periodization of History: A Comparative and Theoretic-mathematical Approach. / Social Evolution & History. Volume 6, Number 2 / September 2007

[22] Hole, Frank., A Reassessment of the Neolithic Revolution, Paléorient, Volume 10, Issue 10-2, pp. 49-60, 1984.

[23] Zohary, D., The mode of domestication of the founder crops of Southwest Asian agriculture. pp. 142-158 in D. R. Harris (ed.) The Origins and Spread of Agriculture and Pastoralism in Eurasia. UCL Press Ltd, London, 1996

[24] Zohary, D., Monophyletic vs. polyphyletic origin of the crops on which agriculture was founded in the Near East. Genetic Resources and Crop Evolution 46 (2) pp. 133-142

[25] Hillman, G. C. and M. S. Davies., Domestication rate in wild wheats and barley under primitive cultivation: preliminary results and archaeological implications of field measurements of selection coefficient, pp. 124-132 in P. Anderson-Gerfaud (ed.) Préhistoire de l'agriculture: nouvelles approches expérimentales et ethnographiques. Monographie du CRA 6, Éditions Centre Nationale Recherches Scientifiques: Paris, 1992

[26] Weiss, Ehud; Kislev, Mordechai E.; Hartmann, Anat (2006). "Autonomous Cultivation Before Domestication". *Science* **312** (5780): 1608–1610. doi:10.1126/science.1127235. PMID 16778044.

[27] "Tamed 11,400 Years Ago, Figs Were Likely First Domesticated Crop".

[28] Brown, T. A.; Jones, M. K.; Powell, W.; Allaby, R. G. (2009). "The complex origins of domesticated crops in the Fertile Crescent". *Trends in Ecology & Evolution* **24** (2): 103. doi:10.1016/j.tree.2008.09.008.

[29] Mithen, Steven (2006). *After the ice : a global human history, 20.000 - 5.000 BC* (1. paperback ed.). Cambridge, Mass.: Harvard Univ. Press. p. 517. ISBN 0-674-01570-3.

[30] Compiled largely with reference to: Weiss, E., Mordechai, E., Simchoni, O., Nadel, D., & Tschauner, H. (2008). Plant-food preparation area on an Upper Paleolithic brush hut floor at Ohalo II, Israel. Journal of Archaeological Science, 35 (8), 2400-2414.

[31] Ozkan, H; Brandolini, A; Schäfer-Pregl, R; Salamini, F (October 2002). "AFLP analysis of a collection of tetraploid wheats indicates the origin of emmer and hard wheat domestication in southeast Turkey". *Molecular Biology and Evolution* **19** (10): 1797–801. doi:10.1093/oxfordjournals.molbev.a004002. PMID 12270906.

[32] van Zeist, W. Bakker-Heeres, J.A.H., Archaeobotanical Studies in the Levant 1. Neolithic Sites in the Damascus Basin: Aswad, Ghoraifé, Ramad., Palaeohistoria, 24, 165-256, 1982.

[33] Hopf, Maria., "Jericho plant remains" in Kathleen M. Kenyon and T. A. Holland (eds.) Excavations at Jericho 5, pp. 576-621, British School of Archaeology at Jerusalem, London, 1983.

[34] Jacques Cauvin (27 July 2000). *The Birth of the Gods and the Origins of Agriculture, p. 53*. Cambridge University Press. ISBN 978-0-521-65135-6. Retrieved 15 August 2012.

[35] E. J. Peltenburg; Alexander Wasse; Council for British Research in the Levant (2004). *Maya Haïdar Boustani, Flint workshops of the Southern Beqa' valley (Lebanon): preliminary results from Qar'oun* in Neolithic revolution: new perspectives on southwest Asia in light of recent discoveries on Cyprus*. Oxbow Books. ISBN 978-1-84217-132-5. Retrieved 18 January 2012.

[36] L. Copeland; P. Wescombe (1966). *Inventory of Stone-Age Sites in Lebanon: North, South and East-Central Lebanon, p. 89*. Impr. Catholique. Retrieved 3 March 2011.

[37] Fuller, D. Q. (2007). "Contrasting Patterns in Crop Domestication and Domestication Rates: Recent Archaeobotanical Insights from the Old World". *Annals of Botany* **100** (5): 903–924. doi:10.1093/aob/mcm048. PMC 2759199. PMID 17495986.

[38] Siddiqi, Mohammad Rafiq. *Tylenchida: Parasites of Plants and Insects*. New York: CABI Pub. 389. p. (2001).

[39] Thacker, Christopher (1985). *The history of gardens*. Berkeley: University of California Press. p. 57. ISBN 978-0-520-05629-9.

[40] Webber, Herbert John (1967–1989). Chapter I. History and Development of the Citrus Industry in *ORIGIN OF CITRUS*, Vol. 1. University of California

[41] Diamond, Jared (1999). *Guns, Germs, and Steel*. New York: Norton Press. ISBN 0-393-31755-2.

[42] The Cambridge History of Africa

[43] Smith, Philip E.L., Stone Age Man on the Nile, Scientific American Vol. 235 No. 2, August 1976: "With the benefit of hindsight we can now see that many Late Paleolithic peoples in the Old World were poised on the brink of plant cultivation and animal husbandry as an alternative to the hunter-gatherer's way of life".

[44] Denham, Tim et al. (received July 2005) "Early and mid Holocene tool-use and processing of taro (*Colocasia esculenta*), yam (*Dioscorea* sp.) and other plants at Kuk Swamp in the highlands of Papua New Guinea" (Journal of Archaeological Science, Volume 33, Issue 5, May 2006)

[45] Hoy, Thomas & Matthew Springs (1992), " Direct evidence for human use of plants 28,000 years ago: starch residues on stone artefacts from the northern Solomon Islands" (Antiquity Volume: 66 Number: 253 Page: 898–912)

[46] McGourty, Christine (2002-11-22). "Origin of dogs traced". BBC News. Retrieved 2006-11-29.

[47] Fleisch, Henri., Notes de Préhistoire Libanaise : 1) Ard es Saoude. 2) La Bekaa Nord. 3) Un polissoir en plein air. BSPF, vol. 63.

[48] *Guns, Germs, and Steel: The Fates of Human Societies*. Jared Diamond (1997).

[49] Hermanussen, Michael; Poustka, Fritz (July–September 2003). "Stature of early Europeans". *Hormones (Athens)* **2** (3): 175–8. doi:10.1159/000079404. PMID 17003019.

[50] Shermer, Michael (2001) The Borderlands of Science, Oxford University Press p.250

[51] Eagly, Alice H. & Wood, Wendy (June 1999). "The Origins of Sex Differences in Human Behavior: Evolved Dispositions Versus Social Roles". *American Psychologist* **54** (6): 408–423. doi:10.1037/0003-066x.54.6.408.

[52] Jared Diamond: "The Worst Mistake in the History of the Human Race," Discover Magazine, May 1987, pp. 64-66.

[53] Sherratt 1981

[54] Furuse, Y.; Suzuki, A.; Oshitani, H. (2010). "Origin of measles virus: Divergence from rinderpest virus between the 11th and 12th centuries". *Virology Journal* **7**: 52. doi:10.1186/1743-422X-7-52. PMC 2838858. PMID 20202190.

[55] *Guns, Germs, and Steel: The Fates of Human Societies* - Jared Diamond, 1997

[56] Halcrow, S.; E., Harris, N. J., Tayles, N., Ikehara-Quebral, R., & Pietrusewsky, M. (2013). "From the mouths of babes: Dental caries in infants and children and the intensification of agriculture in mainland Southeast Asia". *American Journal of Physical Anthropology* **150** (3): 409–420. doi:10.1002/ajpa.22215. PMID 23359102.

[57] Marshall, P. J. Ed. (1996), *Cambridge illustrated History: British Empire*, Cambridge University Press, ISBN 0-521-00254-0, p. 142

[58] Semino, O; et al. (2004). "Origin, Diffusion, and Differentiation of Y-Chromosome Haplogroups E and J: Inferences on the Neolithization of Europe and Later Migratory Events in the Mediterranean Area". *American Journal of Human Genetics* **74** (5): 1023–34. doi:10.1086/386295. PMC 1181965. PMID 15069642.

[59] Lancaster, Andrew (2009). "Y Haplogroups, Archaeological Cultures and Language Families: a Review of the Multidisciplinary Comparisons using the case of E-M35" (PDF). *Journal of Genetic Genealogy* **5** (1).

31.7 Bibliography

- Bailey, Douglass. (2001). *Balkan Prehistory: Exclusions, Incorporation and Identity.* Routledge Publishers. ISBN 0-415-21598-6.

- Bailey, Douglass. (2005). *Prehistoric Figurines: Representation and Corporeality in the Neolithic.* Routledge Publishers. ISBN 0-415-33152-8.

- Balter, Michael (2005). *The Goddess and the Bull: Catalhoyuk, An Archaeological Journey to the Dawn of Civilization.* New York: Free Press. ISBN 0-7432-4360-9.

- Bellwood, Peter. (2004). *First Farmers: The Origins of Agricultural Societies.* Blackwell Publishers. ISBN 0-631-20566-7

- Bocquet-Appel, Jean-Pierre, editor and Ofer Bar-Yosef, editor, *The Neolithic Demographic Transition and its Consequences*, Springer (October 21, 2008), hardcover, 544 pages, ISBN 978-1402085383, trade paperback and Kindle editions are also available.

- Cohen, Mark Nathan (1977)*The Food Crisis in Prehistory: Overpopulation and the Origins of Agriculture.* New Haven and London: Yale University Press. ISBN 0-300-02016-3.

- Jared Diamond, *Guns, germs and steel. A short history of everybody for the last 13'000 years*, 1997.

- Diamond, Jared (2002). "Evolution, Consequences and Future of Plant and Animal Domestication". *Nature*, Vol 418.

- Harlan, Jack R. (1992). *Crops & Man: Views on Agricultural Origins* ASA, CSA, Madison, WI. http://www.hort.purdue.edu/newcrop/history/lecture03/r_3-1.html

- Wright, Gary A. (1971). "Origins of Food Production in Southwestern Asia: A Survey of Ideas" Current Anthropology, Vol. 12, No. 4/5 (Oct.–Dec., 1971), pp. 447–477

- Bartmen, Jeff M. (2008). *Disease.*

- Evidence for food storage and predomestication granaries 11,000 years ago in the Jordan Valley .

- Co-Creators How our ancestors used Artificial Selection during the Neolithic Revolution

31.8 See also

- Çatalhöyük, a Neolithic site in southern Anatolia

- Aşıklı Höyük, in Anatolia

- Natufians, a settled culture preceding agriculture

- Behavioral modernity

- Original affluent society

- Haplogroup G (Y-DNA)

- Haplogroup J2 (Y-DNA)

- Haplogroup J (mtDNA)

- Haplogroup K (mtDNA)

- Neolithic tomb

- Surplus product

- Göbekli Tepe

- Mehrgarh, a Neolithic site in Balochistan

31.9 External links

- The Agricultural Revolution on YouTube: Crash Course World History #1

Chapter 32

Domestication

This article is about the process of artificial selection in animals or plants. For other uses, see Domestication (disambiguation).

Domestication (from the Latin *domesticus*: "of the

Dogs and sheep were among the first animals to be domesticated.

home") is the cultivating or taming[1] of a population of organisms in order to accentuate traits that are desirable to the cultivator or tamer. The desired traits may include a particular physical appearance, behavioral characteristic, individual size, litter size, hair/fur quality or color, growth rate, fecundity, lifespan, ability to use marginal grazing resources, production of certain by-products, and many others.[2] Domesticated organisms may become dependent on humans or human activities, since they sometimes lose

their ability to survive in the wild.[3]

Domestication differs from taming in that it may refer not simply to a change in organisms' behaviors or environmental socialization, but also potentially even in their phenotypical expressions and genotypes. The word *domestication* also is more commonly used to mean a change within whole populations, while *taming* is more commonly used to mean a change within individuals. Furthermore, taming typically applies only to animals and their becoming habituated to human presence, while domestication is a broader term and can include plants, fungi, and other types of organisms.

Plants domesticated primarily for aesthetic enjoyment in and around the home are usually called *house plants* or *ornamentals*, while those domesticated for large-scale food production are generally called *crops*. A distinction can be made between those domesticated plants that have been deliberately altered or selected for special desirable characteristics (see cultigen) and those plants that are used for human benefit, but are essentially no different from the wild populations of the species. Animals domesticated for home companionship are usually called *pets*, while those domesticated for food or work are called *livestock* or *farm animals*.

32.1   Background

Charles Darwin was the first to describe the connection between domestication, selection and evolution.[4] Darwin described how the process of domestication can involve both unconscious and methodical elements. Routine human interactions with animals and plants create selection pressures that cause adaptation to human presence, use or cultivation. Deliberate selective breeding has also been used to create desired changes, often after initial domestication. These two forces, unconscious natural selection and methodical selective breeding, may have both played roles in the processes of domestication throughout history.[2] Both have been described from human perspective as processes of artificial selection.

The domestication of wheat provides an example. Wild wheat falls to the ground to reseed itself when ripe, but domesticated wheat stays on the stem for easier harvesting. There is evidence that this change was possible because of a random mutation that happened in the wild populations at the beginning of wheat's cultivation. Wheat with this mutation was harvested more frequently and became the seed for the next crop. Therefore, without realizing, early farmers selected for this mutation, which may otherwise have died out. The result is domesticated wheat, which relies on farmers for its own reproduction and dissemination.[5]

Selective breeding may best explain how continuing processes of domestication often work. Evidence of the power of selective breeding comes from the Farm-Fox Experiment by the Russian scientist, Dmitri K. Belyaev, in the 1950s. His team bred the domesticated silver fox (*Vulpes vulpes*) and selected the individuals that showed the least fear of humans. Then Belyaev's team selected only those that showed the most positive response to humans. He ended up with a population of grey foxes whose behavior and appearance was significantly changed. They no longer showed any fear of humans and often wagged their tails and licked their human caretakers to show affection. These foxes had floppy ears, smaller skulls, rolled tails and other traits commonly found in dogs.[6] Despite the success of this experiment, it appears that selective breeding cannot always achieve domestication. Attempts to domesticate many kinds of wild animals have been unsuccessful. Although the four species of zebra can interbreed with the horse and the donkey, attempts at domestication have failed.[7] Factors such as temperament, social structure and ability to breed in captivity play a role in determining whether a species can be successfully domesticated.[2] In human history to date, only a few species of large animal have been domesticated. In approximate order of their earliest domestication these are: dog, sheep, goat, pig, ox, yak,[8] reindeer,[9] water buffalo, horse, donkey, llama, alpaca, Bactrian camel and Arabian camel.[10]

32.1.1 Animals

Archaeozoology has identified 3 classes of animal domesticates: (1) commensals, adapted to a human niche (e.g., dogs, cats, guinea pigs); (2) prey animals sought for food (e.g., cows, sheep, pig, goats); and (3) targeted animals for draft and nonfood resources (e.g., horse, camel, donkey).[11]

According to evolutionary biologist Jared Diamond, animal species must meet six criteria in order to be considered for domestication:[12]

1. Flexible diet – Creatures that are willing to consume a

Hereford cattle, domesticated for beef production.

wide variety of food sources and can live off less cumulative food from the food pyramid (such as corn or wheat), particularly food that is not utilized by humans (such as grass and forage) are less expensive to keep in captivity. Carnivores by definition feed primarily or only on flesh, which requires the domesticators to raise additional animals just to feed them, though they may exploit sources of meat not utilized by humans, such as scraps and vermin.

2. Reasonably fast growth rate – Fast maturity rate compared to the human life span allows breeding intervention and makes the animal useful within an acceptable duration of caretaking. Some large animals require many years before they reach a useful size.

3. Ability to be bred in captivity – Creatures that are reluctant to breed when kept in captivity do not produce useful offspring, and instead are limited to capture in their wild state. Creatures such as the panda, antelope and giant forest hog are territorial when breeding and cannot be maintained in crowded enclosures in captivity.

4. Pleasant disposition – Large creatures that are aggressive toward humans are dangerous to keep in captivity. The African buffalo has an unpredictable nature and is highly dangerous to humans; similarly, although the American bison is raised in enclosed ranges in the Western United States, it is much too dangerous to be regarded as truly domesticated. Although similar to the domesticated pig in many ways, Africa's warthog and bushpig are also dangerous in captivity.

5. Temperament which makes it unlikely to panic – A creature with a nervous disposition is difficult to keep in captivity as it may attempt to flee whenever startled. The gazelle is very flighty and it has a powerful leap that allows it to escape an enclosed pen. Some animals, such as the domestic sheep, still have a strong tendency to panic when their flight zone is encroached upon. However, most sheep also show a flocking instinct, whereby they stay close together when pressed.

Livestock with such an instinct may be herded by people and dogs.

6. Modifiable social hierarchy – Social creatures whose herds occupy overlapping ranges and recognize a hierarchy of dominance can be raised to recognize a human as the pack leader:

(a) Tapirs and rhinoceroses are solitary and do not tolerate being penned with each other.

(b) Antelope and deer—except for reindeer—are territorial when breeding and live in herds only for the rest of the year.

(c) Bighorn sheep and peccaries have non-hierarchical herd structures and do not follow any definite leader; instead, males fight continuously with each other for mating opportunities.

(d) Musk ox herds, although they have a defined leader, maintain mutually exclusive territories and two herds will fight if kept together.

Cow domestication in North India for milk production.

However, this list is of limited use because it fails to take into account the profound changes that domestication has on a species. While it is true that some animals retain their wild instincts even if born in captivity, e.g. laying hens,[13] pigs[14] and laboratory mice,[15] some factors must be taken into consideration.

Number (5) may not be a prerequisite for domestication, but rather a natural consequence of a species' having been domesticated. In other words, wild animals are naturally timid and flighty because they are constantly faced by predators; domestic animals do not need such a nervous disposition, as they are protected by their human owners. The same holds true for number (4) – aggressive temperament is an adaptation to the danger from predators. A Cape buffalo can kill even an attacking lion, but most modern large domestic animals were descendants of aggressive ancestors. The wild boar, ancestor of the domestic pig, is certainly renowned for its ferocity; other examples include the aurochs (ancestor of modern cattle), horse, Bactrian camels and yaks, all of which are no less dangerous than their undomesticated wild relatives such as zebras and buffalos. Others have argued that the difference lies in the ease with which breeding can improve the dispositions of wild animals, a view supported by the failure to domesticate the kiang and onager. On the other hand, for thousands of years humans have managed to tame dangerous species like bears and cheetahs whose failed domestications had little to do with their aggressiveness.

Number (6), while it does apply to most domesticated species, also has exceptions, most notably in the domestic cat and ferret, which are both descended from strictly solitary wild ancestors but which tolerate and even seek out social interaction in their domestic forms. Feral domestic cats, for example, naturally form colonies around concentrated food sources, and will even share prey and rear kittens communally, while wildcats remain solitary even in the presence of such food sources.[16] Zoologist Marston Bates devoted a chapter on domestication in his 1960 book *The Forest and the Sea*, in which he talks a great deal about how domestication alters a species. Dispersal mechanisms tend to disappear for the reason stated above, and additionally, because people provide transportation for them, domestic chickens and turkeys have a greatly reduced ability to fly. Similarly, domestic animals cease to have a definite mating season, so the need to be territorial when mating loses its value, and if some of the males in a herd are castrated, the problem is reduced even further. What he says suggests that the process of domestication can itself make a creature domesticable. Besides, the first steps towards agriculture may have involved hunters keeping young animals, who are always more impressionable than the adults, after killing their mothers.

Another strong factor in deciding whether a species will be considered for domestication is quite simply the availability of more suitable (or, better yet, already domesticated) alternatives. For example, a community that had been introduced to domestication by neighboring peoples will generally find it much more practical, economical, and time saving to import already domesticated species than experiment with wild animals, even if they are of the same species. Generally speaking, the species of animals originally domesticated by early humans in the interconnected landmasses of Eurasia and Africa were far superior, both in working capacity and in food production, than the species found in the other continents, namely the Americas and Oceania.

32.1.2 Plants

See also: Cultigen and Horticulture

The earliest human attempts at plant domestication occurred in South-Western Asia. There is early evidence for conscious cultivation and trait selection of plants by pre-Neolithic groups in Syria: grains of rye with domestic traits have been recovered from Epi-Palaeolithic (c. 11,050 BCE) contexts at Abu Hureyra in Syria,[17] but this appears to be a localised phenomenon resulting from cultivation of stands of wild rye, rather than a definitive step towards domestication.[17]

By 10,000 BCE the bottle gourd (*Lagenaria siceraria*) plant, used as a container before the advent of ceramic technology, appears to have been domesticated. The domesticated bottle gourd reached the Americas from Asia by 8000 BCE, most likely due to the migration of peoples from Asia to America.[18]

Cereal crops were first domesticated around 9000 BCE in the Fertile Crescent in the Middle East. The first domesticated crops were generally annuals with large seeds or fruits. These included pulses such as peas and grains such as wheat. The Middle East was especially suited to these species; the dry-summer climate was conducive to the evolution of large-seeded annual plants, and the variety of elevations led to a great variety of species. As domestication took place humans began to move from a hunter-gatherer society to a settled agricultural society. This change would eventually lead, some 4000 to 5000 years later, to the first city states and eventually the rise of civilization itself.

Continued domestication was gradual, a process of trial and error that occurred intermittently. Over time perennials and small trees began to be domesticated including apples and olives. Some plants were not domesticated until recently such as the macadamia nut and the pecan.

In other parts of the world very different species were domesticated. In the Americas squash, maize, beans, and perhaps manioc (also known as cassava) formed the core of the diet. In East Asia millet, rice, and soy were the most important crops. Some areas of the world such as Southern Africa, Australia, California and southern South America never saw local species domesticated.

Domesticated plants often differ from their wild relatives in the way they spread to a more diverse environment and have a wider geographic range;[19] they may also have a different ecological preference; flower and fruit simultaneously; may lack shattering or scattering of seeds, and may have lost their dispersal mechanisms completely; have larger fruits and seeds, and so lower efficiency of dispersal; may have been converted from a perennial to annual; have

lost seed dormancy and photoperiodic controls; lack normal pollinating organs; may have a different breeding system; may lack defensive adaptations such as hairs, spines and thorns, protective coverings and sturdiness; may have better palatability and chemical composition, rendering them more likely to be eaten by animals; may be more susceptible to diseases and pests; may develop seedless parthenocarpic fruits; may have undergone selection for double flowers, which may involve conversion of stamens into petals; may have become sexually sterile and therefore only reproduce vegetatively.

32.2 Degrees

The boundaries between surviving wild populations and domestic clades can be vague. A classification system that can help solve this confusion surrounding animal populations might be set up on a spectrum of increasing domestication:

- **Wild**: These populations experience their full life cycles without deliberate human intervention.

- **Raised in captivity/captured from wild** (in zoos, botanical gardens, or for human gain): These populations are nurtured by humans but (except in zoos) not normally bred under human control. They remain as a group essentially indistinguishable in appearance or behaviour from their wild counterparts. Examples include Asian elephants, animals such as sloth bears and cobras used by showmen in India, and animals such as Asian black bears (farmed for their bile), and zoo animals, kept in captivity as examples of their species. (It should be noted that zoos and botanical gardens sometimes exhibit domesticated or feral animals and plants such as camels, mustangs, and some orchids)

- **Raised commercially** (captive or semidomesticated): These populations are ranched or farmed in large numbers for food, commodities, or the pet trade, commonly breed in captivity, but as a group are not substantially altered in appearance or behavior from their wild cousins. Examples include the ostrich, various deer, alligator, cricket, honeybees, pearl oyster, raptors used in falconry and ball python. (These species are sometimes referred to as *partially domesticated*.)

- **Domesticated**: These populations are bred and raised under human control for many generations and are substantially altered as a group in appearance or behaviour. Examples include sweet potato, garlic, pigs, ferrets, turkeys, canaries, domestic pigeons, budgerigars, goldfish, koi carp, silkworms, dogs, cats, sheep, cattle, chickens, llamas, guinea pigs, laboratory mice, horses, goats and (silver) foxes.[6]

This classification system does not account for several complicating factors: genetically modified organisms, feral populations, and hybridization. Many species that are farmed or ranched are now being genetically modified. This creates a unique category because it alters the organisms as a group but in ways unlike traditional domestication. Feral organisms are members of a population that was once raised under human control, but is now living and multiplying outside of human control. Examples include mustangs. Hybrids can be wild, domesticated, or both: a liger is a hybrid of two wild animals, a mule is a hybrid of two domesticated animals, and a beefalo is a cross between a wild and a domestic animal.

32.3   Tame or domesticated

A herd of Pryor Mustangs

A great difference exists between a tame animal and a domesticated animal. The term "domesticated" refers to an entire species or variety while the term "tame" can refer to just one individual within a species or variety. Humans have tamed many thousands of animals that have never been truly domesticated. These include the elephant, giraffes, and bears and lions. There is debate over whether some species have been domesticated or just tamed. Some state that the elephant has been domesticated, while others argue that the cat has never been domesticated. Dividing lines include whether a specimen born to wild parents would differ in appearance or behavior from one born to domesticated parents. For instance a dog is certainly domesticated because even a wolf (which genetically shares a common ancestor with all dogs) raised from a pup would be very different from a dog, in both appearance and behaviour.[20] Similar problems of definition arise when domesticated cats go feral.

Many other languages use the same word for both concepts.

32.4   Negative aspects

Selection of animals for visible "desirable" traits may make them unfit in other, unseen, ways. The consequences for the captive and domesticated animals were reduction in size, piebald color, shorter faces with smaller and fewer teeth, diminished horns, weak muscle ridges, and less genetic variability. Poor joint definition, late fusion of the limb bone epiphyses with the diaphyses, hair changes, greater fat accumulation, smaller brains, simplified behavior patterns, extended immaturity, and more pathology are a few of the defects of domestic animals. All of these changes have been documented in direct observations of the rat in the 19th century, by archaeological evidence, and confirmed by animal breeders in the 20th century.[21] A 2014 commentary published in *Genetics* proposed that many of these features may arise due to mild neural crest deficits that also cause tameness; hence, selectively breeding tame animals also selects for these negative traits.[22]

One side effect of domestication has been zoonotic diseases. For example, cattle have given humanity various viral poxes, measles, and tuberculosis; pigs and ducks have given influenza; and horses have given the rhinoviruses. Humans share over sixty diseases with dogs . Many parasites also have their origins in domestic animals.[2] The advent of domestication resulted in denser human populations which provided ripe conditions for pathogens to reproduce, mutate, spread, and eventually find a new host in humans.

Other negative aspects of domestication have been explored. For example, Paul Shepherd writes "Man substitutes controlled breeding for natural selection; animals are selected for special traits like milk production of passivity, at the expense of overall fitness and naturewide relationships...Though domestication broadens the diversity of forms – that is, increases visible polymorphism – it undermines the crisp demarcations that separate wild species and cripples our recognition of the species as a group. Knowing only domestic animals dulls our understanding of the way in which unity and discontinuity occur as patterns in nature, and substitutes an attention to individuals and breeds. The wide variety of size, color, shape, and form of domestic horses, for example, blurs the distinction among different species of *Equus* that once were constant and meaningful."[23]

Going further, some anarcho-primitivist authors describe domestication as the process by which previously nomadic human populations shifted towards a sedentary or settled existence through agriculture and animal husbandry. They claim that this kind of domestication demands a totalitarian relationship with both the land and the plants and animals being domesticated. They say that whereas, in a state of wildness, all life shares and competes for resources, domestication destroys this balance. Domesticated landscape (e.g. pastoral lands/agricultural fields and, to a lesser degree, horticulture and gardening) ends the open sharing of resources; where "this was everyone's," it is now "mine." Anarcho-primitivists state that this notion of ownership laid the foundation for social hierarchy as property and power emerged. It also involved the destruction, enslavement, or assimilation of other groups of early people who did not make such a transition.[24]

To primitivists, domestication enslaves both the domesticated species as well as the domesticators. Advances in the fields of psychology, anthropology, and sociology allows humans to quantify and objectify themselves, until they too become commodities.[25]

32.5 Dates and places

Early domestication: cow being milked in ancient Egypt.

Since the process of domestication inherently takes many generations over a long period of time, and the spread of breed and husbandry techniques is also slow, it is not meaningful to give a single "date of domestication". However, it is believed that the first attempt at domestication of both animals and plants were made in the Old World by peoples of the Mesolithic Period. The tribes that took part in hunting and gathering wild edible plants, started to make attempts to domesticate dogs, goats, and possibly sheep, which was as early as 9000 BC. However, it was not until the Neolithic Period that primitive agriculture appeared as a form of social activity, and domestication was well under way. The great majority of domesticated animals and plants that still serve humans were selected and developed during the Neolithic Period, a few other examples appeared later. The rabbit for example, was not domesticated until the Middle Ages, while the sugar beet came under cultivation as a sugar-yielding agricultural plant in the 19th century. As recently as the 20th century, mint became an object of agricultural production, and animal breeding programs to produce high-quality fur were started in the same time period.[26]

The methods available to estimate domestication dates introduce further uncertainty, especially when domestication has occurred in the distant past. So the dates given here should be treated with caution; in some cases evidence is scanty and future discoveries may alter the dating significantly.

Dates and places of domestication are mainly estimated by archaeological methods, more precisely archaeozoology. These methods consist of excavating or studying the results of excavation in human prehistorical occupation sites. Animal remains are dated with archaeological methods, the species they belong to is determined, the age at death is also estimated, and if possible the form they had, that is to say a possible domestic form. Various other clues are taken advantage of, such as slaughter or cutting marks. The aim is to determine if they are game or raised animal, and more globally the nature of their relationship with humans. For example, the skeleton of a cat found buried close to humans is a clue that it may have been a pet cat. The age structure of animal remains can also be a clue of husbandry, in which animals were killed at the optimal age.

New technologies and especially mitochondrial DNA, which are simple DNA found in the mitochondria that determine its function in the cell provide an alternative angle of investigation, and make it possible to reestimate the dates of domestication based on research into the genealogical tree of modern domestic animals.

It is admitted for several species that domestication occurred in several places distinctly. For example, research on mitochondrial DNA of the modern cattle Bos taurus supports the archaeological assertions of separate domestication events in Asia and Africa. This research also shows that Bos taurus and Bos indicus haplotypes are all descendants of the extinct wild ox Bos primigenius.[27][28] However, this does not rule out later crossing inside a species; therefore it appears useless to look for a separate wild ancestor for each domestic breed.

The dog was the first domesticated animal dating 18,000-32,000 years ago, which supports the hypothesis that dog domestication preceded the emergence of agriculture and occurred in the context of European hunter-gatherer cultures.[29] This preceded the domestication of other species by several millennia. In the Neolithic a number of important species such as goats, sheep, pigs and cattle were domesticated, as part of the spread of farming which characterises this period. The goat, sheep and pig in particular were domesticated independently in the Levant and Asia.

There is early evidence of beekeeping, in the form of rock paintings, dating to 13,000 BC.

Recent archaeological evidence from Cyprus indicates domestication of a type of cat by perhaps 9500 BC.[30][31][32]

The earliest secure evidence of horse domestication, bit wear on horse molars at Dereivka in Ukraine, dates to around 4000 BC. The *unequivocal* date of domestication and use as a means of transport is at the Sintashta chariot burials in the southern Urals, c. 2000 BC. Local equivalents and smaller species were domesticated from the 26th century BC.

The availability of both domesticated vegetable and animal species increased suddenly following the voyages of Christopher Columbus and the contact between the Eastern and Western Hemispheres. This is part of what is referred to as the Columbian Exchange.

32.5.1 Approximate dates and locations of original domestication

Main article: List of domesticated animals

Additional domestications:

32.5.2 Modern instances

Researchers at the Max Planck institute in Germany are attempting to find a genetic basis for the processes of taming and domestication. They have obtained two strains of grey rats which were bred by Dmitry Konstantinovich Belyaev at the Institute of Cytology and Genetics in Novosibirsk, Russia, research which was later continued by Irina Plyusnina. One strain had been selected for aggressiveness while the other had been selected for tameness, mimicking the process by which neolithic farmers are thought to have first domesticated animals. A similar experiment studying silver foxes has been ongoing at the same institute since 1959.[49] Richard Wrangham of Harvard suggests that similar genes could be involved in human self-domestication.[49]

32.5.3 Former instances

Some species are said to have been domesticated, but are not any more, either because they have totally disappeared, or since their domestic form no longer exists. Examples include the jaguarundi,[50] the kakapo, the ring-tailed cat, caracal and *Bos aegyptiacus*.

32.5.4 Hybrid domestic animals

- Alpaca: DNA evidence shows that alpacas are a llama/vicuña hybrid

- Beefalo
- Bengal cat
- Cama (animal)
- Chausie
- Coydog
- Dzo
- Domesticated hedgehog: A cross between the Algerian hedgehog and the four-toed hedgehog.
- Sheep-goat hybrid
- Hinny
- Huarizo
- Iron Age pig
- Mule
- Savannah (cat)
- Wolfdog
- Wolphin
- Yakalo
- Zeedonk
- Zorse
- Zony
- Zubron

32.6 Genetic pollution

Main article: Genetic pollution

Animals of domestic origin and feral ones sometimes can produce fertile hybrids with native, wild animals which leads to genetic pollution in the naturally evolved wild gene pools, many times threatening rare species with extinction. Cases include the mallard duck, wildcat, wild boar, the rock dove or pigeon, the red junglefowl (*Gallus gallus*) (ancestor of all chickens), carp, and more recently salmon. Another example is the dingo, itself an early feral dog, which hybridizes with dogs of European origin. On the other hand, genetic pollution seems not to be noticed for rabbits. There is much debate over the degree to which feral hybridization compromises the purity of a wild species. In the case of the mallard, for example, some claim there are no populations which are completely free of any domestic ancestor.

32.7 Notes and references

[1] "Domesticate". *Oxford Dictionaries*. Oxford University Press. 2014.

[2] Diamond, Jared (1999). *Guns, Germs, and Steel*. New York: Norton Press. ISBN 0-393-31755-2.

[3] "Domestication." Dictionary.com. Based on the Random House Dictionary (Random House, Inc. 2013). http://dictionary.reference.com/browse/domesticate

[4] Darwin, Charles (1868). *The Variation of Animals and Plants under Domestication*. London: John Murray. OCLC 156100686.

[5] Zohary, D. & Hopf, M. (2000). *Domestication of Plants in the Old World* Oxford: Oxford Univ. Press.

[6] Lyudmila N. Trut (1999). "Early Canid Domestication: The Farm-Fox Experiment" (PDF). *American Scientist* (Sigma Xi, The Scientific Research Society) **87** (March–April): 160–169. Bibcode:1999AmSci..87.....T. doi:10.1511/1999.20.813. Retrieved June 25, 2011.

[7] Clutton-Brock, J. (1981) *Domesticated Animals from Early Times*. Austin: Univ. Texas Press.

[8] Ning L., Jinge G. and Aireti. 1997. "Yak in Xinjiang", in Miller D.G., Craig S.R. and Rana G.M. (eds), *Proceedings of a workshop on conservation and management of yak genetic diversity held at ICIMOD, Kathmandu, Nepal*, October 29–31, 1996. ICIMOD (International Centre for Integrated Mountain Development), Kathmandu, Nepal. pp. 115–122.

[9] Cronin, M.A.; Renecker, L; Pierson, B.J. and Patton, J.C.; "Genetic variation in domestic reindeer and wild caribou in Alaska"; *Animal Genetics*, volume 26, Issue 6 (December 1995), pp. 427–434

[10] Diamond, Jared; *Guns, Germs, and Steel: The Fates of Human Societies*; p. 147. ISBN 0-393-31755-2

[11] doi:10.1086/659964
This citation will be automatically completed in the next few minutes. You can jump the queue or expand by hand

[12] Diamond, Jared (1998). *Guns, Germs, and Steel*. Vintage. pp. 169–174. ISBN 978-0-09-930278-0.

[13] McBride, G., Parer, I.P. and Foenander, F., (1969). The social organization and behaviour of the feral domestic fowl. Animal Behaviour Monographs, 2:125–181

[14] Stolba, A. and Wood-Gush, D.G.M., (1989). The behaviour of pigs in a semi-natural environment. Animal Production, 48: 419-425

[15] Sherwin, C.M. (2002). "Comfortable Quarters for Mice in Research Institutions". Animal Welfare Institute. Retrieved November 6, 2013.

[16] "The Domestication of the Cat". Messybeast.com. October 5, 2009. Retrieved November 5, 2013.

[17] Hillman G, Hedges R, Moore A, Colledge S, Pettitt P; Hedges; Moore; Colledge; Pettitt (2001). "New evidence of Lateglacial cereal cultivation at Abu Hureyra on the Euphrates". *Holocene* **11** (4): 383–393. doi:10.1191/095968301678302823.

[18] Erickson DL, Smith BD, Clarke AC, Sandweiss DH, Tuross N; Smith; Clarke; Sandweiss; Tuross (December 2005). "An Asian origin for a 10,000-year-old domesticated plant in the Americas". *Proc. Natl. Acad. Sci. U.S.A.* **102** (51): 18315–20. Bibcode:2005PNAS..10218315E. doi:10.1073/pnas.0509279102. PMC 1311910. PMID 16352716.

[19] Zeven, A. C.; de Wit, J. M. (1982). *Dictionary of Cultivated Plants and Their Regions of Diversity, Excluding Most Ornamentals, Forest Trees and Lower Plants*. Wageningen, Netherlands: Centre for Agricultural Publishing and Documentation.

[20] Virányi Z, Gácsi M, Kubinyi E, Topál J, Belényi B, Ujfalussy D, Miklósi Á; Gácsi; Kubinyi; Topál; Belényi; Ujfalussy; Miklósi (2008). "Comprehension of human pointing gestures in young human-reared wolves (Canis lupus) and dogs (Canis familiaris)". *Animal Cognition* **11** (3): 373–387. doi:10.1007/s10071-007-0127-y. PMID 18183437.

[21] Berry, R.J. (1969). "The Genetical Implications of Domestication in Animals". In Ucko, Peter J., Dimbleby, G.W. *The Domestication and Exploitation of Plants and Animals*. Chicago: Aldine. pp. 207–217.

[22] Wilkins, Adam S.; Wrangham, Richard W.; Fitch, W. Tecumseh (July 2014). "The 'Domestication Syndrome' in Mammals: A Unified Explanation Based on Neural Crest Cell Behavior and Genetics". *Genetics* **197** (3): 795. doi:10.1534/genetics.114.165423.

[23] Shepherd, Paul (1973). "Ten Thousand Years of Crisis". *The Tender Carnivore*. Athens, GA: University of Georgia Press. pp. 10–11.

[24] Boyden, Stephen Vickers (1992). "Biohistory: The interplay between human society and the biosphere, past and present". *Man and the Biosphere Series* (Pari: UNESCO) **8** (supplement 173): 665. Bibcode:1992EnST...26..665. doi:10.1021/es00028a604.

[25] Moore, John. "A Primitivist Primer: What is anarcho-primitivism?".

[26] Archived June 26, 2012 at the Wayback Machine

[27] Troy CS, MacHugh DE, Bailey JF; et al. (April 2001). "Genetic evidence for Near-Eastern origins of European cattle". *Nature* **410** (6832): 1088–91. doi:10.1038/35074088. PMID 11323670.

[28] Wendorf F., Schild R.; Schild (1998). "Nabta Playa and its role in ortheastern African prehistory". *J. Anthropol. Archaeol* **17** (2): 97–123. doi:10.1006/jaar.1998.0319.

[29] O. Thalmann, B. Shapiro, P. Cui, V. J. Schuenemann, S. K. Sawyer, D. L. Greenfield, M. B. Germonpré, M. V. Sablin, F. López-Giráldez, X. Domingo-Roura, H. Napierala, H-P. Uerpmann, D. M. Loponte, A. A. Acosta, L. Giemsch, R. W. Schmitz, B. Worthington, J. E. Buikstra, A. Druzhkova, A. S. Graphodatsky, N. D. Ovodov, N. Wahlberg, A. H. Freedman, R. M. Schweizer, K.-P. Koepfli, J. A. Leonard, M. Meyer, J. Krause, S. Pääbo, R. E. Green, R. K. Wayne - Complete Mitochondrial Genomes of Ancient Canids Suggest a European Origin of Domestic Dogs - Science 15 November 2013: Vol. 342 no. 6160 pp. 871-874 DOI: 10.1126/science.1243650 Full Text available here

[30] "Oldest Known Pet Cat? 9500-Year-Old Burial Found on Cyprus". *National Geographic News*. April 8, 2004. Retrieved March 6, 2007.

[31] Muir, Hazel (April 8, 2004). "Ancient remains could be oldest pet cat". *New Scientist*. Retrieved November 23, 2007.

[32] Walton, Marsha (April 9, 2004). "Ancient burial looks like human and pet cat". *CNN*. Retrieved November 23, 2007.

[33] Druzhkova AS, Thalmann O, Trifonov VA, Leonard JA, Vorobieva NV, et al. (2013) Ancient DNA Analysis Affirms the Canid from Altai as a Primitive Dog. PLoS ONE 8(3): e57754. doi:10.1371/journal.pone.0057754

[34] MSNBCE : World's first dog lived 31,700 years ago, ate big

[35] Krebs, Robert E. & Carolyn A. (2003). *Groundbreaking Scientific Experiments, Inventions & Discoveries of the Ancient World*. Westport, CT: Greenwood Press. ISBN 0-313-31342-3.

[36] Simmons, Paula; Carol Ekarius (2001). *Storey's Guide to Raising Sheep*. North Adams, MA: Storey Publishing LLC. ISBN 978-1-58017-262-2.

[37] Giuffra E, Kijas JM, Amarger V, Carlborg O, Jeon JT, Andersson L; Kijas; Amarger; Carlborg; Jeon; Andersson (April 2000). "The origin of the domestic pig: independent domestication and subsequent introgression". *Genetics* **154** (4): 1785–91. PMC 1461048. PMID 10747069.

[38] G. Larson, K. Dobney, U. Albarella, M. Fang, E. Matisso-Smith, J. Robins, S. Lowden, H. Finlayson, T. Brand, E. Willerslev, P. Rowley-Conwy, L. Andersson, A. Cooper; Dobney; Albarella; Fang; Matisoo-Smith; Robins; Lowden; Finlayson; Brand; Willerslev; Rowley-Conwy; Andersson; Cooper (March 2005). "Worldwide Phylogeography of Wild Boar Reveals Multiple Centers of Pig Domestication" (PDF). *Science* **307** (5715): 1618–21. Bibcode:2005Sci...307.1618L. doi:10.1126/science.1106927. PMID 15761152.

[39] Melinda A. Zeder, Goat busters track domestication (Physiologic changes and evolution of goats into a domesticated animal), April 2000, (English) (summarizing research done in Ganj Dareh).

[40] Source : Laboratoire de Préhistoire et Protohistoire de l'Ouest de la France , (French). Archived October 5, 2006 at the Wayback Machine

[41] , domestication of the cat on Cyprus, National Geographic.

[42] West B, Zhou B-X.; Zhou (1989). "Did chickens go north? New evidence for domestication" (PDF). *World's Poultry Science Journal* **45** (3): 205–218. doi:10.1079/WPS19890012. Archived from the original (PDF) on October 14, 2006.

[43] *History of the Guinea Pig* (Cavia porcellus) *in South America, a summary of the current state of knowledge*

[44] Beja-Pereira A, England PR, Ferrand N; et al. (June 2004). "African origins of the domestic donkey". *Science* **304** (5678): 1781. doi:10.1126/science.1096008. PMID 15205528. [*New Scientist* Donkey domestication began in Africa Lay summary] Check |url= scheme (help).

[45] Roger Blench, *The history and spread of donkeys in Africa* PDF (235 KB) (English).

[46] The Domestication of the Horse; see also Domestication of the horse Archived August 11, 2006 at the Wayback Machine

[47] Domestication of Reindeer

[48] Geese: the underestimated species

[49] Nicholas Wade (July 25, 2006). "Nice Rats, Nasty Rats: Maybe It's All in the Genes". *NY times*.

[50] .Sometimes it is because these animals don't breed well in captivity

32.8   Bibliography

- Darwin, Charles. *The Variation of Animals and Plants under Domestication*, 1868.

- Diamond, Jared. *Guns, germs and steel. A short history of everybody for the last 13,000 years*, 1997.

- Hobgood-Oster, Laura. *A Dog's History of the World: Canines and the Domestication of Humans*, 2014.

- Ryden, Hope. *Out of the Wild: The Story of Domesticated Animals*, 1995.

32.9 Further reading

- Halcrow, S. E., Harris, N. J., Tayles, N., Ikehara-Quebral, R. and Pietrusewsky, M. (2013), From the mouths of babes: Dental caries in infants and children and the intensification of agriculture in mainland Southeast Asia. Am. J. Phys. Anthropol., 150: 409–420. doi: 10.1002/ajpa.22215

- Hayden, B. (2003). Were luxury foods the first domesticates? Ethnoarchaeological perspectives from Southeast Asia. World Archaeology, 34(3), 458-469.

- Marciniak, Arkadiusz (2005). *Placing Animals in the Neolithic: Social Zooarchaeology of Prehistoric Farming Communities.* London: UCL Press. ISBN 1844720926.

32.10 See also

- Animal husbandry
- Anthrozoology
- Columbian Exchange
- Domesticated silver fox
- Domestication of the horse
- Domestication theory
- Experimental evolution
- Genetic engineering
- Genetic erosion
- Genetic pollution
- Genomics of domestication
- History of plant breeding
- Horticulture
- Lion taming
- List of domesticated animals
- List of domesticated fungi and microorganisms
- List of domesticated plants
- Marker assisted selection
- Selective breeding
- Self-domestication
- Timeline of agriculture and food technology
- Ethnozoology

32.11 External links

- Crop Wild Relative Inventory and Gap Analysis: reliable information source on where and what to conserve ex-situ, for crop genepools of global importance

- Discussion of animal domestication

- *Guns, Germs and Steel* by Jared Diamond (ISBN 0-393-03891-2)

- News story about an early domesticated cat find

- Belyaev experiment with the domestic fox

- Use of Domestic Animals in Zoo Education

- The Initial Domestication of Cucurbita pepo in the Americas 10,000 Years Ago

- Cattle domestication diagram

- Major topic "domestication": free full-text articles (more than 100 plus reviews) in National Library of Medicine

- *Why don't we ride zebras?* an online children's film about animal domestication

- Isidro A. T. Savillo and Villaluz, Elizabeth A. 2013 this introduces a proposed Domesticity Scale for Wild Birds

Chapter 33

Pottery

A potter at work in Bangalore, India

Unfired "green ware" pottery on a traditional drying rack at Conner Prairie living history museum

Pottery is the ceramic material which makes up potterywares,[1] of which major types include earthenware, stoneware and porcelain. The place where such wares are made is also called a *pottery* (plural "potteries"). Pottery also refers to the art or craft of a potter or the manufacture of pottery.[2][3] A dictionary definition is simply objects of fired clays.[4] The definition of *pottery* used by the American Society for Testing and Materials (ASTM) is "all fired ceramic wares that contain clay when formed, except technical, structural, and refractory products."[5]

Pottery originated before the Neolithic period, with ceramic objects like the Gravettian culture Venus of Dolní Věstonice figurine discovered in the Czech Republic date back to 29,000–25,000 BC,[6] and pottery vessels that were discovered in Jiangxi, China, which date back to 20,000 BC.[7] Early Neolithic pottery have been found in places such as Jomon Japan (10,500 BC),[8] the Russian Far East (14,000 BC),[9] Sub-Saharan Africa and South America.

Pottery is made by forming a clay body into objects of a required shape and heating them to high temperatures in a kiln which removes all the water from the clay, which induces re-

Pottery workshop reconstruction in the Museum of traditional crafts and applied arts, Troyan, Bulgaria

232

actions that lead to permanent changes including increasing their strength and hardening and setting their shape. A clay body can be decorated before or after firing. Prior to some shaping processes, clay must be prepared. Kneading helps to ensure an even moisture content throughout the body. Air trapped within the clay body needs to be removed. This is called de-airing and can be accomplished by a machine called a vacuum pug or manually by wedging. Wedging can also help produce an even moisture content. Once a clay body has been kneaded and de-aired or wedged, it is shaped by a variety of techniques. After shaping it is dried and then fired.

33.1 Production stages

A potter at work in Jaura, Madhya Pradesh, India.

Clay ware takes on varying physical characteristics during the making of pottery.

- Greenware refers to unfired objects. At sufficient moisture content, bodies at this stage are in their most plastic form (they are soft and malleable, and hence can be easily deformed by handling).

- Leather-hard refers to a clay body that has been dried partially. At this stage the clay object has approximately 15% moisture content. Clay bodies at this stage are very firm and only slightly pliable. Trimming and handle attachment often occurs at the leather-hard state.

- Bone-dry refers to clay bodies when they reach a moisture content at or near 0%. It is now ready to be bisque fired.

- Bisque [10][11] refers to the clay after the object is shaped to the desired form and fired in the kiln for the first time, known as "bisque fired" or "biscuit fired". This firing changes the clay body in several ways. Mineral components of the clay body will undergo chemical changes that will change the colour of the clay.

- Glaze fired is the final stage of some pottery making. A glaze may be applied to the bisque form and the object can be decorated in several ways. After this the object is "glazed fired", which causes the glaze material to melt, then adhere to the object. The glaze firing will also harden the body still more as chemical processes can continue to occur in the body.

33.2 Clays bodies and mineral contents

There are several materials that are referred to as clay. The properties of the clays differ, including: Plasticity, the malleability of the body; the extent to which they will absorb water after firing; and shrinkage, the extent of reduction in size of a body as water is removed. Different clay bodies also differ in the way in which they respond when fired in the kiln. A clay body can be decorated before or after firing. Prior to some shaping processes, clay must be prepared. Each of these different clays are composed of different types and amounts of minerals that determine the characteristics of resulting pottery. There can be regional variations in the properties of raw materials used for the production of pottery, and this can lead to wares that are unique in character to a locality. It is common for clays and other materials to be mixed to produce clay bodies suited to specific purposes. A common component of clay bodies is the mineral kaolinite. Other mineral compounds in the clay may act as fluxes which lower the vitrification temperature of bodies. Following is a list of different types of clay used for pottery.[12]

Preparation of Clay for Pottery in India

A man shapes pottery as it turns on a wheel. (Cappadocia, Turkey)

- Kaolin, is sometimes referred to as China clay because it was first used in China. Used for Porcelain.

- Ball clay An extremely plastic, fine grained sedimentary clay, which may contain some organic matter. Small amounts can be added to porcelain to increase plasticity.

- Fire clay A clay having a slightly lower percentage of fluxes than kaolin, but usually quite plastic. It is highly heat resistant form of clay which can be combined with other clays to increase the firing temperature and may be used as an ingredient to make stoneware type bodies.

- Stoneware clay Suitable for creating stoneware. This clay has many of the characteristics between fire clay and ball clay, having finer grain, like ball clay but is more heat resistant like fire clays.

- Common red clay and Shale clay have vegetable and ferric oxide impurities which make them useful for bricks, but are generally unsatisfactory for pottery except under special conditions of a particular deposit.[13]

- Bentonite An extremely plastic clay which can be added in small quantities to short clay to increase the plasticity.

33.3 Methods of shaping

Pottery can be shaped by a range of methods that include:

- Hand-building. This is the earliest forming method. Wares can be constructed by hand from coils of clay, combining flat slabs of clay, or pinching solid balls of clay or some combination of these. Parts of hand-built vessels are often joined together with the aid of

slip, an aqueous suspension of clay body and water. A clay body can be decorated before or after firing. Prior to some shaping processes, clay must be prepared such as tablewares although some studio potters find hand-building more conducive to create one-of-a-kind works of art.

A potter shapes a piece of pottery on an electric-powered potter's wheel

- The potter's wheel. In a process called "throwing" (coming from the Old English word *thrawan* which means to twist or turn,[14]) a ball of clay is placed in the centre of a turntable, called the wheel-head, which the potter rotates with a stick, with foot power or with a variable-speed electric motor.

  During the process of throwing, the wheel rotates while the solid ball of soft clay is pressed, squeezed and pulled gently upwards and outwards

Classic potter's kick wheel in Erfurt, Germany

into a hollow shape. The first step of pressing the rough ball of clay downward and inward into perfect rotational symmetry is called *centring* the clay—a most important skill to master before the next steps: *opening* (making a centred hollow into the solid ball of clay), *flooring* (making the flat or rounded bottom inside the pot), *throwing* or *pulling* (drawing up and shaping the walls to an even thickness), and *trimming* or *turning* (removing excess clay to refine the shape or to create a *foot*).

Considerable skill and experience are required to throw pots of an acceptable standard and, while the ware may have high artistic merit, the reproducibility of the method is poor.[15] Because of its inherent limitations, throwing can only be used to create wares with radial symmetry on a vertical axis. These can then be altered by impressing, bulging, carving, fluting, and incising. In addition to the potter's hands these techniques can use tools, including paddles, anvils & ribs, and those specifically for cutting or piercing such as knives, fluting tools and wires. Thrown pieces can be further modified by the attachment of handles, lids, feet and spouts.

- Granulate pressing: As the name suggests, this is the operation of shaping pottery by pressing clay in a semi-

dry and granulated condition in a mould. The clay is pressed into the mould by a porous die through which water is pumped at high pressure. The granulated clay is prepared by spray-drying to produce a fine and free-flowing material having a moisture content of between about 5 and 6 per cent. Granulate pressing, also known as *dust pressing*, is widely used in the manufacture of ceramic tiles and, increasingly, of plates.

- Injection moulding: This is a shape-forming process adapted for the tableware industry from the method long established for the forming of thermoplastic and some metal components.[16] It has been called *Porcelain Injection Moulding*, or *PIM*.[17] Suited to the mass production of complex-shaped articles, one significant advantage of the technique is that it allows the production of a cup, including the handle, in a single process, and thereby eliminates the handle-fixing operation and produces a stronger bond between cup and handle.[18] The feed to the mould die is a mix of approximately 50 to 60 percent unfired body in powder form, together with 40 to 50 percent organic additives composed of binders, lubricants and plasticisers.[17] The technique is not as widely used as other shaping methods.[19]

- Jiggering and jolleying: These operations are carried out on the potter's wheel and allow the time taken to bring wares to a standardized form to be reduced. *Jiggering* is the operation of bringing a shaped tool into contact with the plastic clay of a piece under construction, the piece itself being set on a rotating plaster mould on the wheel. The jigger tool shapes one face while the mould shapes the other. Jiggering is used only in the production of flat wares, such as plates, but a similar operation, *jolleying*, is used in the production of hollow-wares such as cups. Jiggering and jolleying have been used in the production of pottery since at least the 18th century. In large-scale factory production, jiggering and jolleying are usually automated, which allows the operations to be carried out by semi-skilled labour.

- Roller-head machine: This machine is for shaping wares on a rotating mould, as in jiggering and jolleying, but with a rotary shaping tool replacing the fixed profile. The rotary shaping tool is a shallow cone having the same diameter as the ware being formed and shaped to the desired form of the back of the article being made. Wares may in this way be shaped, using relatively unskilled labour, in one operation at a rate of about twelve pieces per minute, though this varies with the size of the articles being produced. Developed in the UK just after World War II by the company *Service*

Shaping on a potter's kick wheel; Gülşehir, Turkey

Engineers, roller-heads were quickly adopted by manufacturers around the world; they remain the dominant method for producing flatware.[20]

- Pressure casting: Specially developed polymeric materials allow a mould to be subject to application external pressures of up to 4.0 MPa–so much higher than slip casting in plaster moulds where the capillary forces correspond to a pressure of around 0.1 - 0.2 MPa. The high pressure leads to much faster casting rates and, hence, faster production cycles. Furthermore, the application of high pressure air through the polymeric moulds upon demoulding the cast means a new casting cycle can be started immediately in the same mould, unlike plaster moulds which require lengthy drying times. The polymeric materials have much greater durability than plaster and, therefore, it is possible to achieve shaped products with better dimensional tolerances and much longer mould life. Pressure casting was developed in the 1970s for the production of sanitaryware although, more recently, it has been applied to tableware.[21][22][23][24]

- RAM pressing: This is used to shape ware by pressing a bat of prepared clay body into a required shape between two porous moulding plates. After pressing, compressed air is blown through the porous mould plates to release the shaped wares.

- Slipcasting: This ideally suited to the making of wares that cannot be formed by other methods of shaping. A slip, made by mixing clay body with water, is poured into a highly absorbent plaster mould. Water from the slip is absorbed into the mould leaving a layer of clay body covering its internal surfaces and taking its internal shape. Excess slip is poured out of the mould, which is then split open and the moulded object removed. Slipcasting is widely used in the production of sanitary wares and is also used for making smaller articles, such as intricately detailed figurines.

33.4   Decorating and glazing

Contemporary pottery from the State of Hidalgo, Mexico

Pottery may be decorated in many different ways. Some decoration can be done before or after the firing.

33.4.1   Decoration

- Painting has been used since early prehistoric times, and can be very elaborate. The painting is often applied to pottery that has been fired once, and may then be overlaid with a glaze afterwards. Many pigments change colour when fired, and the painter must allow for this.

- Ceramic glaze Perhaps the most common form of decoration, that also serves as protection to the pottery, by being tougher and keeping liquid from penetrating the pottery. Glaze may be clear, especially over painting, or coloured and opaque. There is more detail in the section below.

This is an Italian red earthenware vase covered with a mottled pale blue glaze. It has large blue and gold-coated flowers and a scalloped gold-coated rim.

- Carving Pottery vessels may be decorated by shallow carving of the clay body, typically with a knife or similar instrument used on the wheel. This is common in Chinese porcelain of the classic periods.

- Burnishing the surface of pottery wares may be *burnished* prior to firing by rubbing with a suitable instrument of wood, steel or stone to produce a polished finish that survives firing. It is possible to produce very highly polished wares when fine clays are used or when the polishing is carried out on wares that have been partially dried and contain little water, though wares in this condition are extremely fragile and the risk of breakage is high.

- Additives can be worked into the clay body prior to forming, to produce desired effects in the fired wares. Coarse additives such as sand and grog (fired clay which has been finely ground) are sometimes used to give the final product a required texture. Contrasting coloured clays and grogs are sometimes used to produce patterns in the finished wares. Colourants, usually metal oxides and carbonates, are added singly or in combination to achieve a desired colour. Combustible particles can be mixed with the body or pressed into

the surface to produce texture.

- Lithography also call litho, although the alternative names of transfer print or "*decal*" are also common. These are used to apply designs to articles. The litho comprises three layers: the colour, or image, layer which comprises the decorative design; the cover coat, a clear protective layer, which may incorporate a low-melting glass; and the backing paper on which the design is printed by screen printing or lithography. There are various methods of transferring the design while removing the backing-paper, some of which are suited to machine application.

- Banding is the application by hand or by machine of a band of colour to the edge of a plate or cup. Also known as "lining", this operation is often carried out on a potter's wheel.

- Agateware is named after its resemblance to the quartz mineral agate which has bands or layers of colour that are blended together, agatewares are made by blending clays of differing colours together but not mixing them to the extent that they lose their individual identities. The wares have a distinctive veined or mottled appearance. The term "agateware" is used to describe such wares in the United Kingdom; in Japan the term "*neriage*" is used and in China, where such things have been made since at least the Tang Dynasty, they are called "*marbled*" wares. Great care is required in the selection of clays to be used for making agatewares as the clays used must have matching thermal movement characteristics.

- Engobe: This is a clay slip, that is used to coat the surface of pottery, usually before firing. Its purpose is often decorative though it can also be used to mask undesirable features in the clay to which it is applied. Engobe slip may be applied by painting or by dipping to provide a uniform, smooth, coating. Engobe has been used by potters from pre-historic times until the present day and is sometimes combined with sgraffito decoration, where a layer of engobe is scratched through to reveal the colour of the underlying clay. With care it is possible to apply a second coat of engobe of a different colour to the first and to incise decoration through the second coat to expose the colour of the underlying coat. Engobes used in this way often contain substantial amounts of silica, sometimes approaching the composition of a glaze.

- Gold: Decoration with gold is used on some high quality ware. Different methods exist for its application, including:

An ancient Armenian urn

33.4.2 Glaze

Main article: Ceramic glaze

Glaze is a glassy coating on pottery, the primary purposes

Two panels of earthenware tiles painted with polychrome glazes over a white glaze. Iran. First half of the 19th century.

1. *Best gold* – a suspension of gold powder in essential oils mixed with a flux and a mercury salt extended. This can be applied by a painting technique. From the kiln, the decoration is dull and requires burnishing to reveal the full colour

2. *Acid Gold* – a form of gold decoration developed in the early 1860s at the English factory of Mintons Ltd, Stoke-on-Trent. The glazed surface is etched with diluted hydrofluoric acid prior to application of the gold. The process demands great skill and is used for the decoration only of ware of the highest class.

3. *Bright Gold* – consists of a solution of gold sulphoresinate together with other metal resonates and a flux. The name derives from the appearance of the decoration immediately after removal from the kiln as it requires no burnishing

4. *Mussel Gold* – an old method of gold decoration. It was made by rubbing together gold leaf, sugar and salt, followed by washing to remove solubles

of which are decoration and protection. One important use of glaze is to render porous pottery vessels impermeable to water and other liquids. Glaze may be applied by dusting the unfired composition over the ware or by spraying, dipping, trailing or brushing on a thin slurry composed of the unfired glaze and water. The colour of a glaze after it has been fired may be significantly different from before firing. To prevent glazed wares sticking to kiln furniture during firing, either a small part of the object being fired (for example, the foot) is left unglazed or, alternatively, special refractory "*spurs*" are used as supports. These are removed and discarded after the firing.

Some specialised glazing techniques include:

- Salt-glazing, where common salt is introduced to the kiln during the firing process. The high temperatures cause the salt to volatize, depositing it on the surface of the ware to react with the body to form a sodium aluminosilicate glaze. In the 17th and 18th centuries, salt-glazing was used in the manufacture of domestic pottery. Now, except for use by some studio potters, the process is obsolete. The last large-scale application before its demise in the face of environmental clean air restrictions was in the production of salt-glazed sewer-pipes.[25][26]

- Ash glazing – ash from the combustion of plant matter has been used as the flux component of glazes. The source of the ash was generally the combustion waste from the fuelling of kilns although the potential of ash derived from arable crop wastes has been investigated.[27] Ash glazes are of historical interest in the Far East although there are reports of small-scale use in other locations such as the Catawba Valley Pottery in the United States. They are now limited to small numbers of studio potters who value the unpredictability arising from the variable nature of the raw material.[28]

- Underglaze decoration (in the manner of many blue and white wares). Underglaze may be applied by brush strokes, air brush, or by pouring the underglaze into the mold, covering the inside, creating a swirling effect, then the mold is filled with slip.

- In-glaze decoration

- On-glaze decoration

- Enamel

33.5 Firing

The pottery firing process in Kalabougou, Mali, using a firing mound.

Firing produces irreversible changes in the body. It is only after firing that the article or material is pottery. In lower-fired pottery, the changes include sintering, the fusing together of coarser particles in the body at their points of contact with each other. In the case of porcelain, where

A kiln at a pottery in Bardon Mill, UK.

different materials and higher firing-temperatures are used, the physical, chemical and mineralogical properties of the constituents in the body are greatly altered. In all cases, the object of firing is to permanently harden the wares and the firing regime must be appropriate to the materials used to make them. As a rough guide, earthenwares are normally fired at temperatures in the range of about 1,000°C (1,830 °F) to 1,200 °C (2,190 °F); stonewares at between about 1,100 °C (2,010 °F) to 1,300 °C (2,370 °F); and porcelains at between about 1,200 °C (2,190 °F) to 1,400 °C (2,550 °F).

Firing pottery can be done using a variety of methods, with a kiln being the usual firing method. Both the maximum temperature and the duration of firing influences the final characteristics of the ceramic. Thus, the maximum temperature within a kiln is often held constant for a period of time to *soak* the wares to produce the maturity required in the body of the wares.

The atmosphere within a kiln during firing can affect the appearance of the finished wares. An oxidising atmosphere, produced by allowing air to enter the kiln, can cause the oxidation of clays and glazes. A reducing atmosphere, produced by limiting the flow of air into the kiln, can strip oxygen from the surface of clays and glazes. This can affect the appearance of the wares being fired and, for example, some glazes containing iron fire brown in an oxidising atmosphere, but green in a reducing atmosphere. The atmosphere within a kiln can be adjusted to produce complex effects in glaze.

Kilns may be heated by burning wood, coal and gas or by electricity. When used as fuels, coal and wood can introduce smoke, soot and ash into the kiln which can affect the appearance of unprotected wares. For this reason, wares fired in wood- or coal-fired kilns are often placed in the kiln in saggars, lidded ceramic boxes, to protect them. Modern kilns powered by gas or electricity are cleaner and more easily controlled than older wood- or coal-fired kilns and often allow shorter firing times to be used. In a Western adaptation of traditional Japanese Raku ware firing, wares are removed from the kiln while hot and smothered in ashes, paper or woodchips which produces a distinctive carbonised appearance. This technique is also used in Malaysia in creating traditional *labu sayung*.[29][30]

In Mali, a firing mound is used rather than a brick or stone kiln. Unfired pots are first brought to the place where a mound will be built, customarily by the women and girls of the village. The mound's foundation is made by placing sticks on the ground, then:

> [...]pots are positioned on and amid the branches and then grass is piled high to complete the mound. Although the mound contains the pots of many women, who are related through their husbands' extended families, each women is responsible for her own or her immediate family's pots within the mound.
>
> When a mound is completed and the ground around has been swept clean of residual combustible material, a senior potter lights the fire. A handful of grass is lit and the woman runs around the circumference of the mound touching the burning torch to the dried grass. Some mounds are still being constructed as others are already burning.[31]

33.6 History

Main article: Ceramic art § History

A great part of the history of pottery is prehistoric, part of past pre-literate cultures. Therefore, much of this history can only be found among the artifacts of archaeology. Because pottery is so durable, pottery and sherds from pottery survive from millennia at archaeological sites.

Before pottery becomes part of a culture, several conditions must generally be met.

- First, there must be usable clay available. Archaeological sites where the earliest pottery was found were near deposits of readily available clay that could be properly shaped and fired. China has large deposits of a variety of clays, which gave them an advantage in early development of fine pottery. Many countries have large deposits of a variety of clays.

- Second, it must be possible to heat the pottery to temperatures that will achieve the transformation from raw clay to ceramic. Methods to reliably create fires hot

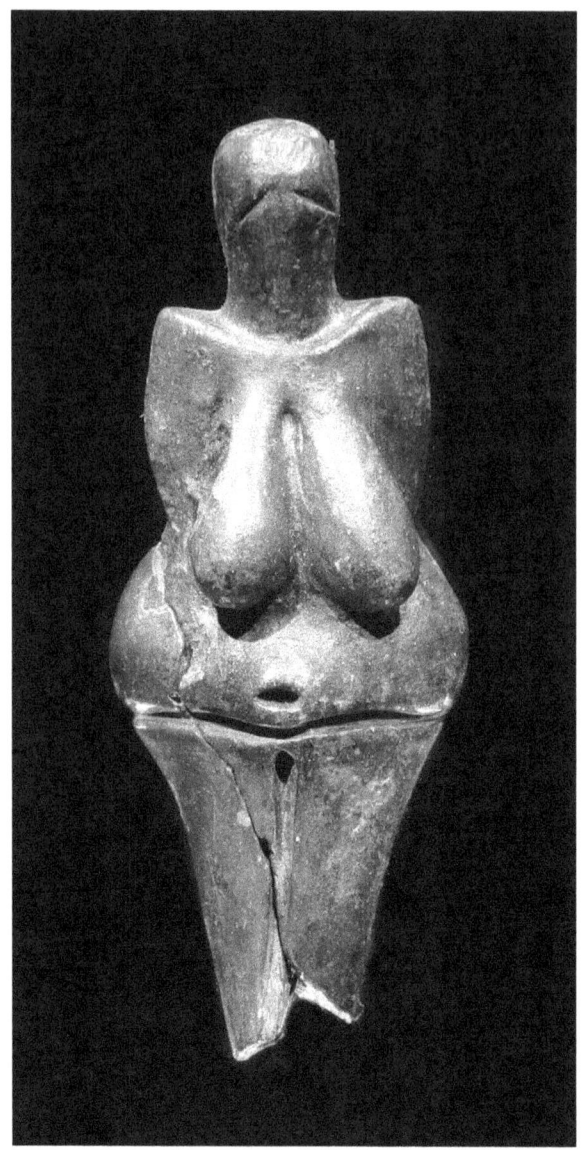

Earliest known ceramics are the Gravettian figurines that date to 29,000 to 25,000 BC

enough to fire pottery did not develop until late in the development of cultures.

- Third, the potter must have time available to prepare, shape and fire the clay into pottery. Even after control of fire was achieved, humans did not seem to develop pottery until a sedentary life was achieved. It has been hypothesized that pottery was developed only after humans established agriculture, which led to permanent settlements. However, the oldest known pottery is from China and dates to 20,000 BC, at the height of the ice age, long before the beginnings of agriculture.

- Fourth, there must be a sufficient need for pottery in order to justify the resources required for its production.[32]

An Incipient Jōmon pottery vessel reconstructed from fragments (10,000-8,000 BC), Tokyo National Museum, Japan

33.6.1 Early pottery

- Methods of forming: Hand-shaping was the earliest method used to form vessels. This included the combination of pinching and coiling.

- Firing: The earliest method for firing pottery wares was the use of bonfires Pit fired pottery. Firing times were short but the peak-temperatures achieved in the fire could be high, perhaps in the region of 900 °C (1,650 °F), and were reached very quickly.[33]

- Clay: Early potters used whatever clay was available to them in their geographic vicinity. However, the lowest quality common red clay was adequate for low-temperature fires used for the earliest pots. Clays tempered with sand, grit, crushed shell or crushed pottery were often used to make bonfire-fired ceramics because they provided an open-body texture that allowed water. The coarser particles in the clay also acted to restrain shrinkage during drying, and hence reduce the risk of cracking.

- Form: In the main, early bonfire-fired wares were made with rounded bottoms to avoid sharp angles that might be susceptible to cracking.

- Glazing: the earliest pots were not glazed.

- The potter's wheel was invented in Mesopotamia sometime between 6,000 and 4,000 BC (Ubaid period) and revolutionised pottery production.

- Biscuit moulds were used to a limited extent as early as the 5th and 6th century BC by the Etruscans[34] and more extensively by the Romans.[35]

- Slipcasting, a popular method for shaping irregular shaped articles. It was first practised, to a limited extent, in China as early as the T'ang dynasty[36]

- Transition to kilns: The earliest intentionally constructed were pit-kilns or trench-kilns—holes dug in the ground and covered with fuel. Holes in the ground provided insulation and resulted in better control over firing.[37]

- Kilns: Pit fire methods were adequate for creating simple earthenware, but other pottery types needed more sophisticated kilns (see below kilns).

33.6.2 History of pottery types

Earthenware

Main article: Earthenware

The earliest forms of pottery were made from clays that were fired at low temperatures in pit-fires or in open bonfires. They were hand formed and undecorated. Because the biscuit form of earthenware is porous, it has limited utility for storage of liquids. However, earthenware has a continuous history from the Neolithic period to today. It can be made from a wide variety of clays. The development of ceramic glaze which makes it impermeable makes it a popular and practical form of pottery making. The addition of decoration has evolved throughout its history.

Stoneware

Main article: Stoneware

Glazed Stoneware was being created as early as the 15th century BC in China. This achievement coincided with kilns that could be fired at higher temperatures.[38]

Porcelain

Main article: Porcelain

Porcelain was first made in China during the Tang Dynasty (AD 618-906). Porcelain was also made in Korea and Japan around the 16th and 17th century AD after suitable kaolin was located in those countries. It was not created outside of the area until the 18th century.[39]

33.6.3 History by region

A sherd or fragment of a vessel, from 20,000 BP. Found in Xianrendong cave in Jiangxi, China.

The earliest-known ceramic objects are Gravettian figurines such as those discovered at Dolní Věstonice in the modern-day Czech Republic. The Venus of Dolní Věstonice (Věstonická Venuše in Czech) is a Venus figurine, a statuette of a nude female figure dated to 29,000–25,000 BC (Gravettian industry).[6] The earliest pottery vessels date back to 20,000 BP and were discovered in Xianrendong cave in Jiangxi, China.[7][40] The pottery may have been used as cookware.[7] Other early pottery vessels include those excavated from the Yuchanyan Cave in southern China, dated from 16,000 BC,[41] and those found in the Amur River basin in the Russian Far East, dated from 14,000 BC.[9][42]

Other early pottery vessels include those made by the Incipient Jōmon people of Japan from around 10,500 BC have also been found.[8][43] The term "Jōmon" means "cord-marked" in Japanese. This refers to the markings made on the vessels and figures using sticks with cords during their production. Recent discovery places the Incipient Jōmon period start to 15,000 to 11,800 cal bp. Nature: Earliest evidence for the use of pottery[44]

It appears that pottery was independently developed in Sub-Saharan Africa during the 11,000-10,000 BC[45] and in South America during the 10,000s BC.[46]

Far East Asia

See also: Chinese ceramics and Jomon pottery

Sherds have been found in China and Japan from a period between 12,000 and perhaps as long as 18,000 years ago.[9][41] As of 2012, the earliest pottery found anywhere in the world,[47] dating to 20,000 to 19,000 years before the present, was found at Xianrendong Cave in the Jiangxi province of China.[7][48] In Japan, the Jōmon period has a long history of development of Jōmon Pottery which was characterized by impressions of rope on the surface of the pottery created by pressing rope into the clay before firing. Glazed Stoneware was being created as early as the 15th century BC in China. Porcelain became a renowned Chinese export during the Tang Dynasty (AD 618-906) and subsequent dynasties.[49] Korean potters produced porcelain as early as the 14th century AD.[50] Koreans brought the art of porcelain to Japan in the 17th century AD.[51]

The secret of making such porcelain was sought in the Islamic world and later in Europe when examples were imported from the East. Many attempts were made to imitate it in Italy and France. However it was not produced outside of the Orient until 1709 in Germany.[52]

South Asia

Pottery was in use in ancient India, including areas now forming Pakistan and northwest India, during the Mehrgarh Period II (5,500-4,800 BC) and Merhgarh Period III (4,800-3,500 BC), known as the ceramic Neolithic and chalcolithic. Pottery, including items known as the ed-Dur vessels, originated in regions of the Saraswati River / Indus River and have been found in a number of sites in the Indus Civilization.[53][54]

Near East

The earliest history of pottery production in the Near East can be divided into four periods, namely: the Hassuna period (7000-6500 BC), the Halaf period (6500-5500 BC), the Ubaid period (5500-4000 BC), and the Uruk period (4000-3100 BC).

Pottery making began in the Fertile Crescent from the 7th millennium BC. The earliest forms, which were found at the Hassuna site, were hand formed from slabs, undecorated, unglazed low-fired pots made from reddish-brown clays.[37] Within the next millennium, wares were decorated

enabled new possibilities and new preparation of clays. Production was now carried out by small groups of potters for small cities, rather than individuals making wares for a family. The shapes and range of uses for ceramics and pottery expanded beyond simple vessels to store and carry to specialized cooking utensils, pot stands and rat traps.[55] As the region developed, new organizations and political forms, pottery became more elaborate and varied. Some wares were made using moulds, allowing for increased production for the needs of the growing populations. Glazing was commonly used and pottery was more decorated.[56]

Aegean region

A potter and his apprentice

Greek red-figure vase in the krater shape, between 470 and 460 BC, by the Altamura Painter

Main articles: Minoan pottery, Pottery of ancient Greece and Ancient Roman pottery

Civilization developed concurrently with the Fertile Crescent in the ancient Mediterranean islands around Greece from about 3200 to 1000 BC and carried to Ancient Greece and Ancient Rome that is considered the Classical era in the Western world. The arts of these cultures eventually became a hallmark for Europe and the New World.

The Minoan pottery was characterized by elaborate painted decoration with natural themes.[57]

The classical Greek culture began to emerge around 1000 BC featuring a variety of well crafted pottery which now included the human form as a decorating motif. The pottery

Persian pottery from the city of Isfahan, Iran, 17th century.

with elaborate painted designs and natural forms, incising and burnished.

The invention of the potter's wheel in Mesopotamia sometime between 6,000 and 4,000 BC (Ubaid period) revolutionized pottery production. Newer kiln designs could fire wares to 1,050 °C (1,920 °F) to 1,200 °C (2,190 °F) which

wheel was now in regular use. Although glazing was known to these potters, it was not widely used. Instead, a more porous clay slip was used for decoration. A wide range of shapes for different uses developed early and remained essentially unchanged during the Greek history.[58]

In the Mediterranean, during the Greek Dark Ages (1,100–800 BC), amphoras and other pottery were decorated with geometric designs such as squares, circles and lines. In the Chalcolithic period in Mesopotamia, Halafian pottery achieved a level of technical competence and sophistication, not seen until the later developments of Greek pottery with Corinthian and Attic ware.

The Etruscan pottery carried on the Greek pottery with its own variations.

The Ancient Roman pottery started by copying Greek and Etruscan styles but soon developed a style of its own.[35]

The distinctive Red Samian ware of the Early Roman Empire was copied by regional potters throughout the Empire.

Islamic pottery

Tajine potter, making tajines

Main article: Islamic pottery

Early Islamic pottery followed the forms of the regions which the Muslims conquered. Eventually, however, there was cross-fertilization between the regions. This was most notable in the Chinese influences on Islamic pottery. Trade between China and Islam took place via the system of trading posts over the lengthy Silk Road. Islamic nations imported stoneware and later porcelain from China. China imported the minerals for Cobalt blue from the Islamic ruled Persia to decorate their blue and white porcelain, which they then exported to the Islamic world.

Likewise, Islamic art contributed to a lasting pottery form identified as Hispano-Moresque in Andalucia (Islamic Spain). Unique Islamic forms were also developed,

Spherical Hanging Ornament, *1575-1585, Ottoman Period. Brooklyn Museum.*

including fritware, lusterware and specialized glazes like tin-glazing, which led to the development of the popular maiolica.[59]

One major emphasis in ceramic development in the Muslim world was the use of tile and decorative tilework.

Europe

Main article: Linear Pottery culture

The early inhabitants of Europe developed pottery at about the same time as in the Near East, circa 5500–4500 BC. These cultures and their pottery were eventually shaped by new cultural influences and technology with the invasions of Ancient Rome and later by Islam. The Renaissance art of Europe was a melding of the art of Classical era and Islamic art.

Americas

Main article: Ceramics of indigenous peoples of the Americas

Most evidence points to an independent development of pottery in the Amerindian cultures, starting with their Archaic Era (3500–2000 BC), and into their Formative period (2000 BC – AD 200). These cultures did not develop the stoneware, porcelain or glazes found in the Old World.

A potter at work, 1605

Pottery in Székely Land, in Romania

Africa

In 2007, Swiss archaeologists discovered pieces of the oldest pottery in Africa in Central Mali, dating back to at least 9,500 BC.[45] The relationship of the introduction of pot-making in many parts of Sub-Saharan Africa with the spread of Bantu languages has been long recognized, al-

though the details remain controversial and awaiting further research, and no consensus has been reached.[60]

Northern Africa includes Egypt, which had several distinct phases of development in pottery. During the early Mediterranean civilizations of the fertile crescent, Egypt developed a unique non-clay-based ceramic which has come to be called Egyptian faience.[note 1]

The other major phase came during the Umayyad Caliphate of Islam, Egypt was a link between early center of Islam in the Near East and Iberia which led to the impressive style of pottery.

It is, however, still valuable to look into pottery as an archaeological record of potential interaction between peoples, especially in areas where little or no written history exists. Because Africa is primarily heavy in oral traditions, and thus lacks a large body of written historical sources, pottery has a valuable archaeological role. When pottery is placed within the context of linguistic and migratory patterns, it becomes an even more prevalent category of social artifact.[60] As proposed by Olivier P. Gosselain in his article, "Materializing Identities: An African Perspective," when comparing pottery techniques in the sub-Saharan region of Africa, it is possible to understand ranges of cross-cultural interaction by looking closely at "chaines operatoires," or production sequences.[61] The methodologies used to produce pottery in early Sub-Saharan Africa are divisible into three categories: techniques visible to the eye (decoration, firing and post-firing techniques), techniques related to the materials (selection or processing of clay, etc.), and techniques of molding or fashioning the clay.[61] We can use these three categories to consider the implications of the reoccurrence of a particular sort of pottery in different areas. Generally, the techniques that are easily visible (the first category of those mentioned above) are thus readily imitated, and may indicate a more distant connection between groups, such as trade in the same market or even relatively close proximity in settlements.[61] Techniques that require more studied replication (i.e., the selection of clay and the fashioning of clay) may indicate a closer connection between peoples, as these methods are usually only transmissible between potters and those otherwise directly involved in production.[61] Such a relationship requires the ability of the involved parties to communicate effectively, implying pre-existing norms of contact or a shared language between the two. Thus, the patterns of technical diffusion in pot-making that are visible via archaeological findings also reveal patterns in societal interaction. This relationship, as referred to by Gosselain, deems pottery in the African context as a "socio-technical aggregate."[61]

Oceania

Polynesia, Melanesia and Micronesia

Pottery has been found in archaeological sites across the islands of Oceania. It is attributed to an ancient archaeological culture called the Lapita. A form of pottery called **Plainware** is found throughout sites of Oceania. The relationship between Lapita pottery and Plainware is not altogether clear.

The Indigenous Australians never developed pottery.[62] After Europeans came to Australia and settled, they found deposits of clay which were analysed by English potters as excellent for making pottery. Less than 20 years later, Europeans came to Australia and began creating pottery. Since then, ceramic manufacturing, mass-produced pottery and studio pottery have flourished in Australia.[63]

33.7 Archaeology

Pottery found at Çatal Höyük - sixth millennium BC

For archaeologists, anthropologists and historians the study of pottery can help to provide an insight into past cultures. Pottery is durable, and fragments, at least, often survive long after artefacts made from less-durable materials have decayed past recognition. Combined with other evidence, the study of pottery artefacts is helpful in the development of theories on the organisation, economic condition and the cultural development of the societies that produced or acquired pottery. The study of pottery may also allow inferences to be drawn about a culture's daily life, religion, social relationships, attitudes towards neighbours, attitudes to their own world and even the way the culture understood the universe.

Chronologies based on pottery are often essential for dating non-literate cultures and are often of help in the dating of historic cultures as well. Trace-element analysis, mostly by neutron activation, allows the sources of clay to be accurately identified and the thermoluminescence test can be used to provide an estimate of the date of last firing. Examining fired pottery shards from prehistory, scientists learned that during high-temperature firing, iron materials in clay record the exact state of Earth's magnetic field at that exact moment.

33.8 Environmental issues in production

Pots in Punjab, Pakistan

Although many of the environmental effects of pottery production have existed for millennia, some of these have been amplified with modern technology and scales of production. The principal factors for consideration fall into two categories: (a) effects on workers, and (b) effects on the general environment. Within the effects on workers, chief impacts are indoor air quality, sound levels and possible over-illumination. Regarding the general environment, factors of interest are fuel consumption, off-site water pollution, air pollution and disposal of hazardous materials.

Historically, "plumbism" (lead poisoning) was a significant health concern to those glazing pottery. This was recognised at least as early as the nineteenth century, and the first legislation in the United Kingdom to limit pottery workers' exposure was introduced in 1899.[64] While the risk to those working in ceramics is now much reduced, it can still not be ignored. With respect to indoor air quality, workers can be exposed to fine particulate matter, carbon monoxide and certain heavy metals. The greatest health risk is the potential to develop silicosis from the long-term exposure to crystalline silica. Proper ventilation can reduce the risks, and the first legislation in the United Kingdom to govern ventilation was introduced in 1899.[64] Another, more recent, study at Laney College, Oakland, California suggests that all these factors can be controlled in a well-designed workshop environment.[65]

33.9 Other usages

The English city of Stoke-on-Trent is widely known as **The Potteries** because of the large number of pottery factories or, colloquially, **Pot Banks**. It was one of the first industrial cities of the modern era where, as early as 1785, two hundred pottery manufacturers employed 20,000 workers.[66] For the same reason, the largest football club in the city is known as **The Potters**.[67]

33.10 See also

- Glossary of pottery terms
- Anagama kiln
- Arts and Crafts Movement
- Asbestos-ceramic
- Barro negro pottery
- Celadon
- Chinese ceramics including porcelain
- Delftware
- Dipped ware
- Faience
- Fiesta (dinnerware) Fiestaware
- Franciscan Ceramics
- Glaze defects
- History of ceramic art
- History of pottery in Palestine
- Ironstone ware
- Jasperware
- Korean ceramics
- Kakiemon pottery
- Longquan celadon
- Maiolica of Renaissance Italy
- Native American pottery
- Persian pottery
- Poole Pottery
- Rockingham Pottery
- Royal Doulton — Henry Doulton, John Doulton
- Sancai
- Sea pottery
- Slipware
- Staffordshire Potteries
- Talavera (pottery), Mexican maiolica
- Thai ceramics
- Wedgwood — Enoch Wedgwood, Josiah Wedgwood
- Victorian majolica
- Vietnamese ceramics
- Yixing pottery

33.11 Notes

[1] The non-clay ceramic called Egyptian faience should not be confused with faience, which is a type of glaze.

33.12 References

[1] 'Pottery Science: materials, process and products.' Allen Dinsdale. Ellis Horwood Limited, 1986.

[2] "Merriam-Webster.com". Merriam-Webster.com. 2010-08-13. Retrieved 2010-09-04.

[3] 'An Introduction To The Technology Of Pottery. 2nd edition. Paul Rado. Institute Of Ceramics & Pergamon Press, 1988

[4] Pottery, meaning 3, mass noun, Oxford English Dictionary, Oxford University Press, 2015

[5] 'Standard Terminology Of Ceramic Whitewares And Related Products.' ASTM C 242–01 (2007.) *ASTM International.*

[6] "No. 359: The Dolni Vestonice Ceramics". Uh.edu. 1989-11-24. Retrieved 2010-09-04.

[7] "Early Pottery at 20,000 Years Ago in Xianrendong Cave, China". *Science* **336** (6089): 1696–1700. June 29, 2012. doi:10.1126/science.1218643. PMID 22745428. Retrieved June 29, 2012.

[8] Diamond, Jared (June 1998). "Japanese Roots". *Discover* (Discover Media LLC). Retrieved 2010-07-10.

[9] 'AMS 14C Age Of The Earliest Pottery From The Russian Far East; 1996-2002.' Derevianko A.P., Kuzmin Y.V., Burr G.S., Jull A.J.T., Kim J.C. Nuclear Instruments And Methods In Physics Research. B223-224 (2004) 735-739.

[10] "The Fast Firing Of Biscuit Earthenware Hollow-Ware In a Single-Layer Tunnel Kiln." Salt D.L. Holmes W.H RP737. *Ceram Research*.

[11] "New And Latest Biscuit Firing Technology". Porzellanfabriken Christian Seltmann GmbH. *Ceram.Forum Int.*/Ber.DKG 87,No.1/2, p.E33-E34,E36. 2010

[12] Ruth M. Home, 'Ceramics for the Potter', Chas. A. Bennett Co., 1952

[13] Home, 1952, p. 16

[14] Dennis Krueger, *Why On Earth Do They Call It Throwing?*, in Ceramics Today

[15] "Whitewares: Production, Testing And Quality Control." W.Ryan & C.Radford. *Pergamon Press.* 1987

[16] "Novel Approach To Injection Moulding." M.Y.Anwar, P.F. Messer, H.A. Davies, B. Ellis. Ceramic Technology International 1996. *Sterling Publications Ltd.*, London, 1995. pg.95-96,98.

[17] "Injection Moulding Of Porcelain Pieces." A. Odriozola, M.Gutierrez, U.Haupt, A.Centeno. *Bol. Soc. Esp. Ceram.* Vidrio 35, No.2, 1996. pg.103-107

[18] "Injection Moulding Of Cups With Handles." U.Haupt. *International Ceramics.* No.2, 1998, pg. 48-51.

[19] "Injection Moulding Technology In Tableware Production." *Ceramic World Review.* 13, No.54, 2003. pg94, 96-97.

[20] An Introduction To The Technology Of Pottery. Paul Rado. Pergamon Press. 1969

[21] 'Sanitaryware Technology'. Domenico Fortuna. Gruppo Editoriale Faenza Editrice S.p.A. 2000.

[22] "DGM-E.pdf" (PDF). Retrieved 2010-09-04.

[23] "Ceramicindustry.com". Ceramicindustry.com. 2000-11-21. Retrieved 2010-09-04.

[24] *Dictionary Of Ceramics.* Arthur Dodd & David Murfin. 3rd edition. The Institute Of Minerals. 1994.

[25] "Clay Sewer Pipe Manufacture. Part II - The Effect Of Variable Alumina, Silica And Iron Oxide In Clays On Some Properties Of Salt Glazes." H.G.Schurecht. *The Journal of the American Ceramic Society.* Volume 6. Issue 6, Pg. 717 – 729.

[26] "Dictionary Of Ceramics." Arthur Dodd & David Murfin. 3rd edition. *The Institute Of Minerals.* 1994.

[27] "Ash Glaze Research." C. Metcalfe. *Ceramic Review* No.202. 2003. pg.48-50.

[28] "Glaze From Wood Ashes And Their Colour Characteristics." Y-S. Han, B-H. Lee. *Korean Ceramic Society* 41. No.2. 2004.

[29] "History of Pottery". Brothers-handmade.com. Retrieved 2010-09-04.

[30] Malaxi Teams. "Labu Sayong, Perak". Malaxi.com. Retrieved 2010-09-04.

[31] Goldner, Janet (Spring 2007). "The women of Kalabougou". *African Arts* **40** (1): 74–79. doi:10.1162/afar.2007.40.1.74.

[32] William K. Barnett and John W. Hoopes, *The Emergence of Pottery: Technology and Innovation in Ancient Society*, Smithsonian Institution Press, 1995, p. 19

[33] Metropolitan Museum of Art http://www.metmuseum.org/toah/hd/jomo/hd_jomo.htm

[34] Glenn C. Nelson, Ceramics: A Potter's Handbook,1966,Holt, Rinehart and Winston, Inc.,p.251

[35] Cooper(2010)

[36] Nelson(1966),p.251

[37] Cooper(2010),p.16

[38] Cooper(2010), p.54

[39] Cooper (2010), pp.72-79,160-179

[40] Stanglin, Douglas (2012-06-29). "Pottery found in China cave confirmed as world's oldest". *USA Today*.

[41] "Chinese pottery may be earliest discovered." Associated Press. 2009-06-01.

[42] 'Radiocarbon Dating Of Charcoal And Bone Collagen Associated With Early Pottery At Yuchanyan Cave, Hunan Province, China.' Boaretto E, Wu X, Yuan J, Bar-Yosef O, Chu V, Pan Y, Liu K, Cohen D, Jiao T, Li S, Gu H, Goldberg P, Weiner S. Proceeding Of The National Academy of Science USA. June 2009. 16;106(24):9595-600.

[43] Kainer, Simon (September 2003). "The Oldest Pottery in the World" (PDF). *Current World Archaeology* (Robert Selkirk). pp. 44–49. Archived from the original (PDF) on 2006-04-23. Retrieved 2006-03-23. (Link currently not functional. 2010-04-09.)

[44] http://www.nature.com/nature/journal/v496/n7445/full/nature12109.html Nature: Earliest evidence for the use of pottery O. E. Craig, H. Saul, A. Lucquin, Y. Nishida, K. Taché, L. Clarke, A. Thompson, D. T. Altoft, J. Uchiyama, M. Ajimoto, K. Gibbs, S. Isaksson, C. P. Heron & P. Jordan

[45] A Swiss-led team of archaeologists has discovered pieces of the oldest African pottery in central Mali.- swissinfo

[46] Barnett & Hoopes 1995:211

[47] "Remnants of an Ancient Kitchen Are Found in China". *The New York Times.*

[48] "Harvard, BU researchers find evidence of 20,000-year-old pottery". *The Boston Globe.*

[49] Emmanuel Cooper, 10,000 Years of Pottery, 2010, University of Pennsylvania Press, p.54

[50] Cooper(2010), p.75

[51] Cooper(2010), p.79

[52] Cooper(2010), p.160-162

[53] Proceedings, American Philosophical Society (vol. 85, 1942). ISBN 1-4223-7221-9

[54] Archaeology of the United Arab Emirates: Proceedings of the First International Conference on the Archaeology of the U.A.E. By Daniel T. Potts, Hasan Al Naboodah, Peter Hellyer. Contributor Daniel T. Potts, Hasan Al Naboodah, Peter Hellyer. Published 2003. Trident Press Ltd. ISBN 1-900724-88-X

[55] Cooper(2010),p.19-20

[56] Cooper(2010),p.20-24

[57] Cooper(2010),p. 36-37

[58] Cooper(2010),p.42

[59] Nelson(1966),pp.23-26

[60] See Koen Bostoen, "Pots, Words and the Bantu Problem: On Lexical Reconstruction and Early African History", *Journal of African History*, **48** (2007), pp. 173-199 for a recent discussion of the issues, and links to further literature.

[61] See Olivier P. Gosselain, [Gosselain, Olivier P. "Materializing Identities: An African Perspective." The Journal of Archaeological Method and Theory 7.3 (2000): 187-217.] for further discussion and sources.

[62] Aboriginal Culture: Introduction

[63] History of Australian Pottery

[64] Health Risks In A Victorian Pottery

[65] *Indoor air quality evaluation for the Butler Building Ceramics Laboratory, Laney College, Oakland, California*, Earth Metrics Incorporated, Alameda County Schools Insurance Association, December, 1989

[66] Patterns of Labour - Work and Social Change in the Pottery Industry. Richard Whipp. Routlidge 1990

[67] "Stokecityfc.com". Stokecityfc.com. 2010-05-13. Retrieved 2010-09-04.

33.13 Further reading

- ASTM Standard C 242-01 *Standard Terminology of Ceramic Whitewares and Related Products*

- Ashmore, Wendy & Sharer, Robert J., (2000). *Discovering Our Past: A Brief Introduction to Archaeology Third Edition*. Mountain View, California: Mayfield Publishing Company. ISBN 978-0-07-297882-7

- Barnett, William & Hoopes, John (Eds.) (1995). *The Emergence of Pottery*. Washington: Smithsonian Institution Press. ISBN 1-56098-517-8

- Childe, V. G., (1951). *Man Makes Himself*. London: Watts & Co.

- Rice, Prudence M. (1987). *Pottery Analysis – A Sourcebook*. Chicago: University of Chicago Press. ISBN 0-226-71118-8.

- Tschegg, C., Hein, I., Ntaflos, Th., 2008. State of the art multi-analytical geoscientific approach to identify Cypriot Bichrome Wheelmade Ware reproduction in the Eastern Nile delta (Egypt). Journal of Archaeological Science 35, 1134-1147.

33.14 External links

- Ancient pottery in Canada

- Ceramic Collection at the Royal Military College of Canada Museum

- Pottery manufacture in recent past

- Stoke-on-Trent Museums - Ceramics Collections Online

Chapter 34

Chalcolithic

Painting of a Copper Age walled city, Los Millares, Iberia

The **Chalcolithic** (English /ˌkælkəlˈlɪθɪk/;[1] Greek: χαλ-κός *khalkós*, "copper" and λίθος *líthos*, "stone")[1] period or **Copper Age**,[1] also known as the **Eneolithic**[1] or **Æne-olithic** (from Latin *aeneus* "of bronze"), is a phase of the Bronze Age before it was discovered that adding tin to copper formed the harder bronze. The Copper Age was originally defined as a transition between the Neolithic and the Bronze Age. However, because it is characterized by the use of metals, the Copper Age is considered a part of the Bronze Age rather than the Stone Age.

The archaeological site of Belovode on the Rudnik mountain in Serbia contains the world's oldest securely dated evidence of copper making at high temperature, from 5,000 BCE.[2][3]

34.1 Origin of name

The multiple names result from multiple recognitions of the period. Originally the term "Bronze Age" meant that either copper or bronze was being used as the chief hard substance for the manufacture of tools and weapons. In 1881, John Evans, recognizing that the use of copper often preceded the use of bronze, distinguished between a transitional Cop-per Age and the Bronze Age proper. He did not include this transitional period in the tripartite system of Early, Middle and Late Bronze Age but placed it at the beginning outside of it. He did not, however, present it as a fourth age, but chose to retain the traditional three-age system.

In 1884, Gaetano Chierici, perhaps following the lead of Evans, renamed it in Italian as the *Eneo-litica*, or "Bronze-stone" transition. This phrase was never intended to mean that the period was the only one in which both bronze and stone were used. The Copper Age features the use of copper, excluding bronze; moreover, stone continued to be used throughout both the Bronze Age and the Iron Age. "Litica" simply names the Stone Age as the point from which the transition began and is not another -lithic age. The Eneolithic was never part of the Stone Age, which ended conclusively the moment the first smelter succeeded in obtaining copper from copper ore for the first time.

Subsequently British scholars used either Evans's "Copper Age" or the term "Eneolithic" (or Aeneolithic), a transla-tion of Chierici's *eneo-litica*. After several years, a num-ber of complaints appeared in the literature that "Ene-olithic" seemed to the untrained eye to be produced from e-neolithic, "outside the Neolithic," clearly not a definitive characterization of the Copper Age. About the year 1900, many writers began to substitute "Chalcolithic" for Ene-olithic, to avoid the false segmentation. It was at this time that the misunderstanding began among those who had not understood the Italian. The -lithic was seen as a new -lithic age, a part of the Stone Age in which copper was used, which may appear paradoxical. Today Copper Age, Eneolithic and Chalcolithic are used synonymously to mean Evans's original definition of Copper Age.

The period is a transitional one, but does not stand outside the traditional three-age system. The analysing stone tool assemblages from sites on the Tehran Plain in Iran has il-lustrated the effects of the introduction of copper working technologies on the in place systems of lithic craft special-ists and raw materials. Networks of exchange and special-ized processing and production that had evolved during the Neolithic seem to have collapsed by the Middle Chalcolithic (*c* 4500-3500 BCE) and been replaced by the use of local materials by a primarily household base production of stone tools.[4] It appears that copper was not widely exploited at first, and that efforts in alloying it with tin and other metals began quite soon, making it difficult to distinguish the distinct Chalcolithic cultures from later periods. The boundary between the Copper and Bronze Ages is indistinct, since alloys faded in and out of use due to the erratic supply of tin.

The emergence of metallurgy may have occurred first in the Fertile Crescent, where it gave rise to the Bronze Age in the 4th millennium BCE (the traditional view), though finds from the Vinča culture in Europe have now been securely dated to slightly earlier than those of the Fertile Crescent. There was an independent invention of copper and bronze smelting first by Andean civilizations in South America extended later by sea commerce to the Mesoamerican civilization in West Mexico (see Metallurgy in pre-Columbian America and Metallurgy in pre-Columbian Mesoamerica).

The literature of European archaeology, in general, avoids the use of 'chalcolithic' (the term 'Copper Age' is preferred), whereas Middle Eastern archaeologists regularly use it. The Copper Age in the Middle East and the Caucasus began in the late 5th millennium BCE and lasted for about a millennium before it gave rise to the Early Bronze Age. The transition from the European Copper Age to Bronze Age Europe occurs about the same time, between the late 5th and the late 3rd millennia BCE.

According to Parpola,[5] ceramic similarities between the Indus Civilization, southern Turkmenistan, and northern Iran during 4300–3300 BCE of the Chalcolithic period (Copper Age) suggest considerable mobility and trade.

34.2   Europe

Main articles: Chalcolithic Europe and Metallurgy during the Copper Age in Europe

An archaeological site in Serbia contains the oldest securely dated evidence of copper making at high temperature, from 7,500 years ago. The find in June 2010 extends the known record of copper smelting by about 800 years, and suggests that copper smelting may have been invented in separate parts of Asia and Europe at that time rather than spreading from a single source.[3] In Serbia, a copper axe was found at Prokuplje, which indicates that humans were using metals in Europe by 7,500 years ago (~5,500 BCE), many years earlier than previously believed.[6] Knowledge of the use of copper was far more widespread than the metal itself. The European Battle Axe culture used stone axes modeled on copper axes, even with imitation "mold marks" carved in

the stone.[7] Ötzi the Iceman, who was found in the Ötztal Alps in 1991 and whose remains were dated to about 3300 BCE, was found with a Mondsee copper axe.

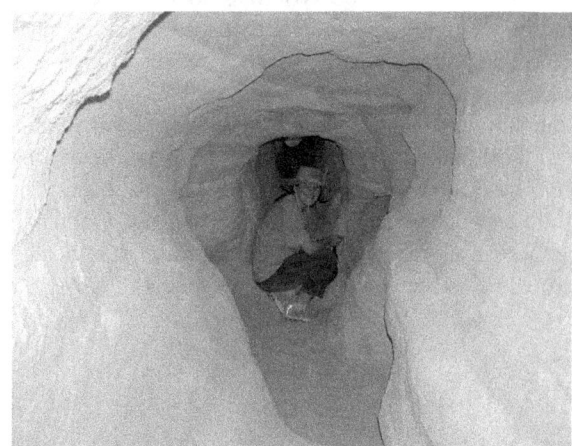

Chalcolithic copper mine in Timna Valley, Negev Desert, Israel

Examples of Chalcolithic cultures in Europe include Vila Nova de São Pedro and Los Millares on the Iberian Peninsula.[8] Pottery of the Beaker people has been found at both sites, dating to several centuries after copper-working began there. The Beaker culture appears to have spread copper and bronze technologies in Europe, along with Indo-European languages.[9]

34.3   South Asia

The inhabitants of Mehrgarh in Pakistan fashioned tools with local copper ore (ore used as pigment) between 7700–3300 BCE.[10] Nausharo site dated to 4500 years ago, a pottery workshop in province of Baluchistan, Pakistan, unearthed 12 blades or blade fragments. The dimensions of these blades are: length 12–18 cm, width 12–20 mm and relatively thin in thickness. The archaeological experiments show that these blades were made with copper indentor, and functioned as potter's tool to trim and shape unfired pottery. The petrographic analysis indicates the local pottery manufacturing, but also reveals that the existence of few exotic black-slipped pottery from Indus Valley.[11]

34.4   East Asia

5th millennium BCE copper artifacts start to appear in East Asia, such as Jiangzhai and Hongshan culture, but those metal artifacts were not widely used.

34.5   Middle East

Timna Valley contains evidence of copper mining 9,000 to 7,000 years ago. The process of transition from Neolithic to Chalcolithic in the Middle East is characterized in archaeological stone tool assemblages by a decline in high quality raw material procurement and use. This dramatic shift is seen throughout the region, including the Tehran Valley, Iran. Here, analysis of six archaeological sites determined a marked downward trend in not only material quality, but also in aesthetic variation in the lithic artefacts. Fazeli et al. use these results as evidence of the loss of craft specialisation caused by increased use of copper tools.[12]

34.6   Africa

Main articles: Copper metallurgy in Africa and Iron Metallurgy in Africa

North Africa and the Nile Valley imported its iron technology from the Near East and followed the Near Eastern course of Bronze Age and Iron Age development. However the Iron Age and Bronze Age occurred simultaneously in much of Africa. The earliest dating of iron in Sub-Saharan Africa is 2500 BCE at Egaro, west of Termit, making it contemporary to the Middle East.[13] The Egaro date is debatable with archaeologists, due to the method used to attain it.[14] The Termit date of 1500 BCE is widely accepted.

In the region of the Aïr Mountains in Niger, we have the development of independent copper smelting between 3000–2500 BCE. The process was not in a developed state, indicating smelting was not foreign. It became mature about 1500 BCE.[15] d

34.7   Americas

Main articles: Metallurgy in pre-Columbian Mesoamerica and Metallurgy in pre-Columbian America

The term is also applied to American civilizations that already used copper and copper alloys thousands of years before the European migration. Besides cultures in the Andes and Mesoamerica, the Old Copper Complex, centered in the Upper Great Lakes region – present-day Michigan and Wisconsin in the United States – mined and fabricated copper as tools, weapons, and personal ornaments.[16] The evidence of smelting or alloying that has been found is subject to some dispute and a common assumption by archaeologists in that objects were cold-worked into shape. Artifacts from some of these sites have been dated from 4000 to 1000 BCE, making them some of the oldest Chalcolithic sites in the entire world.[17] Furthermore, some archaeologists find artifactual and structural evidence of casting by Hopewellian and Mississippian peoples to be demonstrated in the archaeological record.[18]

34.8   See also

- Copper metallurgy in Africa

- Metallurgy during the Copper Age in Europe

- Synoptic table of the principal old world prehistoric cultures

- Three age system

34.9   Notes

[1] The New Oxford Dictionary of English (1998) ISBN 0-19-861263-X, p. 301: "**Chalcolithic** /ˌkælkəˈlɪθɪk/ **adjective** *Archaeology* of, relating to, or denoting a period in the 4th and 3rd millennium BCE, chiefly in the Near East and SE Europe, during which some weapons and tools were made of copper. This period was still largely Neolithic in character. Also called **Eneolithic**... Also called **Copper Age** - *Origin* early 20th cent.: from Greek *khalkos* 'copper' + *lithos* 'stone' + **-ic**".

[2] "Serbian site may have hosted first copper makers". UCL Institute of Archaeology. 23 September 2010.

[3] "Serbian site may have hosted first copper makers". *Science-News*. July 17, 2010.

[4] Fazeli, H.; Donahue, R.E.; Coningham, R.A.E. (2002). "Stone Tool Production, Distribution and Use during the Late Neolithic and Chalcolithic on the Tehran Plain, Iran". *Journal of Persian Studies* **40**: 1–14. JSTOR 4300616.

[5] A.Parpola, 2005

[6] http://www.thaindian.com/newsportal/india-news/ancient-axe-find-suggests-copper-age-began-earlier-than-believed_100105122.html

[7] J. Evans, 1897

[8] C.M.Hogan, 2007

[9] D.W.Anthony, *The Horse, The Wheel and Language: How Bronze-Age riders from the Eurasian steppes shaped the modern world* (2007).

[10] Possehl, Gregory L. (1996)

[11] Méry, S; Anderson, P; Inizan, M.L.; Lechavallier, M; Pelegrin, J (2007). "A pottery workshop with flint tools on blades knapper with copper at Naushaaro (Indus civilisation ca. 2500 BCE)". *Journal of Archaeological Sciences* **34** (7): 1098–1116. doi:10.1016/j.jas.2006.10.002.

[12] Fazeli, H.; Donahue, R.E; Coningham, R.A.E (2002). "Stone Tool Production, Distribution and use during the Late Neolithic and Chalcolithic on the Tehran Plain, Iran". *Iran* **40**: 1–14. doi:10.2307/4300616. Retrieved 27 November 2014.

[13] IRON IN AFRICA: REVISING THE HISTORY(2002). Unesco.

[14] Iron in Sub-Saharan Africa — by Stanley B. Alpern (2005). pp. 71

[15] Ehret, Christopher (2002). The Civilizations of Africa. Charlottesville: University of Virginia, pp. 136, 137 ISBN 0-8139-2085-X.

[16] R. A. Birmingham and L. E. Eisenberg. *Indian Mounds of Wisconsin*. (Madison, Univ Wisconsin Press. 2000.) pp.75-77.

[17] T.C.Pleger, 2000

[18] Neiburger, E. J. 1987. Did Midwest Pre-Columbia Indians Cast Metal? A New Look. *Central States Archaeological Journal* 34(2), 60-74.

34.10   References

- Parpola, Asko (2005). "Study of the Indus script". *Transactions of the 50th International Conference of Eastern Studies* (PDF). Tokyo: The Tôhô Gakkai. pp. 28–66..

- Bogucki, Peter (2007). "Copper Age of Eastern Europe". *The Atlas of World Archaeology*. London: Sandcastle Books. p. 66..

- Evans, John (1897). *The Ancient Stone Implements, Weapons and Ornaments of Great Britain*. London: Longmans, Green, and Company. p. 197..

- Hogan, C. Michael (2007) *Los Silillos*, The Megalithic Portal, ed. A. Burnham

- Pleger, T. C. (2002). "A Brief Introduction to the Old Copper Complex of the Western Great Lakes: 4000-1000 BCE". *Proceedings of Twenty-seventh Annual Meeting of Forest History Association of Wisconsin* (Oconto, Wisconsin: Forest History Association of Wisconsin).

- Possehl, Gregory L. (1996). *Mehrgarh* in *Oxford Companion to Archaeology*, edited by Brian Fagan. Oxford University Press.

34.11   External links

- 'Chalcolithic Era' ; Elizabeth F. Henrickson . Encyclopædia Iranica 1991 .

Chapter 35

Epipaleolithic

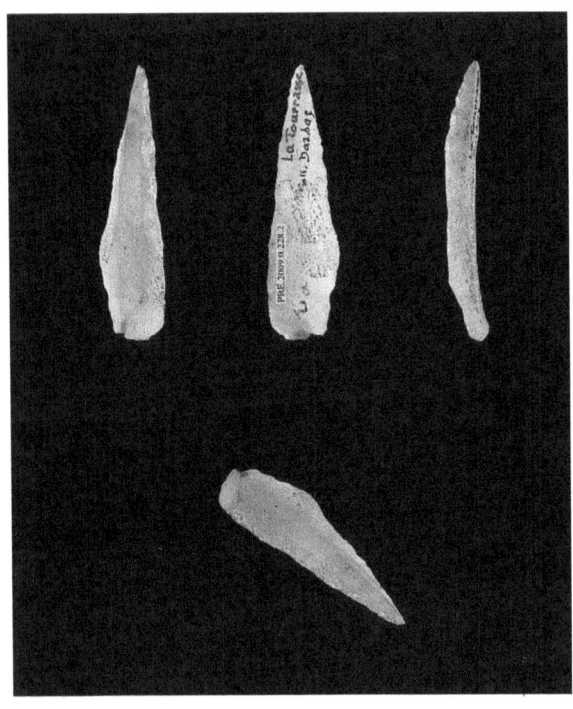

Azilian points, microliths from epipaleolithic northern Spain and southern France.

"**Epipaleolithic**" is a term used for the "final Upper Palaeolithic industries occurring at the end of the final glaciation which appear to merge technologically into the Mesolithic".[1] The period is generally dated from 20,000 BP to about 10,500 BP, having emerged from the Palaeolithic era. [2] The term is sometimes used as a synonym of "Mesolithic". When a distinction is made, "Epipaleolithic" stresses the continuity with the Upper Paleolithic and Mesolithic as we understand it today, whilst "Protoneolithic" stresses a subsequent transition to the Neolithic. [3] Alfonso Moure says in this respect:

> In the language of Prehistorical Archaeology, the most extended trend is to use the term "Epipaleolithic" for the industrial complexes of the post-glacial hunter-gatherer groups. Inversely,

those that are in transitional ways towards artificial production of food are inscribed in the "Mesolithic".[4]

Epipalaeolithic hunter-gatherers made relatively advanced tools from small flint or obsidian blades, known as microliths that were hafted in wooden implements. They were generally nomadic.

Some authors reserve the term "Mesolithic" for the cultures of Europe, where the extinction of the Megafauna had a great impact on the Paleolithic populations at the end of the Ice Age, from about 8000 BCE until the advent of the Neolithic (Sauveterrian, Tardenoisian, Maglemosian, etc.).

35.1 The Epipaleolithic period and animal food sources

The Epipaleolithic is best understood when discussing the southern Levant, as the period is well documented due to good preservation at the site. The most prevalent animal food sources during this period were:[2]

- Gazelle
- Wild equids
- Wild boar
- Deer
- Wild cattle
- Ibex
- Wild goat
- Wild sheep

These were most likely the main food sources through the PPNA. Of these animals, it is likely that only the equids were migrational.[2]

35.2 References

[1] Bahn, Paul, *The Penguin Archaeology Guide*, Penguin, London, pp. 141. ISBN 0-14-051448-1

[2] Byrnes, Andie. "Epipalaeolithic." Epipalaeolithic. N.p., 2005. Web. Dec. 2014.

[3] "The Scandinavian Ice Sheet itself started to retreat northward about 8300 bce, and the period between then and the origins of agriculture (at various times in the 7th to 4th millennia, depending on location) was one of great environmental and cultural change. It is termed the Mesolithic Period (Middle Stone Age) to emphasize its transitional importance, but the alternative term Epipaleolithic, used mostly in eastern Europe, stresses the continuity with processes begun earlier." history of Europe at Encyclopedia Britannica online (accessed April 2013)

[4] A. Moure *El Origen del Hombre*, 1999. ISBN 84-7679-127-5

Chapter 36

Paleolithic diet

This article is about a type of diet. For information on the dietary practices of Paleolithic humans, see Paleolithic#Diet and nutrition.

The **paleolithic diet** is a diet based on the foods ancient

Wild food is an important feature of the diet.

ancestors might likely have eaten, such as meat, nuts, and berries,[1] and excludes food to which they had not yet become familiar, like dairy. The Paleolithic era was a period lasting around 2.5 million years that ended about 10,000 years ago with the advent of farming. It was characterized by the use of flint, stone, and bone tools, hunting, fishing, and the gathering of plant foods.[2]

The diet is based on the premise that Paleolithic humans evolved nutritional needs specific to the foods available at that time, and that the nutritional needs of modern humans remain best adapted to the diet of their Paleolithic ancestors. Proponents argue that this is true because modern human metabolism has been unable to adapt fast enough to handle many of the foods that have become available since the advent of agriculture. Thus, they believe modern humans are maladapted to eating foods such as grain, legumes, and dairy, and in particular the high-calorie processed foods that are a staple of most modern diets. Proponents claim that modern humans' inability to properly metabolize these comparatively new types of food has led to modern-day

problems such as obesity, heart disease, and diabetes. They claim that followers of the Paleolithic diet may enjoy a longer, healthier, more active life.

Critics of the Paleolithic diet have raised a number of objections, including that: paleolithic humans *did* eat grains and legumes;[3] humans are much more nutritionally flexible than Paleolithic diet advocates claim; Paleolithic humans were not genetically adapted to specific local diets; the Paleolithic period was extremely long, and saw a wide variety of natural items that humans used for sustenance; and/or, that very little is even known for certain about exactly what Paleolithic humans ate.

36.1 Health effects

A 2015 systematic review on paleolithic nutrition and its effects on measurable components of metabolic syndrome found that the paleo diets used in the studies led to short-term improvements that had statistical significance for waist circumference, triglycerides, and blood pressure and that there was no statistical significance for changes in HDL cholesterol and fasting blood sugars; it concluded: "Although there is moderate quality evidence from randomized controlled intervention studies to suggest that the Paleolithic diet can improve metabolic syndrome components, we believe that more studies are required before Paleolithic nutrition can be recommended in future guidelines."[4]

As of 2014 there was no good evidence the paleo diet is effective in treating inflammatory bowel disease.[5]

The British Dietetic Association named the paleo diet as among the five worst celebrity-endorsed diets of 2015, saying it risks being "unbalanced, time consuming, [and] socially isolating" and so "a sure-fire way to develop nutrient deficiencies".[6]

David L. Katz and Stephanie Meller have written that the paleolithic diet presents a "scientific case" in part because of its anthropological basis, but that there is comparatively limited evidence supporting its health benefit over other popular contemporary diets.[7]

According to evolutionary biologist Marlene Zuk of the University of Minnesota: "Those who follow the [paleo] diet may be missing out on vital nutrients, and it is believed that could create long term health problems, in particular for adolescent girls who may be at risk of developing osteoporosis later in life as a result of not getting enough calcium."[8]

36.2 History and terminology

The idea of a paleolithic diet can be traced to the work in the 1970s by gastroenterologist Walter Voegtlin.[9] The idea was later developed by Stanley Boyd Eaton and Melvin Konner, and popularized by Loren Cordain in his 2002 book *The Paleo Diet*.[10]

The terms *caveman diet* and *stone-age diet* are also used,[9][11] as is *Paleo Diet*, trademarked by Loren Cordain.[12]

36.3 Foods

Cordain has said the diet requires:[13]

Seeds such as walnuts are rich sources of protein and micronutrients

- More protein and meat: Meat, seafood, and other animal products represent the staple foods of modern-day Paleo diets, since advocates claim protein constitutes 19–35% of the calories in hunter-gatherer diets.[14] The Centers for Disease Control and Prevention, the national public health institute of the United States, recommends that 10–35% of calories come from protein.[15] Advocates recommend, relative to modern diets, that the Paleolithic diet have moderate to higher fat intake dominated by monounsaturated

and polyunsaturated fats and omega-3 fats, but avoiding trans fats, and omega-6 fats.[14] It should be noted that the increased uptake of meat and proteins should consist of meats from grass-fed animals. Livestock raised on a grass diet are able to incorporate omega-3 fatty acids from grass into their tissue, as opposed to a grain fed diet high in omega-6 rich corn. This includes grass fed meats that are "finished with grains."[16]

- Fewer carbohydrates: Non-starchy vegetables. The diet recommends the consumption of non-starchy fresh fruits and vegetables to provide 35–45% daily calories and be the main source of carbohydrates.[14] According to the United States Department of Agriculture, the acceptable macronutrient distribution range for carbohydrates is 45 to 65 percent of total calories.[17] A typical modern diet gets a lot of carbohydrates from dairy products and grains, but these are excluded in the Paleolithic diet.

- High fiber: High fiber intake not from grains, but from non-starchy vegetables and fruits.[14]

36.3.1 Exclusions

Food groups that advocates claim were rarely or never consumed by humans before the Neolithic agricultural revolution are excluded from the diet. These include:

- dairy products

- grains, for example wheat, rye, canary seed, and barley, which make it a gluten-free diet

- legumes, for example beans and peanuts

- processed oils

- refined sugar

- salt

- Neither alcohol[18] nor coffee is considered "paleo" as human ancestors could not produce these drinks.

36.4 Rationale and counter-arguments

36.4.1 Adaptation

The rationale for the Paleolithic diet derives from evolutionary medicine.[19] Advocates of the diet state that humans were genetically adapted to eating specifically

did not live long enough to develop them.[27] Based on the data from recent hunter-gatherer populations, it is estimated that at age 15, life expectancy was an additional 39 years, for a total age of 54.[28] At age 45, it is estimated that average life expectancy was an additional 19 years, for a total age of 64 years.[29][30] Food energy excess, relative to energy expended, rather than the consumption of specific foods may underlie the diseases of affluence. "The health concerns of the industrial world, where calorie-packed foods are readily available, stem not from deviations from a specific diet but from an imbalance between the energy humans consume and the energy humans spend."[31]

Paleolithic carving of a mammoth. Hunting by humans may have been a factor in its extinction, causing resource scarcity which may in turn have contributed to the development of agriculture.

those foods that were readily available to them in their local environments. These foods therefore shaped the nutritional needs of Paleolithic humans. The physiology and metabolism of modern humans have changed little, if at all, since the time of their Paleolithic ancestors.[20] Natural selection took time and the cultural and lifestyle changes to westernized culture occurred too quickly for the gene pool to evolve with the environmental changes.[21] The agricultural revolution brought the addition of grains and dairy to the diet.[22]

According to the evolutionary discordance hypothesis, "many chronic diseases and degenerative conditions evident in modern Western populations have arisen because of a mismatch between Stone Age genes and recently adopted lifestyles."[23] Advocates of the modern Paleolithic diet, including Loren Cordain, take the evolutionary discordance hypothesis for granted, and form their dietary recommendations on its basis. They argue that modern humans should follow a diet that is as nutritionally close to that of their Paleolithic ancestors as possible.

The validity of the evolutionary discordance hypothesis has been brought into doubt by recent research.[24] Studies of traditionally living populations show that humans can live healthily with a wide variety of diets. Humans have evolved to be flexible eaters.[25] Lactose tolerance is an example of how humans have adapted to the introduction of dairy. While the introduction of grains, dairy, and legumes has not necessarily been easy for the modern human, especially the Westernized one, it is safe to say that if humans could only survive in environments similar to that of their ancestors, then the society that humans have would not be in existence today.[26]

36.4.2   Diseases of affluence

Advocates of the diet argue that the increase in diseases of affluence after the dawn of agriculture was caused by the change in diet, but it may be that pre-agricultural foragers did not suffer from the diseases of affluence because they

36.4.3   Historical diet

Adoption of the diet assumes that we can reproduce the hunter-gatherer diet. Molecular biologist Marion Nestle argues that "knowledge of the relative proportions of animal and plant foods in the diets of early humans is circumstantial, incomplete, and debatable and there are insufficient data to identify the composition of a genetically determined optimal diet. The evidence related to Paleolithic diets is best interpreted as supporting the idea that diets based largely on plant foods promote health and longevity, at least under conditions of food abundance and physical activity".[32] Ideas about Paleolithic diet and nutrition are at best hypothetical.[33]

Brassica oleracea, an edible wild plant

The data for Cordain's book only came from six groups, mainly living in marginal habitats.[34] One of the studies was on the !Kung, whose diet was recorded for a single month,[35] and one was on the Eskimos.[36] Due to these limitations, the book has been criticized as painting an incomplete picture of what the diets of Paleolithic ancestors may have looked like.[34] It has been noted that the rationale for the diet does not take adequate account of the fact that, due to the pressures of artificial selection, most modern domesticated plants and animals differ drastically from their Paeleolithic ancestors, whose nutritional profiles often differed drastically from their modern counterparts. For example, wild almonds produce potentially fatal levels of cyanide, but this harmful poison has been bred out of domesticated varieties by artificial selection. Many vegetables like Broccoli "did not ... exist in the Paleolithic period".[37] Broccoli and many other genetically similar vegetables (like cabbage, cauliflower, kale, etc.) are in fact modern cultivars of the ancient species *Brassica oleracea*, a wild plant also known as wild mustard.

Trying to devise an ideal diet by studying contemporary hunter-gatherers is difficult because of the great disparities that exist, for example with the animal-derived calorie percentage ranging from 25% in the Gwi people of southern Africa to 99% in Alaskan Nunamiut.[38] Recommendations to restrict starchy vegetables may not be an accurate representation of the diet of relevant Paleolithic ancestors.[39]

Not all processed foods are a post agricultural introduction, there is evidence early humans processed plant food and possibly prepared flour 30,000 years ago.[40] Researchers have proposed that cooked starches met the energy demands of an increasing brain size, based on variations in the copy number of genes encoding for amylase.[41][42]

36.5 See also

- Inuit diet

- Modern primitive

- Nutritional genomics

- Paleolithic lifestyle

- Low carbohydrate diet

36.6 References

[1] Zimmer, Carl (August 13, 2015). "For Evolving Brains, a 'Paleo' Diet Full of Carbs". *New York Times*. Retrieved August 14, 2015.

[2] "Definition: Paleolithic". Collins. n.d. Retrieved 17 March 2015.

[3] Henry, Amanda; Brooks, Alison; Piperno, Dolores (2014). "Plant foods and the dietary ecology of Neanderthals and early modern humans". *Journal of Human Evolution* **69**: 44–54. doi:10.1016/j.jhevol.2013.12.014. PMID 24612646.

[4] Manhiemer, Eric W; van Zuuren, Esther J; Fedorowicz, Zbys; Pijl, Hanno (August 12, 2015). "Paleolithic nutrition for metabolic syndrome: systematic review and meta-analysis". *Am J Clin Nutr* **102** (4): 922. doi:10.3945/ajcn.115.113613. PMID 26269362.

[5] Hou JK, Lee D, Lewis J; Lee; Lewis (October 2014). "Diet and inflammatory bowel disease: review of patient-targeted recommendations". *Clin. Gastroenterol. Hepatol.* (Review) **12** (10): 1592–600. doi:10.1016/j.cgh.2013.09.063. PMC 4021001. PMID 24107394. Even less evidence exists for the efficacy of the SCD, FODMAP, or Paleo diets. Furthermore, the practicality of maintaining these interventions over long periods of time is doubtful.

[6] "Top 5 Worst Celebrity Diets to Avoid in 2015". British Dietetic Association. 8 December 2014. Retrieved February 2015. An unbalanced, time consuming, socially isolating diet, which this could easily be, is a sure-fire way to develop nutrient deficiencies, which can compromise health and your relationship with food.

[7] Katz DL, Meller S (2014). "Can we say what diet is best for health?". *Annu Rev Public Health* **35**: 83–103. doi:10.1146/annurev-publhealth-032013-182351. PMID 24641555.

[8] Scientists argue that the Paleo diet could be doing more harm than good, 'ignores basic biology'

[9] Fitzgerald M (2014). *Diet Cults: The Surprising Fallacy at the Core of Nutrition Fads and a Guide to Healthy Eating for the Rest of US*. Pegasus Books. p. 38. ISBN 978-1-60598-595-4.

[10] "The modern take on the Paleo diet: is it grounded in science?". *Environmental Nutrition* (7). 2010.

[11] Shariatmadari, David (22 October 2014). "What language tells us about the roots of the stone age diet". Guardian. Retrieved 17 March 2015.

[12] Lowe K (20 July 2014). "A dissenting view on the Paleo Diet". *The Seattle Times*. Retrieved 17 March 2015.

[13] Cordain, Loren (2010). *The Paleo diet Revised*. Houghton Mifflin Harcourt. p. 10. ISBN 978-0470913024.

[14] "THE PALEO DIET PREMISE". *The Paleo Diet*. Retrieved 14 June 2014.

[15] "Protein". *CDC*. US Government. Retrieved 14 June 2014.

[16] Duckett, S. K.; Neel, J. P. S.; Fontenot, J. P.; Clapham, W. M. (2009-05-27). "Effects of winter stocker growth rate and finishing system on: III. Tissue proximate, fatty acid, vitamin, and cholesterol content". *Journal of Animal Science* **87** (9): 2961–2970. doi:10.2527/jas.2009-1850. PMID 19502506.

[17] "Carbohydrates". USDA. Retrieved 14 June 2014.

[18] Cordain, Loren. "ONE TEQUILA, TWO TEQUILA, THREE TEQUILA... PRIMAL!". Retrieved 14 June 2014.

[19] Konner M.; Eaton, S. Boyd (2010). "Paleolithic Nutrition: Twenty-Five Years Later". *Nutrition in Clinical Practice* **25** (6): 594–602. P. 594.

[20] Konner M.; Eaton, S. Boyd (2010). "Paleolithic Nutrition: Twenty-Five Years Later". *Nutrition in Clinical Practice* **25** (6): 594–602. Pp. 594–95.

[21] Carrera-Bastos, P., Fontes-Villalba, M., O'Keefe, J., Lindeberg, S., Cordain, L. 2011. The western diet and lifestyle and diseases of civilization. Research Reports in Clinical Cardiology. doi:10.2147/RRCC.S16919

[22] Ramsden, C.; Faurot, K.; Carrera-Bastos, P.; Cordain, L.; De Lorgeril, M.; Sperling, L. (2009). "Dietary Fat Quality and Coronary Heart Disease Prevention: A Unified Theory Based on Evolutionary, Historical, Global, and Modern Perspectives". *Current Treatment Options in Cardiovascular Medicine* **11** (4): 289–301. doi:10.1007/s11936-009-0030-8. PMID 19627662.

[23] Elton, S (2008). "Environments, Adaptation, and Evolutionary Medicine: Should We be Eating a Stone Age Diet?". In S. Elton, P. O'Higgins (ed.), *Medicine and Evolution: Current Applications, Future Prospects*. Boca Raton, FL: CRC Press. P. 9. ISBN 978-1-4200-5134-6.

[24] Turner, Bethany L; Thompson, Amanda L (August 2013). "Beyond the Paleolithic prescription: incorporating diversity and flexibility in the study of human diet evolution". *Nutrition Reviews* **71** (8): 501–510. doi:10.1111/nure.12039. PMID 23865796.

[25] Leonard, William R. "Food for Thought: Dietary change was a driving force in human evolution". Scientific American.

[26] http://www.scientificamerican.com/article/why-paleo-diet-half-baked-how-hunter-gatherer-really-eat/

[27] Ungar, Peter S.; Grine, Frederick E.; & Teaford, Mark F. (October 2006). "Diet in Early *Homo*: A Review of the Evidence and a New Model of Adaptive Versatility" (PDF). *Annual Review of Anthropology* **35** (1): 209–228. doi:10.1146/annurev.anthro.35.081705.123153.

[28] Hillard Kaplan, Kim Hill, Jane Lancaster, and A. Magdalena Hurtado (2000). "A Theory of Human Life History Evolution: Diet, Intelligence and Longevity" (PDF). *Evolutionary Anthropology* **9** (4): 156–185. doi:10.1002/1520-6505(2000)9:4<156::AID-EVAN5>3.0.CO;2-7. Retrieved September 12, 2010.

[29] Gurven, Michael; Kaplan, Hillard (2007). "Longevity Among Hunter- Gatherers: A Cross-Cultural Examination". *Population and Development Review* **33** (2): 321–365. doi:10.1111/j.1728-4457.2007.00171.x. ISSN 0098-7921.

[30] Osborne, Daniel L.; Hames, Raymond (2014). "A life history perspective on skin cancer and the evolution of skin pigmentation". *American Journal of Physical Anthropology* **153** (1): 1–8. doi:10.1002/ajpa.22408. ISSN 0002-9483. PMID 24459698.

[31] Leonard, William R. (December 2002). "Food for thought: Dietary change was a driving force in human evolution" (PDF). *Scientific American* **287** (6): 106–15. PMID 12469653.

[32] Nestle, Marion (March 2000). "Paleolithic diets: a sceptical view". *Nutrition Bulletin* **25** (1): 43–7. doi:10.1046/j.1467-3010.2000.00019.x.

[33] Milton, Katharine (2002). "Hunter-gatherer diets: wild foods signal relief from diseases of affluence (PDF)" (PDF). In Ungar, Peter S. & Teaford, Mark F. *Human Diet: Its Origins and Evolution*. Westport, CT: Bergin and Garvey. pp. 111–22. ISBN 0-89789-736-6.

[34] Peter S. Ungar; Mark Franklyn Teaford (1 January 2002). *Human Diet: Its Origin and Evolution*. Greenwood Publishing Group. pp. 67–. ISBN 978-0-89789-736-5.

[35] Lee, Richard (1969). "Kung Bushmen Subsistence: An Input-Output Analysis". *Contributions to Anthropology: Ecological Essays. Ottawa: National Museums of Canada* (230): 73–94.

[36] Eaton, M.D., S. Boyd; Shostak, Marjorie; Konner, M.D., Ph.D., Melvin (1988). *The Paleolithic Prescription: A Program of Diet and Exercise and a Design for Living*. Harper and Row. p. 79. ISBN 978-0060916350.

[37] C. Warinner (2013), "Debunking the Paleo Diet", *TEDxOU*, 25 January 2013, https://www.youtube.com/watch?v=BMOjVYgYaG8, accessed 21 August 2014.

[38] Kolbert, Elizabeth. "Flesh of Your Flesh", *The New Yorker*, November 9, 2009, accessed January 27, 2011.

[39] Gibbons, Ann (September 2014). "The Evolution of Diet". *National Geographic*. Retrieved 2014-09-04.

[40] Thirty thousand-year-old evidence of plant food processing

[41] "For Evolving Brains, a 'Paleo' Diet Full of Carbs". New York Times. 13 August 2015. Retrieved 14 August 2015.

[42] Hardy, Karen; Brand-Miller, Jennie; Brown, Katherine D.; Thomas, Mark G.; Copeland, Les (September 2015). "The Importance of Dietary Carbohydrate in Human Evolution". *The Quarterly Review of Biology* **90** (3): 251–268. doi:10.1086/682587. JSTOR 682587.

36.7 Further reading

• Bijlefeld M, Zoumbaris SK (2014). *Paleo Diet. Encyclopedia of Diet Fads: Understanding Science and Society* (2nd ed.) (ABC-CLIO). pp. 164–166. ISBN 978-1-61069-760-6.

• Gorski D (18 March 2013). "It's a part of my paleo fantasy, it's a part of my paleo dream". Science-Based Medicine. Retrieved February 2015.

- Zuk M (2013). *Paleofantasy: What Evolution Really Tells Us about Sex, Diet, and How We Live*. W.W. Norton & Co. ISBN 978-0-393-08137-4.

- George Bryant (2014). *The Paleo Kitchen: Finding Primal Joy in Modern Cooking*. Victory Belt. ISBN 978-1-628-60010-0.

36.8 External links

- "Paleo isn't a fad diet, it's an ideology that selectively denies the modern world"

36.9 Text and image sources, contributors, and licenses

36.9.1 Text

- **Stone Age** *Source:* https://en.wikipedia.org/wiki/Stone_Age?oldid=688863470 *Contributors:* Vicki Rosenzweig, Bryan Derksen, Tarquin, Taw, Alex.tan, Rmhermen, SimonP, Peterlin~enwiki, BryceHarrington, Olivier, Frecklefoot, Patrick, Michael Hardy, Paul Barlow, Gdarin, Mic, Ixfd64, Delirium, Paul A, Todd, Looxix~enwiki, Den fjättrade ankan~enwiki, Glenn, Marteau, Harry Potter, Mxn, Pizza Puzzle, Reddi, The Anomebot, DJ Clayworth, Tpbradbury, Itai, Nv8200pa, Phoebe, Penfold, Owen, Jni, Branddobbe, Robbot, RedWolf, Arkuat, Mayooranathan, Academic Challenger, Ojigiri~enwiki, Sunray, UtherSRG, Alan Liefting, Marc Venot, Gtrmp, Jyril, Tom harrison, Aphaia, Chowbok, Gadfium, Andycjp, DocSigma, Antandrus, Beland, Kusunose, Adamsan, Rdsmith4, Maximaximax, Bodnotbod, Icairns, Karl-Henner, J0m1eisler, Lumidek, Neutrality, Okapi~enwiki, Robin klein, Ratiocinate, Demiurge, Zondor, Trevor MacInnis, Eisnel, Mike Rosoft, EugeneZelenko, Discospinster, Rhobite, Mazi, Vsmith, Lulu of the Lotus-Eaters, Dbachmann, Grutter, Stbalbach, Bender235, ESkog, Kbh3rd, Fenice, Eric Forste, Ben Webber, RoyBoy, Syats, Bobo192, Comtebenoit, Janna Isabot, Stesmo, Smalljim, Func, BrokenSegue, R. S. Shaw, Foobaz, Jolomo, ParticleMan, Giraffedata, Syd1435, PeterisP, Geos, Haham hanuka, Krellis, Nsaa, Alansohn, Gary, JYolkowski, Foant, LtNOWIS, Arthena, Jeltz, Ricky81682, Riana, Mailer diablo, Differentgravy, Bart133, Dhartung, Grenavitar, Sciurinæ, Lerdsuwa, Zxcvbnm, Jguk, Itsmine, New Age Retro Hippie, YixilTesiphon, Oleg Alexandrov, Woohookitty, Mindmatrix, TigerShark, PoccilScript, Brunnock, Ganeshk, TomTheHand, WadeSimMiser, Howabout1, Damicatz, Torqueing, Eras-mus, CharlesC, Wayward, Phlebas, Yst, LeoO3, Mandarax, Graham87, BD2412, Josh Parris, Rjwilmsi, Evin290, Palpatine, Bhadani, Olessi, MapsMan, DirkvdM, Falphin, Wragge, Osprey39, CDThieme, RobertG, Pavlo Shevelo, NekoDaemon, Shadow007, Nwatson, RobyWayne, Gurubrahma, Phoenix2~enwiki, King of Hearts, Chobot, Bornhj, DVdm, NSR, Roboto de Ajvol, Sortan, Stan2525, Jimp, Anglius, Redbaron302000~enwiki, Briaboru, Exir, Adalger, Lexi Marie, Anache, Gardar Rurak, SpuriousQ, Sporks of Mass Destruction, Stephenb, Gaius Cornelius, Eleassar, Rsrikanth05, Cryptic, Ugur Basak, Burek, N2ChristTheKing, SEWilcoBot, Grafen, Erielhonan, Jaxl, Terfili, Brandon, Ezeu, PM Poon, Dbfirs, Deckiller, Zirland, Gadget850, Barnabypage, CLW, Botteville, Evryman, Wknight94, Wardog, Igiffin, Deville, Phgao, Theodolite, Mike Dillon, Closedmouth, Arthur Rubin, E Wing, Anclation~enwiki, Junglecat, Maxamegalon2000, Saltmarsh, Ajdebre, Serendipodous, Mejor Los Indios, That Guy, From That Show!, P. B. Mann, MacsBug, SmackBot, Haymaker, Vald, Bomac, Delldot, Yamaguchi⿰⿰, Gilliam, Skizzik, Kevinalewis, Rmosler2100, Wigren, Chris the speller, Tito4000, MK8, Snori, SchfiftyThree, Hongooi, Oatmeal batman, Scwlong, Javier Arambel, Can't sleep, clown will eat me, Sephiroth BCR, Yidisheryid, Rizzi, Rrburke, Addshore, KnowledgeLord, Smooth O, Underbar dk, Dreadstar, Yom, Evlekis, Kahuroa, Risker, Risssa, ArglebargleIV, Serein (renamed because of SUL), Rukario639, John, Euchiasmus, Scientizzle, Kipala, Heimstern, Thanos5150, Gobonobo, Jimd, Jas131, IronGargoyle, Judge Howarth, A. Parrot, Mr Stephen, Dukemeiser, Skinsmoke, Dl2000, BranStark, Nonexistant User, Loki0074, BananaFiend, Iridescent, Dekaels~enwiki, Laurens-af, Joseph Solis in Australia, Shoeofdeath, CapitalR, Courcelles, Tawkerbot2, Randroide, Fvasconcellos, Farouk92, JForget, Ale jrb, Sir Vicious, Dr.Bastedo, Baiji, AshLin, Ken Gallager, Lokal Profil, MrFish, Mike 7, Mato, Hooded sonny, Gogo Dodo, Siberian Husky, JFreeman, Alexfrance250291, Marssociety, Tawkerbot4, Doug Weller, Christian75, DumbBOT, Kozuch, Emils9, Aazn, Omicronpersei8, UberScienceNerd, Maziotis, Epbr123, Qwyrxian, N5iln, Mojo Hand, Marek69, A3RO, James086, Wildthing61476, Joymmart, J. W. Love, Kohlrabi, Natalie Erin, Escarbot, KevinWho, AntiVandalBot, Milton Stanley, Majorly, Luna Santin, Seaphoto, Crabula, Pro crast in a tor, TimVickers, Tmopkisn, Modernist, Âme Errante, Credema, Fireice, Labongo, Myanw, Redsnapper511, JAnDbot, Tigga, Leuko, MER-C, Skomorokh, Fetchcomms, Andonic, East718, PhilKnight, Savant13, Geniac, Magioladitis, 75pickup, Bongwarrior, VoABot II, Silicon retina, JNW, SineWave, Kevinmon, TAP3AH, 28421u2232nfenfcenc, Allstarecho, Schumi555, Cpl Syx, Fang 23, Vssun, DerHexer, Wguy00, FriendlyFred, JosephCampisi, MartinBot, GimliDotNet, Arjun01, Rettetast, Juansidious, Anaxial, Hazzah!, CommonsDelinker, AlexiusHoratius, Dinkytown, J.delanoy, Darin-0, Snow Shoes, Virgil Valmont, Winampman, AlienZen, DarkFalls, LordAnubisBOT, Clerks, Balthazarduju, Canadian Scouter, AntiSpamBot, HiLo48, Loohcsnuf, NewEnglandYankee, Half-Blood Auror, In Transit, SJP, Largoplazo, KylieTastic, Cometstyles, Jamesofur, Bonadea, 619po, Podyte, JavierMC, Specter01010, Devin.Callahan, CardinalDan, Idioma-bot, Oaxaca dan, Ariobarzan, Malik Shabazz, Deor, VolkovBot, ABF, Jlaramee, Skjbe, Temp234, Jeff G., AlnoktaBOT, Jacroe, Ryan032, Philip Trueman, Mike Cline, TXiKiBoT, Oshwah, Technopat, Tonicblue, Rei-bot, Wiikipedian, Dendodge, LeaveSleaves, Rickito, Cremepuff222, Liberal Classic, Itemirus, Falcon8765, Burntsauce, HiDrNick, AlleborgoBot, Spdhf, TheXenocide, SieBot, Arun11, Steorra, WereSpielChequers, Gfglegal, SheepNot-Goats, Hertz1888, Caltas, ConfuciusOrnis, Yintan, The very model of a minor general, Calabraxthis, Keilana, Happysailor, Flyer22 Reborn, UnrivaledShogun, Oda Mari, Micke-sv, Oxymoron83, Harry~enwiki, Hello71, Benea, Steven Crossin, Lightmouse, Ks0stm, Juneythomas, Calatayudboy, Nordic Crusader, PerryTachett, Pinkadelica, Escape Orbit, Randy Kryn, TwinnedChimera, Floorwalker, Mr. Granger, Alfons Åberg, Martarius, Apuldram, ClueBot, Avenged Eightfold, Fyyer, The Thing That Should Not Be, Gaia Octavia Agrippa, R000t, Photouploaded, Drmies, Mild Bill Hiccup, CounterVandalismBot, Parkwells, Thegargoylevine, Phenylalanine, Puchiko, DragonBot, Excirial, Eeekster, Jayantanth, Nownownow, Ember of Light, Tahmasp, Moberg, Thingg, Leungkh, Ranjithsutari, Versus22, Starlemusique, Yozer1, Johnuniq, SoxBot III, DumZiBoT, BarretB, ChickenFURY, BodhisattvaBot, Stickee, Rror, Shoeofdeathisadouche, Avoided, Skarebo, NellieBly, Badgernet, Noctibus, ZooFari, Bobliang345, Stephen Poppitt, King Pickle, Addbot, ConCompS, Freakmighty, DOI bot, Tcncv, DaughterofSun, Astraydagger, Kristinamwood, Fieldday-sunday, Hoboday, CanadianLinuxUser, Fluffernutter, BabelStone, Download, EhsanQ, ChenzwBot, LinkFA-Bot, West.andrew.g, 5 albert square, Numbo3-bot, Tide rolls, Bluebusy, Albert galiza, Megaman en m, Legobot, PlankBot, Luckas-bot, Pink!Teen, Tohd8BohaithuGh1, Sprachpfleger, Fraggle81, Ojay123, Gobbleswoggler, Timothyhouse1, Inemanja, IW.HG, Eric-Wester, Jungle-Joe, Magog the Ogre, Daveosaurus, AnomieBOT, Andrewrp, 1exec1, ThaddeusB, Jim1138, Piano non troppo, AdjustShift, Thewanger, NickK, Cillian flood, Csigabi, Flewis, Materialscientist, ImperatorExercitus, Citation bot, Gilderoy8, Nika 243, Bob Burkhardt, Yelloeyes, Maxis ftw, Frankenpuppy, Obersachsebot, MauritsBot, Xqbot, Timir2, Intelati, Cureden, Addihockey10, Capricorn42, Sungmanitu, Thermoproteus, Ultimation, J04n, GrouchoBot, Abce2, Brandon5485, Arch27, Doulos Christos, GhalyBot, Halubihalubi, Shadowjams, E0steven, Captain Weirdo the Great, Fortdj33, LucienBOT, Dger, Sebastiangarth, Bukovets, JohnL.Weber, Robo37, HamburgerRadio, Citation bot 1, Harleh, Pinethicket, I dream of horses, Haaqfun, Jonesey95, MJ94, Calmer Waters, Icemerang, Hoo man, Hantzen, SpaceFlight89, Île flottante, Ronald0216, VinnyXY, Newgrounder, Tim1357, ⿰⿰⿰⿰, Abc518, Gamewizard71, FoxBot, TrickyM, TobeBot, Jonkerz, Comet Tuttle, Pclaplante, Kmw2700, Defender of torch, Cowlibob, Joodeak, Mttcmbs, Xrmach, Suffusion of Yellow, PleaseStand, DARTH SIDIOUS 2, Fellisha123, Bmathews96, Onel5969, TheRealSimmonds, RjwilmsiBot, TjBot, Shiftyfifty, Hajatvrc, DASHBot, EmausBot, John of Reading, WikitanvirBot, JohnXCitizen, Gfoley4, Razor2988, RA0808, Nquinn91, Slightsmile, Tommy2010, Wikipelli, ZéroBot, Ὁ οἶστρος, Zap Rowsdower, David J Johnson, Jay-Sebastos, TyA, L Kensington, Photofem, MonoAV, Donner60, CountMacula, Puffin, Quadruplum, ChuispastonBot, VictorianMutant, DASHBotAV, ClueBot NG, Prohistorygeek, Uziw, This lousy T-shirt, Tideflat, McGrowski, Muon, Rezabot, Widr, Wllmevans, PatHadley,

Darrend67, 0987oiuy, Helpful Pixie Bot, Qaewsd, Editking612, Electriccatfish2, Titodutta, Calabe1992, Gob Lofa, Hopekatienom, Seistho, Bellardoo, BG19bot, Smallerjim, Goddamcaptchacode, MusikAnimal, தென்காசி சுப்பிரமணியன், Sowsnek, FiveColourMap, CimanyD, Cowsgobob, Jinglearceus, Lmarcell14, Chip123456, A Timelord, Teammm, Zhaofeng Li, ChrisGualtieri, Codeh, TacticalTurtleneck, EuroCarGT, Titchybear, Verryniceguy2, Dexbot, T v shah, Lugia2453, Ak5791, Frosty, Schiltron, Darth Sitges, PizzaHutCreeper, Ruby Murray, Tentinator, AnthonyJ Lock, Emily2117, Ugog Nizdast, Laurenmathews123, Ginsuloft, Pratheshsum, Xspike15, Manul, Lizia7, Kimboslicee, JaconaFrere, Dinorexcoolio, Monkbot, Supermariolink777, Braden12345, Iluhrs, Owen minns, Flingbong, HMSLavender, 115ash, Alpha Monarch, Oinksgiant2000, Ninafundisha, Speedytacos, Surajbryan, Jessica Simpson03, ⁇⁇⁇⁇⁇, Johanna, Guillame Heavensburg, MacPoli1, Addemf, Aset34, GeneralizationsAreBad, Nokia loomia 630, KasparBot, 3 of Diamonds, Yusdu14, Happypoos, Basuza.majumdar, Wikiass123, Mannermauler, N120pA, Calipachanguero and Anonymous: 1061

- **Paleolithic** *Source:* https://en.wikipedia.org/wiki/Paleolithic?oldid=689605309 *Contributors:* Brion VIBBER, Rickyrab, Gabbe, Sebastian-Helm, Stan Shebs, Jebba, Glenn, Llull, Cimon Avaro, Qqq, Emperorbma, Timwi, Zarggg, Gutza, The Anomebot, Tpbradbury, SEWilco, Joy, Penfold, Pumpie, Lumos3, Robbot, Chris 73, Romanm, Arkuat, Cornellier, UtherSRG, Wereon, Casito, ManuelGR, Nagelfar, MPF, Tourguide, Inter, Hagedis, Yak, Everyking, Wronkiew, Jabowery, Alanl, Chowbok, Gadfium, Pgan002, Andycjp, Geni, Antandrus, Adamsan, Discospinster, Rich Farmbrough, Kdammers, Vsmith, Roybb95~enwiki, Dbachmann, Bender235, ESkog, Malkin, Silentlight, Andrewlunch, Nabla, Eric Forste, Shanes, Bobo192, Vervin, Janna Isabot, Wipe, Jjk, SpeedyGonsales, Beetle B., Nsaa, Mdd, Ranveig, Storm Rider, Alansohn, Philip Cross, Dmismir, Bart133, Snowolf, Burwellian, Wtmitchell, Velella, Jobe6, Amorymeltzer, CloudNine, Sciurinæ, Japanese Searobin, Sheynhertz-Unbayg, Zntrip, Nuno Tavares, Velho, Firsfron, Woohookitty, Mindmatrix, Sandius, Mazca, MatthewJ, Jørgen Holm, Chochopk, Jeff3000, Phlebas, Palica, Stevey7788, Mandarax, Graham87, BD2412, Deadcorpse, Mendaliv, Dwarf Kirlston, Crzrussian, Rjwilmsi, Koavf, Sman~enwiki, Lordkinbote, Kalogeropoulos, Muj0, Nihiltres, Itinerant1, NekoDaemon, RexNL, ChongDae, Jrtayloriv, Losecontrol, Eric.dane~enwiki, Chobot, K2wiki, DVdm, VolatileChemical, Ahpook, Jimbobsween, Kummi, YurikBot, Retaggio, Jimp, Peter G Werner, Phantomsteve, Fabartus, Yamara, Stephenb, Gaius Cornelius, Anomalocaris, NawlinWiki, Grafen, FritzG, Ghidra99, Irishguy, Patrick MMA Bringmans, Ospalh, Stefeyboy, Kkmurray, Botteville, Wknight94, Igiffin, Phgao, 2over0, Snuppy, Lt-wiki-bot, Smoggyrob, Closedmouth, Zorgrian, CWenger, JLaTondre, David Biddulph, Kungfuadam, Luk, SmackBot, Snickersnee, InverseHypercube, NaiPiak, Unyoyega, Mountain Goat, Eskimbot, Alsandro, Athinaios, Yamaguchi⁇⁇, Gilliam, Betacommand, Kinhull, Amatulic, Chris the speller, Bluebot, Silly rabbit, Bazonka, DHN-bot~enwiki, Rama's Arrow, Tsca.bot, Skoglund, Wwelles14, Normxxx, Khoikhoi, Flyguy649, Paul H., Ster~enwiki, The PIPE, BryanG, Cosmix, Takowl, Zeneky, Mitsuki152, Thor Dockweiler, SashatoBot, Swatjester, Straif, Grumpy444grumpy~enwiki, Korean alpha for knowledge, Locutus Borg~enwiki, Mgiganteus1, IronGargoyle, Fandalaar, A. Parrot, S kitahashi, Geologyguy, Anonymous anonymous, Iridescent, SkyWalker, CmdrObot, Richard Keatinge, Utsav.schurmans, Cydebot, Pais, A D 13, Mato, Gogo Dodo, Doug Weller, Phydend, Kozuch, Daven200520, Maziotis, Malleus Fatuorum, Thijs!bot, Epbr123, Parsa, Dasani, Marek69, SGGH, NorwegianBlue, Second Quantization, Chrisdab, Nick Number, Blathnaid, Escarbot, AntiVandalBot, Jj137, Modernist, Darklilac, Spencer, Sluzzelin, Deadbeef, JAnDbot, Samar, Chicken Wing, Maybejames, Michig, Hut 8.5, Jkalm, Geniac, Magioladitis, Connormah, WolfmanSF, Pedro, Bongwarrior, VoABot II, Jerome Kohl, Rivertorch, ClovisPt, Fang 23, DerHexer, Austin luce, Snackattack22, Vigyani, BetBot~enwiki, Ugajin, CommonsDelinker, Nono64, N4nojohn, J.delanoy, Pharaoh of the Wizards, Zorakoid, Uncle Dick, Nigholith, Geog1, McSly, NewEnglandYankee, 83d40m, Juliancolton, Summer35, SoCalSuperEagle, Squids and Chips, VolkovBot, Johan1298~enwiki, DOHC Holiday, Thisisborin9, Macedonian, Jeff G., Krauq, AlnoktaBOT, Stefan Kruithof, Philip Trueman, TXiKiBoT, Zidonuke, Altruism, Technopat, Raven rs, Aymatth2, Wiikipedian, Slysplace, Broadbot, LeaveSleaves, Howryn, Starlists, Billinghurst, Burntsauce, AlleborgoBot, Nagy, Jgsack, Logan, NHRHS2010, EmxBot, SieBot, BotMultichill, Jauerback, Typritc, Isf23, Bentogoa, Flyer22 Reborn, Tiptoety, Micke-sv, SPACKlick, Allmightyduck, Mimihitam, Oxymoron83, Prof. Thomas, Lightmouse, Californiajeff, Diego Grez-Cañete, JohnSawyer, StaticGull, Monroetransfer, Sean.hoyland, Susan118, CP2002, Denisarona, Escape Orbit, Randy Kryn, Velvetron, Invertzoo, Atif.t2, ClueBot, Phoenix-wiki, Hippo99, The Thing That Should Not Be, Witchwooder, Arakunem, Mild Bill Hiccup, Boing! said Zebedee, OccamzRazor, Thegargoylevine, Phenylalanine, WestwoodMatt, Chris Kutler, DragonBot, Jusdafax, Mr squelch, Kanguole, Muenda, Peter.C, Peachypoh, Regainfo, Antiquary, SchreiberBike, Audaciter, Woojitsu, Kakofonous, Chaosdruid, Thingg, SoxBot III, Chhe, DumZiBoT, Bussech, Skarebo, Aunt Entropy, Addbot, Willking1979, DOI bot, Betterusername, Captain-tucker, Richardf630, Fluffernutter, Millerjb, Devrit, AnnaFrance, ChenzwBot, LinkFA-Bot, West.andrew.g, Tassedethe, Tom2wheatley, Tide rolls, BrianKnez, OlEnglish, AlexJFox, CountryBot, Micke, Margin1522, Legobot, Luckas-bot, Yobot, 2D, Apollonius 1236, DisillusionedBitterAndKnackered, THEN WHO WAS PHONE?, Absolutely Trustworthy, Knownot, Untrue Believer, AnomieBOT, IRP, Frank Ridollen, Chuckiesdad, Kingpin13, Colutowe, Flewis, Materialscientist, Hunnjazal, Citation bot, LightCMM, YardsGreen, ArthurBot, Guitarman23, LilHelpa, Xqbot, Anime lover2, Rimonh, Gigemag76, Thermoproteus, GrouchoBot, Addingrefs, Franco3450, Amaury, Philip72, Moxy, Joaquin008, Samwb123, Tobby72, Altg20April2nd, Levalley, Perdman, VS6507, Westcoaston, Weetoddid, Commit charge, Citation bot 1, DrilBot, Pinethicket, I dream of horses, Icemerang, Jschnur, RedBot, MastiBot, Hessamnia, Utility Monster, Trappist the monk, HelenOnline, Vrenator, Moy tuachi 4009, Alexsaba96, Geyuan08, Fireisme, Davish Krail, Gold Five, Maltaydavid, Bobalabalu, Jeffrd10, Stephen MUFC, Brokendrumpedal, Reach Out to the Truth, Ultimate alliance, DARTH SIDIOUS 2, Aviv007, RjwilmsiBot, NameIsRon, Ripchip Bot, Agent Smith (The Matrix), WildBot, Androstachys, Mordekai wiki, Deagle AP, Giorgiogp2, EmausBot, WikitanvirBot, Immunize, Gfoley4, Look2See1, ScottyBerg, Racerx11, Slightsmile, Denhetreil, ZéroBot, Fæ, Josve05a, MithrandirAgain, Moparman147, Cobaltcigs, H3llBot, Wayne Slam, Aidarzver, L Kensington, Nyappy001, Donner60, DASHBotAV, Ebehn, Miradre, Special Cases, Xanchester, Gwen-chan, ClueBot NG, This lousy T-shirt, Deedee96, -sche, EnekoGotzon, Frietjes, Cntras, Tlainen, Marechal Ney, CaroleHenson, Widr, Helpful Pixie Bot, Calabe1992, Gob Lofa, DBigXray, BG19bot, CityOfSilver, MusikAnimal, AvocatoBot, Mark Arsten, ChidemK, Cadiomals, Hello99998, Joshua Jonathan, Nahlaugh, Shay21d, EricTheGreat1999, BattyBot, A Timelord, Pratyya Ghosh, Acadēmica Orientālis, Angela MacLean, Soulbust, Bill.lovell, BrightStarSky, Dexbot, Mogism, Lugia2453, Tary123, UNC fan-4 life, Minecraft1234567890, Peepeelovaa, BobDaWurf, Eyesnore, EvergreenFir, PurserSmith, DavidLeighEllis, Ebmc5, Ginsuloft, Iceman784, Dutchdreams, Zambelo, Fafnir1, Monkbot, SantiLak, TheQ Editor, Adam nieblas-haly, Peter238, Saurusaurus, Aconn91, TranquilHope, Ilovemysister777, Jlb2011, Bro3333, Drkimwells, Russelsouthard, LBUSER13, Bruhlify.3, KasparBot and Anonymous: 677

- **Lower Paleolithic** *Source:* https://en.wikipedia.org/wiki/Lower_Paleolithic?oldid=681229891 *Contributors:* AxelBoldt, Edward, Ixfd64, SebastianHelm, Glenn, Joy, Chrisjj, Casito, Nagelfar, Ancheta Wis, Adamsan, Ushishir, Neutrality, Noisy, Pmsyyz, Vsmith, Barista, Dbachmann, Ben Standeven, Eric Forste, Molecular b, Miaow Miaow, Phlebas, Palica, NekoDaemon, YurikBot, Hairy Dude, Wiki alf, Zwobot, Botteville, Igiffin, Smoggyrob, Xaxafrad, SmackBot, Vald, Alsandro, Hibernian, Sgt Pinback, Chlewbot, Aleator, Retromaniac, A. Parrot, Succubus MacAstaroth, Clf99, Floris V, CieloEstrellado, Thijs!bot, Chrisdab, AntiVandalBot, Czj, Peer Gynt, JAnDbot, Fang 23, NatureA16, Johnbod, MikeEagling, Russianname, 83d40m, VolkovBot, Fences and windows, Raven rs, Jpeeling, SieBot, Randy Kryn, Soporaeternus, Auntof6, Alexbot, Perkeleperkele, SchreiberBike, SilvonenBot, Addbot, Elvire, Asfreeas, Tide rolls, Luckas-bot, TreytonP, Aboalbiss, AnomieBOT,

Hairhorn, Xqbot, Coffeetalkh, Tsinfandel, The Utahraptor, TjBot, EmausBot, Veda784, BartlebytheScrivener, ClueBot NG, LordAkaros, The-Conduqtor, Zackzing, BG19bot, BattyBot, Will Sandberg, OccultZone, Monkbot, Ninafundisha, AnatuZeder, Wyatt242, KasparBot, Jjello and Anonymous: 44

- **Oldowan** *Source:* https://en.wikipedia.org/wiki/Oldowan?oldid=682921427 *Contributors:* Rickyrab, Infrogmation, DopefishJustin, Skysmith, Glenn, Andres, Dimadick, Fredrik, Fifelfoo, Sam Spade, Merovingian, UtherSRG, Michael Snow, Lizard King, Stirling Newberry, Ancheta Wis, Obli, Coldacid, Andycjp, Adamsan, Noisy, Rich Farmbrough, Guanabot, Dbachmann, Bender235, Geos, Storm Rider, Eric Kvaalen, Tainter, Stemonitis, Brunnock, Miaow Miaow, WadeSimMiser, Magister Mathematicae, Mendaliv, Josh Parris, Rjwilmsi, NekoDaemon, Chobot, Bgwhite, Hairy Dude, RussBot, Gaius Cornelius, Anomalocaris, Welsh, Moe Epsilon, Ezeu, Bota47, Botteville, Tekana, SmackBot, Herostratus, Vald, Jagged 85, IstvanWolf, Bluebot, Snori, Hibernian, Bazonka, Colonies Chris, Julius Sahara, Stevenmitchell, Yahewe, Yom, Thanos5150, JorisvS, Locutus Borg~enwiki, Midnightblueowl, Clf99, JoeBot, ErWenn, Courcelles, Bearingbreaker92, CGWhitsett, JForget, Poethical, Floris V, Cydebot, Mattisse, JamesAM, Thijs!bot, Vertium, Philippe, I20, Schlim, Darklilac, Spencer, JAnDbot, T L Miles, Father Goose, The Anomebot2, Fang 23, WLU, R'n'B, MacAuslan, Lizbetann, Johnbod, Chiswick Chap, Blueshifter, Trilobitealive, Santh, VolkovBot, Tubbienine, Aymatth2, Milkbreath, Enigmaman, Pre-ski18888, Closenplay, SieBot, YTS275, Gfglegal, Flyer22 Reborn, Goustien, Randy Kryn, ClueBot, Avenged Eightfold, Ashashyou, Orpheus1989, Garing, BOTarate, Versus22, Keilmesser, SilvonenBot, Jbeans, Aunt Entropy, Addbot, Oryxgazella, Queenmomcat, AndersBot, Numbo3-bot, Zorrobot, CountryBot, Legobot, Luckas-bot, Yobot, WikiDan61, Ptbotgourou, Untrue Believer, Papercup47, AnomieBOT, Citation bot, Xqbot, J04n, Speednat, Levalley, Archaeodontosaurus, AstaBOTh15, Pinethicket, Trappist the monk, Xook1kai Choa6aur, Jonkerz, Kmw2700, RjwilmsiBot, John of Reading, Laszlovszky András, John756, JoeSperrazza, Wakebrdkid, ClueBot NG, TheConduqtor, CaroleHenson, Widr, PatHadley, Andreygeo, Werseuch, BG19bot, Hamish59, Estevezj, BattyBot, Sylvia253, SomeGuyWhoRandomlyEdits, Heyluckysandwich, Dexbot, Hsizang, Lungtikchuanren, SparkE303, Kaficek, Badietz, Debouch, Wikieditingphile, Melcous, Monkbot, Johnsoniensis, Ninafundisha, Lady Miyazawa, Giraffacamelopardalis, A mzungu, Losskakel, JLebowski94, Coolgrandma420, Abedisbatmannow, Liying15, SeriouslySerious, ElphabaThropp13 and Anonymous: 100

- **Acheulean** *Source:* https://en.wikipedia.org/wiki/Acheulean?oldid=686622536 *Contributors:* Llywrch, Stevenj, Glenn, Andres, Jengod, Gutsul, Samsara, Bevo, Penfold, UtherSRG, LX, Lizard King, Nunh-huh, Cantus, Fjarlq, 20040302, PDH, Adamsan, Michaeltaft, Chmod007, Smitha, Cagliost, Dbachmann, Kwamikagami, Jonegn, Ptyxs, SteinbDJ, Sheynhertz-Unbayg, Woohookitty, Miaow Miaow, Palica, Marudubshinki, Lusitana, Rjwilmsi, PinchasC, Mike s, FlaBot, Ian Pitchford, RJP, NekoDaemon, Hal Bird, YurikBot, RussBot, Gustavb, Anomalocaris, Nahallac Silverwinds, Dysmorodrepanis~enwiki, Badagnani, Albedo, Pyroclastic, Ezeu, Bota47, Botteville, Drallim, SmackBot, Vald, Eskimbot, Ohnoitsjamie, Cattus, MalafayaBot, Hibernian, Bazonka, Colonies Chris, Mike hayes, Krsont, Savidan, Geoffr, Ligulembot, Potosino, Locutus Borg~enwiki, Agathoclea, Twas Now, Soneten, CmdrObot, Pankajjain, Ruslik0, Floris V, CieloEstrellado, Thijs!bot, Julia Rossi, JAnDbot, Deflective, T L Miles, Magioladitis, MSisk, Warren Dew, CommonsDelinker, Hans Dunkelberg, 88888, Dispenser, Johnbod, Idioma-bot, Dslmagenta, VolkovBot, AlnoktaBOT, Fences and windows, Tubbienine, TXiKiBoT, VonRichthofen, AlleborgoBot, Arun11, Micke-sv, OsamaBinLogin, Randy Kryn, SallyForth123, Boneyard90, Alexbot, Crywalt, Chefallen, Stephen Poppitt, Addbot, La Fuente, Lightbot, مانی, Karatorian, Luckas-bot, Yobot, Fraggle81, קרול ישראל, Gongshow, AnomieBOT, Xufanc, Citation bot, Jeffrey Mall, GrouchoBot, Speednat, MerlLinkBot, Archaeodontosaurus, D'ohBot, MastiBot, Trappist the monk, Zoeperkoe, Cowlibob, DexDor, DennisIsMe, ChuispastonBot, ClueBot NG, MIKHEIL, Justlettersandnumbers, CaroleHenson, Widr, PatHadley, BG19bot, Jahani65, Nirjhara, MangoWong, தென்காசி சுப்பிரமணியன், Achowat, BattyBot, Ggal2, JYBot, SomeGuyWhoRandomlyEdits, Dexbot, Wyken Seagrave, Lugia2453, Danny Sprinkle, Stamptrader, Ninafundisha, Wyatt242, A mzungu, 120 and Anonymous: 67

- **Homo** *Source:* https://en.wikipedia.org/wiki/Homo?oldid=689003630 *Contributors:* Josh Grosse, Toby Bartels, Azhyd, Montrealais, D, Talshiarr, Gabbe, Julesd, Bogdangiusca, Llull, CarlKenner, Denny, Timwi, Fvw, Jason Potter, Ghouston, Kizor, Rursus, Radomil, Hadal, UtherSRG, GerardM, Giftlite, MPF, Fennec, Curps, Geoffspear, Chowbok, Gdr, Ran, Antandrus, OverlordQ, JohnArmagh, Sonett72, Discospinster, Florian Blaschke, Silence, Arthur Holland, Dbachmann, SamEV, Kbh3rd, Evice, CanisRufus, Livajo, Hayabusa future, Felagund, RoyBoy, Thor Andersen, Themusicgod1, Bobo192, Whosyourjudas, Fremsley, Giraffedata, Nk, Ardric47, Haham hanuka, Ranveig, Alansohn, Gary, Eric Kvaalen, ABCD, Riana, Ahruman, Theodore Kloba, Knowledge Seeker, Dinoguy2, Rhialto, BDD, Chamaeleon, Jackhynes, Zntrip, Mindmatrix, LOL, Rocastelo, WadeSimMiser, Duncan.france, KFan II, GregorB, Eras-mus, Waldir, Banpei~enwiki, Phlebas, GSlicer, Ashmoo, Ryoung122, Cuchullain, Drbogdan, Rjwilmsi, Sacresd, Jake Wartenberg, Quiddity, Ems57fcva, DoubleBlue, Ucucha, Dionyseus, Eubot, Crazycomputers, Nivix, RexNL, Gurch, John Maynard Friedman, Neutrinoman, Gareth E Kegg, SGreen~enwiki, Chobot, DVdm, Bdelisle, Gdrbot, Hede2000, Limulus, Shell Kinney, Rsrikanth05, Dysmorodrepanis~enwiki, Bachrach44, Snek01, Test-tools~enwiki, Voyevoda, Ptcamn, Aufidius, RazorICE, Apokryltaros, Ngorongoro, E rulez, Ctobola, Bucketsofg, Brat32, DeadEyeArrow, Wknight94, WAS 4.250, Andrew Lancaster, Terry Longbaugh, Smoggyrob, Pb30, Dspradau, Emc2, Nixer, AndrewWTaylor, Luk, True Pagan Warrior, SmackBot, Reedy, Herostratus, KnowledgeOfSelf, TestPilot, Hydrogen Iodide, Jrockley, Delldot, Frymaster, Boris Barowski, Timotheus Canens, Gilliam, Ohnoitsjamie, Amatulic, Hitman012, Rkitko, AndrewRT, Enkyklios, Master of Puppets, Behaafarid, Epastore, Quena@sympatico.ca, Darth Panda, Verrai, Modest Genius, King of the Dancehall, Can't sleep, clown will eat me, Abyssal, Alphathon, XQ fan, Addshore, TheLateDentarthurdent, Theonlyedge, "alyosha", Adrigon, DMacks, Metamagician3000, DDima, GameKeeper, John, LinuxDude, Mgiganteus1, Bjankuloski06en~enwiki, Tlesher, Mr. Lefty, A. Parrot, Soulkeeper, Slakr, Optakeover, Iridescent, Kaarel, Chadpeters, Tawkerbot2, Robinhw, JForget, Alexei Kouprianov, Snorkelman, FunPika, Wafulz, Bonás, JohnCD, Yopienso, Richard Keatinge, Ark-pl, Gogo Dodo, Drur93, Chasingsol, Shirulashem, Doug Weller, Ameliorate!, Dyanega, Shall I Make Wikipedia A Boom Town?, Thijs!bot, Epbr123, Nowimnthing, Mojo Hand, Marek69, James086, Yettie0711, Bob the Wikipedian, Sean William, Natalie Erin, Escarbot, Hmrox, KrakatoaKatie, AntiVandalBot, Luna Santin, Mark t young, Seaphoto, Gnixon, Smartse, Farosdaughter, Credema, Petedavo, Myanw, Sluzzelin, JAnDbot, Barek, Denn333, 100110100, WolfmanSF, Bongwarrior, VoABot II, Soulbot, The truth must be told, Afu1111, Zephyr2k~enwiki, Wormcast, BrianGV, Branka France, Ksanyi, BatteryIncluded, Edmundwoods, 28421u2232nfenfcenc, E104421, Lenticel, Jackson Peebles, MartinBot, IgorSF, Rettetast, Anaxial, Keith D, WelshMatt, Julli321, LedgendGamer, J.delanoy, Captain panda, Pharaoh of the Wizards, Bogey97, Natasia, Octopus-Hands, It Is Me Here, McSly, Hunstiger, AntiSpamBot, Vanished User 4517, LittleHow, Juliancolton, Plindenbaum, Prot D, Treisijs, Jakeowns2, RJASE1, Richard New Forest, FeralDruid, Wikieditor06, VolkovBot, Temporarily Insane, HyperSonicBoom, Soliloquial, Philip Trueman, Oshwah, Chaffinc, Imasleepviking, Leafyplant, LeaveSleaves, Seb az86556, UnitedStatesian, Chickyfuzz, Lova Falk, 4444hhhh, Purgatory Fubar, Spinningspark, Insanity Incarnate, Ceranthor, Logan, EmxBot, Coffee, Tommynewman, Tresiden, Rlendog, Dawn Bard, Yintan, Revent, FunkMonk, Qst, SPACKlick, JSpung, Aruton, Tapau6, Avnjay, ReformedXtian, Steven Crossin, Josuf07, Lightmouse, Helikophis, KathrynLybarger, Taggard, Hippie Metalhead, Spotty11222, Pinkadelica, Dolphin51, Randy Kryn, Loren.wilton, ClueBot, The Thing That Should Not Be, Rjd0060, Taxobot, Hafspajen, Otolemur crassicaudatus, Excirial, Orpheus1989, Gnome de plume, Jusdafax, Erebus Morgaine, Eeekster, Gulmammad, Parisian-

eagle, LittleHow, Arodery, Kyle the bot, Mzmadmike, David Condrey, Phil Bridger, Nordic Crusader, Mygerardromance, Randy Kryn, Clue-Bot, Alexbot, Arjayay, Keilmesser, Mpondopondo, Gene Fellner, Addbot, DrHerries, Yobot, Fraggle81, Jim1138, Materialscientist, 3family6, StPernar, Adunbar505, Altg20April2nd, Hchc2009, Axxter99, EmausBot, Dewritech, Donner60, Rocketrod1960, ClueBot NG, CherryX, BG19bot, Ninafundisha, DiSchamelrider, AnatuZeder, Abedisbatmannow, Oleaster and Anonymous: 25

- **Neanderthal** *Source:* https://en.wikipedia.org/wiki/Neanderthal?oldid=689406940 *Contributors:* Dreamyshade, Brion VIBBER, Vicki Rosen-zweig, Bryan Derksen, The Anome, F. Lee Horn, Jeronimo, Clasqm, Fnielsen, Kowloonese, Rmhermen, Ortolan88, Zoe, Azhyd, Jaknouse, B4hand, Olivier, OlofE~enwiki, Vik-Thor, Michael Hardy, Alan Peakall, DopefishJustin, Dan Koehl, Gabbe, Tannin, Ixfd64, Zeno Gantner, Skysmith, Kosebamse, Shimmin, Ahoerstemeier, Ronz, Jimfbleak, Yaronf, Kingturtle, Rlandmann, Julesd, Glenn, Marco Krohn, Mykdavies, Sray, Nikai, Susurrus, Llull, Jiang, Vargenau, Jerryb1961, Crusadeonilliteracy, Timwi, Dcoetzee, Zarggg, Wikiborg, Alex n, Jwrosenzweig, StAkAr Karnak, Wik, Timc, Tpbradbury, Furrykef, Samsara, Topbanana, Joy, Oaktree b, AnonMoos, Wetman, Penfold, Chl, Bcorr, Frazzydee, Jeffq, Owen, 80.255, Jason Potter, Gromlakh, Bearcat, Branddobbe, Robbot, Frank A, Sander123, Astronautics~enwiki, ChrisO~enwiki, Chrism, Kizor, Tomchiukc, Donreed, Altenmann, Peak, Nurg, Naddy, Lowellian, Orthogonal, Mirv, Postdlf, Rursus, Sunray, Hadal, UtherSRG, Wikibot, GerardM, Aetheling, Jor, Casito, Cyrius, Lizard King, Marvelite, Giftlite, DocWatson42, MPF, Awolf002, Abiola Lapite, Wolfkeeper, Nunh-huh, Tom harrison, IRelayer, Aphaia, Orpheus, Binadot, Yak, Peruvianllama, Dratman, Henry Flower, FeloniousMonk, Kenneth Alan, RScheiber, Duncharris, Alensha, Kpalion, J-V Heiskanen, Matthead, Dolfin~enwiki, Bobblewik, Golbez, Erik Carson, Woggly, Gdr, Sonjaaa, Quadell, Blankfaze, Antandrus, Alteripse, Beland, Cjewell, Oscar, Kaldari, Murple, Anythingyouwant, OwenBlacker, DragonflySixtyseven, Martin.komunide.com, Ensrifraff, Sonett72, Trevor MacInnis, Grunt, Freakofnurture, JTN, RedWordSmith, Diagonalfish, Blanchette, Discospinster, Rich Farmbrough, KillerChihuahua, Vague Rant, Vsmith, Iswm, Florian Blaschke, Smyth, Dave souza, Francis Davey, Dbachmann, Michael Zimmermann, Bender235, Rubicon, ESkog, Swid, Dgorsline, El C, Carlon, Cedders, Kwamikagami, Laurascudder, Worldtraveller, Art LaPella, Xed, Ralphounet, Bobo192, Comtebenoit, Janna Isabot, BW, Smalljim, Viriditas, CloudSurfer, Cohesion, Mdkarazim, NickSchweitzer, Sleske, Maxl, Hesperian, MPerel, Haham hanuka, Pharos, Pearle, Paul McMahon, HasharBot~enwiki, Earthbound01, Shadoks, Espoo, Jumbuck, Anthony Appleyard, ThePedanticPrick, 119, Eric Kvaalen, Arthena, Plumbago, Ahruman, Mailer diablo, Wanderingstan, Brentford, Avenue, Katefan0, DreamGuy, Ombudsman, Angelic Wraith, Wtmitchell, Velella, TaintedMustard, Danaman5, Stephan Leeds, Shashark, Cmapm, Pfahlstrom, Guthrie, Ndteegarden, Deathphoenix, Chamaeleon, Walshga, Oleg Alexandrov, Brookie, Sheynhertz-Unbayg, Jackhynes, Woohookitty, Bjones, Mindmatrix, FeanorStar7, LOL, DoctorWho42, Politas, Yansa, Spettro9, Mark K. Jensen, Ekem, Unixer, Venik~enwiki, MGTom, Encyclopedist, Hfarmer, GeorgeTSLC, Eleassar777, T D Thomas, Sengkang, GregorB, Macaddct1984, Arrkhal, Eyreland, Funhistory, Doco, Doric Loon, Prashanthns, Ae7flux, Mtloweman, Phlebas, Volatile, Paxsimius, Ashmoo, Graham87, Flypaper, Ryoung122, WBardwin, FreplySpang, Charmii, Myonlyemotion27, Jclemens, Melesse, Djanvk, Drbogdan, Vberger~enwiki, Rjwilmsi, Koavf, Smoe, DeadlyAssassin, OneWeirdDude, XP1, Josiah Rowe, Feydey, Palpatine, Yug, Cheesy123456789, Ucucha, Yamamoto Ichiro, Dionyseus, Ravidreams, Dinosaurdarrell, FlaBot, Eubot, KarlFrei, Ewlyahoocom, Gurch, Arctic.gnome, Quuxplusone, Tomer Ish Shalom, EronMain, CJLL Wright, Chobot, Hatch68, Gdrbot, Bgwhite, Wjfox2005, Sus scrofa, YurikBot, Hairy Dude, Rtkat3, Waitak, Huw Powell, Ste1n, Davehatcher911, Brandmeister (old), Luis Fernández García, TheDoober, Witan, Warmaster, Pigman, Chris Capoccia, Chuck Carroll, Groogle, Gerbil, Scott5834, Stephenb, Gaius Cornelius, Kyorosuke, Sentausa, NawlinWiki, Muntuwandi, Dysmorodrepanis~enwiki, Wiki alf, Complainer, The Ogre, Voyevoda, TVilkesalo~enwiki, Nebogipfel~enwiki, Joel7687, Sylvain1972, Nascigl, Apokryltaros, Patrick MMA Bringmans, Cenedi, PhilipC, GracieLizzie, Hypehuman, Butterflyvertigo, Dbfirs, DryaUnda, BOT-Superzerocool, Kortoso, DeadEyeArrow, Rdos, Private Butcher, Botteville, Rob117, Thegreyanomaly, Maunus, Narsamson, Wknight94, Atiemann, Igiffin, Pawyilee, FF2010, J S Ayer, Orioane, Paul Magnussen, Citynoise, 2over0, Zzuuzz, Hoopshank, Open2universe, Aremisasling, RWFanMS, RDF, Arthur Rubin, Th1rt3en, Grmagne, Sarefo, BorgQueen, Petri Krohn, Shawnc, Smurrayinchester, Andyluciano~enwiki, HereToHelp, Katieh5584, Otto ter Haar, GrinBot~enwiki, Elliskev, Kf4bdy, Jbull, NickelShoe, NetRolller 3D, Luk, SmackBot, BaKanale, Darthmix, Mjposner, Harbinger22, KnowledgeOfSelf, TestPilot, McGeddon, Giraldusfaber, Speight, WilyD, Davewild, Anastrophe, Declare, Jefs, Antrophica, ASarnat, Athaler, AnOddName, Vugluskr, Francisco Valverde, KED, Yamaguchi[?][?], Gilliam, Universal1300, ERcheck, Smeggysmeg, Andy M. Wang, Durova, Welwitschia, Chris the speller, Daemonward, Nativeborncal, Bluebot, Qwasty, Autarch, Alamaison, Persian Poet Gal, RDBrown, NCurse, Postoak, MK8, Cattus, Tisthammerw, Jayanta Sen, Ryan Paddy, SchfiftyThree, Hibernian, Moshe Constantine Hassan Al-Silverburg, Hooriaj, Jfsamper, Ikiroid, Baa, CMacMillan, Quena@sympatico.ca, Aridd, Zachorious, Scwlong, Peter Campbell, Zsinj, King of the Dancehall, Kotra, Trekphiler, Derekt75, Hartebeest, Scott3, Shalom Yechiel, AltGrendel, Chlewbot, OrphanBot, Avb, XQ fan, Slau, Addshore, Edivorce, Yahewe, Jmlk17, FrankBlissett, Downwards, Sei Shonagon~enwiki, Nakon, SnappingTurtle, Dreadstar, TrogdorPolitiks, Alexandra lb, So cool~enwiki, NickPenguin, DMacks, Ladlergo, Kendrick7, Where, Epf, Captainbeefart, Nasz, Bcasterline, Quendus, Icelandic Hurricane, JzG, 3dnatureguy, Kuru, Titus III, Richard L. Peterson, Euchiasmus, Drewby9000, J 1982, Tazmaniacs, SilkTork, Thanos5150, Gobonobo, Soumyasch, Tim Q. Wells, Mgiganteus1, Ocatecir, Abotnick, Extremophile, AbuAmir, Niroht, The Man in Question, Petter73, Pendragon39, Helzagood, MarkSutton, Principessa, Smith609, BillFlis, Deceglie, Rainwarrior, Munita Prasad, Freyr35, Giant Blue Anteater, InedibleHulk, Mets501, Mbuk, Ryulong, Psj333, Sims2789, MTSbot~enwiki, Digsdirt, SmokeyJoe, Peyre, Xionbox, MrDolomite, Zepheus, KJS77, Jgm22, Iridescent, JMK, TerryE, SanchoEDLP, Kagome717, Brandizzi, Joseph Solis in Australia, JoeBot, Newone, Cela~enwiki, Twas Now, Omulan, Shoshonna, DavidOaks, Happy-melon, Majora4, Trialsanderrors, Courcelles, Ziusudra, Tawkerbot2, CalebNoble, Robinhw, CmdrObot, Tanthalas39, Huehauh, Ale jrb, Andrew E. Drake, Mattbr, Wafulz, Agathman, Scohoust, Makeemlighter, Homeslice640, Ossanha, Rickpedia, Rogerborg, Hakluyt bean, Ruslik0, Djus, Rabid Lemur, Casper2k3, Richard Keatinge, JettaMann, MrFish, Skybon, Utsav.schurmans, Yaris678, Cydebot, Slickpicker, A D 13, Slp1, Rehash84, Anonymi, Anthonyhcole, Drur93, Anonymous44, DangApricot, Dusty relic, Capedia, Strangelv, B, Michael C Price, Tawkerbot4, Quibik, Luccas, Doug Weller, Jeannbean, DumbBOT, Jmasalle, Narayanese, Omicronpersei8, Zalgo, Jahiliyyah, Maziotis, Saintrain, HJJHolm, Jdlyall, Scottaylor, Thijs!bot, Brstahl, Epbr123, Nowimnthing, Tmkeesey, The Sausage Knight, Marek69, Ufwuct, Lewallen, Warfwar3, Khb, EdJohnston, CharlotteWebb, Dawkeye, Edhubbard, Rotundo, Rriegs, I20, Natalie Erin, Escarbot, Thadius856, Hires an editor, AntiVandalBot, BokicaK, Mark t young, Blue Tie, Turlo Lomon, Nagara373, Dbrodbeck, Opelio, Quintote, Jbaranao, Julia Rossi, Jj137, Tjmayerinsf, Mack2, Jason12345, Modernist, Darklilac, RedCoat10, SkoreKeep, Alller, Petedavo, Lantios, Bjenks, Arx Fortis, Eleos, Mike-Lynch, GWhitewood, JAnDbot, Narssarssuaq, Dogru144, Deflective, Barek, Quentar~enwiki, Smiddle, Andonic, Dcooper, Noobeditor, Vultur~enwiki, Dream Focus, Savant13, .anacondabot, Acroterion, Wildhartlivie, Magioladitis, WolfmanSF, Pedro, Grigri, Bongwarrior, VoABot II, MartinDK, Nyq, Wikideman, Bhar100101, Apollyon48, JNW, JamesBWatson, Miurajose, Davez1138, Chchan0, Pugetbill, Jim Douglas, The Anomebot2, Avicennasis, Majestic Lizard, Theroadislong, Gen us, Michaelmc, BatteryIncluded, 28421u2232nfenfcenc, LookingGlass, Rupert Nichol, Icbskateamsk8r, Doughboy914, Fang 23, Ripogenus77, DerHexer, Khalid Mahmood, Stupefaction, Mateoee, WLU, TimidGuy, NatureA16, Hdt83, MartinBot, Agricolae, Arjun01, Aladdin Sane, Rettetast, Speck-Made, Anaxial, Azalea pomp, R'n'B, MerryXIV, CommonsDelinker, AlexiusHoratius, Bloody bug, Nono64, Pdeitiker, Fconaway, MacAuslan, Srielity, Proabivouac, Dudley Miles, Petter Bøckman,

Mepiston, DreamWalker~enwiki, DrKay, Trusilver, Boghog, JoDonHo, Uncle Dick, MistyMorn, Rob Burbidge, Mike.lifeguard, FinnBjork, Octopus-Hands, Bot-Schafter, Stan J Klimas, McSly, QuasiAbstract, Samtheboy, AntiSpamBot, RenniePet, Hm john morse, HiLo48, Plasticup, Margareta, Tparameter, Chauncyo9, LittleHow, Belovedfreak, Marcus1234, NewEnglandYankee, Literacola, ArmadilloProcess, Regalshadow, JackobyLad, Gr8white, Cometstyles, Jamesontai, Tricksterson, JavierMC, Dorftrottel, Useight, Gibmetal77, Iceageman, Spellcast, Signalhead, ACSE, Hugo999, Littleolive oil, VolkovBot, Isak.Swahn, Johan1298~enwiki, Macedonian, Jeff G., Stefan Kruithof, Fences and windows, Chienlit, Tubbienine, Rokus01, Philip Trueman, TXiKiBoT, Cosmic Latte, Planetary Chaos, Rei-bot, Zephyr axiom, Ducky98, Kflester, Qxz, Vanished user ikijeirw34iuaeolaseriffic, Tsarevna, Locogato, Sgs SARVER, SGA SARVER, Graham hajosy, Leafyplant, Jackfork, Raryel, Amog, PDFbot, Cremepuff222, Madeup, Autodidactyl, Onore Baka Sama, Steve3849, Robert1947, Maxim, LBehounek, Doug, Dirkbb, Staka, Thinot, MDfoo, Falcon8765, Burntsauce, Declanbwrox, Angai001, Hannahrae, Cacaoatl, Hunting knife, Schnurrbart, Munci, ZBrannigan, Stringman5, Sylius, EmxBot, Jboutchard, SieBot, Bastiche, Pontificateus, Gfglegal, BotMultichill, ToePeu.bot, GuilhermeB, Jauerback, Philprime, Julain Barbarosa, Stephendcole, Yintan, Majakwe, Roidhrigh, Gravitan, LeadSongDog, Calusarul, Atkindave, FunkMonk, Radon210, Not Andrew, Mimihitam, Oxymoron83, Benea, KoshVorlon, Ausfdsaover439520, Lightmouse, Vasconicus, Thorrstein, IdreamofJeanie, Hatmatbbat10, G.-M. Cupertino, Ashnazgul, JohnSawyer, TB1983, Spartan-James, Chrisrus, Lanny5591, Hank52, Mr. Stradivarius, Jza84, Kalidasa 777, Nergaal, Denisarona, Randy Kryn, Llywelyn2000, Mr. Granger, Lenerd, Ratemonth, Martarius, AerosmithNirvana, ClueBot, NickCT, Binksternet, The Thing That Should Not Be, Helenabella, Taharley, GeneCallahan, Podsednik22, Quinxorin, Nnemo, Pi zero, Pairadox, Arkardx, CounterVandalismBot, Weredozen~enwiki, Parkwells, TarzanASG, Neverquick, Q27, Solar-Wind, Wildspell, Rockfang, Boneyard90, Fillosaurus, F-402, Deselliers, No such user, Excirial, 11jgiese, Alexbot, Jusdafax, Crywalt, Adimovk5, Arcot, SpikeToronto, JonatasM, ParisianBlade, Ashras10, NuclearWarfare, Oneelephantpickle, Kaeso Dio, Arjayay, Eleanor J Miller, Cinncyfan85, Iohannes Animosus, Strpker7734, Hans Adler, CowboySpartan, Elizium23, EGMichaels, Audaciter, Gnip, Ulso, EgraS, Thingg, Flower Priest, Tibbets74, Keilmesser, Amaltheus, DumZiBoT, Ano-User, Bridies, Whoelius, Mpondopondo, Umuntuwandi, GordonUS, Bobbob007, Dthomsen8, Denzel Ingram, Ost316, Bussech, Avoided, IAMTrust, Ncouture, Doc9871, SilvonenBot, Aunt Entropy, Wiki Greek Basketball, Airplaneman, RyanCross, White aasian, Cam1113, KirbyManiac, Kajabla, Cuptea, Addbot, Man with one red shoe, Jbsmith614, Some jerk on the Internet, Besh Saab, DOI bot, Non-dropframe, Trasman, Snakeyes74, Fieldday-sunday, CanadianLinuxUser, Leszek Jańczuk, Fluffernutter, Skyezx, MrOllie, Funky Fantom, Glane23, Bassbonerocks, Bahamut Star, Debresser, FCSundae, Timcali140, Oxlos69, Peter Napkin Dance Party, Setwisohi, 84user, Numbo3-bot, Ehrenkater, Keds0, Craigsjones, Tide rolls, Lightbot, Totorotroll, Teles, Jarble, Ettrig, SaintHammett, Ben Ben, Stuffdoing, Luckas-bot, Yobot, AzureFury, JohnnyCalifornia, Sprachpfleger, Legobot II, Rsquire3, Anypodetos, Professortimithy, KamikazeBot, Kjaer, Universal Life, AnakngAraw, Fjsfjs, Againme, Zomen19, AnomieBOT, DemocraticLuntz, 71.201.241.2, Senor Freebie, Archon 2488, VX, Jim1138, Grey3k, Nilenbert, Collieuk, LlywelynII, Shambalala, Alexikoua, Pingpeng, Materialscientist, Elmmapleoakpine, Wirrad, Citation bot, Vdr826, Xqbot, Historian of the arab people, St.nerol, Shashamula, Wapondaponda, Khajidha, Sellyme, Grim23, Teamjenn, Nneer8, Tyrol5, Aa77zz, Alkibiades231, Almabot, Homyakchik, RibotBOT, Amaury, Pereant antiburchius, Conquistador, Ozzie13, James henrick, Miyagawa, WebCiteBOT, Gnu Ordure, Prari, Dailycare, Collini, Nicolas Perrault III, Komitsuki, Paine Ellsworth, Tobby72, Lothar von Richthofen, Gouerouz, HJ Mitchell, Solaricon, Argumzio, Rgvis, Cannolis, Citation bot 1, Galmicmi, Pinethicket, Nmatavka, Abductive, Kakorot84, Pmokeefe, A8UDI, Moonraker, RedBot, LegendFPS, Alan LeHun, Egudahl, Hardwigg, Gerda Arendt, Kgrad, TobeBot, Trappist the monk, Belchman, Wotnow, Fama Clamosa, Xook1kai Choa6aur, ItsZippy, Jonkerz, Cirrus Editor, Diannaa, Stephen MUFC, Innotata, Bobby122, Lord of the Pit, DARTH SIDIOUS 2, Vladlen666, Onel5969, Updatehelper, RjwilmsiBot, NameIsRon, Androstachys, Billare, Slon02, EmausBot, John of Reading, WikitanvirBot, Milkunderwood, Distal24, Katherine, Super48paul, Faolin42, RenamedUser01302013, The Mysterious El Willstro, Tuxedo junction, Trybald, Kp grewal, Daonguyen95, Fæ, Dolovis, Elspooky, Ziggyseventh, Eilbertn, Medeis, Zloyvolsheb, Gniniv, Wayne Slam, Archaeomoonwalker, HammerFilmFan, Jeff Hunter Ackerman, NightSerf, Kindzmarauli, Abmin, StasMalyga, Vibsrelm, Abtalion, Brandmeister, Jbradley7203, Couloir007, L Kensington, Donner60, Terpentin1984, CountMacula, Puffin, Peter Karlsen, MrHiatus, DASHBotAV, Udahliudsy, Mjbmrbot, Poisonyourmind, Khestwol, Xanchester, ClueBot NG, NobuTamura, Paul Schliesser, This lousy T-shirt, Jareddiamond, Anagoria, LPOG1, Hectonichus, Errantsignal, Katsaridakos, Jen6jen6, Widr, Kraftiga, Cyrrk, SEE-SCAN, Clorenzoga, Galvatron311, Helpful Pixie Bot, Asdfjkl1235, Mira3z, Bibcode Bot, Technical 13, Ephert, BG19bot, PUECH P.-F., Birbasu, Kendall-K1, Marcocapelle, Cold Season, Compfreak7, AdventurousSquirrel, Max braddy, Jitka J, Aranea Mortem, Ayh123, Cgx8253, P'tit Pierre, Harizotoh9, NotWith, Hergilfs, Cornishpasty 2.0, Bruins360360, Chris Bowden, Henry McClean, GreenUniverse, Justincheng12345-bot, Darorcilmir, Hergilei, Jimw338, Bobjoe0987654321, Ekroon, Hsp90, ChrisGualtieri, Embrittled, Healablemarrow4, EuroCarGT, JCJC777, Ekren, Dexbot, Caroline1981, Sminthopsis84, Mogism, Cicero Moraes, Leo Fyllnet, Dustierer, RexRowan, Frosty, Parcival6, Sfgiants1995, CarlesMillan, Zyma, Corinne, Passwordbanana, Lupus Bellator, Muktaka Joshipura, Hillbillyholiday, Meganhams, I am One of Many, Jasmine Sansaverino, Lingzhi, Learnsean, Custard mustard sandwich, Alexandre Candalaft, PhantomTech, UnluckyJohnThomas, Cypork, Krankes-kind, NNGAdultServices, Dirtydirtyretro, DavidLeighEllis, Wikirictor, Macho9000, Karimatris, NottNott, George8211, Shenxingyang, Atotalstranger, Wikifan2744, Anarchistdy, Pruthweesha, Monkbot, Adrià Millán, Filedelinkerbot, Adriano G. V. Esposito, John Santana, Paleolithic Man, Joemangele, TheQ Editor, Izotz Aro, BioRuins, Cyrussmart, Saurusaurus, Sociocerebral, 1up Cypher, Editor abcdef, Datingjanebook, Meakin.2, Abedisbatmannow, Loracon, Pornstar1919, Artillias, The BaconNation, DalekSupreme, Meghanhauser, JISDA, 17b ext2014, Rsinghzoolgpu, Ben1148, ProfessorDorsy1978, Banakin900, Pearlcopse, Mysticalismeathome, Gregory Hopkins, Suksesi, Supdiop, Prinsgezinde, KasparBot, Sky.reid, Jjabrams119, Coolidon, DonaldGolden, ProprioMe OW, User2020202020, CyberWarfare, Sangpolar, Gearldwish, Muke af always1010, Keesleygirl and Anonymous: 1380

- **Archaic humans** *Source:* https://en.wikipedia.org/wiki/Archaic_humans?oldid=685625609 *Contributors:* AnonMoos, UtherSRG, Kaldari, Florian Blaschke, Dbachmann, Bender235, Guettarda, Axeman89, Jackhynes, Paxsimius, Drbogdan, Rjwilmsi, Smoe, Ian Pitchford, Idaltu, WriterHound, RussBot, Gerbil, Muntuwandi, Aeusoes1, S. Neuman, Closedmouth, Arthur Rubin, DerekOYeah, SmackBot, Durova, Scwlong, EOZyo, Paul H., Mander74, SkyWriter, Epf, Drunken Pirate, Titus III, Mgiganteus1, Grumpyyoungman01, Dave420, FairuseBot, Neelix, Cydebot, Metanoid, Ttiotsw, Dusty relic, Juansempere, Barticus88, Pstanton, Nowimnthing, John254, Escarbot, Hires an editor, Seaphoto, Alphachimpbot, GWhitewood, T@nn, Rivertorch, Agricolae, Tgeairn, Colincbn, Olegwiki, Jevansen, Mcewan, TXiKiBoT, Falcon8765, SieBot, Rlr3, Yerpo, Avnjay, BartekChom, Slovenski Volk, Chrisrus, Friendly Cave, WikipedianMarlith, ClueBot, R000t, Spark240, Neko-Nico, Lartoven, Newsroom hierarchies, Thingg, GordonUS, Richard-of-Earth, Jbeans, Elrodriguez, Addbot, OlEnglish, Krano, HerculeBot, Amirobot, AnomieBOT, Rubinbot, Citation bot, Wapondaponda, Altontacoma, Earlypsychosis, Chronus, BenzolBot, Citation bot 1, Pinethicket, O.anatinus, MondalorBot, Jonkerz, Autoreplay, Sulhan, Innotata, Onel5969, WikitanvirBot, Slightsmile, Fanyavizuri, ChuispastonBot, ClueBot NG, CaroleHenson, Helpful Pixie Bot, LITTLEej, Gob Lofa, Plantdrew, Jeraphine Gryphon, Cold Season, Honesty32, Undesignated, Demoniccathandler, Divyang.ch and Anonymous: 77

- **Recent African origin of modern humans** *Source:* https://en.wikipedia.org/wiki/Recent_African_origin_of_modern_humans?oldid=

688922989 *Contributors:* Kpjas, Mav, Zundark, Slrubenstein, William Avery, OlofE~enwiki, Mrwojo, Boud, Michael Hardy, Paul Barlow, Lexor, Tannin, 168..., VeryVerily, AnonMoos, Huangdi, Donarreiskoffer, Altenmann, Sam Spade, UtherSRG, Saforrest, Wereon, Raeky, Nagelfar, Cokoli, Jyril, Avsa, Andycjp, Gdr, Robert Brockway, Vina, Sonett72, Pasquale, Rich Farmbrough, Dave souza, Dbachmann, Dmr2, Bender235, Laurascudder, Rhysn, Reinyday, John Vandenberg, Cohesion, Acjelen, Saccade, Espoo, Ranveig, Danski14, Jhertel, DimaY2K, ThePedanticPrick, Titanium Dragon, Velella, RichBlinne, Reaverdrop, LukeSurl, KTC, Zanaq, Woohookitty, Cryptico, Spettro9, Duncan.france, Funhistory, Doric Loon, Ryoung122, Cuchullain, BD2412, Drbogdan, Rjwilmsi, Koavf, Fred Hsu, Eyu100, Wobble, Jamankowitz, Preslethe, Andreparis, Emiao, Mordicai, YurikBot, Wavelength, JWB, RussBot, Bhny, Groogle, Gerbil, Eleassar, Muntuwandi, Ytcracker, Badagnani, Expensivehat, Sir48, Daniel Mietchen, Davemck, Lockesdonkey, Kortoso, Skepticsteve, MrSativa, Maunus, 2over0, Jrajesh, Andrew Lancaster, Bhumiya, Shawnc, Amalthea, SmackBot, Ashill, Od Mishehu, Ramdrake, Nickst, Athaler, Kintetsubuffalo, Hmains, Durova, Bluebot, Silly rabbit, Hibernian, Neo-Jay, Nbarth, Epastore, Grandmasterka, VikSol, Gerkinstock, Cybercobra, Decltype, Al-Zaidi, Kntrabssi, John D. Croft, Dreadstar, Yom, BRSM, Epf, Torst, Saerain, Anlace, Titus III, Khazar, John, J 1982, JorisvS, Capmo, Extremophile, Ben Moore, Smith609, Keycard, Keith-264, Ossipewsk, Iridescent, Rhillman, CmdrObot, Moreschi, Floris V, Cydebot, Ntsimp, Jonathan Tweet, Arthurian Legend, Mind flux, Michael C Price, Doug Weller, Juansempere, DumbBOT, Bumba claat, Jahuda, Hhmb, Keraunos, Headbomb, Marek69, Itsmejudith, The Proffesor, Futurebird, Hires an editor, Doc Tropics, Storkk, T L Miles, Mcorazao, Ericoides, OhanaUnited, Igodard, Hurmata, Pharillon, SDas, Mdma2, JamesBWatson, PelleSmith, Theroadislong, BatteryIncluded, Eoganan, David Eppstein, JaGa, Manka, Warren Dew, Boynamedsue, Agricolae, Mattjs, Paracel63, Pdeitiker, Dudley Miles, Adavidb, Boghog, Tripleinfidel, Tom Schmal, Octopus-Hands, Maproom, Balthazarduju, Anonywiki, Olegwiki, Prhartcom, Tweisbach, WLRoss, ACSE, Master z0b, Fences and windows, Teledildonix314, Davidwr, Irish Pearl, JayEsJay, A4bot, Metasailor, Autodidactyl, Cosmos416, Penarc, HDThomas, Insanity Incarnate, Pan-ethnic, Gamsbart, Humboldt, Simplifier, Gerry Wachovsky, Jsc83, Skylark42, ConfuciusOrnis, Flyer22 Reborn, Permacultura, Chhandama, Shoebill2, Johntobey, Archaeogenetics, Sunrise, Stfg, Smilo Don, RhiannonAmelie, Varanwal, MoritzB, ImageRemovalBot, Afiya27, Martarius, Khumbula, General Epitaph, Nrkpan~enwiki, Epsilon60198, Parkwells, Rotational, Deselliers, Thgoiter, Excirial, NuclearWarfare, Hans Adler, Mikaey, Johnuniq, DumZiBoT, Ano-User, Mpondopondo, Aunt Entropy, Truthwillprevail12, Santasa99, DOI bot, Patchouli Princess, Debresser, Tobus, Longbowman, ScienceApe, Yobot, Denispir, DropShadow, AnomieBOT, Senor Freebie, In the government, Shambalala, Hunnjazal, Infiniti28, Citation bot, LilHelpa, Madalibi, Gianreali, Alexlange, JimVC3, Wapondaponda, Maulucioni, Regnir, Meshin0, Brout8, Omnipaedista, Moxy, Nitpyck, Paine Ellsworth, Lothar von Richthofen, Zygryk, Citation bot 1, Åkebråke, Pinethicket, Jonesey95, Blubro, Full-date unlinking bot, Trappist the monk, Litawor, Fama Clamosa, Xook1kai Choa6aur, HelenOnline, Livingrm, Anu826, Tbhotch, Jesse V., RjwilmsiBot, Beyond My Ken, Becritical, John of Reading, GoingBatty, Dcirovic, APayan, Tuxedo junction, Ὁ οἶστρος, Glennconti, MisterGugaruz, Andrea Jagher, ResidentAnthropologist, TitaniumCarbide, ClueBot NG, Widr, Saberus, Helpful Pixie Bot, Tdimhcs, Bibcode Bot, Greengrounds, BG19bot, MangoWong, Cold Season, Yowanvista, Cadiomals, Harizotoh9, Aloysius Sebastian, PersianEvolution, Lucy126, Cerabot~enwiki, FonsScientiae, Nottruelosa, Supersaiyen312, Kohelet, Atotalstranger, Monkbot, Trackteur, Lizzy8127, Rjmath, Liann2009, Jackrabbit1234, Liying15, Prinsgezinde, Datbubblegumdoe, DimensionQualm, Tbexbix and Anonymous: 177

- **Upper Paleolithic** *Source:* https://en.wikipedia.org/wiki/Upper_Paleolithic?oldid=686399207 *Contributors:* Rickyrab, SebastianHelm, Glenn, Kaihsu, Jeandré du Toit, Mxn, SEWilco, Joy, Opus33, Chrisjj, Lumos3, Casito, Xanzzibar, DocWatson42, Andycjp, Quadell, OverlordQ, Adamsan, Lacrimosus, Pasquale, Discospinster, Vsmith, Florian Blaschke, Dbachmann, Bender235, Carlon, Art LaPella, Comtebenoit, Wdshu, Sugaar, Fritzpoll, Wdfarmer, Computerjoe, LordAmeth, Yansa, Miaow Miaow, Polyparadigm, Pol098, Cbdorsett, Phlebas, Halcatalyst, Offtherails, BD2412, NekoDaemon, Wjfox2005, Blueyez941, Roboto de Ajvol, YurikBot, Hairy Dude, Peter G Werner, Gardar Rurak, Stephenb, Anomalocaris, Muntuwandi, Dysmorodrepanis~enwiki, Snek01, Nick C, Botteville, Arthur Rubin, Extraordinary, Xaxafrad, SMcCandlish, Carnaval~enwiki, Serendipodous, Hiddekel, KnightRider~enwiki, SmackBot, Vald, Phaldo, Donama, Anthonysenn, Jayanta Sen, Hibernian, Colonies Chris, Can't sleep, clown will eat me, Skoglund, Abyssal, AMK152, Rrburke, Anthon.Eff, BullRangifer, The PIPE, Metamagician3000, Tydus Arandor, Thanos5150, Edwy, Physis, Ckatz, Willy turner, Tapatio, Tawkerbot2, The Letter J, Ken Gallager, Floris V, Doctormatt, MC10, Rrotenbe, Dancter, Doug Weller, CieloEstrellado, HJJHolm, Thijs!bot, Chrisdab, Dmitri Lytov, I20, KrakatoaKatie, Seaphoto, Peer Gynt, JAnDbot, SineWave, MSisk, ClovisPt, EagleFan, Robotman1974, Nat, Fang 23, Ugajin, Lord Einar, Alro, Em Mitchell, J.delanoy, Geog1, It Is Me Here, Stan J Klimas, Johnbod, LordAnubisBOT, M-le-mot-dit, Kraftlos, Inwind, Funandtrvl, CWii, Raven rs, LeaveSleaves, Wingedsubmariner, Zuchinni one, MCTales, Assasin Joe, Hughey, SieBot, Cymi, Hugh16, Crash Underride, Mankar Camoran, Pocopocopocopoco, Randy Kryn, Someone the Person, Sfan00 IMG, Elassint, Perkinsonre, Parkwells, CohesionBot, CowboySpartan, Perkeleperkele, Flavallee, Alex144, Chronicler~enwiki, Mpondopondo, Facts707, Addbot, Steveporet, Asfreeas, Omnipedian, Tassedethe, Tide rolls, Legobot, Luckas-bot, Yobot, KamikazeBot, SwisterTwister, PoizonMyst, Maldek2, VanishedUser sdu9aya9fasdsopa, Hairhorn, Richardlord50, Chuckiesdad, Shambalala, Furby2010, Phthinosuchusisancestor, Maulucioni, Natural Cut, FrescoBot, Fortdj33, Edward44516, Madden88, HRoestBot, Reinfield, MastiBot, Labrynthia9856, White Shadows, Fama Clamosa, SeoMac, Stroppolo, RjwilmsiBot, EmausBot, Veda784, Immunize, T H MITCHELL, GoingBatty, Wikipelli, Q433, Tuxedo junction, BartlebytheScrivener, ChuispastonBot, Cefisher, ClueBot NG, Deedee96, Bped1985, Widr, Weseo, Roymarth, Gob Lofa, Bismaydash, Iselilja, Whitehex, Ladytimide, Joshua Jonathan, SystemsAlliance, Luciebalakova, Wangzilv1998, Leoesb1032, DragonCelery, Paleolithic Man, Daß Wölf, Eryk Norse, KasparBot, Thestoneageman, SpreadItOut and Anonymous: 153

- **Later Stone Age** *Source:* https://en.wikipedia.org/wiki/Later_Stone_Age?oldid=685551952 *Contributors:* Anthere, Dbachmann, Woohookitty, Rjwilmsi, TeaDrinker, Muntuwandi, Grafen, SmackBot, Richard75, I20, Hugo999, VolkovBot, Randy Kryn, Alexbot, Mpondopondo, MystBot, Addbot, LarryJeff, Shambalala, Citation bot, LilHelpa, Axxter99, Look2See1, ClueBot NG, Helpful Pixie Bot, Mogism, Ninafundisha, A mzungu, Omo Obatalá and Anonymous: 5

- **Behavioral modernity** *Source:* https://en.wikipedia.org/wiki/Behavioral_modernity?oldid=689210475 *Contributors:* Michael Hardy, Ronz, Ciphergoth, Feedmecereal, Charles Matthews, Dysprosia, AnonMoos, Xanzzibar, BalthCat, Onco p53, Vanished user 1234567890, TJSwoboda, Moxfyre, Dbachmann, Ben Standeven, RJHall, Lycurgus, Chbarts, Ranveig, 119, GeorgeTSLC, RichardWeiss, Cuchullain, BD2412, Ketiltrout, Drbogdan, Rjwilmsi, ErikHaugen, Mordicai, JWB, Muntuwandi, Aeusoes1, Ospalh, Maunus, Maelgwn, Igiffin, SmackBot, EncycloPetey, Thumperward, Hibernian, Epastore, Scwlong, Anthon.Eff, Kendrick7, JohnI, JorisvS, Mgiganteus1, Fig wright, Ossipewsk, Senorelroboto, Cydebot, Jonathan Tweet, Keraunos, Magioladitis, JamesBWatson, PelleSmith, Stassa, Olivierchaussavoine, R'n'B, Dudley Miles, LittleHow, DASonnenfeld, BlackJar72, Wingedsubmariner, Kyle112, Quietbritishjim, RW Marloe, Randy Kryn, Afiya27, Someone the Person, Epsilon60198, Leadwind, Boneyard90, SchreiberBike, Xsf24, Mpondopondo, MystBot, Addbot, DOI bot, Ludovika26, Jarble, Luckas-bot, Legobot II, Denispir, Speedy la cucaracha, Mr. Muntuwandi, AnomieBOT, Shambalala, Citation bot, Wapondaponda, J JMesserly, Mark Schierbecker, Djchazradio, Citation bot 1, Brionthorpe, Belchman, Xook1kai Choa6aur, Callanecc, The Utahraptor, Dewritech, SwordsmanRyan, Finn Bjørklid, RememberingLife, Gnaussor, ChuispastonBot, ClueBot NG, Gareth Griffith-Jones, Braincricket, Helpful Pixie Bot, Gob

Lofa, Bibcode Bot, BG19bot, Iselilja, Philvish, Aymankamelwiki, Unquiet pasts, Happycactuz, Soranoch, Krotera, Monkbot, GalenMillerAtkins and Anonymous: 63

- **Origin of the domestic dog** *Source:* https://en.wikipedia.org/wiki/Origin_of_the_domestic_dog?oldid=689530023 *Contributors:* William Avery, Rickyrab, Dogface, Lumos3, Coldacid, Beland, Eregli bob, Kevyn, Chris Howard, Rich Farmbrough, Vsmith, Dbachmann, R. S. Shaw, Of~enwiki, Maxl, Anthony Appleyard, Wtmitchell, RPH, Deacon of Pndapetzim, Pappa, Pauli133, Gene Nygaard, Dennis Bratland, Marasmusine, Woohookitty, LOL, Robert K S, Chris Buckey, Bytor, Graham87, BD2412, Rjwilmsi, Nightscream, Georgelazenby, Gurch, Choess, Bgwhite, Wavelength, Khisanth, Hairy Dude, RJC, Groogle, NawlinWiki, Muntuwandi, Grafen, ONEder Boy, Rjensen, Ashwinr, Leontes, Epipelagic, Aaron Schulz, Cmskog, Shawnc, Allens, Patiwat, Bibliomaniac15, SmackBot, Herostratus, Lawrencekhoo, Jacek Kendysz, Eskimbot, JoeMarfice, Skizzik, Rmosler2100, Bluebot, Thumperward, Snori, Hibernian, Suicidalhamster, Liontooth, Gaddy1975, John D. Croft, Risssa, GameKeeper, SirIsaacBrock, Lambiam, Gloriamarie, Kuru, Joelmills, Vanished user v8n3489h3tkjnsdkq30u3f, Ian Dalziel, Dan Gluck, Røed, JForget, Hirokazu, Searles2sels, Ken Gallager, Myasuda, W.F.Galway, Metanoid, GordonE, Doug Weller, DumbBOT, Monsieur Fou, PKT, CieloEstrellado, Epbr123, PetePassword, Hollomis, Rossab, Steve Dufour, Northumbrian, Mmyers1976, Seaphoto, Uvaphdman, QuiteUnusual, Nipisiquit, Epeefleche, Black Mamba, Igodard, PhilKnight, Autangelist, Bongwarrior, VoABot II, Steven Walling, Theroadislong, Nat, MartinBot, Morgan Wright, Anaxial, Mooni, CommonsDelinker, RockMFR, Roly Perera, Vanished User 4517, Toon05, Speciate, VolkovBot, Jalwikip, Mr. Shean, Gorank4, Jobberone, Wavehunter, PericlesofAthens, Deconstructhis, SylviaStanley, EJF, SieBot, Coffee, Calliopejen1, FunkMonk, Zucchini Marie, StaticGull, ZioPera, WikiLaurent, ImageRemovalBot, Hafwyn, Sfan00 IMG, ClueBot, Unbuttered Parsnip, Shinpah1, Hafspajen, Leodmacleod, Edude2, William Ortiz, Jeremiestrother, Thedugganaut, Awickert, Excirial, Naleh, Hharren, Sun Creator, Arjayay, JasonAQuest, Dark Mage, Dthomsen8, WP-Vandal, HexaChord, Addbot, Grey Geezer, DOI bot, Jcknowles, Ronhjones, FearAsylum, Download, Gsmileyv, Lightbot, Cesiumfrog, Zorrobot, Jarble, Yobot, Donfbreed, Darx9url, Edwin Luciano, AnomieBOT, IRP, OpenFuture, Citation bot, Hermie016, Mariomassone, Annabanana59876, Donpayette, Aa77zz, Champlax, Miyagawa, Erik9, Ngkimvy, FrescoBot, Riventree, Glo rida, Spindocter123, Craig Pemberton, Citation bot 1, Pinethicket, Chris814, Pmokeefe, Full-date unlinking bot, Trappist the monk, Moscow Connection, Hamfist, Jonkerz, Pbrower2a, DARTH SIDIOUS 2, RjwilmsiBot, Superk1a, EmausBot, John of Reading, Az29, Wikipelli, Brandmeister, Puffin, ChuispastonBot, SemanticMantis, Strangetruther, ClueBot NG, Gilderien, Snotbot, CopperSquare, Widr, DrChrissy, Helpful Pixie Bot, Calabe1992, Bibcode Bot, Lowercase sigmabot, BG19bot, Emayv, Winfredtheforth, NotWith, Hergilfs, Dontreader, BattyBot, Hergilei, Rivalnator, The Illusive Man, Commander v99, Onepebble, Dexbot, Dissident93, Webclient101, Mogism, Sidelight12, AWalker21, The Anonymouse, Mandruss, OccultZone, Monkbot, William Harris, Nblund, Editor abcdef, Snodgrass.370, Iannelli.4, Frevel8093, Nøkkenbuer and Anonymous: 178

- **Spear-thrower** *Source:* https://en.wikipedia.org/wiki/Spear-thrower?oldid=688738812 *Contributors:* Infrogmation, JohnOwens, Stormwriter, Llywrch, Ixfd64, Arthur3030, Skysmith, Usedbook, Glenn, Emperorbma, Zoicon5, SEWilco, Hajor, Wikibot, Yak, Bradeos Graphon, Ds13, Rick Block, Stevietheman, Chowbok, Adamsan, Sam Hocevar, Bhugh, Reflex Reaction, Guanabot, Hydrox, Dbachmann, Harriv, Kjoonlee, RJHall, Kwamikagami, Lkmorlan, Bgeer, John Fader, Alansohn, Anthony Appleyard, Pappa, Carioca, Throbblefoot, TriNotch, Dismas, A D Monroe III, Polyparadigm, GregorB, Marudubshinki, Dysepsion, Fbkintanar, Graham87, Rjwilmsi, Quale, TitaniumDreads, S Chapin, JohnGH, Lignomontanus, CJLL Wright, YurikBot, Wavelength, Red Slash, Gaius Cornelius, Ptcamn, JPMcGrath, Doncram, Mongrel, Maunus, Bhumiya, Streltzer, Vicarious, Scoutersig, JJL, Lavintzin, Eskimbot, Yamaguchi☐☐, Peter Isotalo, Gilliam, Bluebot, Keegan, Thumperward, Rorybowman, Madman2001, Brainhell, Nakon, Trieste, Derek R Bullamore, Hammer1980, The Ungovernable Force, Ser Amantio di Nicolao, Jotamar, John, Robofish, Mathias-S, Rkmlai, Dfred, Patrick Schwemmer, Blehfu, Geeman, Moonwatcher13, Vanisaac, WeggeBot, Richard Keatinge, Cydebot, Marvel3666, Ward3001, Pustelnik, Thijs!bot, Marek69, Natalie Erin, RobotG, Joan-of-arc, Tillman, Zatchmort, MrFlit, Deflective, 100110100, Simon Burchell, VoABot II, ClovisPt, Thunderbirdatlatl, Fang 23, Unclepea, KTo288, AlphaEta, Pharaoh of the Wizards, Jtbarnabas, Jamesmcardle, Johnbod, Drunkenteaparty, STBotD, Pdcook, IceDragon64, Wilhelm meis, Squids and Chips, VolkovBot, DOHC Holiday, Uyvsdi, Philip Trueman, TXiKiBoT, Double Fanucci, Technopat, Mzmadmike, LeaveSleaves, Andreas Carter, Mario1952, AHMartin, BotMultichill, Triwbe, Devonpike, Senor Cuete, Capitalismojo, Jimbovang, DeXXus, Randy Kryn, ClueBot, Hadrianheugh, Tomas e, Der Golem, Sabri76, Pointillist, Excirial, Mikaey, DumZiBoT, XLinkBot, Aaaalllleeeexxxx, MatthewVanitas, Addbot, Dark Lobo~enwiki, Usieeman, Lightbot, Ben Ben, Legobot, Luckas-bot, Yobot, AnomieBOT, Jim1138, Royote, Quebec99, LilHelpa, Xqbot, Peterdx, Mark Schierbecker, Amaury, Doulos Christos, Akagpak, Perdman, Haeinous, Yanajin33, Diremarc, Chenopodiaceous, Pinethicket, Smuckola, Bellaswan987, Wayne Riddock, 19cass20, Animalparty, Fama Clamosa, Nivekc711, Armondof, DARTH SIDIOUS 2, DexDor, Alph Bot, Marie Paradox, In ictu oculi, John of Reading, Nachosan, Tommy2010, K6ka, AvicBot, Bahudhara, Yonameistobie45, Vorziblix, L Kensington, Paname-IV, ClamDip, Peter Karlsen, Chubby muffins, Browniedog101, ClueBot NG, Atlatler, Iiii I I I, Gob Lofa, MerscratianAce, Bmaddenwiki, Ephert, BG19bot, Northamerica1000, Furkaocean, Mark Arsten, Sorcize, Серрей 6662, Murúg, Mpiva, BattyBot, Cyberbot II, Kurper, Starmick, A Certain Lack of Grandeur, Oudemia, GeneralizationsAreBad and Anonymous: 182

- **Mesolithic** *Source:* https://en.wikipedia.org/wiki/Mesolithic?oldid=688761365 *Contributors:* SimonP, Ubiquity, Lir, JohnOwens, Glenn, Llull, Andres, Genie, Emperorbma, Timwi, The Anomebot, Head, Joy, AnonMoos, Penfold, Pumpie, Robbot, ChrisO~enwiki, Romanm, Arkuat, Ojigiri~enwiki, Auric, Casito, Nagelfar, DocWatson42, Tourguide, Yak, Auximines, 1297, Adamsan, Murple, Eep², Stepp-Wulf, Discospinster, Rich Farmbrough, Kdammers, Dave souza, Roybb95~enwiki, Ivan Bajlo, Dbachmann, Pavel Vozenilek, Risacher, Pt, Vervin, Janna Isabot, Russ3Z, DCEdwards1966, Gary, Mark Dingemanse, Eric Kvaalen, Lectonar, Sugaar, Snowolf, Wtmitchell, Cough, Pauli133, Nuno Tavares, Chochopk, Phlebas, Island, Rjwilmsi, Kalogeropoulos, Cethegus, FlaBot, CarolGray, NekoDaemon, Gurch, Sstrader, Chobot, DVdm, Roboto de Ajvol, YurikBot, Bhny, Pigman, Gaius Cornelius, Anomalocaris, Wiki alf, Trovatore, Chakazul, Botteville, Grw, Brianlucas, Mejor Los Indios, DVD R W, Yvwv, SmackBot, Jab843, Gilliam, Ak70g2, Keegan, Persian Poet Gal, Stevage, VikSol, Laslovarga, Snowmanradio, SundarBot, Stevenmitchell, Khoikhoi, Wizardman, Bejnar, Risssa, Nareek, Grumpy444grumpy~enwiki, Mathsci, Nehrams2020, Iridescent, Shoeofdeath, PetaRZ, Eastlaw, Anima Rytak, Kotzker, Dolphinholmer, Zarex, Law soma, Moreschi, Richard Keatinge, Cydebot, Bill "Hemingray" Meier, Kazubon~enwiki, Kotiwalo, Doug Weller, HJJHolm, Thijs!bot, John254, A3RO, James086, Chrisdab, I20, Escarbot, TimVickers, Peer Gynt, Mikenorton, JAnDbot, Repku, Bongwarrior, Fang 23, Gwern, NatureA16, Anaxial, Dudley Miles, J.delanoy, UBeR, Johnbod, Naniwako, Largoplazo, JWetzel, CardinalDan, Idioma-bot, VolkovBot, Macedonian, Chienlit, Rokus01, Roarshocker, TXiKiBoT, Oshwah, Antoni Barau, Hqb, Rei-bot, Charlesdrakew, Raven rs, Autodidactyl, Ftjrwrites, Falcon8765, Fanatix, SieBot, ToePeu.bot, Gerakibot, Micke-sv, Faradayplank, AngelOfSadness, Helikophis, Jons63, Randy Kryn, Velvetron, ClueBot, Phoenix-wiki, CounterVandalismBot, Babul roy, DragonBot, Excirial, Jusdafax, Eeekster, Peter.C, SchreiberBike, Audaciter, Kblday, Nickkk1983~enwiki, Addbot, Tanhabot, Gururaj Nayak, ChenzwBot, Numbo3-bot, Zorrobot, Jerzas, Luckas-bot, II MusLiM HyBRiD II, Gongshow, TestEditBot, Getsardon, AnomieBOT, Rubinbot, Jim1138, Piano non troppo, Chuckiesdad, Kingpin13, Materialscientist, Citation bot, Andrewcottrell, Xqbot, TinucherianBot II, Grim23, Margaret50, Paleodigitalist,

GrouchoBot, Citation bot 1, DrilBot, Pinethicket, 10metreh, Jaysonisidro120495, RedBot, Serols, Jauhienij, Searine, Callanecc, Dinamik-bot, Extra999, DARTH SIDIOUS 2, Dynamicpricing, Alexandrudinu, Androstachys, EmausBot, John of Reading, Look2See1, Ajraddatz, ZéroBot, Qqchose2sucre, Candm8, DASHBotAV, 28bot, Monakhfirst, Dag448, ClueBot NG, Prioryman, Jack Greenmaven, Rtucker913, Deedee96, O.Koslowski, Dream of Nyx, Widr, PatHadley, MerlIwBot, Helpful Pixie Bot, Gob Lofa, WNYY98, BZTMPS, FoxCE, Snow Blizzard, Riley Huntley, SkepticalRaptor, The Illusive Man, YFdyh-bot, Dexbot, Luciebalakova, Jamesx12345, I am One of Many, NavinoEvans, Svoboman, Kebabpizza, Bladesmulti, Jillybean19, TranquilHope, Parikshit2922, Poopcr222, Supdiop, KasparBot, XTremeVG7, WesternWolfFTP and Anonymous: 208

- **Microlith** *Source:* https://en.wikipedia.org/wiki/Microlith?oldid=681740985 *Contributors:* Glenn, SD6-Agent, Bearcat, Robbot, RedWolf, Rholton, DocWatson42, Yak, Pgan002, Adamsan, Joyous!, Rich Farmbrough, Rama, Raverdrew, Longhair, Grutness, Orelstrigo, Grenavitar, Mel Etitis, Woohookitty, Rjwilmsi, Guyd, Bensin, NekoDaemon, YurikBot, Pburka, Pigman, Gadget850, Arthur Rubin, SmackBot, Athinaios, Chris the speller, Locutus Borg~enwiki, 16@r, Pjbflynn, Law soma, Richard Keatinge, Floris V, Verdi1, PKT, Thijs!bot, Dmitri Lytov, Enaidmawr, Fconaway, DarkFalls, Chiswick Chap, Idioma-bot, VolkovBot, Bridget.harrington, WOSlinker, TXiKiBoT, O crandell, SieBot, Calliopejen1, Gerakibot, TubularWorld, Randy Kryn, Beeblebrox, LonelyBeacon, SilvonenBot, Addbot, Sherrold, Asfreeas, Omnipedian, LinkFA-Bot, Lightbot, Zorrobot, Middayexpress, Luckas-bot, Yobot, AnomieBOT, Geregen2, Xqbot, FrescoBot, Archaeodontosaurus, D'ohBot, BillyGambela, MastiBot, English Fig, PleaseStand, EmausBot, GoingBatty, Paul Bedson, Richard asr, ZéroBot, TYelliot, Aparker621, CocuBot, PatHadley, Ne12308, ChrisGualtieri, Dexbot, Monkbot, Skytrekkertw and Anonymous: 24

- **Bow and arrow** *Source:* https://en.wikipedia.org/wiki/Bow_and_arrow?oldid=688773008 *Contributors:* Grouse, Rmhermen, William Avery, Ray Van De Walker, SimonP, KF, Ubiquity, Patrick, Ixfd64, Stevenj, Julesd, Glenn, Gavri, Hashar, Dcoetzee, Janko, Denni, Andrewman327, Tpbradbury, Hyacinth, Wetman, Francs2000, PuzzletChung, Robbot, PBS, Cek, Buster2058, DocWatson42, Pmaguire, Wiglaf, Yak, Muke, Everyking, Per Honor et Gloria, Bobblewik, Golbez, John Abbe, Kudz75, Antandrus, Evertype, Melloss, Pembers, Sanbec, Ultranol, Mrtrey99, Sam Hocevar, Drnnivel, A-giau, Discospinster, Solitude, Rich Farmbrough, Hidaspal, Cnyborg, Dbachmann, JemeL, ESkog, Kbh3rd, Petersam, CanisRufus, Kwamikagami, Triona, Bendono, Thuresson, Bobo192, Yonghokim, Smalljim, AlphBetaFive555111, The strategy freak, Conny, Jumbuck, Storm Rider, Anthony Appleyard, Arthena, Riana, Pion, Titanium Dragon, Wtmitchell, Velella, Ronark, TaintedMustard, Gene Nygaard, HGB, Shrap, Simetrical, Bellhalla, Jonathanbishop, WadeSimMiser, BlaiseFEgan, Tom M, MarcoTolo, Ae7flux, Palica, LeoO3, Mandarax, Gettingtoit, Magister Mathematicae, Josh Parris, Sjö, Koavf, Sdornan, Makaristos, XR-CatD, Yamamoto Ichiro, FlaBot, Lebha, Celestianpower, RexNL, Kolbasz, Chobot, The Rambling Man, YurikBot, Wavelength, Matanya (renamed), GBMorris, LizL05, RussBot, Sputnikcccp, Pigman, RadioFan, Gaius Cornelius, Ksyrie, Pseudomonas, Wiki alf, Grafen, Aaron Brenneman, Anetode, Nate1481, Mysid, Kortoso, Bota47, Craigkbryant, Wknight94, JakkoWesterbeke, Pawyilee, Durak, JereKrischel, Mev532, 21655, WikiY, Closedmouth, JoanneB, Croat Canuck, LeonardoRob0t, JLaTondre, Warreed, Allens, Katieh5584, John Broughton, GrinBot~enwiki, DVD R W, BonsaiViking, SmackBot, Ariedartin, Anastrophe, RedSpruce, Ace of Risk, AKismet, Commander Keane bot, Gilliam, Mr Barndoor, Andy M. Wang, Durova, Tyciol, Chris the speller, Bk22, Jprg1966, Thumperward, Snori, SchfiftyThree, Hibernian, Sadads, L clausewitz, DHN-bot~enwiki, The1exile, Darth Panda, Trekphiler, Can't sleep, clown will eat me, Cplakidas, OrphanBot, Rrburke, SundarBot, Nahum Reduta, Nakon, Xyzzy n, Just plain Bill, Latebird, Where, Skinnyweed, ArglebargleIV, Vriullop, Markjeff, DA3N, DO11.10, Copeland.James.H, Pat Payne, Statsone, IronGargoyle, Mr. Vernon, Vincentpang, Incognit000, Kiver16, Peter Horn, Jose77, Norm mit, Octane, SohanDsouza, Civil Engineer III, Jedihjc, Srain, Mrjahan, Luke c, JForget, TheHerbalGerbil, Ken Gallager, Richard Keatinge, No1lakersfan, Slazenger, Abeg92, ChristTrekker, Gogo Dodo, Tawkerbot4, Nabokov, Knight45, Lysandros, Oyo321, Wandalstouring, Epbr123, Hawkes~enwiki, 54gsze4ghz5, Dfrg.msc, Dpenguinman, RoboServien, Oreo Priest, AntiVandalBot, Luna Santin, CZmarlin, Tigeroo, DWolf2k2, Farosdaughter, Ingolfson, JAnDbot, Tigga, Ndyguy, Chanakyathegreat, Trebor trouble, Leolaursen, Barefact, Kerotan, Bongwarrior, VoABot II, JNW, Dor Posner, Crunchy Numbers, Fang 23, Darkrangerj, Sumitsoren, Glen, DerHexer, Yeahsoo~enwiki, MartinBot, FlieGerFaUstMe262, STBot, Mr Wednesday, Jim.henderson, Osirishotep, GeorgHH, Anaxial, Kostisl, J.delanoy, MRFraga, Trusilver, AstroHurricane001, Bogey97, KiwiBiggles, Rwebb1991, Erik Springelkamp, Thomas Larsen, NewEnglandYankee, Knulclunk, KylieTastic, Inwind, Izno, Llamanator, Phoenix264, Idioma-bot, Ariobarzan, VolkovBot, CWii, Uyvsdi, Nburden, Nik Sage, VasilievVV, Guardian Tiger, TheOtherJesse, Philip Trueman, TXiKiBoT, Mu$hroom, Edchant, Traumrune, Andreas Kaganov, Melsaran, Sirkad, Raymondwinn, Hey jude, don't let me down, Bowzzz, Kehrbykid, Symane, Uifareth Cuthalion, NHRHS2010, TheXenocide, Mjepson, Gaelen S., SieBot, CsikosLo, Tiptoety, Oxymoron83, Antonio Lopez, Ealdgyth, Mm.w0mbat, Dabomb87, DRTllbrg, Randy Kryn, Explicit, ClueBot, The Thing That Should Not Be, Phoenix999, Hostile Amish, Niceguyedc, Lbeben, Excise, Nick19thind, Jagdfeld, Superdummy999, LeoFrank, Phileasson, Gnome de plume, Alexbot, Machoman11, Diaboli, MartinFields, Jotterbot, Dekisugi, Mikaey, Arqueira, Thingg, Aitias, Berean Hunter, Johnuniq, Egmontaz, Slayerteez, Grizzley 118, DumZiBoT, XLinkBot, Bill98762222, Mjangels, Dthomsen8, Sergeant Sevorg, RyanCross, HexaChord, Addbot, Guoguo12, Fyrael, Morning277, Editforfun, Blaylockjam10, Tide rolls, Al3xil, Bobtheglob, Luckas-bot, Yobot, Mmxx, THEN WHO WAS PHONE?, KamikazeBot, AnomieBOT, 732SOUTHPAW, Breadchastick, Materialscientist, The High Fin Sperm Whale, E2eamon, ArthurBot, Anna Fruehwirth, Xqbot, Arghof, Ched, J04n, Abce2, RibotBOT, Wikieditor1988, Amaury, Brutaldeluxe, Wödenhelm, Shadowjams, Iswearius, Green Cardamom, Lateral mu, Fisheatsbear, Perdman, Bowman with microw midas 4, Phoenix000, Orangeace, Hchc2009, Pinethicket, I dream of horses, Rameshngbot, SpaceFlight89, Fumitol, Orenburg1, FoxBot, Firefang95, Fox Wilson, SeoMac, Stalwart111, Weedwhacker128, Mean as custard, Vinnyzz, EmausBot, Opus 113, Katherine, Samuraiantiqueworld, Ilovekgsr, Winner 42, Wikipelli, Alexrocks13, Melakavijay, FunkyCanute, ZéroBot, Josve05a, Imperial Monarch, A Thousand Doors, Zelbava, Erianna, Dfdffddfdffddffdffdfd, Jay-Sebastos, Bill Wilde, Donner60, Hunter4739984, Tboby12, HIST406-10rgomez, Sonicyouth86, Mehmet709, Clue-Bot NG, Satellizer, Wagner Texas Ranger, The High Fin Sock Whale, Rezabot, Widr, Theopolisme, Helpful Pixie Bot, Calabe1992, Gob Lofa, Lowercase sigmabot, Footmikyd, Player017, Carlstak, Altaïr, Aranea Mortem, Earendil56, Tamara Ustinova, Glacialfox, 22hockey, Britehit23, HueSatLum, WOLfan112, OKLR, Comatmebro, A.J.Johnny, Sonarboy, Canadianredneck383, Iammrsmellark, Abdesk2008, Lugia2453, Frosty, Brandvenkatr, BOBBYBOB789, Portelli9, Vladimir Alexiev, Jakec, JoshuaChen, Ugog Nizdast, Z3GY, Snowsuit Wearer, Aaronknowsitall, JaconaFrere, Yoshi24517, Hskoppek, Ianc43, Ivandabomb, Puuuuj, Amortias, Za dom spremi, TranquilHope, Krjohn, Trash bag765927, Emiliemldavidson, Masonsnider, Mason snyder, Markclarkjones, Mousenight, Xovady, Tropicalkitty, RoadWarrior445, Don't belive this dude, Arrow54321 and Anonymous: 616

- **Canoe** *Source:* https://en.wikipedia.org/wiki/Canoe?oldid=687772625 *Contributors:* Marj Tiefert, Bryan Derksen, Rmhermen, William Avery, Heron, Hephaestos, JohnOwens, 7265, Ruckc, Egil, Ahoerstemeier, CatherineMunro, Kingturtle, Harry Wood, Glenn, Habj, Nikai, Smack, Schneelocke, RichardB~enwiki, Tpbradbury, Bevo, Renato Caniatti~enwiki, Pollinator, Stargoat, Robbot, Moriori, Chris 73, RedWolf, Securiger, Cornellier, Academic Challenger, Meelar, Peking Duck, Mark Richards, Mark.murphy, Bkonrad, Alan Au, Hayne, Mendel, Slowking Man, Antandrus, Jareha, Klemen Kocjancic, Jonathanriddell, Dr.frog, A-giau, Chris j wood, Discospinster, Jyp, YUL89YYZ, Harriv, Ben-

- **Neolithic** *Source:* https://en.wikipedia.org/wiki/Neolithic?oldid=687462247 *Contributors:* AxelBoldt, MichaelTinkler, The Epopt, Brion VIB-BER, Eclecticology, SimonP, Heron, Hephaestos, Infrogmation, Michael Hardy, Fred Bauder, Kku, Menchi, Ixfd64, Gnomon42, Sannse, Shoaler, (, Ahoerstemeier, Glenn, Llull, Csernica, Raven in Orbit, Genie, Emperorbma, N-true, Gutza, The Anomebot, Itai, Nv8200pa, SEWilco, EikwaR, Joy, Mackensen, Wetman, Penfold, Finlay McWalter, Pumpie, Nufy8, Robbot, RedWolf, Altenmann, Steeev, Wikibot, Aetheling, Casito, GreatWhiteNortherner, Carnildo, ManuelGR, Nagelfar, Alan Liefting, Usommer~enwiki, DocWatson42, Tourgulde, Martijn faassen, Yak, Bkonrad, Dratman, Hans-Friedrich Tamke, Waltpohl, Rookkey, Dumbo1, Chowbok, Fergananim, Pgan002, Vinay, Antandrus, Beland, Adamsan, PITILai, Neutrality, Adashiel, Brianjd, Discospinster, Rich Farmbrough, Guanabot, Vsmith, Dave souza, Dbachmann, Stbal-bach, SamEV, Bender235, Malkin, Andrejj, Hapsiainen, MBisanz, El C, Kwamikagami, Shanes, Vervin, Janna Isabot, Smalljim, Nectarflowed, Russ3Z, Nesnad, Daf, Tgr, MPerel, Haham hanuka, Alansohn, Gary, Eric Kvaalen, Bathrobe, Calton, Axl, Harburg, Zyqqh, Kfitzgib, Jjhake, Saga City, Jheald, Dave.Dunford, Boyd Steere, Sleigh, Gene Nygaard, Tainter, HenryLi, Kazvorpal, Vanished user dfvkjmet9jweflkmdkcn234, Kfitzner, Roylee, OwenX, Rejs, Rattus, GeorgeTSLC, Tabletop, Twthmoses, JRHorse, GregorB, Eras-mus, Phlebas, Graham87, Noit, BD2412, Rjwilmsi, CristianChirita, Moosh88, Oblivious, Ligulem, Kalogeropoulos, The wub, Ttwaring, FlavrSavr, Ucucha, FlaBot, Pavlo Shevelo, RJP, SeptimusOrcinus, NekoDaemon, RexNL, Gurch, Jrtayloriv, Pevernagie, Codex Sinaiticus, Malhonen, Eric.dane~enwiki, CJLL Wright, Chobot, Sherool, Sbrools, WriterHound, Jimp, Kafziel, Peter G Werner, RussBot, Crazytales, Pigman, Ksyrie, Eleassar, Wimt, Anomalocaris, Dysmoro-drepanis~enwiki, Aeusoes1, Bloodofox, Anetode, Brian Crawford, CecilWard, Deucalionite, Gadget850, Kortoso, Crypteia, Botteville, Ribben-trop, Wknight94, Jkelly, Pawyilee, Sandstein, Bobstopper, Andrew Lancaster, Terry Longbaugh, Theda, Xaxafrad, CWenger, Saukkomies, Moomoomoo, Saltmarsh, GrinBot~enwiki, Selmo, DVD R W, Yvwv, SmackBot, Amcbride, KnowledgeOfSelf, Kilo-Lima, Jagged 85, Dav-ewild, WookieInHeat, AtilimGunesBaydin, Delldot, TharkunColl, Alexisrockcool, Yamaguchi[], Gilliam, Hmains, Chris the speller, Full Shun-yata, Asclepius, Hibernian, TheLeopard, Can't sleep, clown will eat me, OrphanBot, TheKMan, DR04, Anthon.Eff, Stevenmitchell, Khoikhoi, Aldaron, George, SnappingTurtle, G716, Derek R Bullamore, Ozgurgerilla, MaliNorway, FunkyFly, Soap, Vgy7ujm, Shrew, Peterlewis, Jimvin, A. Parrot, Ghelae, InedibleHulk, Richman271, Iridescent, Joseph Solis in Australia, Shoeofdeath, Oussjarrouse, Twas Now, SweetNeo85, Kax, Kevin Murray, JForget, CmdrObot, Ale jrb, Koffie, Ballista, MaxEnt, Cydebot, Future Perfect at Sunrise, Gogo Dodo, MotherFunctor, ST47, Querencia, Guitardemon666, Kotiwalo, Tawkerbot4, Doug Weller, DumbBOT, FastLizard4, Derzak, Kozuch, Abtract, SteveMcCluskey, Maziotis, Renegade MUFC, Mattisse, Malleus Fatuorum, Thijs!bot, Epbr123, Biruitorul, Kablammo, Marek69, Bobblehead, Chrisdab, Joym-mart, Philippe, Klausness, Noclevername, AntiVandalBot, Luna Santin, Seaphoto, Rehnn83, Tangerines, Clamster5, Wayiran, JAnDbot, Leuko, MER-C, Armhaed, Sapphire, Fetchcomms, J-stan, Barefact, GoodDamon, Dukeku, Io Katai, Connormah, Bongwarrior, VoABot II, Dentren, Hasek is the best, JNW, Protodruid, Antiquitas, Branka France, Leks81, Dragfyre, Baileydw@cardiff.ac.uk, Cpl Syx, Fang 23, DerHexer, Austin luce, Baristarim, Patstuart, Ubai1982, Jkaki00, Gwern, MartinBot, BetBot~enwiki, Arjun01, Ben MacDui, Rettetast, R'n'B, CommonsDelinker, AlexiusHoratius, Tgeairn, J.delanoy, Fowler&fowler, Hans Dunkelberg, Uncle Dick, Bohoi, Cocoaguy, .NERGAL, Acalamari, Zipzipzip, Mc-Sly, Naniwako, Eaingshong, Rosenknospe, Potatoswatter, TheNewPhobia, Idioma-bot, Pietru, Deor, VolkovBot, Macedonian, Jeff G., Indu-bitably, Nburden, Arnd Klotz, Tubbienine, Zerpent, Philip Trueman, TXiKiBoT, Zidonuke, Marcus334, Mosmof, Ask123, Srikipedia, Raven rs, Una Smith, Steven J. Anderson, Martin451, LeaveSleaves, Margil~enwiki, Edgehaedjr, Artemis9~enwiki, Y, FinnWiki, Mar vin kaiser, Monty845, Onceonthisisland, Milowent, SieBot, Hertz1888, Dawn Bard, Matthew Yeager, Flyer22 Reborn, Seethaki, Carnun, Zugraga, Nas-sim Abi Chahine, Oxymoron83, Lightmouse, Techman224, OKBot, DancingPhilosopher, Andrij Kursetsky, Prof saxx, Jza84, Wahrmund, Lawrence saliba, Denisarona, Randy Kryn, Kanonkas, Orshick, Scroch, Troy 07, ClueBot, GorillaWarfare, The Thing That Should Not Be, Dean Wormer, Arakunem, Drmies, Niceguyedc, Chris Kutler, Mspraveen, DragonBot, Excirial, AssegaiAli, Bob 1232345324644, Jusdafax, Exact~enwiki, Lartoven, NuclearWarfare, CMW275, Audaciter, Bilbaosr, Dvrtmcc, BOTarate, Redrocketboy, Versus22, GeeAlice, PuraVida2, Crazy Boris with a red beard, Heironymous Rowe, BodhisattvaBot, Opplkj, Feinoha, ErgoSum88, WikHead, Poiuytrewq22, SilvonenBot, Frood, Gene Fellner, MystBot, GwenHitomi, Thatguyflint, Mabalu, Jillpaans, Addbot, Deston johnson, Causteau, Toyokuni3, SpellingBot, Ronhjones, Leszek Jańczuk, Download, Callingchrissy, KenEmerson, Mikenlesley, SpBot, Ginosbot, West.andrew.g, Tobus, Esasus, Ace45954, Ssschhh, VASANTH S.N., Tide rolls, Ssonkris5, Zorrobot, Aday, LuK3, Luckas-bot, Yobot, Apollonius 1236, Tohd8BohaithuGh1, Bballlova99, Boksi, Khanilian, Richigi, AnomieBOT, 1exec1, Kimotori, 9258fahsflkh917fas, AdjustShift, Chuckiesdad, Ivan2007, RandomAct, Flewis, Materi-alscientist, The High Fin Sperm Whale, Citation bot, Jtamad, E2eamon, Roux-HG, Clark89, Sabes3, Madalibi, John Bessa, Xqbot, Romanfall, Capricorn42, Jackiestud, Stars4change, LevenBoy, Ruy Pugliesi, RibotBOT, Geopersona, TarseeRota, Mattis, WebCiteBOT, FrescoBot, Final-ius, Citation bot 1, Arekrishna, Pinethicket, I dream of horses, Macker33, Tanweer Morshed, Jonesey95, Phelbasar, Pmokeefe, Veronica Roberts, RedBot, MastiBot, SpaceFlight89, OMGWEEGEE2, FoxBot, Trappist the monk, Notpietru, Fang524, Saxonman111, Sawhti, Iron0037, Xs-park5, GanqSterzRFlii09, Unrulyevil, Suffusion of Yellow, Tbhotch, Reach Out to the Truth, YyyyyyFyyyyyyyyy 989, DARTH SIDIOUS 2, Zcrules, Bento00, Androstachys, Kiko4564, DASHBot, EmausBot, WikitanvirBot, Look2See1, Niluop, Racerx11, GoingBatty, Empathic-trust, Slightsmile, Tommy2010, Wikipelli, Paul Bedson, Werieth, Its snowing in East Asia, PBS-AWB, Access Denied, H3llBot, Aidarzver, Y-barton, L Kensington, Donner60, Aidan345345, Sgccgs, Discodolphin224, Monakhfirst, Helpsome, ClueBot NG, Kcoxz1, Gareth Griffith-Jones, Deedee96, DoncoN, Braincricket, Rezabot, CaroleHenson, Bbkinky, Qwerty8590, Helpful Pixie Bot, Gob Lofa, BG19bot, Vagobot, City-OfSilver, Northamerica1000, Dzlinker, MusikAnimal, AvocatoBot, Zyxwv99, Mark Arsten, Itswrong, CapitalistOverlord, FoxCE, Yahoow5, MrAmster, AllenBender, Aisteco, Horai 551, Osiris, Jeremy112233, A Timelord, Mdann52, ProudHuman42, EuroCarGT, IjonTichyIjon-Tichy, Dexbot, Hmainsbot1, TwoTwoHello, Lugia2453, Spark123720q, Foonarres, Squirtyjazz, PizzaHutCreeper, Iwondernot, Rockstar798, Eyesnore, Little Runs With Scissors, NottNott, FabForrest, Library Guy, Bonerspam69, Kabahaly, Monkbot, Dan-i-shrabi, Peter238, Sairp, As-dklf;, Makeandtoss, Awesomeguy74847483, Crumblingstatue, Jjosevelezz, Jamutaq, IEditEncyclopedia, KasparBot, Blue Papa Boy, Ncasale, Therealrossgeller, CLCStudent and Anonymous: 782

- **Neolithic Revolution** *Source:* https://en.wikipedia.org/wiki/Neolithic_Revolution?oldid=688900704 *Contributors:* Bryan Derksen, The Anome, William Avery, Paul Barlow, Llywrch, Fred Bauder, Wintran, Ellywa, Ahoerstemeier, Glenn, Raven in Orbit, Guaka, Reddi, IceKarma, AnonMoos, Wetman, Penfold, Pakaran, JorgeGG, AlainV, Goethean, Altenmann, Arkuat, Cornellier, Academic Challenger, GreatWhiteNorth-erner, Alan Liefting, Centrx, DocWatson42, Christopher Parham, Bfinn, Erdal Ronahi, Archie, Foobar, Jackol, Ilikeverin, Antandrus, Beland, Bcameron54, Adamsan, Pat Berry, Karl-Henner, Gscshoyru, Gary D, Redfax, Pm215, Ukexpat, Avihu, Adashiel, Thorwald, Alsocal, An Siarach, Discospinster, Brianhe, Rich Farmbrough, MCBastos, Vsmith, Florian Blaschke, Smyth, LindsayH, Dbachmann, Pavel Vozenilek, Bender235, Aranel, Carlon, Screensaver, Bobo192, Denorris, Nectarflowed, Nyenyec, MPerel, Polylerus, Pharos, Nsaa, Ranveig, Storm Rider, Gow, Alan-sohn, Gary, Albrecht Conz, Atlant, PatrickFisher, Logologist, Zyqqh, Hu, Bart133, A.Kurtz, Sleigh, LordAmeth, AlexTiefling, WilliamKF, Angr, Woohookitty, Webwanderer56, GeorgeTSLC, Jwanders, Eleassar777, Terence, Stefanomione, Phlebas, Joe Roe, Tslocum, Graham87, Magister Mathematicae, BD2412, Psm, Jclemens, Edison, Rjwilmsi, Nightscream, Bill37212, Kalogeropoulos, Lairor, Yamamoto Ichiro, FlaBot, Mar-gosbot~enwiki, Gurch, Losecontrol, Davepetr, CJLL Wright, Ggb667, DVdm, Bgwhite, Debivort, YurikBot, Hairy Dude, Peter G Werner, Russ-

Bitbut, Milowent, Bfpage, SieBot, Swingline 2005, Gerakibot, Da Joe, Smsarmad, Keilana, Not home, Flyer22 Reborn, Yerpo, Boppet, OKBot, Torchwoodwho, JohnnyMrNinja, Rabo3, Mr. Stradivarius, Isaster, Roger D Spencer, ImageRemovalBot, ClueBot, GorillaWarfare, PipepBot, The Thing That Should Not Be, IceUnshattered, Sting au, CounterVandalismBot, Torqtorqtorq, MC Scared of Bees, Reigndream, Excirial, TReidLewis, Eeekster, Coinmanj, Mweites, Drawn Some, Iohannes Animosus, SchreiberBike, Muro Bot, Rui Gabriel Correia, Versus22, Amaltheus, SoxBot III, Tdslk, DumZiBoT, Loranchet, Heironymous Rowe, XLinkBot, Koumz, PervyPirate, Gliderman, Coreylook, Gene Fellner, Addbot, Proofreader77, Heavenlyblue, Some jerk on the Internet, Friginator, Vishnava, MrOllie, Protonk, LaaknorBot, Michael Belisle, Favonian, Tassedethe, Tide rolls, Capitallst Shrugged, Krukouski, Pinus jeffreyi, Weganwock, Luckas-bot, Yobot, 2D, IAmHeron, Amirobot, Kamikaze-Bot, Azcolvin429, Tempodivalse, Synchronism, Mdw0, AnomieBOT, Templatehater, Eteklema-GMU, Jgayoso-GMU, Materialscientist, Swithrow2546, RobertEves92, Bstoopack, Citation bot, Naj-GMU, Dkabban-GMU, Neurolysis, Xqbot, Timir2, Plumpurple, Kindredcharles, Tyrol5, 0wn4g3 41if3, Voksen, GrouchoBot, Abce2, Marshallallensmith, Shattered Gnome, Brambleshire, Etienne4, Seeleschneider, Шуфель, L Seed, نايم, Taka76, LucienBOT, Drakenwolf, SL93, Pinethicket, I dream of horses, HRoestBot, MJ94, Timetuner, Ba dust, Oldsingerman20, Tangles82, Jauhienij, TangoFett, Jonkerz, Bobsteel09, Diannaa, Lord of the Pit, 13smithwalker, DARTH SIDIOUS 2, RjwilmsiBot, Steve03Mills, EmausBot, WikitanvirBot, Gfoley4, Az29, MacDogald, NotAnonymous0, Somebody500, Tommy2010, Thecheesykid, ZéroBot, John Cline, Ashyo99, Zap Rowsdower, Joe Chill 2, Staszek Lem, Augurar, IGeMiNix, GeorgeLyras, Sven Manguard, DASHBotAV, Amrbc, Petrb, Clairegray019, ClueBot NG, Iiii I I I, Davlag, Tideflat, Polskivinnik, Asukite, CaroleHenson, Widr, ساجد, ساجد امجد ساجد, DrChrissy, Helpful Pixie Bot, TheStealthWorker, ViezeRick, Titodutta, Gob Lofa, Bibcode Bot, Plantdrew, BG19bot, KimS012, The Banner Turbo, Puramyun31, AvocatoBot, Asur~enwiki, MKwek, Dhess13, Retroporter3000, Jamesreidel, Eduardofeld, NazmusLabs, Pratyya Ghosh, Jimw338, Cyberbot II, ChrisGualtieri, JYBot, Dexbot, Oughtcover, Hmainsbot1, Mogism, Asadron, Inayity, Jerambam, Lugia2453, Sidelight12, Izak1223, Reatlas, Utkarshsingh.1992, Dontfwithdalions, Number.6.freeman, JackinTrade, Ugog Nizdast, Tpha, Xolani90, Ginsuloft, AfadsBad, Glennonbobo, 5g4g2s1, Vbernau, Tobusfist, Sam.hill7, Fafnir1, Monkbot, William Harris, Lizzy8127, Funnybone224, Frankthetankk, Towering peaks, Galvitir, Gibson.701, Jaydefigurethis, KOtterbeck, Artsyraquel11228, Pyrotle, Dxf vgmds xdmsdgmvx, Lucascrafto2434, OrdinaryComix, KasparBot, HayRas123456789, Alucardatuman and Anonymous: 526

- **Pottery** *Source:* https://en.wikipedia.org/wiki/Pottery?oldid=687605803 *Contributors:* The Anome, Jeronimo, Andre Engels, Rmhermen, Karen Johnson, Heron, Camembert, Hephaestos, Olivier, Renata, Patrick, Llywrch, Fred Bauder, Ixfd64, Ellywa, Ahoerstemeier, Synthetik, Pjamescowie, Snoyes, Darkwind, DropDeadGorgias, Qed, Robertkeller, Genie, Zarius, HolIgor, Reddi, Wik, Tpbradbury, Wetman, Nnh, Sewing, Nickfl, Astronautics~enwiki, Pigsonthewing, ChrisO~enwiki, Altenmann, Lowellian, Mayooranathan, Hadal, Cyrius, Syntax~enwiki, Dmn, GreatWhiteNortherner, Randyoo, Yak, Everyking, Bkonrad, NeoJustin, Leonard G., Yekrats, Per Honor et Gloria, Jorge Stolfi, Node ue, Bobblewik, John Abbe, Wmahan, Stevietheman, Andycjp, Zeimusu, Slowking Man, Antandrus, OverlordQ, JoJan, MacGyverMagic, Neutrality, Tmstapf, Imjustmatthew, Fg2, Rculatta, Grstain, Mike Rosoft, Shipmaster, Discospinster, Rich Farmbrough, KillerChihuahua, Florian Blaschke, Ross Uber, Carptrash, Dbachmann, Bender235, Kbh3rd, Nabla, El C, Walden, Robotje, Smalljim, Viriditas, Jguk 2, Nk, Obradovic Goran, Haham hanuka, Scandium~enwiki, Ranveig, Alansohn, JadziaLover, Calton, Clubmarx, Lerdsuwa, Bsadowski1, Sleigh, LordAmeth, HGB, Roylee, Richard Arthur Norton (1958-), Jeffrey O. Gustafson, Kurmis~enwiki, Brunnock, Benbest, Pbhj, Kelisi, Mandarax, Sparkit, WBardwin, GoldRingChip, BD2412, MauriceJFox3, Cheffie, Sjakkalle, Rjwilmsi, Koavf, Quale, Jmcc150, Remurmur, Sherilyn, FayssalF, Titoxd, G Clark, Loggie, Nihiltres, AJR, RexNL, Oybobby, Kolbasz, Overand, Tysto, Silversmith, TearJohnDown, CStyle, Sharkface217, DVdm, Roboto de Ajvol, The Rambling Man, Wavelength, RussBot, Red Slash, Stephenb, Ksyrie, Wimt, GeeJo, Thane, David R. Ingham, Frost Indri, Wiki alf, Grafen, Yahya Abdal-Aziz, Irishguy, Dhollm, Jpbowen, Moe Epsilon, CLW, Crisco 1492, Zzuuzz, Nelfer, BorgQueen, Vicarious, Saukkomies, Winstonwolfe, Katieh5584, Bridgman, Appleby, CIreland, Luk, SmackBot, Dlc3007~enwiki, Melchoir, Funnugget, Deiaemeth, Liashi, Frymaster, BiT, HeartofaDog, Gilliam, Hmains, Oscarthecat, Skizzik, ParthianShot, Chris the speller, Bluebot, Ian13, Pieter Kuiper, Snori, Fplay, Hibernian, N.Hopton, Colonies Chris, Dethme0w, JonHarder, Rrburke, Elendil's Heir, Gartart, FiveRings, TheLimbicOne, TedE, John D. Croft, Nmotus, Thunk, Lanserj630, Dreadstar, Kapilthakur79, Bidabadi~enwiki, Nmnogueira, Danlina, Hlucho, Gobonobo, Kanuk, Sir Nicholas de Mimsy-Porpington, Edwy, Gregorydavid, IronGargoyle, RomanSpa, Ckatz, MarkSutton, AndyAndyAndy, Viv Hamilton, Klmarcus, Jose77, Hiroe, Swampyank, BranStark, DouglasCalvert, Nonexistant User, Courcelles, PsycheMan, CmdrObot, Raysonho, Basawala, Siyajkak~enwiki, DavidFHoughton, ShelfSkewed, MarsRover, Neelix, Iokseng, Funnyfarmofdoom, Slazenger, KarolS, Gogo Dodo, Amandajm, Doug Weller, Nsaum75, Victoriaedwards, Ereboschi, Epbr123, Parsa, Wheldon Boddy, Oerjan, Mojo Hand, ClosedEyesSeeing, Xaverius, Marek69, Edal, CTZMSC3, Flosseveryday, AntiVandalBot, Derzsi Elekes Andor, Quintote, Goldenrowley, Modernist, Fernando Maia Jr., Hml21st, Chill doubt, Myanw, Armkong, GWhitewood, JAnDbot, SuperLuigi31, Barek, The Transhumanist, Instinct, Jaysweet, Bongwarrior, VoABot II, Kajasudhakarababu, Vito Genovese, Nyttend, Plinth molecular gathered, Catgut, Theroadislong, Indon, Gabriel Kielland, Adrian J. Hunter, Pan Dan, The Sanctuary Sparrow, Chuckwatson, Rickterp, Kownudl, MartinBot, Berneegirl, Arjun01, Poeloq, Anaxial, R'n'B, CommonsDelinker, AlexiusHoratius, Fpbear, Wiki Raja, LedgendGamer, EdBever, Tgeairn, RockMFR, J.delanoy, Trusilver, Garysung168, Hans Dunkelberg, Stephanwehner, Jerry, TrinaLoyd, Sabila1, Shawn in Montreal, Katalaveno, Johnbod, Ihutchesson, Balthazarduju, NewEnglandYankee, SE-V6, Knulclunk, Trilobitealive, SJP, 83d40m, SSSN, Bob, Cometstyles, Uhai, WRoseman, Treisijs, SoCalSuperEagle, Idiomabot, Meaningful Username, Leebo, Jeff G., Tomer T, Apnaraj, Pelarmian, WOSlinker, TXiKiBoT, Theriac, Plenumchamber~enwiki, Gregmy, Lillyundfreya~enwiki, Phillip Rosenthal, Beyond silence, Martin451, ErikWestlund, C.Kent87, Katka193, Jackfork, LeaveSleaves, Muhammad Mahdi Karim, Seb az86556, S. M. Sullivan, Telecineguy, Dough9, Meters, Cantiorix, Falcon8765, Mar vin kaiser, Wikiway, Monty845, Symane, Denisewey, NHRHS2010, SieBot, Steorra, WereSpielChequers, Caltas, Lucasbfrbot, Yintan, Forestarethebest, M.thoriyan, Blago Tebi, Flyer22 Reborn, Radon210, Jjw, Oda Mari, Yerpo, Doncsecz~enwiki, Tdotter, Anchor Link Bot, Precious Roy, ShajiA, ImageRemovalBot, Sagdeep, Keinstein, ClueBot, LAX, SwanSupporter, Jackollie, The Thing That Should Not Be, Torib1234, R000t, Drmies, Mild Bill Hiccup, Hafspajen, Irish hunta, Niceguyedc, Mgcustado06, Estevoaei, Ckeavney, BANZ111, Murdered Jackal, Excirial, GoRight, Jeffg1011, 7&6=thirteen, Tnxman307, Jai Dixit, Grapeguy, Paulmewis, Laurac1636, DumZiBoT, XLinkBot, Kwork2, Wikiuser100, Dthomsen8, Nepenthes, WikHead, PL290, Hightowerpottery, Iamanerd2215, HexaChord, Iranway, Addbot, Brumski, Willking1979, Bloodbath 87, Tcncv, Pattych~enwiki, Ronhjones, Mr. Wheely Guy, Startstop123, Fetʼour, Vchorozopoulos, CanadianLinuxUser, Leszek Jańczuk, Wikisocko, AndersBot, LinkFA-Bot, Pince Nez, 5 albert square, Cezar.ceramics, Numbo3-bot, Tide rolls, Lightbot, OlEnglish, Gail, Zorrobot, Crazyquilter, Jarble, LuK3, Legobot, Middayexpress, Drpickem, Luckas-bot, Timurite, Yobot, Vague, Pink!Teen, TaBOT-zerem, Vaibhav.gupta191, Billlogalneedslove, Yngvadottir, Tea expertchinese, Nallimbot, Taj2008, Potatochippp, Eric-Wester, AnomieBOT, Andrewrp, Jim1138, Materialscientist, Asjlife, Maxis ftw, GB fan, ArthurBot, Bruce Foods, Xqbot, Lovettc, All4art, Date delinker, Anna Frodesiak, Andbrew.downes, Petropoxy (Lithoderm Proxy), J04n, GrouchoBot, Ute in DC, Chris.urs-o, BARNOIN, Luciandrei, FlowerOS, Shadowjams, AlexanderVanLoon, Some standardized rigour, Darwinius, FrescoBot, Tobby72, Colin.camphausen, Riventree, Keithartistsatwork, Moloch09, D'ohBot, Pinethicket, Elockid, Hard Sin, Alonso de Mendoza, Shoulee, Tomi.bojnec, KC130, Jauhienij, Orenburg1, Jugni, Clarkcj12, Defender of torch, Reaper Eternal, Qm museum bot, Ef80,

Linguisticgeek, Suffusion of Yellow, Lolzzlolzz, Satdeep Gill, Tbhotch, Scottdjp, DARTH SIDIOUS 2, Olawlor, Onel5969, Mean as custard, The Utahraptor, RjwilmsiBot, Bento00, Pandjarov, Nuclear Lunch Detected, Yoshih3r0, Slon02, Oddmidge, EmausBot, Orphan Wiki, Wikitanvir-Bot, Vinod rakte, Smtjan0524, Tylerw113, Tommy2010, Nathankimoto, Wikipelli, 4k7a2, Flyguy-is GAY, Tyranny Sue, Italia2006, Josve05a, Wowme12, Jack solomon, Lyk4, Hdgin, Noobkillar, The Nut, Bahudhara, JaLpro, H3llBot, Bugsbunnyyyyy, Bsolutions, Jarodalien, Decoyraven, Frank.Defalco, Erianna, Rxlx, L Kensington, Alborzagros, Spamterrorist3, Donner60, Christopher Daniel Stephanidies Cushion, Mattdude73, Baconage3013, Davidwclee, Donjuan214, Wikiwind, Rocketrod1960, ClueBot NG, Cprl bubbles, Fsbassister, Wendellpineda, MelbourneStar, This lousy T-shirt, Rtucker913, Piast93, Bigaalday, Smeagol321, Potaoesalad, Torontosethian, Wikichangerer, Cntras, HHaeckel, O.Koslowski, Marechal Ney, CaroleHenson, Rurik the Varangian, Karl 334, Berseker7980, Rainaaa13, ساجد, امجد ساجد, Shamto, Darkoonly, BenJChadwick, Zharradaan, Helpful Pixie Bot, Oddski Boddski, HMSSolent, Nightenbelle, Calabe1992, Regulov, Pine, Garrett430, Vilvos, PTJoshua, Gmcbjames, Sanky12123, Europeanhistorian, Lexi33313, OttawaAC, Itzdapotdog, Altaïr, Eman2129, Snow Blizzard, MrBill3, Hergilfs, Evil dooooooodette, Slushy9, HippysRus, Glacialfox, Warriorofdeath, Joethomp289, Ernie8472, A Timelord, Wmscott111, Mdann52, Pdemps01, Heba.amir, EuroCarGT, Eagerptosjdj, JYBot, Winkelvi, Calathea, Dexbot, Johnwhewell, DevinFunk, Webclient101, 331dot, HostileHamster, Ms. tuffy, Lugia2453, Fox2k11, Sriharsh1234, Poopininmapantsin, Corinne, Zackary4504, Sharla20, Pp391, Mugsymusic88, Eyesnore, Cmckain14, Smellyaj, DavidLeighEllis, Babitaarora, Ugog Nizdast, Sam Sailor, Khafesho, Deners884, Mnvitalone, 7Sidz, Embishh, Real7777, Tom.essinger.hileman, Farsinevisaneiran, ODell2001, Carebear2424, Grammar nazi Josh, Blackguyz69, Sok CHHAN, OrbDigital, Msanitam, Kadenjudd, Johncharlesronaldroberts, Supdiop, Wikifacker, KasparBot, Socoolkid123, Macklewhore, Peppy Paneer, PaulEviston, Tylersterry321, Cookiebowl and Anonymous: 643

- **Chalcolithic** *Source:* https://en.wikipedia.org/wiki/Chalcolithic?oldid=689346785 *Contributors:* AxelBoldt, Youssefsan, Rmhermen, Michael Hardy, Ellywa, Glenn, Harry Potter, Genie, Jengod, Jallan, HarryHenryGebel, AnonMoos, Penfold, Robbot, PBS, RedWolf, Altenmann, Arkuat, Sam Spade, Nagelfar, Ancheta Wis, Yak, Jason Quinn, Dumbo1, Antandrus, Mustafaa, Adamsan, Zfr, WpZurp, Rich Farmbrough, Vsmith, Florian Blaschke, Mjpieters, Dbachmann, Bender235, Janna Isabot, Russ3Z, Jumbuck, Larry Grossman, Alansohn, PaulHanson, Monado, Alex '05, Sleigh, KoRnholio8, Paxsimius, RxS, Rjwilmsi, Nightscream, Jquarry, Xosé, FlaBot, RJP, Chobot, Hall Monitor, YurikBot, Hairy Dude, Peterkingiron, Gaius Cornelius, Ksyrie, Anomalocaris, JFD, FourthAve, Countakeshi, Botteville, Closedmouth, Petri Krohn, WIN, Ajdebre, SmackBot, Hmains, Kevinalewis, Koryakov Yuri, Kurykh, Full Shunyata, DMS, Hibernian, Amber388, Grhabyt, SundarBot, Austinfidel, Jeff Wheeler, Drphilharmonic, SashatoBot, Naphureya, Shadowlynk, Ian Dalziel, Masoninman, Joseph Solis in Australia, Darkchun, Dpeters11, Revcasy, Law soma, TheTito, Andyt., Cydebot, Alucard (Dr.), Thijs!bot, Epbr123, Tapir Terrific, Nick Number, Kevphenry, BokicaK, Erin-Howarth, Wayiran, Bèrto 'd Sèra, Uchohan, JAnDbot, 100110100, RebelRobot, Carlwev, Michael Goodyear, Nyttend, Enaidmawr, Tgkohn, CommonsDelinker, Dudley Miles, Fowler&fowler, Love Krittaya, Cometstyles, Idioma-bot, German.Knowitall, Hibbity Dibbity, TXiKiBoT, Rei-bot, Gerrish, Raven rs, Doug, Enviroboy, Insanity Incarnate, Rknasc, SieBot, Frans Fowler, Shaheenjim, Wilson44691, Oxymoron83, Kevincof, Calatayudboy, Prof saxx, Randy Kryn, ClueBot, Mild Bill Hiccup, Audaciter, Mythdon, InternetMeme, BarretB, XLinkBot, EastTN, SilvonenBot, Aunt Entropy, Cewvero, Fyrael, Download, AnnaFrance, SpBot, Lightbot, OlEnglish, Zorrobot, JSR, Sechinsic, Legobot, PlankBot, Luckas-bot, Yobot, AnakngAraw, AnomieBOT, Materialscientist, Viletraveller, Citation bot, ArthurBot, Xqbot, Theonetwo3, TechBot, Omnipaedista, Amaury, Green Cardamom, FrescoBot, Dger, Metricmike, Macker33, Abductive, Moonraker, RedBot, Jopinder, Kgrad, TrickyM, TobeBot, Hanay, Mitchell Powell, Dinamik-bot, EmausBot, Broad Wall, Look2See1, SantosBorb, O'DaveY, Darkfight, AvicBot, Dolovis, Chuispastonbot, NTox, Llightex, ClueBot NG, CocuBot, Joefromrandb, Mémo-ST, Gob Lofa, Voldemort175, DBigXray, Mysthoric, Blake Burba, Zyxwv99, Ardasquin, BattyBot, Lugia2453, Lactasamir, Itc editor2, TruthShallSetTheeFree, Sitzmark, Doktor Wunderbar, Coolgrandma420, Chirt Rockwell, Skytrekkertw, KasparBot and Anonymous: 126

- **Epipaleolithic** *Source:* https://en.wikipedia.org/wiki/Epipaleolithic?oldid=663992424 *Contributors:* Glenn, Emperorbma, Joy, Penfold, Arkuat, Casito, Nagelfar, Yak, Leonard G., PenguiN42, Antandrus, Adamsan, Guanabot, Dbachmann, Janna Isabot, Sugaar, Phlebas, Mana Excalibur, Salix alba, Kalogeropoulos, NekoDaemon, YurikBot, Hairy Dude, Zwobot, Tvdog, SmackBot, Eskimbot, Cattus, Hibernian, Stevenmitchell, SashatoBot, Bcasterline, Law soma, Floris V, Anthonyhcole, Doug Weller, Thijs!bot, Chrisdab, Dmitri Lytov, Escarbot, Magioladitis, Gwern, Million Moments, Idioma-bot, Raven rs, Frans Fowler, YonaBot, Gerakibot, Micke-sv, OsamaBinLogin, Archaeogenetics, Randy Kryn, Skäpperöd, Quercus basaseachicensis, Koro Neil, Legobot, Luckas-bot, Yobot, ArthurBot, Tenofour, Omnipaedista, Archaeodontosaurus, MastiBot, Finn Bjørklid, Paul Bedson, ClueBot NG, PatHadley, BG19bot, Bernorix, Cerabot~enwiki, Nonsenseferret, Lanzente, Eadzz and Anonymous: 12

- **Paleolithic diet** *Source:* https://en.wikipedia.org/wiki/Paleolithic_diet?oldid=689188968 *Contributors:* Rmhermen, Michael Hardy, Ahoerstemeier, Ronz, Glenn, Rfr, Brigman, TonyClarke, Jpspeno, Omegatron, Bevo, Hankwang, RedWolf, Wikikiwi, Goethean, Modulatum, Ashley Y, Cornellier, Auric, Rhombus, Davidcannon, Wolfkeeper, Lethe, Peruvianllama, Michael Devore, GGordonWorleyIII, Antandrus, Beland, MisfitToys, Tharenthel~enwiki, TiMike, Neutrality, Joyous!, O'Dea, Discospinster, Rich Farmbrough, Kdammers, Nina Gerlach, Florian Blaschke, Smyth, Phrost, Arthur Holland, Elwikipedista~enwiki, Kiand, Kwamikagami, The bellman, CDN99, Stesmo, Davidruben, Shenme, Arcadian, Yellowking, Guy Harris, Fourthords, Omphaloscope, Gpvos, Jsmorse47, Kusma, SteinbDJ, Gene Nygaard, Finsternis, Tiger Khan, WayneMokane, Bobrayner, Angr, Pekinensis, Mindmatrix, LOL, Tierlieb, Hdante, MONGO, Alexescalona, GregorB, Dysepsion, Quantum bird, Sjö, Drbogdan, Rjwilmsi, Zbxgscqf, Rkeene0517, Ekspiulo, Salleman, Redecke~enwiki, Ground Zero, Nihiltres, Itinerant1, RexNL, Ralphael, Blue canary, DaGizza, DVdm, Bgwhite, EamonnPKeane, YurikBot, Sceptre, RussBot, Matthendrix, Chris Capoccia, Gaius Cornelius, Lusanaherandraton, Irrevenant, DVirus101, Dialectric, AugieWest, Lepidoptera, Coderzombie, DeadEyeArrow, Asarelah, Thegreyanomaly, Flooey, WAS 4.250, Deville, 2over0, Sotakeit, Colin, Youssef51, Tevildo, Kgf0, SmackBot, Sticky Parkin, InverseHypercube, C.Fred, Arny, IronDuke, Kintetsubuffalo, Sebesta, Septegram, Andrew J. MacDonald, Peter Isotalo, Gilliam, Tyciol, RDBrown, NCurse, Hibernian, Nbarth, Oatmeal batman, Frap, OrphanBot, Rrburke, Cícero, Thrallie, Baiter, Zdavidross, Master Scott Hall, Oanabay04, BullRangifer, MattCutts, Chad.brewbaker, Earthlingme, Leatherbear, Dmh~enwiki, Harryboyles, Valfontis, JzG, Rodney Boyd, AmiDaniel, Gobonobo, Narmical, James.S, Tlesher, CredoFromStart, Noian, Mmmsnouts, SandyGeorgia, Arstchnca, Peter Horn, Hu12, DabMachine, Iridescent, JoeBot, Tarl Cabot, Joshua Lutz, CRGreathouse, CmdrObot, KyraVixen, Nunquam Dormio, Jsmaye, TonySebas, Mt1955, Gran2, Cydebot, Dianathemath, Slp1, Vorlon19, Rracecarr, Doug Weller, HitroMilanese, DumbBOT, Kozuch, NMChico24, Gimmetrow, Mattisse, Epbr123, Rwmnau, Anupam, Ggarron, Aiko, Ufwuct, Nick Number, Dezidor, Heroeswithmetaphors, Mmortal03, Obiwankenobi, Crabula, Barek, Mcorazao, Janejellyroll, Albany NY, Greensburger, Poolboy8, Eburge, Joemacgregor, WolfmanSF, Scholariusx, Amateria1121, VoABot II, Mrund, AuburnPilot, Cadsuane Melaidhrin, Froid, Ariley, LookingGlass, Fang 23, WLU, Nodekeeper, Warren Dew, Jonnyhabenero, Yobol, Axlq, Bus stop, Christian424, Smithma7, J.delanoy, TimBuck2, TyrS, All Is One, Petegranger, Katalaveno, Mccajor, McSly, Brad Mosely, Anonywiki, Plasticup, LittleHow, 83d40m, Jmcw37, Mufka, Cuckooman4, Bonadea, RVJ, Zoezed, OnaTutors, Murderbike, Holme053, Mugander, Fences and windows,

Philip Trueman, Oshwah, GimmeBot, Tirakuna, Rei-bot, Anonymous Dissident, Sukaim6, Una Smith, Steven J. Anderson, Corvus cornix, Sarahjansen, Fortunaa, Kmhkmh, Geprodis, GlassFET, Spinningspark, JeremyT923, Doc James, MrChupon, Sam Hane, Calliopejen1, Spartan, Tiddly Tom, Euryalus, Dawn Bard, Happysailor, Flyer22 Reborn, Alexbrn, Cfrontz, Charles Paladin, JSpung, AngelOfSadness, Avnjay, Lightmouse, Navy.enthusiast, Col. Kernal, Mygerardromance, Fess-it, Maralia, Certayne, Tom Reedy, Joseph Meisenhelder, Prof. Campbell, ClueBot, Vrmlguy, Snigbrook, Hippo99, Wikievil666, Havers, Sural, Supasaru, OccamzRazor, Wikiaway, LonelyBeacon, Parkwells, Historian 1000, Leadwind, Phenylalanine, Paulcmnt, Campoftheamericas, Catfish Jim and the soapdish, Eeekster, Abrech, Lartoven, Xodarap00, Medos?, Holothurion, Jushmater, Kakofonous, Aitias, Johnuniq, Apparition11, CynRN, Against the current, MarmotteNZ, Nathan Johnson, Jytdog, Laser brain, Feinoha, Ragnord, VanishedUser ewrfgdg3df3, Addbot, DOI bot, Faiz40, B2daC, Lavishlova, Happyemo88, Ocdnctx, Rockstarmode, PatrickFlaherty, Daringderi, Mac Dreamstate, MrOllie, Download, LaaknorBot, Eltheodigraeardgesece, Fundamentisto, Scotchleaf, Pietrow, Ben Ben, Yobot, Legobot II, Jugney, Mdw0, AnomieBOT, Jim1138, Crecy99, Hiace28, Materialscientist, Citation bot, Jtamad, Bellemonde, Basilisk4u, ArthurBot, Xqbot, Ulysses elias, Nasnema, Mac520, Thermoproteus, Xasodfuih, The big man of mystery, Wōdenhelm, Shadowjams, Tabledhote, Videoqualia, AlexanderVanLoon, Ellenois, A.amitkumar, FrescoBot, Whazzupdoc, Glider87, Citation bot 1, Pinethicket, HRoestBot, Abductive, Jonesey95, Ahartzog, Rushbugled13, Jaybird vt, ScottMHoward, Jmedwards.uk, Rocalpi, Jandalhandler, SkyMachine, Wikididact, Kelly2357, Jordgette, Yunshui, Treesjm, HelenOnline, Antisoapbox, Editor99999, SarahMalek, Onel5969, Mean as custard, RjwilmsiBot, NameIsRon, Ripchip Bot, Phlegat, EmausBot, John of Reading, Jeffsu350, JteB, BillyPreset, Razor2988, Slightsmile, Winner 42, Bollyjeff, Skmishraindia, H3llBot, Laneways, AManWithNoPlan, Tolly4bolly, Barek-public, EricWesBrown, Alber Holmquist, Donner60, DM4242, LibertyOrDeath, Aldnonymous, Muchachomalonj, Ganerd, Incommand, Kkrueger, Morton16ok, Autodidact1, ClueBot NG, Segolyoda, Jack Greenmaven, Jhenderson8, Rokonn, Jennyfbrown, Rdd0013, Tyuigo, Texasss, Widr, Biomechanist, Helpful Pixie Bot, Gob Lofa, Angel Ayala Torres, Bibcode Bot, BG19bot, TessRose, Lahfoiado, Wikiterpsichore, CityOfSilver, MusikAnimal, Aasmae, Tfnn, Mark Arsten, Paleolithicdiet, Dainomite, Rob7866, Tony Tan, Yahoow5, Happyfang, Sadfang, MrBill3, Bluesky86, A1candidate, BattyBot, Worm12ga, Justincheng12345-bot, Fennfoot, Overspline, Abel James, Khazar2, TylerDurden8823, My 21cent, Eoxenford, LucyintheSkyyye, Kevinc20012, AutomaticStrikeout, Piyaro, Webclient101, Mogism, SpaceCatOnMushrooms, Lugia2453, Everything Is Numbers, Elitedresses, Frosty, Jocktapustheplatypus, Lizzyrose743, Danny Sprinkle, Cellistcat, Informedmama, Another.is.i, Epichomosapien, Wiki-madsen, Jabdou207, Acetotyce, Mrjohn1010, Tentinator, MdntDrgn, Everymorning, GoGoTob2, ChapelofJustice, Fred Biggin, Finnusertop, Ericjwilhelm, Catharine14, Drinkupthewine, Mj9021, Nustaris, OccultZone, NRozakos, GrassHopHer, JaconaFrere, TGHHL, Csara23, TuxLibNit, Htbiker, Monkbot, Mistaspock, Npdavies11, Paleolithic Man, Clothierb01, Lgkkitkat, Kelseeswenn, Upjav, Danatyko, JamesPem, TerryAlex, Roger 8 Roger, Alrich44, KH-1, Roshmax, Editor abcdef, Mazzola.23, Deaththealien, Dr.aviva.hill, Potatoxx69xxpotato, S65520, Zoedonne, Asambrailo, Obergen, Primula veris, Awhitehead3, Russelbsouthard, Exoduspal, Hamedaan, K scheik, Paleogray, Hup1922, Jerodlycett, Nøkkenbuer, Jennerchen, Emmaprizzia, Raddandfabb, Hillsie, Loienhug, YourPaleoPractitioner, ChrystyneO, Eating Nicely, Lth247, Issyella1, Doka07, Tinniestbore, A.beaumont, Diet2s, 208pony, Nouhb, Rheider921, JallDAMNday2002 and Anonymous: 570

36.9.2 Images

- **File:001117_15-44-2002-To-grupper-rosa-Qajar-Fliser2.jpg** *Source:* https://upload.wikimedia.org/wikipedia/commons/9/91/001117_ 15-44-2002-To-grupper-rosa-Qajar-Fliser2.jpg *License:* Public domain *Contributors:* [1] *Original artist:* **English:** Owji

- **File:1998-10-tema-canoe.jpg** *Source:* https://upload.wikimedia.org/wikipedia/commons/0/08/1998-10-tema-canoe.jpg *License:* CC BY 3.0 *Contributors:* Transferred from en.wikipedia to Commons by Lcawte using CommonsHelper. *Original artist:* Robertbody at English Wikipedia

- **File:20070818-0001-strolling_reindeer_cropped.jpg**   *Source:*   https://upload.wikimedia.org/wikipedia/commons/1/15/ 20070818-0001-strolling_reindeer_cropped.jpg *License:* CC BY-SA 3.0 *Contributors:*

- 20070818-0001-strolling_reindeer.jpg *Original artist:* 20070818-0001-strolling_reindeer.jpg: Nattfodd

- **File:A-potter-and-his-apprentice.jpg** *Source:* https://upload.wikimedia.org/wikipedia/commons/e/e6/A-potter-and-his-apprentice.jpg *License:* Public domain *Contributors:* https://archive.org/details/panjabisketches00twofiala *Original artist:* Unknown

- **File:A.afarensis.jpg** *Source:* https://upload.wikimedia.org/wikipedia/commons/3/32/A.afarensis.jpg *License:* Public domain *Contributors:* No machine-readable source provided. Own work assumed (based on copyright claims). *Original artist:* No machine-readable author provided. Esv assumed (based on copyright claims).

- **File:Acheuleanhandaxes.jpg** *Source:* https://upload.wikimedia.org/wikipedia/commons/8/81/Acheuleanhandaxes.jpg *License:* Public domain *Contributors:* ? *Original artist:* ?

- **File:Afghanistan_12.jpg** *Source:* https://upload.wikimedia.org/wikipedia/commons/9/93/Afghanistan_12.jpg *License:* Public domain *Contributors:* Collection personnelle (personal collection) *Original artist:* Davric

- **File:African_LSA_Biface.jpg** *Source:* https://upload.wikimedia.org/wikipedia/commons/8/85/African_LSA_Biface.jpg *License:* Public domain *Contributors:* Own work *Original artist:* José-Manuel Benito Álvarez —> Locutus Borg

- **File:African_Mitochondrial_descent.PNG**   *Source:*   https://upload.wikimedia.org/wikipedia/commons/4/44/African_Mitochondrial_ descent.PNG *License:* CC BY 3.0 *Contributors:* Own work *Original artist:* Maulucioni

- **File:African_cave_paintings.jpg** *Source:* https://upload.wikimedia.org/wikipedia/commons/d/dd/African_cave_paintings.jpg *License:* Public domain *Contributors:* ? *Original artist:* ?

- **File:Afro-Eurasia_location_map.svg** *Source:* https://upload.wikimedia.org/wikipedia/commons/6/60/Afro-Eurasia_location_map.svg *License:* Public domain *Contributors:*

- World_location_map.svg *Original artist:* World_location_map.svg:

- **File:Altamura_Painter_-_Red-Figure_Calyx_Krater_-_Walters_48262_-_Side_A.jpg**   *Source:*   https://upload.wikimedia.org/ wikipedia/commons/1/16/Altamura_Painter_-_Red-Figure_Calyx_Krater_-_Walters_48262_-_Side_A.jpg   *License:*   Public domain *Contributors:*   Walters   Art   Museum: <a href='http://thewalters.org/' data-x-rel='nofollow'><img alt='Nuvola filesystems folder home.svg'   src='https://upload.wikimedia.org/wikipedia/commons/thumb/8/81/Nuvola_filesystems_folder_home.svg/20px-Nuvola_

- **File:Ambox_contradict.svg** *Source:* https://upload.wikimedia.org/wikipedia/commons/2/2e/Ambox_contradict.svg *License:* Public domain *Contributors:* self-made using Image:Emblem-contradict.svg *Original artist:* penubag, Rugby471

- **File:Ambox_important.svg** *Source:* https://upload.wikimedia.org/wikipedia/commons/b/b4/Ambox_important.svg *License:* Public domain *Contributors:* Own work, based off of Image:Ambox scales.svg *Original artist:* Dsmurat (talk · contribs)

- **File:Animal_husbandry.jpeg** *Source:* https://upload.wikimedia.org/wikipedia/commons/1/12/Animal_husbandry.jpeg *License:* CC BY-SA 3.0 *Contributors:* Own work *Original artist:* Utkarshsingh.1992

- **File:AntarcticaDomeCSnow.jpg** *Source:* https://upload.wikimedia.org/wikipedia/commons/b/bd/AntarcticaDomeCSnow.jpg *License:* CC BY 2.5 *Contributors:* Own work *Original artist:* Stephen Hudson

- **File:Aphaia_pediment_polychrome_model_W-XI_Glyptothek_Munich.jpg** *Source:* https://upload.wikimedia.org/wikipedia/commons/e/ ee/Aphaia_pediment_polychrome_model_W-XI_Glyptothek_Munich.jpg *License:* Public domain *Contributors:* User:Bibi Saint-Pol, own work, 2007-02-08 *Original artist:* Unknown

- **File:Apollo-11_stone_slab.jpg** *Source:* https://upload.wikimedia.org/wikipedia/commons/7/7d/Apollo-11_stone_slab.jpg *License:* Public domain *Contributors:* Own work *Original artist:* José-Manuel Benito Álvarez —>Locutus Borg

- **File:Archery_pictogram.svg** *Source:* https://upload.wikimedia.org/wikipedia/commons/8/8e/Archery_pictogram.svg *License:* Public domain *Contributors:* Own work *Original artist:* Thadius856 (SVG conversion) & Parutakupiu (original image)

- **File:Arpón_con_microlitos.png** *Source:* https://upload.wikimedia.org/wikipedia/commons/5/5b/Arp%C3%B3n_con_microlitos.png *License:* CC BY-SA 2.5 *Contributors:* Own work *Original artist:* José-Manuel Benito

- **File:Arrow.svg** *Source:* https://upload.wikimedia.org/wikipedia/commons/3/32/Arrow.svg *License:* CC-BY-SA-3.0 *Contributors:* Transferred from en.wikipedia to Commons. *Original artist:* Wolfmankurd at English Wikipedia

- **File:Arrowhead.jpg** *Source:* https://upload.wikimedia.org/wikipedia/commons/5/54/Arrowhead.jpg *License:* Public domain *Contributors:* http://www.ornl.gov/info/news/pulse/pulse_v44_99.htm *Original artist:* ?

- **File:Ateriense-punta_pedunculada.jpg** *Source:* https://upload.wikimedia.org/wikipedia/commons/d/d7/Ateriense-punta_pedunculada.jpg *License:* CC BY-SA 2.5 *Contributors:* Own work *Original artist:* José-Manuel Benito

- **File:Atlatl.jpg** *Source:* https://upload.wikimedia.org/wikipedia/en/0/08/Aztec_atl-atl_%28Museo_Nacional_de_Antropolog%C3%ADa%29. jpg *License:* PD *Contributors:*
Own work
Original artist:
Fsunoles (talk) (Uploads)

- **File:Atlatl.png** *Source:* https://upload.wikimedia.org/wikipedia/commons/d/d5/Atlatl.png *License:* Public domain *Contributors:* ? *Original artist:* ?

- **File:Awashrivermap.png** *Source:* https://upload.wikimedia.org/wikipedia/commons/9/94/Awashrivermap.png *License:* CC BY-SA 3.0 *Contributors:* Own work, Elevation data from SRTM, drainage basin from GTOPO [1], all other features from Vector Map. Rand McNally "New International Atlas" (1993) used as reference. *Original artist:* Kmusser

- **File:BBC-artefacts.jpg** *Source:* https://upload.wikimedia.org/wikipedia/commons/0/09/BBC-artefacts.jpg *License:* CC BY 2.5 *Contributors:* Transferred from en.wikipedia to Commons. *Original artist:* Chenshilwood at English Wikipedia

- **File:BWCA_Canoe_Outing_-_001.jpg** *Source:* https://upload.wikimedia.org/wikipedia/commons/4/4e/BWCA_Canoe_Outing_-_001.jpg *License:* CC BY-SA 3.0 *Contributors:* Own work *Original artist:* ShakataGaNai

- **File:Babylonlion.JPG** *Source:* https://upload.wikimedia.org/wikipedia/commons/8/88/Babylonlion.JPG *License:* Public domain *Contributors:* ? *Original artist:* ?

- **File:Backing_sheep_at_sheepdog_competition.jpg** *Source:* https://upload.wikimedia.org/wikipedia/commons/0/00/Backing_sheep_at_ sheepdog_competition.jpg *License:* CC BY 2.0 *Contributors:* sheep dog *Original artist:* Peter Shanks from Lithgow, Australia

- **File:Bannerstone.jpg** *Source:* https://upload.wikimedia.org/wikipedia/commons/4/41/Bannerstone.jpg *License:* Public domain *Contributors:* http://www.nps.gov/archeology/visit/ohio/ohTimeline2.htm *Original artist:* National Park Service

- **File:Bardon_mill_kiln.jpg** *Source:* https://upload.wikimedia.org/wikipedia/commons/6/62/Bardon_mill_kiln.jpg *License:* CC BY 2.0 *Contributors:* http://www.flickr.com/photos/johndal/2701245389/sizes/o/ *Original artist:* johndal

- **File:Bhimbetka_rock_paintng1.jpg** *Source:* https://upload.wikimedia.org/wikipedia/commons/b/b6/Bhimbetka_rock_paintng1.jpg *License:* CC-BY-SA-3.0 *Contributors:* ? *Original artist:* ?

- **File:Biface_Cintegabelle_MHNT_PRE_2009.0.201.1_V2.jpg** *Source:* https://upload.wikimedia.org/wikipedia/commons/8/87/Biface_ Cintegabelle_MHNT_PRE_2009.0.201.1_V2.jpg *License:* CC BY-SA 4.0 *Contributors:*
Didier Descouens, 2 October 2010
Original artist: ?

- **File:Sapiens_neanderthal_comparison.jpg** *Source:* https://upload.wikimedia.org/wikipedia/commons/f/f2/Sapiens_neanderthal_comparison.jpg *License:* CC BY-SA 2.0 *Contributors:* http://www.flickr.com/photos/hmnh/3033749380/ *Original artist:* hairymuseummatt

- **File:Sapiens_neanderthal_comparison_en.png** *Source:* https://upload.wikimedia.org/wikipedia/commons/2/2d/Sapiens_neanderthal_comparison_en.png *License:* CC BY-SA 2.0 *Contributors:* http://www.flickr.com/photos/hmnh/3033749380/ (original photo) *Original artist:* hairymuseummatt (original photo), KaterBegemot (derivative work)

- **File:Scythians_shooting_with_bows_Kertch_antique_Panticapeum_Ukrainia_4th_century_BCE.jpg** *Source:* https://upload.wikimedia.org/wikipedia/commons/4/4e/Scythians_shooting_with_bows_Kertch_antique_Panticapeum_Ukrainia_4th_century_BCE.jpg *License:* CC BY-SA 3.0 *Contributors:* Own work, photographed at Musée du Louvre *Original artist:* PHGCOM

- **File:Siddi_Folk_Dancers,_at_Devaliya_Naka,_Sasan_Gir,_Gujarat.jpg** *Source:* https://upload.wikimedia.org/wikipedia/commons/c/cd/Siddi_Folk_Dancers%2C_at_Devaliya_Naka%2C_Sasan_Gir%2C_Gujarat.jpg *License:* CC BY-SA 2.0 *Contributors:* http://www.flickr.com/photos/chromatic_aberration/3360274830/in/set-72157613784288435/ *Original artist:* http://www.flickr.com/photos/chromatic_aberration/

- **File:Skara_Brae_house_1_5.jpg** *Source:* https://upload.wikimedia.org/wikipedia/commons/d/d2/Skara_Brae_house_1_5.jpg *License:* CC-BY-SA-3.0 *Contributors:* Own work *Original artist:* Wknight94

- **File:Skara_Brae_house_9.jpg** *Source:* https://upload.wikimedia.org/wikipedia/commons/c/c7/Skara_Brae_house_9.jpg *License:* CC-BY-SA-3.0 *Contributors:* Own work *Original artist:* Wknight94

- **File:Skeleton_and_restoration_model_of_Neanderthal_La_Ferrassie_1.jpg** *Source:* https://upload.wikimedia.org/wikipedia/commons/3/39/Skeleton_and_restoration_model_of_Neanderthal_La_Ferrassie_1.jpg *License:* CC BY-SA 3.0 *Contributors:* Own work *Original artist:* Photaro

- **File:Speakerlink-new.svg** *Source:* https://upload.wikimedia.org/wikipedia/commons/3/3b/Speakerlink-new.svg *License:* CC0 *Contributors:* Own work *Original artist:* Kelvinsong

- **File:Spherical_Hanging_Ornament,_1575-1585.jpg** *Source:* https://upload.wikimedia.org/wikipedia/commons/e/eb/Spherical_Hanging_Ornament%2C_1575-1585.jpg *License:* Public domain *Contributors:* Brooklyn Museum *Original artist:* Brooklyn Museum

- **File:Spreading_homo_sapiens_la.svg** *Source:* https://upload.wikimedia.org/wikipedia/commons/2/27/Spreading_homo_sapiens_la.svg *License:* Public domain *Contributors:* File:Spreading homo sapiens ru.svg by Urutseg *Original artist:* NordNordWest

- **File:Spy_Skull.jpg** *Source:* https://upload.wikimedia.org/wikipedia/commons/a/a1/Spy_Skull.jpg *License:* CC-BY-SA-3.0 *Contributors:* Own work *Original artist:* We El

- **File:Symbol_book_class2.svg** *Source:* https://upload.wikimedia.org/wikipedia/commons/8/89/Symbol_book_class2.svg *License:* CC BY-SA 2.5 *Contributors:* Mad by Lokal_Profil by combining: *Original artist:* Lokal_Profil

- **File:Székely_Land_-_Great_Market_Hall,_2014.09.12_(26).JPG** *Source:* https://upload.wikimedia.org/wikipedia/commons/d/d1/Sz%C3%A9kely_Land_-_Great_Market_Hall%2C_2014.09.12_%2826%29.JPG *License:* CC BY-SA 4.0 *Contributors:* Own work *Original artist:* Derzsi Elekes Andor

- **File:Sépulture_de_Teviec_Global.jpg** *Source:* https://upload.wikimedia.org/wikipedia/commons/e/e4/S%C3%A9pulture_de_Teviec_Global.jpg *License:* CC BY-SA 4.0 *Contributors:* Didier Descouens *Original artist:* ?

- **File:Tajine_potter.jpg** *Source:* https://upload.wikimedia.org/wikipedia/commons/e/ed/Tajine_potter.jpg *License:* CC BY 2.0 *Contributors:* http://www.flickr.com/photos/clodreno/209188987/ *Original artist:* Renault

- **File:Terra-Amata-Hut.gif** *Source:* https://upload.wikimedia.org/wikipedia/commons/e/e1/Terra-Amata-Hut.gif *License:* Public domain *Contributors:* Own work *Original artist:* José-Manuel Benito

- **File:Text_document_with_red_question_mark.svg** *Source:* https://upload.wikimedia.org/wikipedia/commons/a/a4/Text_document_with_red_question_mark.svg *License:* Public domain *Contributors:* Created by bdesham with Inkscape; based upon Text-x-generic.svg from the Tango project. *Original artist:* Benjamin D. Esham (bdesham)

- **File:The_Cava_dei_Servi_dolmen_(Ragusa-Sicily).jpg** *Source:* https://upload.wikimedia.org/wikipedia/commons/a/ae/The_Cava_dei_Servi_dolmen_%28Ragusa-Sicily%29.jpg *License:* CC BY-SA 4.0 *Contributors:* Own work *Original artist:* Spiccolo

- **File:Timba+1.jpg** *Source:* https://upload.wikimedia.org/wikipedia/commons/4/43/Timba%2B1.jpg *License:* Copyrighted free use *Contributors:* http://www.tamaskan-vom-muensterland.de/ *Original artist:* Kirsten Dieks

- **File:TimnaChalcolithicMine.JPG** *Source:* https://upload.wikimedia.org/wikipedia/commons/a/a5/TimnaChalcolithicMine.JPG *License:* Public domain *Contributors:* self-made; Mark A. Wilson[1] *Original artist:* Wilson44691

- **File:Traditional-pottery-workshop.jpg** *Source:* https://upload.wikimedia.org/wikipedia/commons/b/b7/Traditional-pottery-workshop.jpg *License:* CC BY-SA 3.0 *Contributors:* Own work *Original artist:* Edal Anton Lefterov

- **File:Tree_of_life.svg** *Source:* https://upload.wikimedia.org/wikipedia/commons/0/09/Tree_of_life.svg *License:* CC-BY-SA-3.0 *Contributors:* No machine-readable source provided. Own work assumed (based on copyright claims). *Original artist:* No machine-readable author provided. Vanished user fijtji34toksdcknqrjn54yoimascj assumed (based on copyright claims).

- **File:TrihedralNeolithic.jpg** *Source:* https://upload.wikimedia.org/wikipedia/commons/a/a7/TrihedralNeolithic.jpg *License:* CC0 *Contributors:* Own work *Original artist:* Paul Bedson

- **File:Tripolye_hut.jpg** *Source:* https://upload.wikimedia.org/wikipedia/commons/b/bc/Tripolye_hut.jpg *License:* CC BY-SA 3.0 *Contributors:* ? *Original artist:* ?

- **File:Turkey.Gülşehir001.jpg** *Source:* https://upload.wikimedia.org/wikipedia/commons/2/2e/Turkey.G%C3%BCl%C5%9Fehir001.jpg *License:* CC BY 3.0 *Contributors:* Own work *Original artist:* Georges Jansoone JoJan

- **File:Tværmose_arrow_(Denmark).png** *Source:* https://upload.wikimedia.org/wikipedia/commons/4/4b/Tv%C3%A6rmose_arrow_%28Denmark%29.png *License:* Public domain *Contributors:* CLARK, J. G. D. (1936) **The Mesolithic Settlement of Northern Europe**. Cambridge. Cambridge University Press. *Original artist:* J. G. D. Clark, 1936
- **File:Töpferscheibe.jpg** *Source:* https://upload.wikimedia.org/wikipedia/commons/9/93/T%C3%B6pferscheibe.jpg *License:* Public domain *Contributors:* Self-photographed *Original artist:* Oliver Kurmis
- **File:Unbalanced_scales.svg** *Source:* https://upload.wikimedia.org/wikipedia/commons/f/fe/Unbalanced_scales.svg *License:* Public domain *Contributors:* ? *Original artist:* ?
- **File:Upper_Paleolihic_Art_in_Europe.gif** *Source:* https://upload.wikimedia.org/wikipedia/commons/4/4f/Upper_Paleolihic_Art_in_Europe.gif *License:* Public domain *Contributors:* English wikipedia en:Image:Upper Paleolihic Art in Europe.gif, Based in A. Moure, *El Origen del Hombre*, Historia 16 ed. ISBN 84-7679-127-5 *Original artist:* Sugaar
- **File:Urn.jpg** *Source:* https://upload.wikimedia.org/wikipedia/commons/3/34/Urn.jpg *License:* CC-BY-SA-3.0 *Contributors:* Transferred from en.wikipedia *Original artist:* Original uploader was Dmn at en.wikipedia
- **File:Venus_de_Brassempouy.jpg** *Source:* https://upload.wikimedia.org/wikipedia/commons/2/26/Venus_de_Brassempouy.jpg *License:* Public domain *Contributors:* Own work by uploader, photographed at the en:Musée d'Archéologie Nationale *Original artist:* PHGCOM
- **File:Venus_von_Willendorf_01.jpg** *Source:* https://upload.wikimedia.org/wikipedia/commons/5/50/Venus_von_Willendorf_01.jpg *License:* CC BY 2.5 *Contributors:* Own work *Original artist:* User:MatthiasKabel
- **File:Vestonicka_venuse_edit.jpg** *Source:* https://upload.wikimedia.org/wikipedia/commons/b/b8/Vestonicka_venuse_edit.jpg *License:* CC BY-SA 2.5 *Contributors:* che *Original artist:* che

- **File:Walnuts02.jpg** *Source:* https://upload.wikimedia.org/wikipedia/commons/9/9a/Walnuts02.jpg *License:* GFDL 1.2 *Contributors:* Own work *Original artist:*
  fir0002 | flagstaffotos.com.au

- **File:Wells_Reindeer_Age_articles.png** *Source:* https://upload.wikimedia.org/wikipedia/commons/6/6c/Wells_Reindeer_Age_articles.png *License:* Public domain *Contributors:* Wells, H. G. (1920). The Outline of History. Garden City, New York: Garden City Publishing Co., Inc.. *Original artist:* H. G. Wells
- **File:Wien_NHM_Venus_von_Willendorf.jpg** *Source:* https://upload.wikimedia.org/wikipedia/commons/7/70/Wien_NHM_Venus_von_Willendorf.jpg *License:* CC-BY-SA-3.0 *Contributors:* Own work *Original artist:* Oke
- **File:Wiki_letter_w_cropped.svg** *Source:* https://upload.wikimedia.org/wikipedia/commons/1/1c/Wiki_letter_w_cropped.svg *License:* CC-BY-SA-3.0 *Contributors:*
- Wiki_letter_w.svg *Original artist:* Wiki_letter_w.svg: Jarkko Piiroinen
- **File:Wikiartifact-page-001.jpg** *Source:* https://upload.wikimedia.org/wikipedia/commons/e/e5/Wikiartifact-page-001.jpg *License:* CC BY-SA 4.0 *Contributors:* Own work *Original artist:* Piesquared93
- **File:Wikibooks-logo-en-noslogan.svg** *Source:* https://upload.wikimedia.org/wikipedia/commons/d/df/Wikibooks-logo-en-noslogan.svg *License:* CC BY-SA 3.0 *Contributors:* Own work *Original artist:* User:Bastique, User:Ramac et al.
- **File:Wikiquote-logo.svg** *Source:* https://upload.wikimedia.org/wikipedia/commons/f/fa/Wikiquote-logo.svg *License:* Public domain *Contributors:* ? *Original artist:* ?
- **File:Wikisource-logo.svg** *Source:* https://upload.wikimedia.org/wikipedia/commons/4/4c/Wikisource-logo.svg *License:* CC BY-SA 3.0 *Contributors:* Rei-artur *Original artist:* Nicholas Moreau
- **File:Wikispecies-logo.svg** *Source:* https://upload.wikimedia.org/wikipedia/commons/d/df/Wikispecies-logo.svg *License:* CC BY-SA 3.0 *Contributors:* Image:Wikispecies-logo.jpg *Original artist:* (of code) cs:User:-xfi-
- **File:Wiktionary-logo-en.svg** *Source:* https://upload.wikimedia.org/wikipedia/commons/f/f8/Wiktionary-logo-en.svg *License:* Public domain *Contributors:* Vector version of Image:Wiktionary-logo-en.png. *Original artist:* Vectorized by Fvasconcellos (talk · contribs), based on original logo tossed together by Brion Vibber
- **File:Wiktionary-logo.svg** *Source:* https://upload.wikimedia.org/wikipedia/commons/e/ec/Wiktionary-logo.svg *License:* CC BY-SA 3.0 *Contributors:* ? *Original artist:* ?
- **File:Wine_grapes03.jpg** *Source:* https://upload.wikimedia.org/wikipedia/commons/5/5e/Wine_grapes03.jpg *License:* GFDL 1.2 *Contributors:* Own work *Original artist:* Fir0002
- **File:Women_C-2.jpg** *Source:* https://upload.wikimedia.org/wikipedia/commons/f/f9/Women_C-2.jpg *License:* Public domain *Contributors:* Own work *Original artist:* Daniel Gauthier
- **File:Zhoukoudian_Caves_July2004.jpg** *Source:* https://upload.wikimedia.org/wikipedia/commons/a/a3/Zhoukoudian_Caves_July2004.jpg *License:* CC-BY-SA-3.0 *Contributors:* ? *Original artist:* ?
- **File:Каменный_век_(1).jpg** *Source:* https://upload.wikimedia.org/wikipedia/commons/f/fa/%D0%9A%D0%B0%D0%BC%D0%B5%D0%BD%D0%BD%D1%8B%D0%B9_%D0%B2%D0%B5%D0%BA_%281%29.jpg *License:* Public domain *Contributors:* http://www.picture.art-catalog.ru/picture.php?id_picture=3316 *Original artist:* Viktor M. Vasnetsov

36.9.3 Content license

- Creative Commons Attribution-Share Alike 3.0

www.ingramcontent.com/pod-product-compliance
Lightning Source LLC
Chambersburg PA
CBHW081433170526
45166CB00008B/2197